ALGEBRA AND TRIGONOMETRY

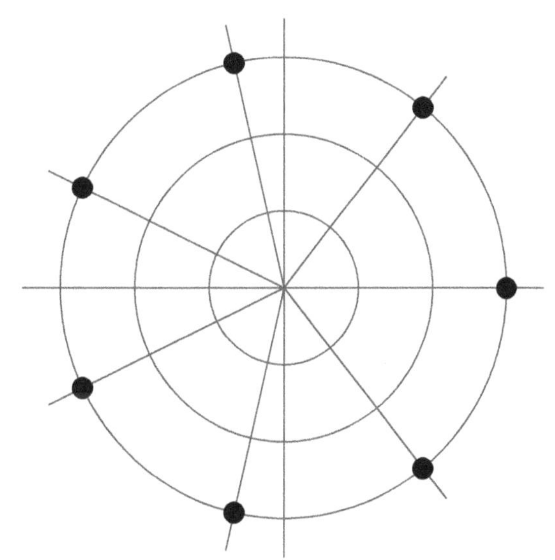

D.R. OLSON

SECOND EDITION

ISBN: 978-1-7358126-5-6

Preface to the Second Edition

The second edition of *Algebra and Trigonometry* differs little from the first. I have corrected a few errors and clarified a few sentences, added examples in several sections, and compiled an index, but I have left most of the text alone.

In a more significant way, I have also remedied a deficiency of a different kind in the book. In the preface to the first edition, I wrote that "the number of exercises in this book is not adequate for someone to learn algebra well, even if he should complete them all." While this statement remains true for the second edition, I have since published another book entitled *Exercises in Algebra and Trigonometry* that serves as a separate source of exercises that the student can complete for deeper understanding of the subjects.

D.R. Olson
26 January 2024

Preface to the First Edition

Philosophy

The purpose of studying algebra is to discipline the mind. We perform mental exercise when we solve algebraic problems with logical analysis and systematic procedures. While we can also use technology or friends to find solutions, such shortcuts will impart little benefit to us. If we want physical exercise, we should not ride a bus over a route we might run. If we want mental exercise, we should not accept mindlessly a solution to a problem through which we might think.

Relatively few of us will ever need to use algebra to solve real problems. Hence for most of us, algebraic problems will have no direct practical significance. Therefore, we should regard them as purely theoretical exercises that derive their value from the mental stimulation they induce. The correct solution to such a problem then is not a single number at the bottom of a page of intermediate steps, but the aggregate of those steps themselves.

Objectives

I wrote this book with the broad objective of helping anyone who wants to learn algebra, but with several narrower objectives as well. First, I wanted to help instructors and students of algebra to use their time in the classroom more efficiently. Term after term in classes worldwide, instructors repeat identical lectures and students copy reams of identical mathematical material from blackboards, whiteboards, and screens. The format of this book frees both student and instructor from the tedium of writing definitions and theorems and allows them more time to engage in other activities.

Second, I wanted to provide a means by which a person could learn algebra without the aid of a teacher. Someone with sufficient motivation who can follow a recipe or instructional manual can probably follow the examples in this book.

Third, I wanted to present traditional methods of solving algebraic problems to students who are confused by the modern methods currently taught in some schools.

Description

Each section of the book consists of information and examples related to that information. The kinds of information include definitions, properties, and procedures. The examples are typical algebraic exercises, but some are finished and some are unfinished.

Finished examples, which appear mainly on the left halves of the pages, provide the reader with models to follow as he completes the unfinished examples, which appear mainly on the right halves of the pages. For optimum benefit, the reader should complete all unfinished examples by himself without the use of a calculator, except where indicated otherwise. He should try to follow the format of the models as closely as possible as he works.

The number of exercises in this book is not adequate for someone to learn algebra well, even if he should complete them all. The reader should find and solve additional (perhaps twenty or twenty-five per section on average) problems from other sources.

Prerequisites

Before a person attempts to learn algebra, he should be able to:
- add, subtract, multiply, and divide two integers or terminating decimals;
- apply proper order of operations to evaluate long numerical expressions;
- evaluate numerical expressions with parentheses;
- write an integer as a product of two factors and determine its prime factorization;
- find the Least Common Multiple (LCM) of two integers;
- find the Greatest Common Factor (GCF) of two integers;
- add, subtract, multiply, and divide two fractions;
- reduce a fraction to lowest terms;
- convert a fraction into a decimal and vice versa;
- convert mixed numerals into fractions and vice versa;
- convert decimals to percentages and vice versa;
- evaluate simple expressions involving exponents and roots;
- find percentages of a given number;
- round a number to a specific place value or number of decimal places;
- substitute numbers into formulas and evaluate the resulting expressions;
- calculate characteristics (such as perimeter and area) of basic geometric shapes.

Furthermore, he should understand:
- the symbols $<$, $>$, and $=$;
- terminology regarding fractions (numerator, denominator, reciprocal);
- place values of a number (such as the thousand's, ten's, and hundredth's digits);
- negative numbers and absolute value;
- measurements of angles in degrees;
- types of angles (right, acute, obtuse);
- types of triangles (scalene, isosceles, equilateral; right, acute, obtuse);
- the meaning of parallel and perpendicular.

Final Note

Learning algebra takes hard work and persistence, for which there are no substitutes.

D.R. Olson
27 July 2021

Symbols

The table below describes symbols that the reader may encounter in the following pages.

Symbol	Meaning
$\mathbf{N}$	the set of natural numbers
$\mathbf{Z}$	the set of integers
$\mathbf{Q}$	the set of rational numbers
$\mathbf{R}$	the set of real numbers
$\mathbf{C}$	the set of complex numbers
$=$	is equal to, equals
$<$	is less than
$>$	is greater than
$\leq$	is less than or equal to
$\geq$	is greater than or equal to
$\approx$	is approximately
$\therefore$	therefore
$\Rightarrow$	implies
$\Leftrightarrow$	if and only if
$\exists$	there exists
$\forall$	for all
$\ni$	such that
$\in$	in (as an element of a set)
$\subset$	subset (of a set)
$\cup$	union (of sets)
$\cap$	intersection (of sets)
$\setminus$	difference (of sets)
$\emptyset$	empty (or null) set
$\parallel$	parallel
$\perp$	perpendicular
∞	infinity
$\to$	approaches
$\mapsto$	translates to
$\checkmark$	valid or correct
X	invalid or incorrect
$\square$	end of example
$\boxdot$	end of proof

Contents

Part One

Part Two

Part One

Section 1: Algebraic Expressions

Definitions

- A *variable* is a letter that can represent any number.
- A *constant* is a number or a letter that represents a specific number.
- An *algebraic expression* is a variable, a constant, or a mathematical phrase that connects variables and constants by mathematical operators.
- A *zero* of an algebraic expression in one variable is a number that causes the value of the expression to equal zero when substituted for the variable.

Notation

The symbols we use in algebra to indicate addition, subtraction, and division are the same as those we use in arithmetic, but we rarely use the multiplicative symbol "$\times$" in algebra because it resembles closely the variable x. The expressions below all denote the product of the variables a and b:

$$ab \qquad a \cdot b \qquad a(b) \qquad (a)(b) \qquad (a)b$$

Procedure A

To evaluate a numerical expression that does not contain grouping symbols (such as parentheses, brackets, or fraction bars), we apply operations in the following order:

1. simplify powers from left to right;
2. multiply and divide numbers connected by the symbols $\times$ and $\div$ from left to right;
3. add and subtract numbers connected by the symbols $+$ and $-$ from left to right.

Procedure B

To evaluate a numerical expression that contains parentheses or brackets, we:

1. evaluate the expression within the innermost set of grouping symbols in the manner described in procedure A;
2. remove the grouping symbols around the value determined in step (1);
3. repeat steps (1) and (2) as many times as necessary.

Procedure C

To evaluate a fraction, we:

1. evaluate the expressions in the numerator and denominator separately in the manners described in procedures A and B;
2. simplify the fraction resulting from step (1).

Procedure D

To evaluate an algebric expression, we:

1. substitute known values for each variable in the expression;
2. apply procedure A, B, or C to evaluate the numerical expression resulting from step (1).

Comment

We cannot use the multiplicative form ab after we substitute numbers for both variables. For example, if $a = 2$ and $b = 3$, then we cannot write $ab = 23$ because 23 means "twenty-three", not "two times three". Instead, we might write $ab = 2 \cdot 3$.

Example 1

A. n, x, and y are variables.
B. 5 and π are constants.
C. $9x$, $x - 2$, and $8m + n/5 - 10$ are algebraic expressions.

□

Example 2

Evaluate each expression using the given values of the variables.

A. $x + y$ for $x = 3$ and $y = 9$

Solution
For $x = 3$ and $y = 9$:

$$\begin{aligned} x + y &= 3 + 9 \\ &= 12 \end{aligned}$$

□

B. $a + b$ for $a = 7$ and $b = -2$

C. $9m - 2n$ for $m = 5$ and $n = 4$

Solution
For $m = 5$ and $n = 4$:

$$\begin{aligned} 9m - 2n &= 9(5) - 2(4) \\ &= 45 - 8 \\ &= 37 \end{aligned}$$

□

D. $x + 5y$ for $x = 11$ and $y = 5$

E. $-x + y - z$
for $x = -2/3$, $y = 5/12$, and $z = 7/4$

Solution
For $x = -2/3$, $y = 5/12$, and $z = 7/4$:

$$\begin{aligned} &-x + y - z \\ &= -(-2/3) + 5/12 - 7/4 \\ &= 2/3 + 5/12 - 7/4 \\ &= (2/3)(4/4) + 5/12 - (7/4)(3/3) \\ &= 8/12 + 5/12 - 21/12 \\ &= (8 + 5 - 21)/12 \\ &= -8/12 \\ &= -2/3 \end{aligned}$$

□

F. $(ab - c)(a + c)$
for $a = 5$, $b = 1/3$, and $c = 2$

G. $8xy - 5xz + 3yz$
for $x = 2$, $y = 3$ and $z = 7$

Solution
For $x = 2$, $y = 3$, and $z = 7$:

$$\begin{aligned}8xy - 5xz + 3yz &= 8(2)(3) - 5(2)(7) + 3(3)(7)\\ &= 48 - 70 + 63\\ &= 41\end{aligned}$$

□

H. $(w + 29)/5 + (24 - w)/9$ for $w = 6$

I. $(x + z)(x - z)$ for $x = 5$ and $z = 8$

Solution
For $x = 5$ and $z = 8$:

$$\begin{aligned}(x + z)(x - z) &= (5 + 8)(5 - 8)\\ &= (13)(-3)\\ &= -39\end{aligned}$$

□

J. $47 - 4[2(m + 5)]$ for $m = -8$

Example 3
Evaluate the expression $[h(b_1 + b_2)]/2$ using the given values of the variables.

A. $h = 7$, $b_1 = 5$, $b_2 = 3$

Solution
For $h = 7$, $b_1 = 5$, and $b_2 = 3$:

$$\begin{aligned}[h(b_1 + b_2)]/2 &= [7(5 + 3)]/2\\ &= [7(8)]/2\\ &= 56/2\\ &= 28\end{aligned}$$

□

B. $h = 5$, $b_1 = 10$, $b_2 = 3$

Example 4
For $x = 2$:

$$\begin{aligned}3x - 6 &= 3(2) - 6\\ &= 6 - 6\\ &= 0\end{aligned}$$

Therefore, the number 2 is a zero of the expression $3x - 6$.

□

Example 5
For $n = 5$:

$$\begin{aligned}n + 2 &= 5 + 2\\ &= 7\\ &\neq 0\end{aligned}$$

Therefore, the number 5 is not a zero of the expression $n + 2$.

□

Section 2: Sets

Definitions

- A *set* is a collection of objects.
- An *element* is an object in a set.
- The set A is a *subset* of the set B if every element of A is also an element of B.
- The *empty set* or *null set*, denoted by $\emptyset$, is the set with no elements.

Operations on Sets

- The *union* ($\cup$) of multiple sets is the set of elements contained in at least one of the sets.
- The *intersection* ($\cap$) of multiple sets is the set of elements contained in every one of the sets.
- The *difference* ($\setminus$) of sets A and B is the set of elements contained in A but not in B.

Notation

- The symbols $\in$ and $\notin$ indicate whether or not an element is in a set:
  - $b \in S$ means that b is an element of the set S;
  - $b \notin S$ means that b is not an element of the set S.
- The symbols $\subset$ and $\not\subset$ indicate whether or not a set is a subset of another set:
  - $A \subset B$ means that the set A is a subset of the set B;
  - $A \not\subset B$ means that the set A is not a subset of the set B;
  - $A = B$ means that the sets A and B contain the same elements.
- The symbols $\{$ and $\}$ enclose the list of elements contained in a set.
- The symbol $\ldots$ (called an "ellipsis") indicates the continuation of a pattern.

Set Builder Notation

Set builder notation is a notation for indicating the elements in a set without listing them explicitly. Set builder notation can take one of two forms:

- $\{\, x \in S \mid \text{property } P \,\}$ means the set of all elements x in S that satisfy P;
- $\{\, \text{expression } E \mid x_1 \in S_1,\ x_2 \in S_2,\ \ldots,\ x_n \in S_n \,\}$ means the set of all values that E can assume if $x_1 \in S_1$, $x_2 \in S_2$, $\ldots$, $x_{n-1} \in S_{n-1}$, and $x_n \in S_n$.

Comments

- $A = B$ if $A \subset B$ and $B \subset A$
- $\emptyset$ is equivalent to $\{\ \}$, but $\emptyset$ is not equivalent to 0 (zero).
- We may omit the " $\in S$" in the first form of set builder notation described above and write simply $\{\, x \mid \text{property } P \,\}$ if the kind of elements under consideration is clear from context.
- In any list of the elements in a set, no single element should appear more than once. The order in which elements of a set are listed does not matter.
- The symbol $\in$ signifies a relationship between an element and a set, while the symbol $\subset$ signifies a relationship between two sets.

Example 1

The notation "$S = \{1, 2, 3, \mathrm{x}, \mathrm{y}, \mathrm{z}\}$" means that S is the set that contains the elements 1, 2, 3, x, y, and z.

□

EXAMPLE 2
Suppose $A = \{1, 2, 3, 4, 5\}$, $B = \{1, 2, 3\}$, $C = \{2, 4, 6\}$, and $D = \{2\}$.
Then:

- $3 \in B$
- $8 \notin A$
- $B \subset A$
- $\{3, 4\} \subset A$
- $C \not\subset A$
- $6 \in C$
- $\{6\} \subset C$
- $D \subset B \subset A$

□

EXAMPLE 3
Suppose the set S consists of the colors on the Italian flag.
Then:

- $S = \{\text{Green, Red, White}\}$
- Green $\in S$
- Brown $\notin S$

□

EXAMPLE 4
Suppose the set R consists of all the letters in the English alphabet
and the set T consists of the vowels in the English alphabet.
Then:

- $R = \{\text{a, b, c, d, e}, \ldots, \text{x, y, z}\}$ (assuming the customary sequence of the letters)
- $T = \{\text{a, e, i, o, u}\}$
- $\text{e} \in T$
- $T \subset R$
- $\text{m} \notin T$
- $R \not\subset T$

□

EXAMPLE 5
Suppose $F = \{1, 3, 5, 7, 9\}$, $G = \{2, 4, 6, 8, 10\}$, and $H = \{1, 2, 3, 4, 5, 6\}$. List the elements in the indicated sets.

A. $F \cup G$, $H \cap F$, $G \cup H$, $F \backslash H$

SOLUTION

- $F \cup G = \{1, 2, 3, 4, 5, 6, 7, 8, 9, 10\}$
- $H \cap F = \{1, 3, 5\}$
- $G \cup H = \{1, 2, 3, 4, 5, 6, 8, 10\}$
- $F \backslash H = \{7, 9\}$

□

B. $F \cup H$, $G \cap H$, $F \cap G$, $H \backslash G$

EXAMPLE 6
The set $\{1, 5, 8\}$ is equivalent to the set $\{8, 1, 5\}$.

□

EXAMPLE 7
Suppose A is the set of all the names of the states in the U.S.A.,

$R = \{\text{a, b, c, d, e}, \ldots, \text{x, y, z}\}$,
$S = \{3, 5, 7\}$,
and $T = \{2, 4\}$.

Then:

- $\{\, x \in R \mid x \text{ is a vowel} \,\} = \{\text{a, e, i, o, u}\}$
- $\{\, x \mid x \text{ is an even number between 7 and 19} \,\} = \{8, 10, 12, 14, 16, 18\}$
- $\{\, x \in A \mid x \text{ begins with the letter I} \,\} = \{\text{Idaho, Illinois, Indiana, Iowa}\}$
- $\{\, a/b \mid a \in S,\ b \in T \,\} = \{3/2,\ 5/2,\ 7/2,\ 3/4,\ 5/4,\ 7/4\}$

□

Section 3: Real Numbers

Definitions

- The *natural numbers*, denoted by $\mathbf{N}$, are the elements of the set $\{1, 2, 3, 4, 5, \ldots\}$.
- The *whole numbers* are the elements of the set $\{0, 1, 2, 3, 4, 5, \ldots\}$.
- The *integers*, denoted by $\mathbf{Z}$, are the elements of the set $\{\ldots, -3, -2, -1, 0, 1, 2, 3, \ldots\}$.
- The *rational numbers*, denoted by $\mathbf{Q}$, are the elements of the set $\{\, a/b \mid a, b \in \mathbf{Z}, b \neq 0 \,\}$.
- The *irrational numbers* are the numbers whose decimal representations neither terminate nor repeat.
- The *real numbers*, denoted by $\mathbf{R}$, are the elements in the union of the sets of rational and irrational numbers.

Notation

- The Greek letter π represents the number equal to the ratio of the circumference of a circle to the diameter of that circle. The value of π is approximately 3.14159.
- The letter e represents the number approached by the value of the expression $(1 + 1/n)^n$ as n increases without bound. The value of e is approximately 2.71828.

Comments

- The natural numbers are also called *counting numbers*.
- The decimal representations of rational numbers either terminate or repeat.
- We can denote the whole numbers by $\mathbf{N} \cup \{0\}$.
- We can denote the irrational numbers by $\mathbf{R} \backslash \mathbf{Q}$.
- $\mathbf{N} \subset \mathbf{Z} \subset \mathbf{Q} \subset \mathbf{R}$.
- When writing symbols by hand for the sets of numbers described above, it is customary to write $\mathbb{N}$, $\mathbb{Z}$, $\mathbb{Q}$, and $\mathbb{R}$ rather than use bold face.
- The numbers that π and e represent are both irrational.
- We can find an approximation to the value of e by evaluating the expression $(1+1/n)^n$ for a large number n; the larger the value of n, the better the approximation, as the table below shows:

n	$(1 + 1/n)^n$
1	2.0000000
10	2.5937425
100	2.7048138
1000	2.7169239
10000	2.7181459
100000	2.7182682
1000000	2.7182805
10000000	2.7182817

Section 4: Interval Notation

DEFINITIONS

Suppose $a, b \in \mathbf{R}$ with $a \leq b$.

- The *open interval* from a to b is the set of real numbers greater than a and less than b.
- The *closed interval* from a to b is the set of real numbers greater than or equal to a and less than or equal to b.
- The *lower bound* of an interval from a to b is the number a.
- The *upper bound* of an interval from a to b is the number b.

NOTATION

Notations for intervals of various kinds are shown below. Observe that a bound next to a parenthesis is excluded from the set and a bound next to a bracket is included in the set.

$$
\begin{aligned}
(a,b) &= \{\, x \in \mathbf{R} \mid a < x < b \,\} & \qquad (a,\infty) &= \{\, x \in \mathbf{R} \mid a < x \,\} \\
[a,b) &= \{\, x \in \mathbf{R} \mid a \leq x < b \,\} & \qquad [a,\infty) &= \{\, x \in \mathbf{R} \mid a \leq x \,\} \\
(a,b] &= \{\, x \in \mathbf{R} \mid a < x \leq b \,\} & \qquad (\infty,b) &= \{\, x \in \mathbf{R} \mid x < b \,\} \\
[a,b] &= \{\, x \in \mathbf{R} \mid a \leq x \leq b \,\} & \qquad (\infty,b] &= \{\, x \in \mathbf{R} \mid x \leq b \,\}
\end{aligned}
$$

COMMENTS

- Intervals of the kind $[a,b)$ or $(a,b]$ are called *half-open* (or *half-closed*) intervals.
- The infinity (∞) and negative infinity ($-\infty$) symbols denote the absence of a bound; one should not place a square bracket next to either symbol.
- The interval $(-\infty, \infty)$ is the set of all real numbers.

EXAMPLE 1

Write each set in interval notation.

A. $\{\, x \in \mathbf{R} \mid -3 \leq x < 4 \,\}$

SOLUTION

$\{\, x \in \mathbf{R} \mid -3 \leq x < 4 \,\} = [-3, 4)$

□

B. $\{\, x \in \mathbf{R} \mid 0 < x \leq 7 \,\}$

C. $\{\, x \in \mathbf{R} \mid x > 2 \,\}$

SOLUTION

$\{\, x \in \mathbf{R} \mid x > 2 \,\} = (2, \infty)$

□

D. $\{\, x \in \mathbf{R} \mid x \leq -88 \,\}$

E. $\{\, x \in \mathbf{R} \mid -10 \leq x \leq -1 \,\}$

SOLUTION

$\{\, x \in \mathbf{R} \mid -10 \leq x \leq -1 \,\} = [-10, -1]$

□

F. $\{\, x \in \mathbf{R} \mid 6 < x \,\}$

EXAMPLE 2
Write each set in set builder notation.

A. $(5, \infty)$

SOLUTION
$(5, \infty) = \{ x \in \mathbf{R} \mid x > 5 \}$ □

B. $(-6, 8]$

C. $(0, 19)$

SOLUTION
$(0, 19) = \{ x \in \mathbf{R} \mid 0 < x < 19 \}$ □

D. $(-\infty, -2]$

E. $[-11, -4)$

SOLUTION
$[-11, -4) = \{ x \in \mathbf{R} \mid -11 \leq x < -4 \}$ □

F. $[7, 16]$

EXAMPLE 3
Simplify each expression.

A. $[-8, 6) \cap (3, 10)$

SOLUTION
$[-8, 6) \cap (3, 10) = (3, 6)$ □

B. $(-5, 2] \cup (0, 7]$

C. $(4, \infty) \cup [1, 13]$

SOLUTION
$(4, \infty) \cup [1, 13] = [1, \infty)$ □

D. $(-\infty, -6] \cap (-7, 38)$

E. $(-\infty, 27] \cup (20, \infty)$

SOLUTION
$(-\infty, 27] \cup (20, \infty) = \mathbf{R}$ □

F. $[2, 9] \cap (9, 11)$

Section 5: The Real Number Line

Definitions

- The *real number line* is a line on which each point represents a real number.
- A *coordinate* of a point is a number that indicates the location of the point.
- A *graph* is a locus of points.
- The *origin* on the real number line is the point that represents the number zero.
- To *plot* a point means to make a dot in an appropriate location.

Procedure

To depict points with given coordinates on the real number line, we:

1. construct a portion of the real number line by drawing a horizontal line segment;
2. place tick marks along the line segment at regular intervals according to a scale where:
   - the value at the leftmost mark is slightly less than the lowest coordinate and
   - the value at the rightmost mark is slightly greater than the highest coordinate;
3. identify for each point with coordinate a its proper location on the number line, where the point should lie:
   - $|a|$ units to the right of the origin if $a \geq 0$ or
   - $|a|$ units to the left of the origin if $a < 0$;
4. plot each point on the line segment in reference to the tick marks.

Procedure

To draw the graph of an interval, we:

1. construct a portion of the real number line;
2. place tick marks along the number line;
3. place on the line at each point that corresponds to a bound of the interval either:
   - a dot, which indicates the bound is included in the interval, or
   - a circle, which indicates the bound is excluded from the interval;
4. plot the points that correspond to the numbers between the bounds of the interval by shading the portion of the number line between the dots or circles made in step (3).

Notation

The notation $P\colon a$ means that the point P has the coordinate a.

Comments

- The coordinate of a point on the number line is equal to the real number it represents.
- The graph of:
  - a finite set of real numbers is a locus of isolated points;
  - an interval is a line segment;
  - the union of two sets consists of the points on at least one of the graphs of the sets;
  - the intersection of two sets consists of the points on both of the graphs of the sets.
- The number line extends without termination in two directions, though we can depict physically only a piece of it.
- Before we plot a point with a coordinate that is not in the form of a decimal, we should approximate that coordinate with a decimal.

Example 1
Plot on the real number line the points P: $\sqrt{3}$, Q: -2.5, and R: π.

Solution
We first determine by calculator that $\pi \approx 3.14$ and $\sqrt{3} \approx 1.73$. Since the given coordinates all lie between -3 and 4, we will make tick marks on our number line from -4 to 4 (a range that allows us to center the number line nicely about the origin) in increments of one unit. After forming the number line and placing these tick marks, we may plot the points that correspond to our given numbers:

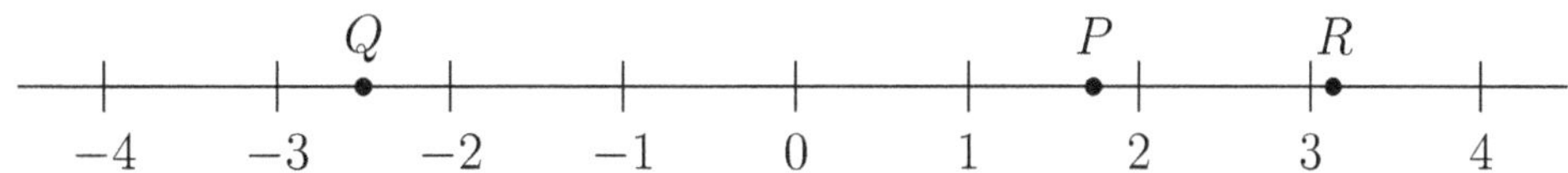

□

Example 2
Draw the graph of each set of real numbers on the number line provided.

A. $\{-12, -3, 7\}$

Solution

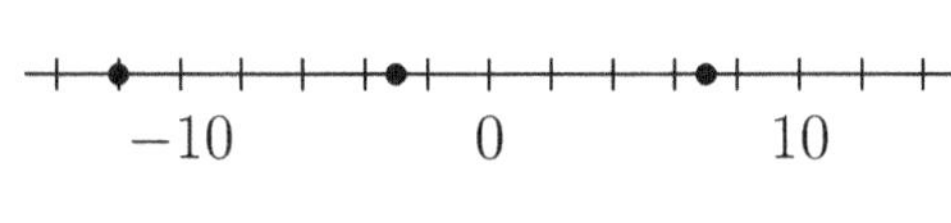

□

B. $\{5.2, 6.5, 6.8\}$

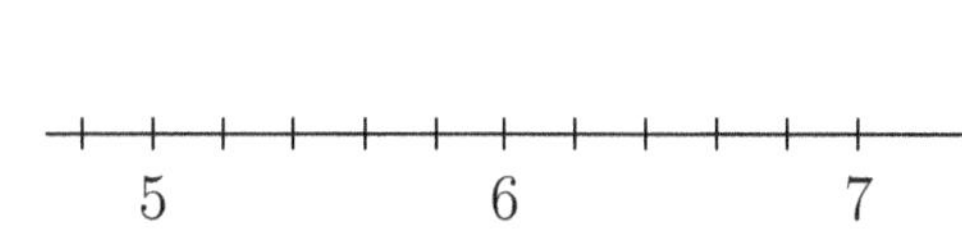

Example 3
Draw the graph of each interval.

A. $[-3, 4)$

Solution

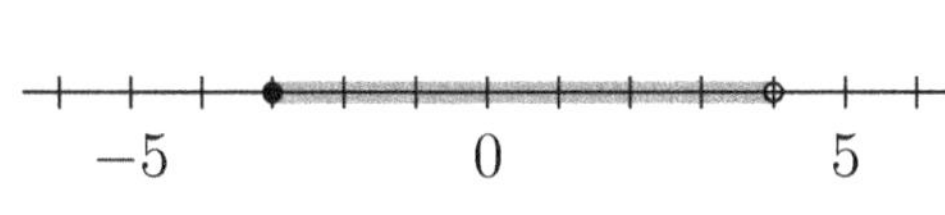

□

B. $(0, 7]$

C. $(8, \infty)$

Solution

5 10 15

□

D. $(-\infty, -1]$

EXAMPLE 4
Given the graphs of sets of real numbers A and B, draw the graphs of $A \cup B$ and $A \cap B$.

A. Graph of A:

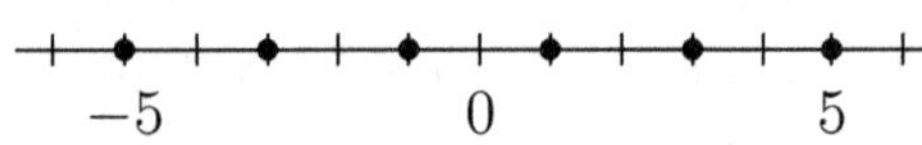

Graph of B:

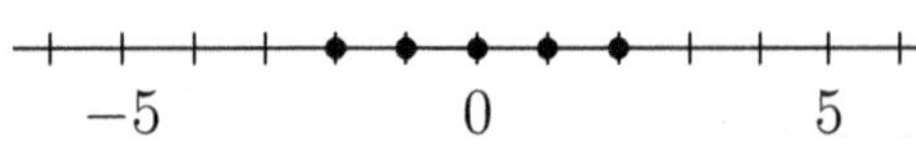

SOLUTION

- Graph of $A \cup B$:

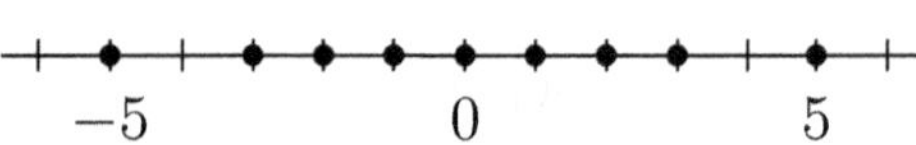

- Graph of $A \cap B$:

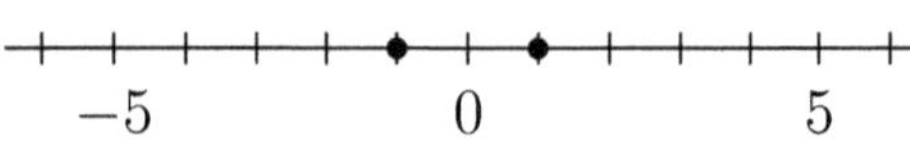

□

B. Graph of A:

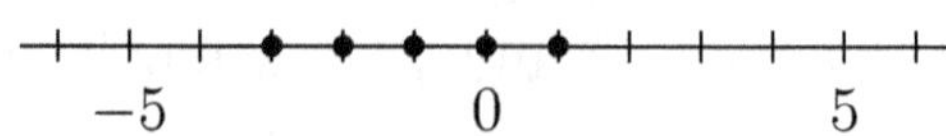

Graph of B:

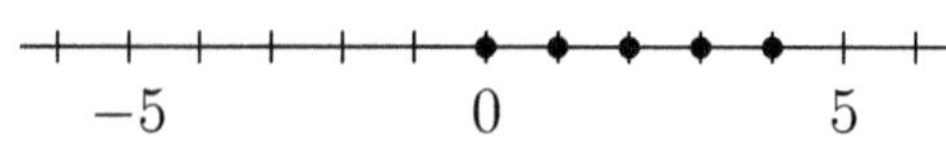

C. Graph of A:

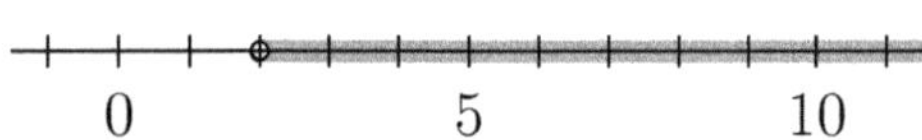

Graph of B:

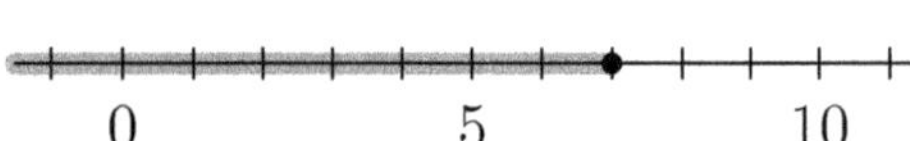

SOLUTION

- Graph of $A \cup B$:

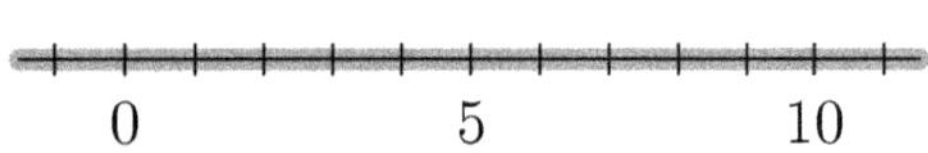

- Graph of $A \cap B$:

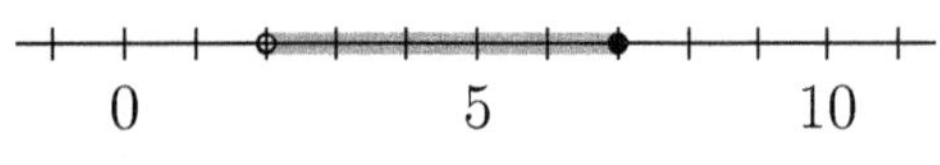

□

D. Graph of A:

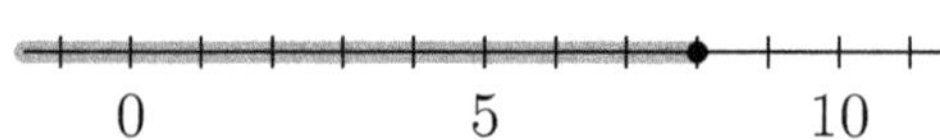

Graph of B:

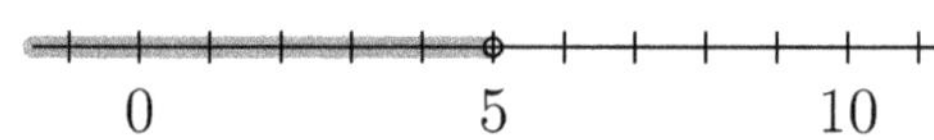

Section 6: Algebraic Statements

Definitions

- An *algebraic statement* is a mathematical sentence that asserts a relationship between algebraic expressions.
- An *equation* is an algebraic statement that asserts that the value of one expression is equal to the value of another expression.
- An *inequality* is an algebraic statement that asserts that the value of one expression is less than or greater than the value of another expression.
- A *solution* of an algebraic statement in one variable is a number that makes the statement true when substituted for the variable.
- To *solve* an algebraic statement means to determine the set of solutions of that statement.
- Algebraic statements are *algebraically equivalent* if they have the same set of solutions.

Comments

- The symbol $\leq$ means "less than or equal to".
- The symbol $\geq$ means "greater than or equal to".
- Algebraic statements can be either true or false. The veracity of an equation or inequality that involves a variable usually depends on the value of that variable; the statement may be true for some values but false for others.
- The set of solutions of an inequality usually contains an infinite number of elements. Thus we usually state the set of solutions of an inequality in set builder or interval notation.

Example 1

Determine whether the given equation is true or false, or if its veracity depends on the value of the variables it contains.

A. $2 + 3 = 5$
B. $9 \cdot 7 = 40$
C. $6x - 4 = 14$
D. $A = l \cdot w$,
where A is the area of a rectangle,
l is the length of the rectangle,
and w is the width of the rectangle

Solutions

A. true
B. false
C. veracity depends on the value of x
D. true

□

Example 2

Determine whether the given inequality is true or false, or if its veracity depends on the value of the variables it contains.

A. $4 > 5$
B. $2 \cdot 5 \leq 11$
C. $n^2 + 2n \geq 3$
D. $a + b < c$
where a, b, and c represent the lengths of the sides of a triangle

Example 3
Determine whether the given numbers are solutions of the given equation.

A. $x + 7 = 10$; 3, 5

Solution
Notes
- $3 + 7 = 10$
- $5 + 7 = 12$
  $\neq 10$

Conclusion
3 is a solution but 5 is not a solution of the equation $x + 7 = 10$. □

B. $2m - 5 > 9$; 1, 13

Example 4
A. The real number 2 is the only solution of the equation $6x - 4 = 8$. Therefore, the set of solutions of this equation is $\{2\}$.

B. The real numbers -3 and 1 are the only solutions of the equation $n^2 + 2n = 3$. Therefore, the set of solutions of this equation is $\{-3, 1\}$.

C. The real numbers less than 5 are the only solutions to the inequality $3x < 15$. Therefore, the set of solutions of this inequality is $\{\, x \in \mathbf{R} \mid x < 5 \,\}$ or $(-\infty, 5)$.

□

Section 7: Properties of Operations on Real Numbers

Definitions

Suppose $\diamond$ represents an operation on real numbers.

- The operation $\diamond$ is *commutative* if $a \diamond b = b \diamond a$ for all $a, b \in \mathbf{R}$.
- The operation $\diamond$ is *associative* if $(a \diamond b) \diamond c = a \diamond (b \diamond c)$ for all $a, b, c \in \mathbf{R}$.
- An *identity* element for $\diamond$ is a real number d for which $a \diamond d = d \diamond a = a$ for all $a \in \mathbf{R}$.
- The *inverse* of a real number a under $\diamond$ is a real number $a^\star$ for which $a \diamond a^\star = a^\star \diamond a = d$, where d represents the identity element for $\diamond$.

Properties of Addition$(+)$

- Addition is commutative since $a + b = b + a$ for all $a, b \in \mathbf{R}$.
- Addition is associative since $(a + b) + c = a + (b + c)$ for all $a, b, c \in \mathbf{R}$.
- 0 serves as an identity element for addition since $a + 0 = 0 + a = a$ for all $a \in \mathbf{R}$.
- The inverse of a real number a under addition is $-a$ since $a + (-a) = (-a) + a = 0$.

Properties of Multiplication$(\cdot)$

- Multiplication is commutative since $a \cdot b = b \cdot a$ for all $a, b \in \mathbf{R}$.
- Multiplication is associative since $(a \cdot b) \cdot c = a \cdot (b \cdot c)$ for all $a, b, c \in \mathbf{R}$.
- 1 serves as an identity element for multiplication since $a \cdot 1 = 1 \cdot a = a$ for all $a \in \mathbf{R}$.
- The inverse of a real number a under multiplication is $1/a$ since $a \cdot (1/a) = (1/a) \cdot a = 1$.

Distributive Property

Suppose $a, b, c \in \mathbf{R}$.
Then:

- $a \cdot (b + c) = a \cdot b + a \cdot c$
- $(b + c) \cdot a = b \cdot a + c \cdot a$

Comments

- The multiplicative inverse of 0 does not exist.
- The multiplicative inverse of a nonzero real number is equivalent to its reciprocal.
- Subtraction $(-)$ is merely addition of an additive inverse and division $(\div)$ is merely multiplication by a multiplicative inverse. The convertibility of expressions from the form "$a - b$" to the form "$a + (-b)$" and from the form "$a \div b$" to the form "$a \cdot (1/b)$" removes all need for properties of subtraction and division.
- The operations subtraction and division are neither commutative nor associative.
- Neither an identity element nor inverses exist under subtraction or division.

Example 1

Use an example to show that subtraction is not commutative.

Solution

$9 - 5 \neq 5 - 9$

□

Example 2

Use an example to show that division is not associative.

EXAMPLE 3

Complete the table to identify the additive and multiplicative inverses of each real number.

A.

Number	Additive Inverse	Multiplicative Inverse
3	−3	1/3
−10	10	−1/10
0	0	does not exist
5/11	−5/11	11/5
−1/8	1/8	−8

☐

B.

Number	Additive Inverse	Multiplicative Inverse
−7	☐	☐
π	☐	☐
1	☐	☐
1/12	☐	☐
−5/3	☐	☐

EXAMPLE 4

Identify the property that justifies each statement.

A. $7 \cdot (3 \cdot 9) = (7 \cdot 3) \cdot 9$
B. $0 + 64 = 64$
C. $(3)(10) = (10)(3)$
D. $5 + (-5) = 0$

SOLUTIONS
A. associative property of multiplication
B. property of additive identity element
C. commutative property of multiplication
D. property of additive inverses

☐

E. $5 \cdot (20 + 3) = (5 \cdot 20) + (5 \cdot 3)$
F. $\frac{1}{2} \cdot 2 = 1$
G. $7 + \sqrt{3} = \sqrt{3} + 7$
H. $6 \cdot 1 = 6$

EXAMPLE 5

Simplify each numerical expression using properties of addition and multiplication. Justify each step and do not use a calculator.

A. $-54 + [54 + (-10)]$

SOLUTION

$-54 + [54 + (-10)]$
$= [-54 + 54] + (-10)$
(by associative prop. of addition)
$= [0] + (-10)$
(by property of additive inverses)
$= -10$
(by property of additive identity)

☐

B. $2 \cdot (767 \cdot 50)$

C. $33(-73) + 33(-27)$

Solution

$33(-73) + 33(-27)$

$= 33(-73 + (-27))$

(by distributive property)

$= 33(-100)$

(by addition)

$= -3{,}300$

(by multiplication)

$\square$

D. $(49)(131) + (-48)(131)$

Section 8: Powers and Exponents

Definition

Suppose $b \in \mathbf{R}$ and $n \in \mathbf{N}$.

- The nth *power* of b is the number b^n, where $b^n = \overbrace{b \cdot b \cdot b \cdot \ldots \cdot b}^{n \text{ factors of } b}$.
- The *exponent* of the nth power of b is the number n.
- The *base* of the nth power of b is the number b.

Terminology

- The *square* of b means the 2nd power of b.
- The *cube* of b means the 3rd power of b.

Comments

- The term "power" is also often used as a synonym for "exponent".
- An exponent applies to a single number or variable in the absence of parentheses.

Example 1

A. $-3^4 = -(3 \cdot 3 \cdot 3 \cdot 3)$
B. $(-3)^4 = (-3) \cdot (-3) \cdot (-3) \cdot (-3)$
C. $2/7^5 = 2/(7 \cdot 7 \cdot 7 \cdot 7 \cdot 7)$
D. $(2/7)^5 = (2/7) \cdot (2/7) \cdot (2/7) \cdot (2/7) \cdot (2/7)$
E. $8x^3 = 8 \cdot (x \cdot x \cdot x)$
F. $(8x)^3 = (8x) \cdot (8x) \cdot (8x)$

□

Example 2

Evaluate each numerical expression.

A. 2^7

Solution

$$\begin{aligned} 2^7 &= 2 \cdot 2 \cdot 2 \cdot 2 \cdot 2 \cdot 2 \cdot 2 \\ &= 128 \end{aligned}$$

□

B. -5^2

C. $(-3)^3 - 3^2 - 6^2$

Solution

$$\begin{aligned} &(-3)^3 - 3^2 - 6^2 \\ &\quad = (-3)(-3)(-3) - (3)(3) - (6)(6) \\ &\quad = -27 - 9 - 36 \\ &\quad = -72 \end{aligned}$$

□

D. $(3/2)^3 - 5(3)^2/4$

EXAMPLE 3

Use exponents to write each expression in a more compact form.

A. $4 \cdot z \cdot z \cdot z$

SOLUTION

$4 \cdot z \cdot z \cdot z = 4z^3$

□

B. $9 \cdot 2 \cdot a \cdot a \cdot a \cdot a \cdot b \cdot b$

EXAMPLE 4

Evaluate each expression using the given values of the variables.

A. $-a^3 + 3b^2$ for $a = -3$ and $b = -2$

SOLUTION

$$\begin{aligned} -a^3 + 3b^2 &= -(-3)^3 + 3(-2)^2 \\ &= -(-27) + 3(4) \\ &= 27 + 12 \\ &= 39 \end{aligned}$$

□

B. $x^2 + 2xy - y^2$ for $x = -\frac{3}{2}$ and $y = 4$

Section 9: Like Terms

Definitions

- A *factor* is an algebraic expression that is multiplied to another expression within a product.
- A *term* is a variable, a constant, or a product of constants and variables.
- A *coefficient* is the numerical factor of a term.
- *Like terms* are terms that contain the same variables raised to the same exponents.

Procedure

To simplify a sum of like terms, we rewrite the expression as the product of the sum of the coefficients of the terms and the other factors of the terms.

Comments

- We cannot simplify a sum of terms that are not like.
- We use implicitly the distributive property when we combine like terms.

Example 1

The table below provides examples of terms and their coefficients:

Term:	$2x^2$	$-3xyz$	$7mn$	8	$ab/3$	y	$-c$
Coefficient:	2	-3	7	8	$1/3$	1	-1

Note that:

- the coefficient of $ab/3$ is $1/3$ since $ab/3 = (1/3) \cdot (ab)$;
- the coefficient of y is 1 since $y = 1 \cdot y$;
- the coefficient of $-c$ is -1 since $-c = (-1)c$.

□

Example 2

A. The terms $3x^2$ and $-5x^2$ are like.
B. The terms $-2x$ and $10y$ are not like; the first contains an x but the second does not.
C. The terms $3ab^2c$, $-7ab^2c$, and $8ab^2c$ are like.
D. The terms $4xy$ and $5xy^2$ are not like; the exponents on the y differ.

□

Example 3

Simplify each expression.

A. $9m + 3m + 5m$

Solution

$$\begin{aligned} 9m + 3m + 5m &= (9 + 3 + 5)m \\ &= 17m \end{aligned}$$

□

B. $4ab + 7ab$

C. $2mn^2q + (-5mn^2q)$

SOLUTION

$$\begin{aligned} 2mn^2q + (-5mn^2q) &= (2 + (-5))mn^2q \\ &= -3mn^2q \end{aligned}$$

□

D. $19vw^4 - 4vw^4 - 8vw^4$

Section 10: Simplifying Algebraic Expressions

Procedure

To simplify a term, we:

1. group constants and powers of the same variables together using the commutative and associative properties of multiplication;
2. evaluate the product of the constants;
3. rewrite each product of powers of the same variable as a single power of that variable.

Procedure

To simplify an algebraic expression with multiple terms, we:

1. convert all differences to sums of additive inverses;
2. eliminate grouping symbols using the distributive property;
3. simplify all terms;
4. group like terms together using the commutative and associative properties of addition;
5. combine like terms.

Comment

We should normally simplify expressions immediately after we perform algebraic operations on them.

Example 1

Simplify each term.

A. $(-5c^2)(3c^3)$

Solution

$$\begin{aligned}(-5c^2)(3c^3) &= (-5 \cdot c^2)(3 \cdot c^3)\\ &= -5 \cdot c^2 \cdot 3 \cdot c^3\\ &= -5 \cdot 3 \cdot c^2 \cdot c^3\\ &= (-5 \cdot 3)(c \cdot c \cdot c \cdot c \cdot c)\\ &= -15c^5\end{aligned}$$

□

B. $(2xy)(7x^2y)$

Example 2

Simplify each expression.

A. $4y - 11y + 2y$

Solution

$$\begin{aligned}4y - 11y + 2y &= 4y + (-11y) + 2y\\ &= (4 + (-11) + 2)y\\ &= -5y\end{aligned}$$

□

B. $-4x^2 + 9x^2$

C. $10x - 3y^3 - 4x - 15y^3$

SOLUTION

$$\begin{aligned}
&10x - 3y^3 - 4x - 15y^3\\
&\quad = 10x + (-3y^3) + (-4x) + (-15y^3)\\
&\quad = 10x + (-4x) + (-3y^3) + (-15y^3)\\
&\quad = (10 + (-4))x + (-3 + (-15))y^3\\
&\quad = 6x + (-18)y^3\\
&\quad = 6x - 18y^3
\end{aligned}$$

□

D. $14mn - 7 + 19mn - 5 - 3mn$

E. $7(xz + 6) + 9(xz + 1)$

SOLUTION

$$\begin{aligned}
&7(xz + 6) + 9(xz + 1)\\
&\quad = 7(xz) + 7(6) + 9(xz) + 9(1)\\
&\quad = 7xz + 42 + 9xz + 9\\
&\quad = 7xz + 9xz + 42 + 9\\
&\quad = (7 + 9)xz + (42 + 9)\\
&\quad = 16xz + 51
\end{aligned}$$

□

F. $-4(c + 3) + 6(c + 8) + 4(c - 5)$

EXAMPLE 3

Simplify each expression and then evaluate the result for the given values of the variables.

A. $-(n + 2) - 3(n - 6)$ for $n = 11$

SOLUTION

Simplification

$$\begin{aligned}
&-(n + 2) - 3(n - 6)\\
&\quad = (-1)(n + 2) + (-3)(n + (-6))\\
&\quad = (-1)(n) + (-1)(2)\\
&\qquad\qquad + (-3)(n) + (-3)(-6)\\
&\quad = (-1n) + (-2) + (-3n) + 18\\
&\quad = (-1n) + (-3n) + (-2) + 18\\
&\quad = ((-1) + (-3))n + ((-2) + 18)\\
&\quad = -4n + 16
\end{aligned}$$

Evaluation

For $n = 11$:

$$\begin{aligned}
-4n + 16 &= -4(11) + 16\\
&= -44 + 16\\
&= -28
\end{aligned}$$

□

B. $5(x + y) - 7(x - y)$ for $x = 3$ and $y = -4$

Section 11: Properties of Equality and Inequality

Properties of Equality

Suppose a, b, and c represent algebraic expressions.

- Reflexive property: $a = a$
- Symmetric property: if $a = b$, then $b = a$.
- Transitive property: if $a = b$ and $b = c$, then $a = c$.
- Additive property: $a = b$ if and only if $a + c = b + c$
- Multiplicative property: $a = b$ if and only if $ac = bc$ (provided $c \neq 0$)
- Substitutive property: if $a = b$, then the substitution of b for a in an algebraic statement will not alter the truth of that statement

Properties of Inequality

Suppose a, b, and c represent algebraic expressions.

- Transitive property: if $a < b$ and $b < c$, then $a < c$.
- Additive property: $a < b$ if and only if $a + c < b + c$
- Multiplicative property: $a < b$ if and only if $ac < bc$ when $c > 0$
  $a < b$ if and only if $ac > bc$ when $c < 0$

Example 1

Identify the property that justifies each statement.

A. $5 = 5$
B. if $m = n$ and $n = 11$, then $m = 11$
C. $z = 7$ if and only if $z \cdot 4 = 7 \cdot 4$
D. $x > y$ if and only if $-5x < -5y$
E. $10 > a$ if and only if $10 + b > a + b$
F. if $x = 4$, then $4 = x$
G. $b = -3$ if and only if $b + 8 = -3 + 8$
H. if $y = 2x - 5$ and $6 + y = 19$, then $6 + (2x - 5) = 19$
I. $7 < v$ if and only if $3 \cdot 7 < 3 \cdot v$

Solutions

A. reflexive property of equality
B. transitive property of equality
C. multiplicative property of equality
D. multiplicative property of inequality
E. additive property of inequality

□

Section 12: Linear Equations in One Variable

Definition
A *linear* (or *first-degree*) *equation in one variable* x is an equation that can be written equivalently in the form $Ax = B$ for some constants $A, B \in \mathbf{R}$ with $A \neq 0$.

Procedure
To solve a linear equation in one variable, we:

1. convert all differences to sums of additive inverses and all quotients to products of multiplicative inverses, then simplify both sides of the equation;
2. add additive inverses to both sides of the equation as necessary until one side contains no variables and the other side contains no constant terms, then simplify;
3. multiply both sides of the equation by the multiplicative inverse of the coefficient of the term with the variable, then simplify;
4. state the set of solutions.

Comments

- A linear equation is called a first-degree equation because the highest exponent on the variable in the equation is 1.
- A linear equation in one variable will have one solution.
- Regarding the procedure above:
  - We should simplify both sides of the equation after each of the first three steps.
  - We will convert differences to sums and quotients to products infrequently and apply properties of addition, multiplication, and equality less explicitly within our processes of finding solutions after a little practice.
  - After step (1), the given linear equation should take the form $ax + b = cx + d$.
    After step (2), the equation should take the form $ex = f$.
    After step (3), the equation should take the form $x = g$.
  - The solution to the equation yielded after step (3), namely the number equal to the variable, will also be the solution to the original equation. Since the solution is obvious at this point, it is common practice to omit the somewhat superfluous statement of the set of solutions in step (4) (see the final example of this section).
- After step (2) of the procedure above, if the equation takes the form $0 = f$ where:
  - $f \neq 0$, then the original equation has no solutions, so the set of solutions is $\emptyset$;
  - $f = 0$, then all real numbers are solutions to the equation, so the set of solutions is $\mathbf{R}$.

Notation

- The symbol $\Rightarrow$ means "implies". We place this symbol between two statements when the second statement is true if the first statement is true.
- The symbol $\Leftrightarrow$ means "if and only if". We place this symbol between two statements when those statements are logically equivalent (that is, if one statement is true, then the other statement is also true). In other words, "Statement1 $\Leftrightarrow$ Statement2" means both that "Statement1 $\Rightarrow$ Statement2" and "Statement2 $\Rightarrow$ Statement1."
- The symbol $\therefore$ means "therefore". We place this symbol to the left of a statement that follows logically from the preceding statement or statements.

EXAMPLE 1
Solve each equation.

A. $x + 6 = 9$

SOLUTION
$$\begin{aligned} x + 6 = 9 &\Leftrightarrow (x+6) + (-6) = 9 + (-6) \\ &\Leftrightarrow x + (6 + (-6)) = 3 \\ &\Leftrightarrow x + 0 = 3 \\ &\Leftrightarrow x = 3 \end{aligned}$$
Set of solutions: $\{3\}$

$\square$

B. $n + 18 = 3$

C. $22 = x - 14$

SOLUTION
$$\begin{aligned} &22 = x - 14 \\ &\Leftrightarrow 22 = x + (-14) \\ &\Leftrightarrow 22 + 14 = (x + (-14)) + 14 \\ &\Leftrightarrow 36 = x + ((-14) + 14) \\ &\Leftrightarrow 36 = x + 0 \\ &\Leftrightarrow 36 = x \end{aligned}$$
Set of solutions: $\{36\}$

$\square$

D. $b - 2.7 = 5.2$

E. $8n = 40$

SOLUTION
$$\begin{aligned} 8n = 40 &\Leftrightarrow (1/8) \cdot (8n) = (1/8) \cdot 40 \\ &\Leftrightarrow ((1/8) \cdot 8)(n) = 5 \\ &\Leftrightarrow (1)(n) = 5 \\ &\Leftrightarrow n = 5 \end{aligned}$$
Set of solutions: $\{5\}$

$\square$

F. $-4y = 3/5$

G. $(-4/9)x = 24$

SOLUTION
$$\begin{aligned} &(-4/9)x = 24 \\ &\Leftrightarrow (-9/4) \cdot (-4/9)x = (-9/4) \cdot 24 \\ &\Leftrightarrow ((-9/4) \cdot (-4/9))(x) = -54 \\ &\Leftrightarrow (1)x = -54 \\ &\Leftrightarrow x = -54 \end{aligned}$$
Set of solutions: $\{-54\}$

$\square$

H. $(2/3)x = 10$

I. $m/11 = -9$

SOLUTION

$m/11 = -9$
$\Leftrightarrow (1/11) \cdot m = -9$
$\Leftrightarrow 11 \cdot ((1/11) \cdot m) = 11 \cdot (-9)$
$\Leftrightarrow (11 \cdot (1/11)) \cdot m = -99$
$\Leftrightarrow 1 \cdot m = -99$
$\Leftrightarrow m = -99$

Set of solutions: $\{-99\}$

□

J. $-1.4z = 9.8$

K. $2c + 5 = 19$

SOLUTION

$2c + 5 = 19$
$\Leftrightarrow (2c + 5) + (-5) = 19 + (-5)$
$\Leftrightarrow 2c + (5 + (-5)) = 14$
$\Leftrightarrow 2c + 0 = 14$
$\Leftrightarrow 2c = 14$
$\Leftrightarrow (1/2)(2c) = (1/2)(14)$
$\Leftrightarrow ((1/2) \cdot 2)(c) = 7$
$\Leftrightarrow (1)(c) = 7$
$\Leftrightarrow c = 7$

Set of solutions: $\{7\}$

□

L. $5x - 3 = 22$

M. $17 = -11 - 7a$

SOLUTION

$17 = -11 - 7a$
$\Leftrightarrow 17 = -11 + (-7a)$
$\Leftrightarrow 11 + 17 = 11 + (-11 + (-7a))$
$\Leftrightarrow 28 = (11 + (-11)) + (-7a)$
$\Leftrightarrow 28 = 0 + (-7a)$
$\Leftrightarrow 28 = -7a$
$\Leftrightarrow (-1/7)(28) = (-1/7)(-7a)$
$\Leftrightarrow -4 = ((-1/7)(-7))(a)$
$\Leftrightarrow -4 = (1)(a)$
$\Leftrightarrow -4 = a$

Set of solutions: $\{-4\}$

□

N. $7 - 3x = 7$

O. $-x = 8$

Solution

$$\begin{aligned}-x = 8 &\Leftrightarrow (-1)(x) = 8\\ &\Leftrightarrow (-1)((-1)(x)) = (-1)(8)\\ &\Leftrightarrow ((-1)(-1))(x) = -8\\ &\Leftrightarrow (1)(x) = -8\\ &\Leftrightarrow x = -8\end{aligned}$$

Set of solutions: $\{-8\}$

□

P. $32 - m = 45$

Q. $5.2 = -6.1 - 3.2x$

Solution

$$\begin{aligned}&5.2 = -6.1 - 3.2x\\ &\Leftrightarrow 5.2 = -6.1 + (-3.2x)\\ &\Leftrightarrow 6.1 + 5.2 = 6.1 + (-6.1 + (-3.2x))\\ &\Leftrightarrow 11.3 = (6.1 + (-6.1)) + (-3.2x)\\ &\Leftrightarrow 11.3 = 0 + (-3.2x)\\ &\Leftrightarrow 11.3 = -3.2x\\ &\Leftrightarrow (-1/3.2)(11.3) = (-1/3.2)(-3.2x)\\ &\Leftrightarrow -11.3/3.2 = ((-1/3.2)(-3.2))(x)\\ &\Leftrightarrow (-11.3/3.2)(10/10) = (1)(x)\\ &\Leftrightarrow -113/32 = x\end{aligned}$$

Set of solutions: $\{-113/32\}$

□

R. $z + 5/8 = 3/5$

S. $2m + 3 + 5m = -18$

Solution

$$\begin{aligned}&2m + 3 + 5m = -18\\ &\Leftrightarrow 2m + 5m + 3 = -18\\ &\Leftrightarrow 7m + 3 = -18\\ &\Leftrightarrow (7m + 3) + (-3) = -18 + (-3)\\ &\Leftrightarrow 7m + (3 + (-3)) = -21\\ &\Leftrightarrow 7m + 0 = -21\\ &\Leftrightarrow 7m = -21\\ &\Leftrightarrow (1/7)(7m) = (1/7)(-21)\\ &\Leftrightarrow ((1/7)\cdot 7)(m) = -3\\ &\Leftrightarrow (1)(m) = -3\\ &\Leftrightarrow m = -3\end{aligned}$$

Set of solutions: $\{-3\}$

□

T. $y + 6 + 5y - 3 = 15$

U. $3 + 7x = 5x + 3 + 2x$

SOLUTION

$3 + 7x = 5x + 3 + 2x$
$\Leftrightarrow 3 + 7x = 5x + 2x + 3$
$\Leftrightarrow 7x + 3 = 7x + 3$
$\Leftrightarrow (7x + 3) + (-3) = (7x + 3) + (-3)$
$\Leftrightarrow 7x + (3 + (-3)) = 7x + (3 + (-3))$
$\Leftrightarrow 7x + (0) = 7x + (0)$
$\Leftrightarrow 7x = 7x$
$\Leftrightarrow 7x + (-7x) = 7x + (-7x)$
$\Leftrightarrow 0x = 0x$
$\Leftrightarrow 0 = 0$

Set of solutions: **R**

□

V. $9 - 3x + 2 = 16 - 3x$

W. $-2x = 7 - 2x$

SOLUTION

$-2x = 7 - 2x$
$\Leftrightarrow -2x = 7 + (-2x)$
$\Leftrightarrow -2x + 2x = (7 + (-2x)) + 2x$
$\Leftrightarrow 0x = 7 + ((-2x) + 2x)$
$\Leftrightarrow 0 = 7 + (0x)$
$\Leftrightarrow 0 = 7$

Set of solutions: $\emptyset$

□

X. $m + 9 + 4m = 17 + 5m - 8$

Y. $-6x + 9 = -7x - 1$

SOLUTION

$-6x + 9 = -7x - 1$
$\Leftrightarrow -6x + 9 = -7x + (-1)$
$\Leftrightarrow 7x + (-6x + 9) = 7x + (-7x + (-1))$
$\Leftrightarrow (7x + (-6x)) + 9$
$\quad = (7x + (-7x)) + (-1)$
$\Leftrightarrow (1x) + 9 = (0x) - 1$
$\Leftrightarrow x + 9 = 0 - 1$
$\quad = -1$
$\Leftrightarrow (x + 9) + (-9) = -1 + (-9)$
$\Leftrightarrow x + (9 + (-9)) = -10$
$\Leftrightarrow x + (0) = -10$
$\Leftrightarrow x = -10$

□

Z. $-3x + 6 + 5x = 7x - 9x - 5$

Section 13: From English to Algebra

Terminology

Below are some common words and phrases associated with arithmetic operations:

- Addition (+): plus, sum, increased by, added to, more than
- Subtraction (−): minus, difference, decreased by, subtracted from, less than
- Multiplication (·): times, product, multiplied by
- Division (÷): divided by, quotient, divided into

Comments

- We often represent quotients as fractions. For example, we can write $7 \div 3$ as $\frac{7}{3}$ or $7/3$.
- When preceding a phrase, the word "quantity" indicates that we should enclose the algebraic expression corresponding to the phrase in parentheses.

Example 1

We would translate each phrase below into the algebraic expression $z + 5$:

- the sum of z and 5
- z plus 5
- z increased by 5
- 5 more than z

□

Example 2

We would translate each phrase below into the algebraic expression $7 - x$:

- the difference of 7 and x
- 7 minus x
- 7 decreased by x
- x less than 7

□

Example 3

We would translate each phrase below into the algebraic expression $6 \cdot y$:

- the product of 6 and y
- 6 times y
- 6 multiplied by y

□

Example 4

We would translate each phrase below into the algebraic expression $m/8$:

- the quotient of m and 8
- m divided by 8
- 8 divided into m

□

Example 5

Translate each phrase into an algebraic expression.

A. the difference of m and n
B. eight times the quantity, x plus 3
C. the sum of four times z and six times y
D. the square of x subtracted from 9
E. π divided into C
F. 22 minus the product of a and b
G. the quotient of d and 5, plus negative 4
H. 5 plus the quantity, m minus 5, cubed

Solutions

A. $m - n$
B. $8 \cdot (x + 3)$
C. $4z + 6y$
D. $9 - x^2$

□

EXAMPLE 6
Answer each question with an algebraic expression.

A. Robert is r years old. How old was he two years ago?

SOLUTION
$r - 2$ years old

□

B. Susie is s years old. Her brother is twice as old as she is. What is Susie's brother's age?

C. Suppose z represents an integer. What is the next largest integer?

SOLUTION
$z + 1$

□

D. Suppose w represents an even integer. What is the next lower even integer?

E. James has n nickels worth five cents each. How many cents does he have?

SOLUTION
$5n$ cents

□

F. Tom can run t meters in one minute. How many meters can he run in one hour?

G. Suppose g grams of gold cost d dollars. What is the price of gold per gram?

SOLUTION
d/g dollars per gram

□

H. The width of a rectangle is w feet. The length of the rectangle is one more foot than thrice its width. What is the perimeter of the rectangle?

Section 14: Word Problems: Linear Applications

Procedure

To solve word problems with a single unknown quantity, we:

1. declare a variable to represent the unknown quantity;
2. write an equation in the variable that reflects the information given;
3. solve the equation of step (2);
4. answer the original question.

Comments

- Use the word "let" to declare variables.
- Choose meaningful letters when declaring variables; appropriate choices of variables add clarity to the process of formulating an equation and finding its solution.
- Use subscripts to differentiate between variables that represent quantities of the same type.
- It can be helpful to draw a diagram that depicts the situation in the given problem.
- We can sometimes use the substitutive property of equality to convert an equation with two variables into an equation with one variable.

Example 1

Solve each word problem.

A. The sum of a certain number and 12 is 15. Find the number.

Solution

Let n represent the number in question.

$n + 12 = 15$

$\Leftrightarrow (n + 12) + (-12) = 15 + (-12)$

$\Leftrightarrow n + (-12 + 12) = 3$

$\Leftrightarrow n = 3$

The number is 3.

□

B. When we subtract a certain number from 33, the difference is 25. Find the number.

C. Eight years ago, Bo was fifteen years old. How old is Bo now?

Solution

Let b represent Bo's current age.

$b - 8 = 15$

$\Leftrightarrow b + (-8) = 15$

$\Leftrightarrow (b + (-8)) + 8 = 15 + 8$

$\Leftrightarrow b + ((-8) + 8) = 23$

$\Leftrightarrow b = 23$

Bo is now twenty-three years old.

□

D. In five years, Carl will be 48 years old. How old is Carl now?

E. Marv bought 12 apples for \$10.20. What was the price per apple?

SOLUTION

Let p represent the price per apple.

$12p = 10.20$

$\Leftrightarrow (1/12)(12p) = (1/12)(10.20)$

$\Leftrightarrow p = 0.85$

The price per apple was \$0.85.

□

F. Sixteen is eighty-six less than six times a certain number. Find the number.

G. The total cost of parts and labor for the repair of a bicycle was \$68. The cost of the parts was \$26. The cost of labor was assessed at a rate of \$12 per hour. Find the number of hours of labor involved.

SOLUTION

Let h be the number of hours of labor.

$26 + 12h = 68$

$\Leftrightarrow (-26) + (26 + 12h) = (-26) + 68$

$\Leftrightarrow 12h = 42$

$\Leftrightarrow (1/12)(12h) = (1/12)(42)$

$\Leftrightarrow h = 3.5$

The number of hours of labor was 3.5.

□

H. A graduate in humanities earned \$364 last year. These wages were \$100 less than twice his wages five years ago. What were his wages five years ago?

I. Find two consecutive even numbers whose sum is 74.

SOLUTION

Let m denote the lower even number and n denote the higher even number.

Note that $n = m + 2$.

$m + n = 74$

$\Rightarrow m + (m + 2) = 74$

$\Rightarrow 2m + 2 = 74$

$\Rightarrow (2m + 2) + (-2) = 74 + (-2)$

$\Rightarrow 2m = 72$

$\Rightarrow (1/2)(2m) = (1/2)(72)$

$\Rightarrow m = 36$

$\Rightarrow n = 36 + 2$

$= 38$

The numbers are 36 and 38.

□

J. Find three consecutive odd numbers whose sum is 87.

K. Tom bought a shirt and a tie for $49 in total. The price of the shirt was three times the price of the tie. How much did each article of clothing cost?

SOLUTION
Let t represent the price of the tie and
s represent the price of the shirt.
Note that $s = 3t$.
$s + t = 49$
$\Rightarrow 3t + t = 49$
$\Rightarrow 4t = 49$
$\Rightarrow (1/4)(4t) = (1/4)(49)$
$\Rightarrow t = 12.25$
$\Rightarrow s = 3(12.25)$
$= 36.75$
The price of the tie was $12.25 and the price of the shirt was $36.75. □

L. A computer and a printer cost $430 in total. The price of the computer is $10 more than five times the price of the printer. Find the price of each.

M. A group of friends went bowling. The highest score was eight points less than three times the lowest score, and the sum of the highest and lowest scores was 364. Determine the highest and lowest scores.

SOLUTION
Let h represent the highest score and
l represent the lowest score.
Note that $h = 3l - 8$.
$h + l = 364$
$\Rightarrow (3l - 8) + l = 364$
$\Rightarrow 3l + (-8) + l = 364$
$\Rightarrow 4l + (-8) = 364$
$\Rightarrow (4l + (-8)) + 8 = 364 + 8$
$\Rightarrow 4l = 372$
$\Rightarrow (1/4)(4l) = (1/4)(372)$
$\Rightarrow l = 93$
$\Rightarrow h = 3(93) - 8$
$= 279 - 8$
$= 271$
The highest score was 271 and the lowest score was 93. □

N. An urn contains 230 marbles colored blue, yellow, and red. There are 18 fewer blue marbles than yellow marbles, and there are twice as many red marbles as yellow marbles. How many marbles of each color are in the urn?

O. The difference between the measures of two complementary angles is 58°. Find the measures.

SOLUTION

Declarations

Let a_1 denote the larger measure and a_2 the smaller angle measure.

Substitution

$a_1 + a_2 = 90$
$\Leftrightarrow (a_1 + a_2) + (-a_2) = 90 + (-a_2)$
$\Leftrightarrow a_1 = 90 - a_2$

Values

$a_1 - a_2 = 58$
$\Rightarrow (90 - a_2) - a_2 = 58$
$\Rightarrow 90 + (-a_2) + (-a_2) = 58$
$\Rightarrow 90 + (-2a_2) = 58$
$\Rightarrow -90 + (90 + (-2a_2)) = -90 + 58$
$\Rightarrow -2a_2 = -32$
$\Rightarrow (-1/2)(-2a_2) = (-1/2)(-32)$
$\Rightarrow a_2 = 16$
$\Rightarrow a_1 = 90 - 16$
$= 74$

Conclusion

The angles measure 74° and 16°.

□

P. Two angles are supplementary and one is four times as large as the other. Find the measures of both angles.

Section 15: Linear Equations with Parentheses

Procedure

To solve a linear equation in one variable that contains parentheses, we:

1. eliminate the parentheses using the distributive property;
2. simplify the equation yielded after step (1);
3. solve the equation yielded after step (2).

Example 1

Solve each equation.

A. $4(x-3)=24$

Solution

$4(x-3)=24$

$\Leftrightarrow 4(x+(-3))=24$

$\Leftrightarrow 4(x)+4(-3)=24$

$\Leftrightarrow 4x+(-12)=24$

$\Leftrightarrow 4x+(-12)+12=24+12$

$\Leftrightarrow 4x=36$

$\Leftrightarrow (1/4)(4x)=(1/4)(36)$

$\Leftrightarrow x=9$

Set of solutions: $\{9\}$

□

B. $-5(x+7)=-15$

C. $-(x+7)=-4(x+10)$

Solution

$-(x+7)=-4(x+10)$

$\Leftrightarrow (-1)(x+7)=-4(x+10)$

$\Leftrightarrow (-1)(x)+(-1)(7)$
$=(-4)(x)+(-4)(10)$

$\Leftrightarrow (-1)x+(-7)=-4x+(-40)$

$\Leftrightarrow (-1)x+(-7)+7$
$=-4x+(-40)+7$

$\Leftrightarrow (-1)x=-4x+(-33)$

$\Leftrightarrow 4x+(-1)x=4x+(-4x)+(-33)$

$\Leftrightarrow 3x=-33$

$\Leftrightarrow (1/3)(3x)=(1/3)(-33)$

$\Leftrightarrow x=-11$

Set of solutions: $\{-11\}$

□

D. $-2(x+6)+3(3x-2)=-3(x-5)$

E. $5[3+8(x-1)-4(9+3x)] = 7(10-17x)+22$

Solution

$5[3+8(x-1)-4(9+3x)] = 7(10-17x)+22$
$\Leftrightarrow 5[3+8(x+(-1))+(-4)(9+3x)] = 7(10+(-17x))+22$
$\Leftrightarrow 5[3+8(x)+8(-1)+(-4)(9)+(-4)(3x)] = 7(10)+7(-17x)+22$
$\Leftrightarrow 5[3+8x+(-8)+(-36)+(-12x)] = 70+(-119x)+22$
$\Leftrightarrow 5[8x+(-12x)+3+(-8)+(-36)] = (-119x)+70+22$
$\Leftrightarrow 5[-4x+(-41)] = -119x+92$
$\Leftrightarrow 5(-4x)+5(-41) = -119x+92$
$\Leftrightarrow -20x+(-205) = -119x+92$
$\Leftrightarrow 119x+(-20x)+(-205) = 119x+(-119x)+92$
$\Leftrightarrow 99x+(-205) = 92$
$\Leftrightarrow 99x+(-205)+205 = 92+205$
$\Leftrightarrow 99x = 297$
$\Leftrightarrow (1/99)(99x) = (1/99)(297)$
$\Leftrightarrow x = 3$

Set of solutions: $\{3\}$

$\square$

Example 2

Find two consecutive integers such that three times the larger integer minus the smaller integer equals 37.

Solution

Let m represent the larger number and
n represent the smaller number.
Note that $m = n+1$.

$3m - n = 37$
$\Rightarrow 3(n+1) - n = 37$
$\Rightarrow 3(n) + 3(1) - n = 37$
$\Rightarrow 3n + 3 + (-n) = 37$
$\Rightarrow 3n + (-n) + 3 = 37$
$\Rightarrow 2n + 3 = 37$
$\Rightarrow 2n + 3 + (-3) = 37 + (-3)$
$\Rightarrow 2n = 34$
$\Rightarrow (1/2)(2n) = (1/2)(34)$
$\Rightarrow n = 17$
$\Rightarrow m = n + 1$
$= 17 + 1$
$= 18$

The integers are 17 and 18.

$\square$

Example 3

Find two consecutive odd integers where the larger number plus seven times the smaller number equals 122.

Example 4
In a collection of 421 marbles, there are 35 more yellow ones than red ones and twice as many green ones as yellow ones. Find the number of marbles of each color.

Solution
Declarations
Let y be the number of yellow marbles,
r be the number of red marbles, and
g be the number of green marbles.

Substitutions
- $y = r + 35$
- $g = 2y$

Values
$r + y + g = 421$
$\Rightarrow r + (r + 35) + (2y) = 421$
$\Rightarrow r + r + 35 + (2(r + 35)) = 421$
$\Rightarrow 2r + 35 + (2)(r) + (2)(35) = 421$
$\Rightarrow 2r + 35 + 2r + 70 = 421$
$\Rightarrow 2r + 2r + 35 + 70 = 421$
$\Rightarrow 4r + 105 = 421$
$\Rightarrow 4r + 105 + (-105) = 421 + (-105)$
$\Rightarrow 4r = 316$
$\Rightarrow (1/4)(4r) = (1/4)(316)$
$\Rightarrow r = 79$
$\Rightarrow y = 79 + 35$
$= 114$
$\Rightarrow g = 2(114)$
$= 228$

Conclusion
The collection has 114 yellow marbles, 79 red marbles, and 228 green marbles.

□

Example 5
A jar contains seven fewer 5-cent coins than 25-cent coins, but three times as many 10-cent coins as 25-cent coins.
If the coins of these denominations are worth a total of 625 cents, then how many coins of each type are in the jar?

Section 16: Linear Equations with Fractions

Procedure

To solve a linear equation in one variable that contains fractions, we:

1. identify the least common multiple of the denominators of the fractions;
2. eliminate the fractions by multiplying both sides of the equation by the least common denominator found in step (1);
3. solve the equation yielded after step (2).

Comments

The equations in this section will not contain fractions with variables in their denominators; we will follow a slightly different procedure to solve equations that do contain such.

Example 1

Solve each equation.

A. $(1/2)x - 2/3 = 7/6$

Solution

$(1/2)x - 2/3 = 7/6$
$\Leftrightarrow 6((1/2)x + (-2/3)) = 6(7/6)$
$\Leftrightarrow 6(1/2)(x) + 6(-2/3) = 7$
$\Leftrightarrow 3x + (-4) = 7$
$\Leftrightarrow 3x + (-4) + 4 = 7 + 4$
$\Leftrightarrow 3x = 11$
$\Leftrightarrow (1/3)(3x) = (1/3)(11)$
$\Leftrightarrow x = 11/3$
Set of solutions: $\{11/3\}$

□

B. $m/6 + 3m/8 = 5/12$

C. $(10/3)r + (10/3)(r - 3) = 70$

Solution

$(10/3)r + (10/3)(r - 3) = 70$
$\Leftrightarrow (10/3)r + (10/3)(r + (-3)) = 70$
$\Leftrightarrow (10/3)r + (10/3)(r) + (10/3)(-3) = 70$
$\Leftrightarrow ((10/3) + (10/3))(r) + (-10) = 70$
$\Leftrightarrow (20/3)r - 10 = 70$
$\Leftrightarrow (20/3)r - 10 + 10 = 70 + 10$
$\Leftrightarrow (20/3)r = 80$
$\Leftrightarrow (3/20)(20/3)r = (3/20)(80)$
$\Leftrightarrow r = 12$
Set of solutions: $\{12\}$

□

D. $t + (3/5)t - 3 = 21$

E. $\dfrac{3x-4}{2} - 6 = \dfrac{x-2}{4}$

SOLUTION

$$\frac{3x-4}{2} - 6 = \frac{x-2}{4}$$
$$\Leftrightarrow (1/2)(3x-4) + (-6) = (1/4)(x-2)$$
$$\Leftrightarrow 4[(1/2)(3x-4) + (-6)] = 4[(1/4)(x-2)]$$
$$\Leftrightarrow 4(1/2)(3x-4) + (4)(-6) = x-2$$
$$\Leftrightarrow (2)(3x-4) + (-24) = x-2$$
$$\Leftrightarrow (2)(3x + (-4)) + (-24) = x-2$$
$$\Leftrightarrow (2)(3x) + (2)(-4) + (-24) = x-2$$
$$\Leftrightarrow 6x + (-8) + (-24) = x + (-2)$$
$$\Leftrightarrow 6x + (-32) = x + (-2)$$
$$\Leftrightarrow 6x + (-32) + 32 = x + (-2) + 32$$
$$\Leftrightarrow 6x = x + 30$$
$$\Leftrightarrow (-x) + 6x = (-x) + x + 30$$
$$\Leftrightarrow 5x = 30$$
$$\Leftrightarrow (1/5)(5x) = (1/5)(30)$$
$$\Leftrightarrow x = 6$$

Set of solutions: $\{6\}$

□

F. $\dfrac{x+2}{7} - \dfrac{x-3}{5} = \dfrac{4}{5}$

G. $0.6t + 0.3(20-t) = 0.4(40)$

SOLUTION

$$0.6t + 0.3(20-t) = 0.4(40)$$
$$\Leftrightarrow 0.6t + 0.3(20 + (-t)) = 0.4(40)$$
$$\Leftrightarrow 0.6t + 0.3(20) + 0.3(-t) = 16$$
$$\Leftrightarrow 0.6t + 6 + (-0.3)t = 16$$
$$\Leftrightarrow 0.3t + 6 = 16$$
$$\Leftrightarrow 0.3t + 6 + (-6) = 16 + (-6)$$
$$\Leftrightarrow 0.3t = 10$$
$$\Leftrightarrow (10)(0.3t) = (10)(10)$$
$$\Leftrightarrow 3t = 100$$
$$\Leftrightarrow (1/3)(3t) = (1/3)(100)$$
$$\Leftrightarrow t = 100/3$$

Set of solutions: $\{100/3\}$

(We also could have converted the decimals in the equation to fractions (0.3 to 3/10, 0.4 to 4/10, and 0.6 to 6/10) before we proceeded to solve it.)

□

H. $6x = (7/3)(x + 1/3)$

I. $(x+12)/3+2x=4+7x/3$

SOLUTION

$(x+12)/3+2x=4+7x/3$

$\Leftrightarrow (1/3)(x+12)+2x=4+(7/3)x$

$\Leftrightarrow 3[(1/3)(x+12)+2x] = 3[4+(7/3)x]$

$\Leftrightarrow (3)(1/3)(x+12)+(3)(2x) = (3)(4)+(3)((7/3)x)$

$\Leftrightarrow (x+12)+6x=12+7x$

$\Leftrightarrow 7x+12=12+7x$

$\Leftrightarrow 7x+12+(-7x)=12+7x+(-7x)$

$\Leftrightarrow 12=12$

Set of solutions: **R**

□

J. $(y-8)/2-3y=-5y/2+7$

EXAMPLE 2

The sum of three-eighths of a number and five-sixths of the same number is 7. Find the number.

SOLUTION

Let x represent the number in question.

$(3/8)(x)+(5/6)(x)=7$

$\Leftrightarrow 24[(3/8)(x)+(5/6)(x)]=24[7]$

$\Leftrightarrow 24(3/8)(x)+24(5/6)(x)=168$

$\Leftrightarrow 9x+20x=168$

$\Leftrightarrow 29x=168$

$\Leftrightarrow (1/29)(29x)=(1/29)(168)$

$\Leftrightarrow x=168/29$

The number is 168/29.

□

EXAMPLE 3

The sum of one-fourth of the complement of an angle and one-third of the supplement of the same angle equals 44°. Find the measure of the angle.

Section 17: Subtraction and Division of Terms

PROCEDURE

To this point we have converted differences to sums of additive inverses and quotients to products of multiplicative inverses as we have solved equations. Provided we recognize that neither subtraction nor division are commutative and we combine like terms with care, we can perform these steps less explicitly. That is:

- we may omit the steps of converting equations:
  - of the form $x - a = b$ to the form $x + (-a) = b$ before adding a to both sides and
  - of the form $x/a = b$ to the form $(1/a) \cdot x = b$ before multiplying both sides by a;
- we may solve equations:
  - of the form $x + a = b$ by subtracting a instead of adding $-a$ on both sides and
  - of the form $ax = b$ by dividing by a instead of multiplying by $1/a$ on both sides.

PROPERTIES

Suppose $a, b, c \in \mathbf{R}$.

Then:

- $a \cdot (b - c) = a \cdot b - a \cdot c$
- $(b - c) \cdot a = b \cdot a - c \cdot a$

PROOF (first property only)

$$\begin{aligned} a \cdot (b - c) &= a \cdot (b + (-c)) \\ &= a \cdot b + a \cdot (-c) \\ &= a \cdot b + (-a \cdot c) \\ &= a \cdot b - a \cdot c \end{aligned}$$

⊡

PROPERTIES (proofs left as exercises)

Suppose $a, b \in \mathbf{R}$.

Then:

- $-(a + b) = -a - b$
- $-(a - b) = -a + b$

COMMENTS

- Within the process of solving an equation, we should draw a slash for clarity through terms or factors that we eliminate with the addition of an additive inverse or the multiplication of a multiplicative inverse.
- Extensions of the properties above remain valid where the expressions within parentheses contain more than two terms. For example:
  - $a(b - c + d - e) = a \cdot b - a \cdot c + a \cdot d - a \cdot e$
  - $-(a + b + c + d + e) = -a - b - c - d - e$
- We can prove that an equation is true with a chain of expressions where the first expression is one (usually the less simple) side of the equation, every other expression equates to its predecessor in the chain, and the last expression is the other side of the equation.

EXAMPLE 1
Solve each equation.

A. $m - 5.32 = 11.87$

SOLUTION
$m - 5.32 = 11.87$
$\Leftrightarrow m - \cancel{5.32} + \cancel{5.32} = 11.87 + 5.32$
$\Leftrightarrow m = 17.19$
Set of solutions: $\{17.19\}$

□

B. $k + 7.4 = 23.7$

C. $v/5 = 7.1$

SOLUTION
$v/5 = 7.1 \Leftrightarrow (v/\cancel{5})(\cancel{5}) = (7.1)(5)$
$\Leftrightarrow v = 35.5$
Set of solutions: $\{35.5\}$

□

D. $12p = 30$

E. $t = 39 + 0.7t$

SOLUTION
$t = 39 + 0.7t$
$\Leftrightarrow t - 0.7t = 39 + \cancel{0.7t} - \cancel{0.7t}$
$\Leftrightarrow 0.3t = 39$
$\Leftrightarrow \cancel{0.3}t/\cancel{0.3} = 39/0.3$
$\Leftrightarrow t = 130$
Set of solutions: $\{130\}$

□

F. $0.85 - x = 0.08x$

G. $2(x - 6.91) = 5.22$

SOLUTION
$2(x - 6.91) = 5.22$
$\Leftrightarrow 2(x) - 2(6.91) = 5.22$
$\Leftrightarrow 2x - 13.82 = 5.22$
$\Leftrightarrow 2x - \cancel{13.82} + \cancel{13.82} = 5.22 + 13.82$
$\Leftrightarrow 2x = 19.04$
$\Leftrightarrow \cancel{2}x/\cancel{2} = 19.04/2$
$\Leftrightarrow x = 9.52$
Set of solutions: $\{9.52\}$

□

H. $0.12x + 0.09(100 - x) = 24$

EXAMPLE 2
Suppose $a, b \in \mathbf{R}$.
Prove that $-(a+b) = -a - b$.

PROOF

$$\begin{aligned} -(a+b) &= (-1)(a+b) \\ &= (-1)(a) + (-1)(b) \\ &= -a + (-b) \\ &= -a - b \end{aligned}$$

⊡

EXAMPLE 3
Suppose $a, b \in \mathbf{R}$.
Prove that $-(a-b) = -a + b$.

EXAMPLE 4
Rewrite each repeating decimal in the form a/b where $a \in \mathbf{Z}$ and $b \in \mathbf{N}$.

A. $0.\overline{4}$

SOLUTION
Let $n = 0.\overline{4}$.

$$\begin{aligned} 10n &= 10(0.\overline{4}) \\ &= 10(0.4\overline{4}) \\ &= 4.\overline{4} \end{aligned}$$

$$\begin{aligned} \Leftrightarrow 10n - n &= 4.\overline{4} - 0.\overline{4} \\ &= 4 + 0.\overline{4} - 0.\overline{4} \end{aligned}$$

$\Leftrightarrow 9n = 4$
$\Leftrightarrow 9n/9 = 4/9$
$\Leftrightarrow n = 4/9$

Thus $0.\overline{4} = 4/9$.

□

B. $0.\overline{13}$

C. $5.0\overline{27}$

SOLUTION
Let $n = 5.0\overline{27}$.

$$\begin{aligned} 1000n - 10n &= 1000(5.0\overline{27}) - 10(5.0\overline{27}) \\ &= 1000(5.027\overline{27}) - 50.\overline{27} \\ &= 5{,}027.\overline{27} - 50.\overline{27} \\ &= 4{,}977 \end{aligned}$$

$\Leftrightarrow 990n = 4{,}977$
$\Leftrightarrow 990n/990 = 4{,}977/990$
$\Leftrightarrow n = (553 \cdot 9)/(110 \cdot 9)$
$= 553/110$

Thus $5.0\overline{27} = 553/110$.

□

D. $0.\overline{729}$

Section 18: Linear Inequalities in One Variable

Definition
A *linear inequality in one variable* x is an inequality that can be written equivalently in the form $Ax < B$ for some constants $A, B \in \mathbf{R}$ with $A \neq 0$.

Procedure
To solve a linear inequality in one variable, we:

1. simplify the inequality as much as possible by eliminating parentheses, clearing fractions (assuming positive denominators), and combining like terms;
2. add additive inverses to all sides of the inequality and simplify until one side contains only a variable term and every other side contains only a numerical term;
3. multiply all sides of the inequality by the multiplicative inverse of the coefficient of the variable term and simplify, bearing in mind that we must change the direction of the inequality sign from $<$ to $>$ or vice versa if the multiplicative inverse is negative;
4. state the set of solutions.

Comments

- We also follow the procedure above to solve statements that can be written equivalently in the form $Ax \leq B$.
- The inequality "$c < x < d$" is equivalent to "$c < x$ AND $x < d$" (see page 50).

Example 1
Solve each inequality. State the set of solutions using interval notation.

A. $x - 5 > 11$

Solution
$$x - 5 > 11 \Leftrightarrow x - \cancel{5} + \cancel{5} > 11 + 5$$
$$\Leftrightarrow x > 16$$
Set of solutions: $(16, \infty)$

□

B. $x + 8 \leq 3$

C. $8x - 3 \geq 4x + 9$

Solution
$$8x - 3 \geq 4x + 9$$
$$\Leftrightarrow 8x - \cancel{3} + \cancel{3} \geq 4x + 9 + 3$$
$$\Leftrightarrow 8x \geq 4x + 12$$
$$\Leftrightarrow -4x + 8x \geq \cancel{-4x} + \cancel{4x} + 12$$
$$\Leftrightarrow 4x \geq 12$$
$$\Leftrightarrow (1/4)(4x) \geq (1/4)(12)$$
$$\Leftrightarrow x \geq 3$$
Set of solutions: $[3, \infty)$

□

D. $-x + 4 < x - 3$

E. $72 \le -8n$

Solution
$72 \le -8n \Leftrightarrow (-1/8)(72) \ge (-1/8)(-8n)$
$\Leftrightarrow -9 \ge n$
Set of solutions: $(\infty, -9]$
□

F. $6x > 15$

G. $7x + 5 < 22$

Solution
$7x + 5 < 22$
$\Leftrightarrow 7x + 5 + (-5) < 22 + (-5)$
$\Leftrightarrow 7x < 17$
$\Leftrightarrow (1/7)(7x) < (1/7)(17)$
$\Leftrightarrow x < 17/7$
Set of solutions: $(\infty, 17/7)$
□

H. $3x - 8 - 5x > 6$

I. $-3(m - 4) \ge 2$

Solution
$-3(m - 4) \ge 2$
$\Leftrightarrow (-3)(m) - (-3)(4) \ge 2$
$\Leftrightarrow -3m + 12 \ge 2$
$\Leftrightarrow -3m + 12 + (-12) \ge 2 + (-12)$
$\Leftrightarrow -3m \ge -10$
$\Leftrightarrow (-1/3)(-3m) \le (-1/3)(-10)$
$\Leftrightarrow m \le 10/3$
Set of solutions: $(-\infty, 10/3]$
□

J. $7(x - 2) < 5(x + 2)$

K. $(3/4)x + (1/6)x < 3/2$

Solution
$(3/4)x + (1/6)x < 3/2$
$\Leftrightarrow (12)[(3/4)x + (1/6)x] < (12)[3/2]$
$\Leftrightarrow (12)(3/4)x + (12)(1/6)x < 18$
$\Leftrightarrow 9x + 2x < 18$
$\Leftrightarrow 11x < 18$
$\Leftrightarrow (1/11)(11x) < (1/11)(18)$
$\Leftrightarrow x < 18/11$
Set of solutions: $(-\infty, 18/11)$
□

L. $n \ge 4.2 + 0.4n$

M. $(x+4)/3+(x-3)/5>43/15$

SOLUTION

$(x+4)/3+(x-3)/5>43/15$
$\Leftrightarrow 15[(x+4)/3+(x-3)/5] > 15(43/15)$
$\Leftrightarrow 15(x+4)/3+15(x-3)/5>43$
$\Leftrightarrow (5)(x+4)+(3)(x-3)>43$
$\Leftrightarrow 5(x)+5(4)+3(x)-3(3)>43$
$\Leftrightarrow 5x+20+3x-9>43$
$\Leftrightarrow 8x+11>43$
$\Leftrightarrow 8x+11+(-11)>43+(-11)$
$\Leftrightarrow 8x>32$
$\Leftrightarrow (1/8)(8x)>(1/8)(32)$
$\Leftrightarrow x>4$

Set of solutions: $(4,\infty)$

□

N. $(z-5)/4-(z-4)/3<2$

O. $-3\leq 2x-1<11$

SOLUTION

$-3\leq 2x-1<11$
$\Leftrightarrow -3+1\leq 2x-\not{1}+\not{1}<11+1$
$\Leftrightarrow -2\leq 2x<12$
$\Leftrightarrow (1/2)(-2)\leq(1/2)(2x)<(1/2)(12)$
$\Leftrightarrow -1\leq x<6$

Set of solutions: $[-1,6)$

□

P. $1\leq 3x+7\leq 37$

EXAMPLE 2

Bill earned scores of 68, 77, 56, and 71 on the first four tests he took in algebra. What score must he earn on the fifth test to ensure his average reaches at least 70?

SOLUTION

Let s denote the score on the fifth test.
average ≥ 70
$\Leftrightarrow (68+77+56+71+s)/5\geq 70$
$\Leftrightarrow (5)[(272+s)/5]\geq(5)(70)$
$\Leftrightarrow 272+s\geq 350$
$\Leftrightarrow (-272)+272+s\geq(-272)+350$
$\Leftrightarrow s\geq 78$

Bill's fifth score must be at least 78.

□

EXAMPLE 3

Fred wants to design a triangular window where one side of a window is twice as long as another side, and the third side has a length of 10 inches. Furthermore, the perimeter of the triangle must not exceed 100 inches. Find the maximum length of the first two sides.

Section 19: Compound Inequalities

Definitions

- A *compound statement* is a statement that consists of multiple statements connected by a logical operator.
- A *conjunction* is a compound statement that is true if and only if every one of its componential statements are true.
- A *disjunction* is a compound statement that is true if and only if at least one of its componential statements are true.

Comments

- The logical operator AND connects the componential statements of a conjunction.
- The logical operator OR connects the componential statements of a disjunction.
- If the sets of solutions of the statements ST1 and ST2 are A and B, respectively, then:
  - the set of solutions of the conjunction "ST1 AND ST2" is $A \cap B$, the set of elements for which both statements are true;
  - the set of solutions of the disjunction "ST1 OR ST2" is $A \cup B$, the set of elements for which at least one of the statements is true.

Example 1

State the set of solutions of each compound inequality.

A. $x > 2$ AND $x \leq 7$

Solution
Set of solutions: $(2, 7]$

☐

B. $x < -5$ OR $x > 3$

C. $x > 17$ OR $x \geq 6$

Solution
Set of solutions: $[6, \infty)$

☐

D. $x < 5$ AND $x \leq 8$

E. $x < -8$ OR $x \geq -4$

Solution
Set of solutions: $(-\infty, -8) \cup [-4, \infty)$

☐

F. $x \geq 3$ AND $x < 9$

G. $x \leq 1$ OR $x > -2$

Solution
Set of solutions: $\mathbf{R}$

☐

H. $x \leq 0$ AND $x \geq 11$

Example 2
Solve each compound inequality.

A. $x - 1 < 1$ AND $3x > 21$

Solution
Componential Statements

- $x - 1 < 1 \Leftrightarrow x - 1 + 1 < 1 + 1$
  $\Leftrightarrow x < 2$
- $3x > 21 \Leftrightarrow (1/3)(3x) > (1/3)(21)$
  $\Leftrightarrow x > 7$

Compound Statement
$x - 1 < 1$ AND $3x > 21$
$\Leftrightarrow x < 2$ AND $x > 7$

Set of Solutions
Set of solutions: $\emptyset$
(Since no real number is both less than 2 and greater than 7.)

□

B. $-2x \leq 8$ OR $\frac{1}{5}x \leq 0$

C. $-4x > 12 - x$ OR $5x \geq -8$

Solution
Componential Statements

- $-4x > 12 - x$
  $\Leftrightarrow -4x + x > 12 - x + x$
  $\Leftrightarrow -3x > 12$
  $\Leftrightarrow (-1/3)(-3x) < (-1/3)(12)$
  $\Leftrightarrow x < -4$
- $5x \geq -8$
  $\Leftrightarrow (1/5)(5x) \geq (1/5)(-8)$
  $\Leftrightarrow x \geq -8/5$

Compound Statement
$-4x > 12 - x$ OR $5x \geq -8$
$\Leftrightarrow x < -4$ OR $x \geq -8/5$

Set of Solutions
Set of solutions: $(-\infty, -4) \cup [-8/5, \infty)$

□

D. $x + 3 \geq 2$ AND $5 - 2x < 13 - 3x$

Section 20: Absolute Value

Definition

The *absolute value* of a real number a, denoted by $|a|$, is the distance between 0 and a on the number line.

Properties of Absolute Value

Suppose $n \in \mathbf{N}$ and $a, b, c \in \mathbf{R}$ with $c > 0$.
Then:

- $|x| = c \Leftrightarrow x = -c \text{ OR } x = c$
- $|x| = 0 \Leftrightarrow x = 0$
- $|x| < c \Leftrightarrow -c < x < c$ [Alternative form: $|x| < c \Leftrightarrow -c < x \text{ AND } x < c$]
- $|x| > c \Leftrightarrow x < -c \text{ OR } x > c$
- $|x| \leq c \Leftrightarrow -c \leq x \leq c$ [Alternative form: $|x| \leq c \Leftrightarrow -c \leq x \text{ AND } x \leq c$]
- $|x| \geq c \Leftrightarrow x \leq -c \text{ OR } x \geq c$
- $|a| = |-a|$
- $|ab| = |a||b|$
- $|a/b| = |a|/|b|$ (provided $b \neq 0$)
- $|a^n| = |a|^n$
- $-|a| \leq a \leq |a|$

Proofs

We will not prove formally these properties, but we show graphically the set of solutions common to the algebraic statements of the first, third, and fourth properties.

$|x| = c \Leftrightarrow x = -c \text{ OR } x = c$

$-c$ 0 c

$|x| < c \Leftrightarrow -c < x < c$

$-c$ 0 c

$|x| > c \Leftrightarrow x < -c \text{ OR } x > c$

$-c$ 0 c

⊡

Comments

- Since distances are nonnegative, the absolute value of a positive number will equal the number itself and the absolute value of a negative number will equal the additive inverse (the negative) of that negative number. Thus the definition of absolute value implies that:

$$|a| = \begin{cases} a & \text{if } a \geq 0 \\ -a & \text{if } a < 0 \end{cases}$$

- If c is a negative number, then:
  - the equation $|x| = c$ will have no solution;
  - the equation $|x| < c$ will have no solution;
  - the equation $|x| > c$ will have a set of solutions of all real numbers.

EXAMPLE 1
Find the absolute value of each number and draw a graph that shows its distance from 0.

A. 5

SOLUTION
- $|5| = 5$
-

□

B. -3

SOLUTION
- $|-3| = 3$
-

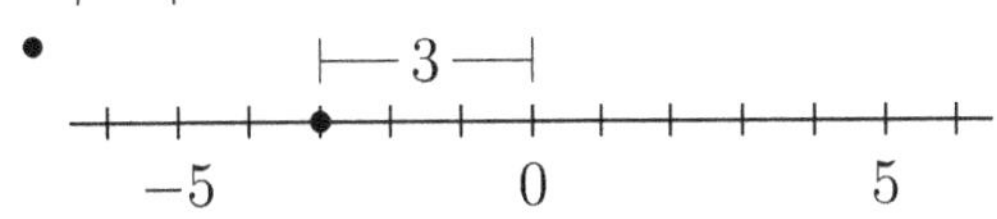

□

EXAMPLE 2
Solve each equation or inequality.

A. $|x| = 7$

SOLUTION
$|x| = 7 \Leftrightarrow x = -7$ OR $x = 7$
Set of solutions: $\{-7, 7\}$

□

B. $|x| < 6$

C. $|x - 7| > -1$

SOLUTION
$|x - 7| > -1$ for all $x \in \mathbf{R}$.
Set of solutions: $\mathbf{R}$

□

D. $|3x| = -\frac{3}{2}$

E. $|2x - 1| = 9$

SOLUTION
$|2x - 1| = 9$
$\Leftrightarrow 2x - 1 = -9$
$\Leftrightarrow 2x = -9 + 1$
$= -8$
$\Leftrightarrow (1/2)(2x) = (1/2)(-8)$
$\Leftrightarrow x = -4$
OR $2x - 1 = 9$
$\Leftrightarrow 2x = 9 + 1$
$= 10$
$\Leftrightarrow (1/2)(2x) = (1/2)(10)$
$\Leftrightarrow x = 5$
Set of solutions: $\{-4, 5\}$

□

F. $|5x + 3| = 13$

G. $|3x+2| \le 4$

SOLUTION

$|3x+2| \le 4$

$\Leftrightarrow -4 \le 3x+2 \le 4$

$\Leftrightarrow -4-2 \le 3x+2-2 \le 4-2$

$\Leftrightarrow -6 \le 3x \le 2$

$\Leftrightarrow (1/3)(-6) \le (1/3)(3x) \le (1/3)(2)$

$\Leftrightarrow -2 \le x \le 2/3$

Set of solutions: $[-2, 2/3]$

□

H. $|-2x-6| < 2$

I. $|-4x+10| > 18$

SOLUTION

$|-4x+10| > 18$

$\Leftrightarrow -4x+10 < -18$

$\Leftrightarrow -4x < -18-10$

$\Leftrightarrow -4x < -28$

$\Leftrightarrow (-1/4)(-4x) > (-1/4)(-28)$

$\Leftrightarrow x > 7$

OR $-4x+10 > 18$

$\Leftrightarrow -4x > 18-10$

$\Leftrightarrow -4x > 8$

$\Leftrightarrow (-1/4)(-4x) < (-1/4)(8)$

$\Leftrightarrow x < -2$

Set of solutions: $(-\infty, -2) \cup (7, \infty)$

□

J. $|6x-1| \ge 1$

EXAMPLE 3

Solve the inequality $|3x+2| \le 4$ using the alternative form of the property for $|x| \le c$.

SOLUTION

$|3x+2| \le 4$

$\Leftrightarrow -4 \le 3x+2$ AND $3x+2 \le 4$

$\Leftrightarrow -4-2 \le 3x+2-2$	$\Leftrightarrow 3x+2-2 \le 4-2$
$\Leftrightarrow -6 \le 3x$	$\Leftrightarrow 3x \le 2$
$\Leftrightarrow (1/3)(-6) \le (1/3)(3x)$	$\Leftrightarrow (1/3)(3x) \le (1/3)(2)$
$\Leftrightarrow -2 \le x$	$\Leftrightarrow x \le 2/3$

Set of solutions: $[-2, 2/3]$

□

Section 21: Distance and Midpoint Formulas, One Dimension

NOTATION

Suppose P and Q are points.

- The expression $d(P, Q)$ denotes the distance between P and Q.
- The expression PQ denotes the line segment that joins P and Q.

DISTANCE AND MIDPOINT FORMULAS

Suppose points P and Q on the number line have the coordinates x_1 and x_2, respectively. Then:

- $d(P, Q) = |x_2 - x_1|$
- the midpoint of line segment PQ has the coordinate $(x_1 + x_2)/2$

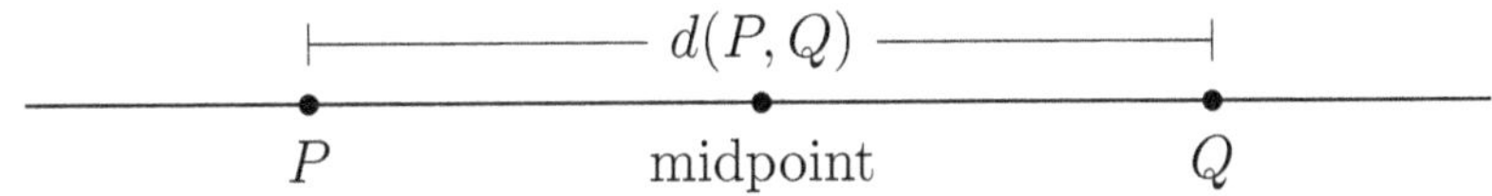

COMMENT

The coordinate of the midpoint of line segment PQ is the arithmetic mean (the average) of the coordinates of P and Q.

EXAMPLE 1

Given the coordinates of points P and Q on the number line, find:

- the distance $d(P, Q)$ between the points and
- the coordinate of the midpoint of the line segment PQ.

A. $P : -3,\ Q : 8$

SOLUTION

- $d(P, Q) = |8 - (-3)|$
  $= |11|$
  $= 11$

- (coordinate of midpoint) $= (-3 + 8)/2$
  $= 5/2$

□

B. $P : 4,\ Q : 20$

Section 22: Geometric Formulas

Definition
A *formula* is a rule expressed as an equation.

Formulas (Areas of Common Two-Dimensional Geometric Objects)
Suppose A represents the area of an object.

- For a rectangle with length l and width w: $A = lw$
- For a circle with radius r: $A = \pi r^2$
- For a triangle with height h and base of length b: $A = (1/2)bh$
- For a parallelogram with height h and base of length b: $A = bh$
- For a trapezoid with height h and bases of lengths b_1 and b_2: $A = (1/2)(b_1 + b_2)h$

Formulas (Perimeters of Common Two-Dimensional Geometric Objects)
Suppose P represents the perimeter of an object.

- For a rectangle with length l and width w: $P = 2l + 2w$
- For a circle with radius r: $P = 2\pi r$
- For a polygon with sides of lengths $s_1, s_2, s_3, \ldots, s_n$: $P = s_1 + s_2 + s_3 + \cdots + s_n$

Formulas (Volumes of Common Three-Dimensional Geometric Objects)
Suppose V represents the volume of an object.

- For a rectangular prism with height h, length l, and width w: $V = lwh$
- For a sphere with radius r: $V = (4/3)\pi r^3$
- For a cylinder with height h and radius r: $V = \pi r^2 h$
- For a cone with height h and radius r: $V = (1/3)\pi r^2 h$
- For a pyramid with height h and base with area B: $V = (1/3)Bh$

Formulas (Surface Areas of Common Three-Dimensional Geometric Objects)
Suppose S represents the surface area of an object.

- For a rectangular prism with height h, length l, and width w: $S = 2lw + 2wh + 2lh$
- For a sphere with radius r: $S = 4\pi r^2$
- For a cylinder with height h and radius r: $S = 2\pi r^2 + 2\pi rh$
- For a cone with height h, radius r, and slant of length s: $S = \pi r^2 + \pi rs$

Terminology
To "solve an equation for a variable" means to find an equation equivalent to the equation given, but with the indicated variable isolated on one side of the equal sign and all other variables and constants on the other side.

Comments

- The perimeter of a circle is called its *circumference*. Hence we usually write the formula for the perimeter of a circle as $C = 2\pi r$ instead of $P = 2\pi r$.
- The process of solving an equation for a variable in terms of other variables does not yield a set of numerical solutions.
- In the following examples, assume the variables we use represent quantities described above.

EXAMPLE 1
Solve each word problem.

A. Find the area of the circle that has a circumference of 12π meters.

SOLUTION
Radius
For $C = 12\pi$:
$$C = 2\pi r \Rightarrow 12\pi = 2\pi r$$
$$\Rightarrow (12\pi)/(2\pi) = (2\pi r)/(2\pi)$$
$$\Rightarrow 6 = r$$

Area
For $r = 6$:
$$A = \pi r^2 \Rightarrow A = \pi(6)^2$$
$$= 36\pi$$
The area is 36π square meters. □

B. Find the surface area of a can of beer that has a radius of three centimeters and a height of twelve centimeters.

C. A rectangular picture is 25 cm long and 15 cm wide. Find the area of a frame of width 3 cm that surrounds the picture.

SOLUTION
Figure

Area
Let F denote the area of the frame.
$$F = \text{(sum of areas of rectangles in figure)}$$
$$= ((3+25+3)\cdot 3)\cdot 2 + (15\cdot 3)\cdot 2$$
$$= 93\cdot 2 + 45\cdot 2$$
$$= 276$$
The area of the frame is 276 cm^2. □

D. A bed of red geraniums takes the shape of a triangle. One side of the triangle is 120 inches long. The altitude to this side is 80 inches long. If a bag of dirt covers 5,000 square inches, will there be enough dirt to cover the bed?

E. If the Pyramid of Khufu in Egypt has a height of 147 meters and a base with length of 230 meters, then what is the volume of the pyramid? (Hint: The base of a pyramid is a square.)

SOLUTION
Area of Base
For $l = 230$ and $w = 230$:
$B = lw \Rightarrow B = (230)(230)$
$= 52900$

Volume
For $h = 147$ and $B = 52900$:
$V = (1/3)Bh \Rightarrow V = (1/3)(52900)(147)$
$= 2592100$
The volume is 2,592,100 cubic meters. □

F. A boy places a scoop of ice cream atop a sugar cone and then injects chocolate syrup into the hollow of the cone until full. If the spherical scoop has a radius of 4 cm, the cone has a height of 8 cm, and exactly half the scoop lies above the top of the cone, then what is the total volume of the ice cream and the syrup?

G. The length of a rectangle is three inches more than twice its width. The perimeter of the rectangle is 60 inches. Find its length and width.

SOLUTION
Substitution
$l = 2w + 3$

Values
$P = 60 \Rightarrow 2l + 2w = 60$
$\Rightarrow 2(2w + 3) + 2w = 60$
$\Rightarrow 2(2w) + 2(3) + 2w = 60$
$\Rightarrow 4w + 6 + 2w = 60$
$\Rightarrow 6w + 6 = 60$
$\Rightarrow 6w + 6 - 6 = 60 - 6$
$\Rightarrow 6w = 54$
$\Rightarrow 6w/6 = 54/6$
$\Rightarrow w = 9$
$\Rightarrow l = 2(9) + 3$
$= 21$

Conclusion
The rectangle has a length of 21 inches and a width of 9 inches. □

H. A triangle has sides labeled A, B, and C. Side B is 2 feet longer than side A, and the length of side C is 27 feet less than three times the length of side B. If the perimeter of the triangle is 41 feet, then find the lengths of all its sides.

EXAMPLE 2
Use the given formula and values to solve for the specified variable.

A. Solve $i = Prt$ for r,
where $i = 50$, $P = 985$, $t = 4$.

SOLUTION
For $i = 50$, $P = 985$, and $t = 4$:
$i = Prt \Rightarrow 50 = (985)(r)(4)$
$\Rightarrow 50 = 3940r$
$\Rightarrow 50/3940 = 3940r/3940$
$\Rightarrow 0.0127 \approx r$

□

B. Solve $A = P + Prt$ for P,
where $A = 1000$, $r = 0.08$, $t = 9$.

EXAMPLE 3
Solve each equation for the specified variable in terms of the other variables.

A. $A = (1/2)bh$ for h

SOLUTION
$A = (1/2)bh \Leftrightarrow 2 \cdot A = 2 \cdot (1/2)bh$
$\Leftrightarrow 2A = bh$
$\Leftrightarrow 2A/b = bh/b$
$\Leftrightarrow 2A/b = h$

□

B. $C = (5/9)(F - 32)$ for F

C. $(x+1)/3 = (y-5)/2$ for y

SOLUTION
$(x+1)/3 = (y-5)/2$
$\Leftrightarrow (6)((x+1)/3) = (6)((y-5)/2)$
$\Leftrightarrow (2)(x+1) = (3)(y-5)$
$\Leftrightarrow (2)(x) + (2)(1) = (3)(y) - (3)(5)$
$\Leftrightarrow 2x + 2 = 3y - 15$
$\Leftrightarrow 2x + 2 + 15 = 3y - 15 + 15$
$\Leftrightarrow 2x + 17 = 3y$
$\Leftrightarrow (1/3)(2x + 17) = (1/3)(3y)$
$= y$
$\Leftrightarrow y = (2x + 17)/3$

□

D. $P = 2l + 2w$ for w

Section 23: Ratios and Proportions

Definitions

- A *ratio* is a comparison of two quantities by division.
- A *proportion* is an algebraic statement that equates two ratios.

Notation

The expressions a to b, $a : b$, and a/b all denote the ratio of quantity a to quantity b.

Property

Suppose $a, b, c, d \in \mathbf{R}$ with $b \neq 0$ and $d \neq 0$.
Then $a/b = c/d$ if and only if $ad = bc$.

Proof

Suppose $a, b, c, d \in \mathbf{R}$ with $b \neq 0$ and $d \neq 0$.

$$
\begin{aligned}
(\Rightarrow)\ a/b = c/d &\Rightarrow (a/b)(bd) = (c/d)(bd) \\
&\Rightarrow (a)(1/b)(b)(d) = (c)(1/d)(d)(b) \\
&\Rightarrow a \cdot 1 \cdot d = c \cdot 1 \cdot b \\
&\Rightarrow ad = bc
\end{aligned}
$$

$$
\begin{aligned}
(\Leftarrow)\ ad = bc &\Rightarrow ad/bd = bc/bd \\
&\Rightarrow (a/b)(d/d) = (b/b)(c/d) \\
&\Rightarrow (a/b)(1) = (1)(c/d) \\
&\Rightarrow a/b = c/d
\end{aligned}
$$

⊡

Comments

- The products ad and bc in the property above are called *cross-products*.
- We will generally use the fractional form a/b to denote the ratio of a to b.
- No number that, when substituted for a variable, causes the denominator of a fraction in an equation to evaluate to zero can serve as a solution to that equation.

Example 1

Solve each equation.

A. $x/3 = 5/7$

Solution

$$
\begin{aligned}
x/3 = 5/7 &\Leftrightarrow (x)(7) = (3)(5) \\
&\Leftrightarrow (x)(7)(1/7) = (3)(5)(1/7) \\
&\Leftrightarrow x = 15/7
\end{aligned}
$$

Set of solutions: $\{15/7\}$

□

B. $(x - 2)/5 = (x + 3)/6$

C. $-8/(x+1) = -3/(x-2)$

Solution

Zeros of the Denominators

- $x+1=0 \Leftrightarrow x+1-1=0-1$
  $\Leftrightarrow x=-1$
- $x-2=0 \Leftrightarrow x-2+2=0+2$
  $\Leftrightarrow x=2$

Set of Solutions

Assuming $x \neq -1$ AND $x \neq 2$:

$-8/(x+1) = -3/(x-2)$
$\Leftrightarrow (-8)(x-2) = (x+1)(-3)$
$\Leftrightarrow -8(x) - (-8)(2) = x(-3) + 1(-3)$
$\Leftrightarrow -8x + 16 = -3x + (-3)$
$\Leftrightarrow -8x + 16 - 16 = -3x - 3 - 16$
$\Leftrightarrow -8x = -3x - 19$
$\Leftrightarrow 3x + (-8x) = 3x + (-3x) - 19$
$\Leftrightarrow -5x = -19$
$\Leftrightarrow (-1/5)(-5x) = (-1/5)(-19)$
$\Leftrightarrow x = 19/5$

Set of solutions: $\{19/5\}$

□

D. $(12-n)/n = 3/5$

E. $(x-3)/5 - (x+4)/7 = -3$

Solution

$(x-3)/5 - (x+4)/7 = -3$
$\Leftrightarrow 35[(x-3)/5 - (x+4)/7] = 35(-3)$
$\Leftrightarrow 35(x-3)/5 - 35(x+4)/7 = -105$
$\Leftrightarrow 7(x-3) - 5(x+4) = -105$
$\Leftrightarrow 7(x-3) + (-5)(x+4) = -105$
$\Leftrightarrow 7(x) - 7(3) + (-5)(x) + (-5)(4) = -105$
$\Leftrightarrow 7x - 21 + (-5x) + (-20) = -105$
$\Leftrightarrow 2x - 41 = -105$
$\Leftrightarrow 2x - 41 + 41 = -105 + 41$
$\Leftrightarrow 2x = -64$
$\Leftrightarrow (1/2)(2x) = (1/2)(-64)$
$\Leftrightarrow x = -32$

Set of solutions: $\{-32\}$

□

F. $x/8 + x/5 = 2$

EXAMPLE 2
Write a proportion that models each situation described.

A. The number of blue balls in an urn is b. The number of red balls in the urn is 35. The ratio of blue balls to red balls in the urn is 1 to 7.

SOLUTION
$b/35 = 1/7$

□

B. On a map, 1 cm represents 750 miles. Two towns are 10 cm apart on the map. The towns are d miles apart in actual distance.

SOLUTION
$1/750 = 10/d$

□

EXAMPLE 3
Solve each word problem.

A. On a map, 1 inch represents 20 miles. If two cities lie 9.5 inches apart on the map, then what is the actual distance between the cities?

SOLUTION
Let m represent the actual distance (in miles) between the cities.
$m/9.5 = 20/1$
$\Leftrightarrow (9.5)(m/9.5) = (9.5)(20/1)$
$\Leftrightarrow m = 190$
The cities are 190 miles apart.

□

B. The ratio of the length of a rectangle to its width is 5 to 3. If the length is 115 feet, then what is the width?

C. Two out of every thirteen people in a town suffer disease of the gums. If the population of the town is 91,000, then how many people there have the disease?

SOLUTION
Let r represent the number of people in the town who have the disease.
$r/91000 = 2/13 \Leftrightarrow (13)(r) = (91000)(2)$
$\Leftrightarrow 13r = 182000$
$\Leftrightarrow 13r/13 = 182000/13$
$\Leftrightarrow r = 14000$
There are 14,000 people with bad gums.

□

D. The ratio of male rats to female rats in a cage is 2 to 3. If the total number of rats in the cage is 400, find the number that are male and the number that are female.

Section 24: Direct and Inverse Variation

Terminology

Suppose x, y, and z represent related quantities.

- The phrase "y varies directly as x" means $y = kx$ for some nonzero constant k.
- The phrase "y varies inversely as x" means $y = k/x$ for some nonzero constant k.
- The phrase "z varies jointly as x and y" means $z = kxy$ for some nonzero constant k.
- The phrase "z varies directly as x and inversely as y" means $z = kx/y$ for some nonzero constant k

Comments

- The constant k in the equations above is called the *constant of proportionality*.
- Alternative terminology:
  - "is directly proportional to" and "is proportional to" mean "varies directly as";
  - "is inversely proportional to" means "varies inversely as".

Example 1

Write the equation that corresponds to each statement.

A. v varies inversely as t
B. V is proportional to h
C. q varies directly as a and inversely as b

Solutions

A. $v = k/t$
B. $V = kh$
C. $q = ka/b$

□

D. C varies directly as r
E. h is inversely proportional to b
F. A varies jointly as l and w

Example 2

Determine the constant of proportionality k using the given pieces of information.

A. A varies jointly as b and h.
If $b = 4$ and $h = 7$, then $A = 14$.

Solution

For $b = 4$, $h = 7$, and $A = 14$:

$$\begin{aligned} A = kbh &\Rightarrow 14 = k(4)(7) \\ &\Rightarrow 14 = 28k \\ &\Rightarrow 14/28 = 28k/28 \\ &\Rightarrow k = 1/2 \end{aligned}$$

□

B. t varies inversely as r.
If $r = 2$, then $t = 5$.

EXAMPLE 3
Suppose y varies directly as x with a constant of proportionality of 10.

A. Find y if $x = 4$.
B. Find x if $y = 15$.

SOLUTION
A. For $x = 4$:
$$y = 10x \Rightarrow y = 10(4) = 40$$

B. For $y = 15$:
$$y = 10x \Rightarrow 15 = 10x \Rightarrow 15/10 = 10x/10 \Rightarrow 3/2 = x \Rightarrow x = 3/2$$

□

EXAMPLE 4
Suppose z varies directly as x and inversely as y with a constant of proportionality of 12.

A. Find z if $x = 2$ and $y = 6$.
B. Find x if $y = 5$ and $z = 3$.
C. Find y if $x = 7$ and $z = 4$.

Section 25: Percentages

Definitions

- *Percent*, denoted by %, means "per one hundred".
- A *percentage* is a number of units per one hundred units.

Comments

- By definition, $p\% = p/100$, or equivalently, $(100p)\% = p$. Thus:
  - to convert a percentage to a decimal we divide by 100;
  - to convert a decimal to a percentage we multiply by 100.
- A quantity that equals p percent of x measures $(p\%)(x)$.
- A quantity that has increased from x by $p\%$ will now measure $x + (p\%)(x)$.
- A quantity that has decreased from x by $p\%$ will now measure $x - (p\%)(x)$.

Example 1

Convert each percentage to a decimal.

A. 20%

Solution

$$\begin{aligned} 20\% &= 20/100 \\ &= 0.2 \end{aligned}$$

□

B. 132%

Example 2

Convert each decimal to a percentage.

A. 0.09

Solution

$$\begin{aligned} 0.09 &= (100 \cdot 0.09)\% \\ &= 9\% \end{aligned}$$

□

B. 0.47

Example 3

Convert each fraction to a percentage.

A. 5/8

Solution

$$\begin{aligned} 5/8 &= 0.625 \\ &= (100 \cdot 0.625)\% \\ &= 62.5\% \end{aligned}$$

□

B. 4/3

Example 4
Convert each percentage to a fraction.

A. 60%

Solution
$60\% = 60/100$
$= (3 \cdot \cancel{20})/(5 \cdot \cancel{20})$
$= 3/5$
□

B. 32%

Example 5
Answer each question. Use a calculator if necessary.

A. What is 25% of 68?

Solution
Let x represent the unknown number.
$x = (25\%)(68)$
$= (25/100)(68)$
$= 17$
So 25% of 68 is 17.
□

B. What is 7% of 183?

C. 72% of what number is 45?

Solution
Let x represent the unknown number.
$(72\%)(x) = 45$
$\Leftrightarrow (72/100)(x) = 45$
$\Leftrightarrow (100/72)(72/100)(x) = (100/72)(45)$
$\Leftrightarrow x = 62.5$
So 72% of 62.5 is 45.
□

D. 120% of what number is 78?

E. 35 is what percent of 53?

Solution
Let p represent the unknown percent.
$35 = (p\%)(53)$
$\Leftrightarrow 35 = (p/100)(53)$
$\Leftrightarrow (100/53)(35) = (100/53)(53/100)p$
$\Leftrightarrow 66.04 \approx p$
So 35 is approximately 66.04% of 53.
□

F. 69 is what percent of 150?

EXAMPLE 6
Solve each word problem. Use a calculator if necessary.

A. The weight of a boy increased from 80 lbs by 21%. What was his final weight?

SOLUTION
Let w denote the final weight (in lbs) of the boy.

$$\begin{aligned} w &= 80 + (21\%)(80) \\ &= 80 + (21/100)(80) \\ &= 80 + 16.8 \\ &= 96.8 \end{aligned}$$

The boy now weighs 96.8 pounds. □

B. The quantity of milk in a refrigerator decreased from 4.07 liters by 65%. How much milk was on hand after the decline?

C. If the rate of flow of raw sewage into a river decreased by 13% to 50,286 gallons per minute, then what was the original rate of flow of the sewage?

SOLUTION
Let r denote the original rate of flow (in gallons per minute) of the sewage.

$$\begin{aligned} 50286 &= r - (13\%)r \\ &= r - 0.13r \\ &= r(1 - 0.13) \\ &= r(0.87) \end{aligned}$$

$$\Leftrightarrow 50286/0.87 = r(0.87)/0.87$$

$$\Leftrightarrow 57800 = r$$

The original rate of flow of the sewage was 57,800 gallons per minute. □

D. The number of children in a condition of mental retardation increased by 3.24% between the years 1995 and 2005. If the number of such children was 443,932 in 2005, then what was the number in 1995?

Section 26: Discount and Profit

Definitions

- A *discount* is a reduction in the price of an item for sale.
- *Profit* is the difference between the price of sale and the cost of purchase of an item.

Formulas

With regard to an item:

- original price – discount = final price
- profit = price of sale – cost of purchase

Comments

- We often express a discount on an item as a percentage of its original price.
- We often express a profit on an item as a percentage of its cost.

Example 1

Solve each word problem. Use a calculator if necessary.

A. Bob purchased a hat at a 20% discount for \$15. What was its original price?

Solution

Let p denote the original price of the hat.

$p - (20\%)(p) = 15$

$\Leftrightarrow p - 0.20p = 15$

$\Leftrightarrow (1 - 0.20)p = 15$

$\Leftrightarrow 0.80p = 15$

$\Leftrightarrow 0.80p/0.80 = 15/0.80$

$\Leftrightarrow p = 18.75$

The original price was \$18.75.

□

B. David bought a radio for 15% less than the listed price. The listed price was \$26. How much did David pay for his radio?

C. A pair of shoes cost a merchant \$20. If he wants to sell them for a profit of 45%, then what price should he charge buyers for the shoes?

Solution

Let s denote the price of the shoes.

$(45\%)(20) = s - 20$

$\Leftrightarrow (0.45)(20) = s - 20$

$\Leftrightarrow (0.45)(20) + 20 = s - 20 + 20$

$\Leftrightarrow 29 = s$

The merchant should charge \$29.

□

D. Suppose it cost you \$4.50 to make three dozen cookies. If you want to garner a profit of 80%, then for how much money should you sell the cookies?

E. Sammy paid \$6.29 for a duffel bag that was originally priced at \$8.50. What rate of discount (as a percentage of the original price) did he receive on the bag?

SOLUTION

Let r denote the rate of discount.

$8.50 - (r)(8.50) = 6.29$

$\Leftrightarrow 8.50(1-r) = 6.29$

$\Leftrightarrow \cancel{8.50}(1-r)/\cancel{8.50} = 6.29/8.50$

$\Leftrightarrow 1 - r = 0.74$

$\Leftrightarrow -1 + 1 - r = -1 + 0.74$

$\Leftrightarrow -r = -0.26$

$\Leftrightarrow (-1)(-r) = (-1)(-0.26)$

$\Leftrightarrow r = 0.26$

$= 26\%$

The rate of discount was 26%. $\square$

F. A boy possesses a cigarette that cost him eight cents. If he sells the cigarette to his friend for thirty cents, then what was his rate of profit based on the cost of the cigarette?

Section 27: Simple Interest

Definitions

- *Interest* is a payment for the use of money.
- *Principal* is an amount of money borrowed.
- *Simple interest* is interest that is calculated only on the original principal.

Formula

Suppose P represents an amount of money borrowed,
r represents the annual rate of simple interest paid by the borrower,
t represents the number of years for which the money is borrowed,
and i represents the total amount of simple interest paid by the borrower.
Then $i = Prt$.

Comments

- We can also use the formula for simple interest above to calculate the total monetary gain i on an investment where the annual rate of return r is a constant percentage of an original amount P of capital (also called the principal) invested for a number of years t.
- The amount of money paid by banks to depositors on savings is also called interest because a deposit at a bank is a loan to the bank.
- Interest on loans is also called *usury*.

Example 1

Solve each word problem. Assume that the variables i, P, r, and t represent the quantities described in the formula above. Use a calculator.

A. A credit card company charges \$171.60 in simple interest for one month on a balance of \$8,580. Find the annual rate of interest charged by the company.

Solution

Substitutions

- $t = 1/12$ (since 1 month = 1/12 year)
- $P = 8580$
- $i = 171.60$

Value

$$\begin{aligned} i = Prt \Rightarrow 171.60 &= (8580)(r)(1/12) \\ &= (8580)(1/12)(r) \\ &= 715(r) \\ \Rightarrow 171.60/715 &= 715r/715 \\ \Rightarrow 0.24 &= r \end{aligned}$$

The rate of interest charged is 24%.

□

B. Suppose Tom wants to receive \$2,500 per month in simple interest. Assuming an annual rate of interest of 5%, how much money does Tom need to lend in order to accomplish his goal?

C. For how long must a person loan \$500 at an annual rate of interest of 4.2% in order to earn \$150 in simple interest?

SOLUTION
For $P = 500$, $i = 150$, and $r = 0.042$:

$$\begin{aligned} i = Prt \Rightarrow 150 &= (500)(0.042)(t) \\ &= 21t \\ \Rightarrow 150/21 &= 21t/21 \\ \Rightarrow 7.14 &\approx t \end{aligned}$$

He must loan the money for approximately 7.14 years, or 7 years 2 months. □

D. A usurer lends \$12,000 to a compulsive gambler at an annual rate of interest of 22%. How much simple interest will he charge in total over a period of six months?

E. Karl had \$600 to invest. He placed some of the money in a savings account at a bank and purchased a junk bond with the remainder. After five years, he found that the money in the savings account had earned simple interest at a annual rate of 1.5%, while the investment in the bond had earned simple interest at a annual rate of 13.3%. If the total of the returns from both investments was \$154.15, then how much money did Karl place initially in each?

SOLUTION
Let i_1, P_1, r_1, and t_1 denote quantities pertaining to the return on the savings account and i_1, P_1, r_1, and t_1 denote quantities pertaining to the return on the junk bond.
Note that $P_1 + P_2 = 600$, or equivalently, $P_1 = 600 - P_2$.
For $r_1 = 0.015$, $t_1 = 5$, $r_2 = 0.133$, and $t_2 = 5$:
(total interest) $= i_1 + i_2$

$$\begin{aligned} \Rightarrow 154.15 &= P_1r_1t_1 + P_2r_2t_2 \\ &= P_1(0.015)(5) + P_2(0.133)(5) \\ &= 0.075P_1 + 0.665P_2 \\ &= 0.075P_1 + 0.665(600 - P_1) \\ &= 0.075P_1 + 0.665(600) - 0.665P_1 \\ &= 0.075P_1 + 399 - 0.665P_1 \\ &= 0.075P_1 - 0.665P_1 + 399 \\ &= -0.59P_1 + 399 \end{aligned}$$

$$\begin{aligned} &\Rightarrow 154.15 - 399 = -0.59P_1 + \cancel{399} - \cancel{399} \\ &\Rightarrow -244.85 = -0.59P_1 \\ &\Rightarrow -244.85/(-0.59) = -0.59P_1/(-0.59) \\ &\Rightarrow 415 = P_1 \\ &\Rightarrow P_2 = 600 - 415 \\ &\qquad = 185 \end{aligned}$$

Karl placed \$415 in the savings account and purchased the junk bond for \$185. □

F. How long will it take for an investment of \$100 to triple in value if the annual yield is 8% of the original principal?

SOLUTION
For $P = 100$ and $r = 0.08$:
$P + i = 3P$
$\Rightarrow P + Prt = 3P$
$\Rightarrow 100 + 100(0.08)t = 3(100)$
$\Rightarrow 100 + 8t = 300$
$\Rightarrow -100 + 100 + 8t = -100 + 300$
$\Rightarrow 8t = 200$
$\Rightarrow 8t/8 = 200/8$
$\Rightarrow t = 25$
The time required is 25 years.

□

G. How long will it take for an investment of \$500 to double in value if the annual yield is 12% of the original investment?

Section 28: Mixture

Terminology

A solution is "$x\%$ a substance" if:

- (volume of substance in solution)/(total volume of solution) $= x\%$
- (volume of substance in solution) $= (x\%)$(total volume of solution)

Example 1

Solve each word problem.

A. How many pints of pure water must a chemist add to 80 pints of a solution that is 30% acid in order to make a solution that is 10% acid?

Solution

Declarations

Let w denote the volume of pure water to add to the original solution,
a denote the volume of acid in the final solution, and
s denote the total volume of the final solution,
where all volumes are in pints.

Substitutions

- $a = (0\%)(w) + (30\%)(80)$
  $= (0)(w) + (0.30)(80)$
  $= 24$
- $s = 80 + w$

Value

$a = (10\%)s$
$\Leftrightarrow 24 = (0.10)(80 + w)$
$= (0.10)(80) + (0.10)(w)$
$= 8 + 0.10w$
$\Leftrightarrow -8 + 24 = -8 + 8 + 0.10w$
$\Leftrightarrow 16 = 0.10w$
$\Leftrightarrow 16/0.10 = 0.10w/0.10$
$\Leftrightarrow 160 = w$

Conclusion

He should add 160 pints of water.

□

B. A pickler mixes 30 liters of a solution that is 55% salt with 25 liters of a solution that is 70% salt. What is the percentage of salt in the final solution?

C. The reservoir for coolant in an automobile contains 10 quarts of a solution that is 40% antifreeze. How much of this solution must a grease monkey drain out and replace with pure antifreeze in order to obtain a solution that is 50% antifreeze?

SOLUTION

Declarations

Let r denote the volume of the original solution to drain out and replace,
a denote the volume of antifreeze in the final solution, and
s denote the total volume of the final solution,
where all volumes are in quarts.

Substitutions

- $a = (40\%)(10) - (40\%)(r) + (100\%)(r)$
  $= (0.40)(10) - (0.40)(r) + (1)(r)$
  $= 4 + (-0.40 + 1.00)r$
  $= 4 + 0.60r$
- $s = 10$

Solution

$a = (50\%)s$
$\Leftrightarrow 4 + 0.60r = (0.50)(10)$
$= 5$
$\Leftrightarrow -4 + 4 + 0.60r = -4 + 5$
$\Leftrightarrow 0.60r = 1$
$\Leftrightarrow 0.60r/0.60 = 1/0.60$
$\Leftrightarrow r = 1/0.60$
≈ 1.67

Conclusion

He should replace roughly 1.67 quarts of solution with pure antifreeze.

□

D. A bootlegger buys one barrel of solution that is 30% alcohol and one barrel of solution that is 60% alcohol. How many liters of each solution must he mix in order to obtain 18 liters of a solution that is 40% alcohol?

Section 29: Adding and Subtracting Polynomials

Definitions

- A *polynomial* in one variable x is a sum of terms of the form $a_i x^i$ where i and a_i are constants with $i \in \mathbf{N} \cup \{0\}$ and $a_i \in \mathbf{R}$.
- The *degree of a term* $a_i x^i$ of a polynomial is the value of the exponent i, assuming $a_i \neq 0$.
- The *degree of a polynomial* is the highest degree of all its nonzero terms.
- The *leading coefficient* of a polynomial is the coefficient on its term of highest degree.
- A *monomial* is a polynomial with exactly one term.
- A *binomial* is a polynomial with exactly two terms.
- A *trinomial* is a polynomial with exactly three terms.

Notation

Given a polynomial P, $\deg(P)$ means "the degree of P".

Comments

- To add and subtract polynomials, we combine like terms.
- Underscores can help us indicate and identify like terms.
- The degree of 0 (the *zero polynomial*) is undefined.
- We usually write a polynomial in one variable with terms in descending order of degree so that it takes the form $a_n x^n + a_{n-1} x^{n-1} + a_{n-2} x^{n-2} + \ldots + a_2 x^2 + a_1 x + a_0$.

Example 1

Complete the table to identify the degree and the leading coefficient of each polynomial.

A.

Polynomial	Degree	Leading Coefficient
$8x^3 - 5x^2 + x + 7$	3	8
$5 - 12x$	1	-12
9	0	9

□

B.

Polynomial	Degree	Leading Coefficient
$x^4 - 3x - 2$	□	□
$7 + 2x^4 - 5x^3$	□	□
$-x$	□	□

Example 2

Simplify each expression.

A. $(3x^2 - 5x - 1) + (x^2 + 2x - 3)$

Solution

$$\begin{aligned}
&(3x^2 - 5x - 1) + (x^2 + 2x - 3) \\
&\quad = 3x^2 + (-5x) + (-1) + x^2 + 2x + (-3) \\
&\quad = \underline{\underline{3x^2}} + \underline{\underline{x^2}} + (\underline{-5x}) + \underline{2x} + (-1) + (-3) \\
&\quad = 4x^2 + (-3x) + (-4) \\
&\quad = 4x^2 - 3x - 4
\end{aligned}$$

□

B. $(-6x^3 + 8x - 7) + (4x^2 + 11)$

C. $(-5x^2 - 3x + 5) - (2x^2 - 7x - 3)$

SOLUTION

$$\begin{aligned}
&(-5x^2 - 3x + 5) - (2x^2 - 7x - 3) \\
&\quad = \underline{\underline{-5x^2}} + (\underline{-3x}) + 5 + (\underline{\underline{-2x^2}}) + \underline{7x} + 3 \\
&\quad = -7x^2 + 4x + 8
\end{aligned}$$

□

D. $(9m + 8) - (11m + 3)$

E. $(x^2 + 4x + 3) + (-2x^3 - 7x^2 - 11)$

SOLUTION

$$\begin{aligned}
&(x^2 + 4x + 3) + (-2x^3 - 7x^2 - 11) \\
&\quad = \underline{\underline{x^2}} + \underline{4x} + 3 + (\underline{\underline{\underline{-2x^3}}}) + (\underline{\underline{-7x^2}}) \\
&\qquad\qquad + (-11) \\
&\quad = -2x^3 + (-6x^2) + 4x + (-8) \\
&\quad = -2x^3 - 6x^2 + 4x - 8
\end{aligned}$$

□

F. $(x^2 + 4x + 3) - (-2x^2 - 7x - 11)$

G. $(4x - 3) + (8x - 5) - (11x - 4)$

SOLUTION

$$\begin{aligned}
&(4x - 3) + (8x - 5) - (11x - 4) \\
&\quad = \underline{4x} + (-3) + \underline{8x} + (-5) + (\underline{-11x}) + 4 \\
&\quad = 1x + (-4) \\
&\quad = x - 4
\end{aligned}$$

□

H. $(-2x + 4) - (-x - 5) - (-5x + 3)$

I. $(-7m + 2) - [-2m + (6m - 3)]$

SOLUTION

$$\begin{aligned}
&(-7m + 2) - [-2m + (6m - 3)] \\
&\quad = (-7m) + 2 + 2m - (6m - 3) \\
&\quad = (\underline{-7m}) + 2 + \underline{2m} + (\underline{-6m}) + 3 \\
&\quad = -11m + 5
\end{aligned}$$

□

J. $-9x - [4x - [3x - (2x - 8)]]$

Section 30: Properties of Exponents within Products

Properties of Exponents

Suppose $a, b \in \mathbf{R}$ and $m, n \in \mathbf{N}$.

Then:

- $b^m \cdot b^n = b^{m+n}$
- $(b^m)^n = b^{mn}$
- $(ab)^n = a^n b^n$

Proofs

Suppose $a, b \in \mathbf{R}$ and $m, n \in \mathbf{N}$.

- $$\begin{aligned} b^m \cdot b^n &= \underbrace{b \cdot b \cdot b \cdot \ldots \cdot b}_{m \text{ factors}} \cdot \underbrace{b \cdot b \cdot b \cdot \ldots \cdot b}_{n \text{ factors}} \\ &= \underbrace{b \cdot b \cdot b \cdot \ldots \cdot b}_{m+n \text{ factors}} \\ &= b^{m+n} \end{aligned}$$

- $$\begin{aligned} (b^m)^n &= \underbrace{b^m \cdot b^m \cdot \ldots \cdot b^m}_{n \text{ factors}} \\ &= \underbrace{\underbrace{b \cdot b \cdot b \cdot \ldots \cdot b}_{m \text{ factors}} \cdot \underbrace{b \cdot b \cdot b \cdot \ldots \cdot b}_{m \text{ factors}} \cdot \ldots \cdot \underbrace{b \cdot b \cdot b \cdot \ldots \cdot b}_{m \text{ factors}}}_{n \text{ sets of } m \text{ factors}} \\ &= \underbrace{b \cdot b \cdot b \cdot \ldots \cdot b}_{m \cdot n \text{ factors}} \\ &= b^{mn} \end{aligned}$$

- $$\begin{aligned} (ab)^n &= \underbrace{ab \cdot ab \cdot ab \cdot \ldots \cdot ab}_{n \text{ factors of } ab} \\ &= \underbrace{a \cdot b \cdot a \cdot b \cdot a \cdot b \cdot \ldots \cdot a \cdot b}_{n \text{ factors of } a \text{ and } n \text{ factors of } b} \\ &= \underbrace{a \cdot a \cdot a \cdot \ldots \cdot a}_{n \text{ factors}} \cdot \underbrace{b \cdot b \cdot b \cdot \ldots \cdot b}_{n \text{ factors}} \\ &= a^n b^n \end{aligned}$$

⊡

Example 1

Simplify each expression.

A. $(5x^3)(8x^2)$

Solution

$$\begin{aligned} (5x^3)(8x^2) &= 5 \cdot 8 \cdot x^3 \cdot x^2 \\ &= 40 \cdot x^{3+2} \\ &= 40x^5 \end{aligned}$$

□

B. $(2m)(3m^5)$

C. $(-4y^2)(-x^4y)$

SOLUTION

$$\begin{aligned}(-4y^2)(-x^4y) &= -4 \cdot y^2 \cdot (-1) \cdot x^4 \cdot y \\ &= -4 \cdot (-1) \cdot x^4 \cdot y^2 \cdot y^1 \\ &= 4 \cdot x^4 \cdot y^{2+1} \\ &= 4x^4y^3\end{aligned}$$

□

D. $(11m^3n)(-5m^2n^7)$

E. $(4a^3b^2)(-3b^4)(-7ab)$

SOLUTION

$$\begin{aligned}&(4a^3b^2)(-3b^4)(-7ab) \\ &= (4)(-3)(-7) \cdot (a^3)(a) \cdot (b^2)(b^4)(b) \\ &= 84 \cdot a^{3+1} \cdot b^{2+4+1} \\ &= 84a^4b^7\end{aligned}$$

□

F. $(-\frac{4}{3}x^4y^3)(\frac{5}{8}xy^2)$

G. $(5x^2)^3$

SOLUTION

$$\begin{aligned}(5x^2)^3 &= (5)^3(x^2)^3 \\ &= 125(x^{2\cdot 3}) \\ &= 125x^6\end{aligned}$$

□

H. $(-6x^3y^4)^2$

I. $(-2a^6b)^4$

SOLUTION

$$\begin{aligned}(-2a^6b)^4 &= (-2)^4(a^6)^4(b)^4 \\ &= (16)(a^{6\cdot 4})(b^4) \\ &= 16a^{24}b^4\end{aligned}$$

□

J. $(-x^2y^5)^9$

K. $-5x(4x^2 - 3)$

SOLUTION

$$\begin{aligned}&-5x(4x^2 - 3) \\ &= -5x(4x^2) - (-5x)(3) \\ &= (-5)(4)(x)(x^2) - (-5)(3)(x) \\ &= -20x^{1+2} + 15x \\ &= -20x^3 + 15x\end{aligned}$$

□

L. $7m(-4m + 2)$

M. $x^2y(5x^2+3x-5)$

SOLUTION

$$\begin{aligned} &x^2y(5x^2+3x-5)\\ &\quad = x^2y(5x^2)+x^2y(3x)-x^2y(5)\\ &\quad = 5x^{2+2}y+3x^{2+1}y-5x^2y\\ &\quad = 5x^4y+3x^3y-5x^2y \end{aligned}$$

□

N. $-6x(4x^2-5x-2)$

O. $2x(3x^2-x+7)-3x(x^2+2x+3)$

SOLUTION

$$\begin{aligned} &2x(3x^2-x+7)-3x(x^2+2x+3)\\ &\quad = 2x(3x^2)-2x(x)+2x(7)\\ &\qquad\quad -[(3x)(x^2)+(3x)(2x)+(3x)(3)]\\ &\quad = 6x^3-2x^2+14x-[3x^3+6x^2+9x]\\ &\quad = 6x^3-2x^2+14x-3x^3-6x^2-9x\\ &\quad = \underline{\underline{\underline{6x^3}}}+(\underline{\underline{-2x^2}})+\underline{14x}\\ &\qquad\qquad +(\underline{\underline{\underline{-3x^3}}})+(\underline{\underline{-6x^2}})+(\underline{-9x})\\ &\quad = 3x^3+(-8x^2)+5x\\ &\quad = 3x^3-8x^2+5x \end{aligned}$$

□

P. $5[2x-(3x+1)]-2(5x-6)$

Q. $(2xy^2)^4(3x^4y)^2$

SOLUTION

$$\begin{aligned} (2xy^2)^4(3x^4y)^2 &= 2^4x^4(y^2)^4\cdot 3^2(x^4)^2y^2\\ &= 16x^4y^8\cdot 9x^8y^2\\ &= (16\cdot 9)(x^4x^8)(y^8y^2)\\ &= 144x^{4+8}y^{8+2}\\ &= 144x^{12}y^{10} \end{aligned}$$

□

R. $(xy^3z^2)^5(-x^2z)^3$

Section 31: Multiplying Polynomials

PROPERTIES (proofs left as exercises)
Suppose $a, b \in \mathbf{R}$.
Then:

- $(a+b)^2 = a^2 + 2ab + b^2$
- $(a-b)^2 = a^2 - 2ab + b^2$
- $(a+b)(a-b) = a^2 - b^2$

COMMENT
To multiply two polynomials in general, we can apply the distributive property repeatedly. The properties above provide shortcuts for this process in three special cases.

EXAMPLE 1
Expand each product.

A. $(x+9)(x-7)$

SOLUTION
$(x+9)(x-7)$
$= (x+9)(x+(-7))$
$= (x+9)(x) + (x+9)(-7)$
$= (x)(x) + (9)(x) + (x)(-7) + (9)(-7)$
$= x^2 + 9x + (-7x) + (-63)$
$= x^2 + 2x - 63$

□

B. $(x+3)(x+5)$

C. $(x-2)(y+8)$

SOLUTION
$(x-2)(y+8)$
$= (x-2)(y) + (x-2)(8)$
$= (x)(y) - (2)(y) + (x)(8) - (2)(8)$
$= xy - 2y + 8x - 16$

□

D. $(y+3)(2x-z+6)$

E. $(4x-3)(5x+1)$

SOLUTION
$(4x-3)(5x+1)$
$= (4x-3)(5x) + (4x-3)(1)$
$= (4x)(5x) - (3)(5x) + (4x)(1) - (3)(1)$
$= 20x^2 - 15x + 4x - 3$
$= 20x^2 - 11x - 3$

□

F. $(2n+1)(12n-9)$

G. $(2b+5)(b^2-3b+2)$

Solution

$$\begin{aligned}
&(2b+5)(b^2-3b+2)\\
&\quad= (2b+5)(b^2+(-3b)+2)\\
&\quad= (2b+5)(b^2)+(2b+5)(-3b)\\
&\qquad\qquad + (2b+5)(2)\\
&\quad= 2b^3+5b^2-6b^2-15b+4b+10\\
&\quad= 2b^3-b^2-11b+10
\end{aligned}$$

□

H. $(x^2-2x-6)(x^2+8x+3)$

I. $(x+4)^3$

Solution

$$\begin{aligned}
&(x+4)^3\\
&\quad= (x+4)(x+4)(x+4)\\
&\quad= [(x+4)(x+4)](x+4)\\
&\quad= [(x+4)(x)+(x+4)(4)](x+4)\\
&\quad= [(x)(x)+(4)(x)+(x)(4)+(4)(4)]\\
&\qquad\qquad \cdot (x+4)\\
&\quad= [x^2+8x+16](x+4)\\
&\quad= [x^2+8x+16](x)+[x^2+8x+16](4)\\
&\quad= (x^2)(x)+(8x)(x)+(16)(x)\\
&\qquad\qquad + (x^2)(4)+(8x)(4)+(16)(4)\\
&\quad= \underline{\underline{\underline{x^3}}}+\underline{\underline{8x^2}}+\underline{16x}+\underline{\underline{4x^2}}+\underline{32x}+64\\
&\quad= x^3+12x^2+48x+64
\end{aligned}$$

□

J. $(3x-1)^3$

Example 2

Expand each product in two ways:

- using the distributive property
- using one of the properties given in this section

A. $(x+8)^2$

Solution

- $$\begin{aligned}
&(x+8)^2\\
&\quad= (x+8)(x+8)\\
&\quad= (x+8)(x)+(x+8)(8)\\
&\quad= (x)(x)+(8)(x)+(x)(8)+(8)(8)\\
&\quad= x^2+16x+64
\end{aligned}$$

- $$\begin{aligned}
(x+8)^2 &= x^2+2(x)(8)+(8)^2\\
&= x^2+16x+64
\end{aligned}$$

□

B. $(n-3)^2$

EXAMPLE 3
Use the properties given in this section to expand each product.

A. $(2m-7)^2$

SOLUTION

$$\begin{aligned}(2m-7)^2 &= (2m)^2 - 2(2m)(7) + (7)^2\\ &= (2)^2(m)^2 - 28m + 49\\ &= 4m^2 - 28m + 49\end{aligned}$$

□

B. $(3x+10y)^2$

C. $(3+4x)(3-4x)$

SOLUTION

$$\begin{aligned}(3+4x)(3-4x) &= (3)^2 - (4x)^2\\ &= 9 - (4)^2x^2\\ &= 9 - 16x^2\end{aligned}$$

□

D. $(7x-2)(7x+2)$

EXAMPLE 4
Suppose $a, b \in \mathbf{R}$.
Prove that $(a+b)^2 = a^2 + 2ab + b^2$.

PROOF

$$\begin{aligned}(a+b)^2 &= (a+b)(a+b)\\ &= (a+b)(a) + (a+b)(b)\\ &= a^2 + ba + ab + b^2\\ &= a^2 + ab + ab + b^2\\ &= a^2 + 2ab + b^2\end{aligned}$$

⊡

EXAMPLE 5
Suppose $a, b \in \mathbf{R}$.
Prove that $(a+b)(a-b) = a^2 - b^2$.

Section 32: Multiplying Algebraic Expressions

PROPERTY

Suppose $a_1, a_2, a_3, \ldots, a_n \in \mathbf{R}$ and $b_1, b_2, b_3, \ldots, b_m \in \mathbf{R}$.
Then $(a_1 + a_2 + a_3 + \ldots + a_n)(b_1 + b_2 + b_3 + \ldots + b_m)$

$$= a_1b_1 + a_1b_2 + \ldots + a_1b_m + a_2b_1 + a_2b_2 + \ldots + a_2b_m + a_3b_1 + a_3b_2 + \ldots + a_3b_m + \ldots + a_nb_1 + a_nb_2 + \ldots + a_nb_m.$$

PROOF

Suppose $a_1, a_2, a_3, \ldots, a_n \in \mathbf{R}$ and $b_1, b_2, b_3, \ldots, b_m \in \mathbf{R}$.

$$\begin{aligned}
&(a_1 + a_2 + a_3 + \ldots + a_n)(b_1 + b_2 + b_3 + \ldots + b_m) \\
&\quad = a_1(b_1 + b_2 + \ldots + b_m) + a_2(b_1 + b_2 + \ldots + b_m) + \ldots + a_n(b_1 + b_2 + \ldots + b_m) \\
&\quad = a_1b_1 + a_1b_2 + \ldots + a_1b_m + a_2b_1 + a_2b_2 + \ldots + a_2b_m + a_3b_1 + a_3b_2 + \ldots + a_3b_m \\
&\qquad + \ldots + a_nb_1 + a_nb_2 + \ldots + a_nb_m
\end{aligned}$$

⊡

COMMENT

In the property above, every term a_i in the expression $a_1 + a_2 + \ldots + a_n$ will be multiplied to every term b_j in the expression $b_1 + b_2 + \ldots + b_m$ in the expansion of the product of the two expressions. In other words, the product of two expressions will be equal to the sum of the products of each term in the first expression and each term in the second expression.

EXAMPLE 1

Expand each product.

A. $(2x + 3y - 5z)(7x - y - 2z)$

SOLUTION

$$\begin{aligned}
&(2x + 3y - 5z)(7x - y - 2z) \\
&\quad = (2x + 3y + (-5z))(7x + (-y) + (-2z)) \\
&\quad = (2x)(7x) + (2x)(-y) + (2x)(-2z) \\
&\qquad + (3y)(7x) + (3y)(-y) + (3y)(-2z) \\
&\qquad + (-5z)(7x) + (-5z)(-y) \\
&\qquad + (-5z)(-2z) \\
&\quad = 14x^2 + (\underline{-2xy}) + (\underline{\underline{-4xz}}) \\
&\qquad + \underline{21xy} + (-3y^2) + (\underline{\underline{\underline{-6yz}}}) \\
&\qquad + (\underline{\underline{-35xz}}) + \underline{\underline{\underline{5yz}}} + 10z^2 \\
&\quad = 14x^2 + 19xy + (-39xz) \\
&\qquad + (-3y^2) + (-yz) + 10z^2 \\
&\quad = 14x^2 + 19xy - 39xz - 3y^2 - yz + 10z^2
\end{aligned}$$

□

B. $(x - 2)(x^3 + 2x^2 + 4x + 8)$

Section 33: The FOIL Method

PROPERTY

Suppose $m, n, p, q, x \in \mathbf{R}$.
Then $(px+m)(qx+n) = pqx^2 + (pn+mq)x + mn$.

PROOF

Suppose $m, n, p, q, x \in \mathbf{R}$.

$$\begin{aligned}(px+m)(qx+n) &= (px)(qx) + (px)(n) + (m)(qx) + (m)(n) \\ &= pqx^2 + pnx + mqx + mn \\ &= pqx^2 + (pn+mq)x + mn\end{aligned}$$

⊡

COMMENT

The method by which we expanded the product of the binomials in the first line of the proof above is sometimes called the "FOIL method", where FOIL is an acronym that indicates the positions of the terms within the two expressions that we multiply together. We may observe that the expansion is the sum of the products of the:

- first terms of the expressions, namely pqx^2 — $(\underline{px}+m)(\underline{qx}+n)$
- outside terms of the expressions, namely pnx — $(\underline{px}+m)(qx+\underline{n})$
- inside terms of the expressions, namely mqx — $(px+\underline{m})(\underline{qx}+n)$
- last terms of the expressions, namely mn — $(px+\underline{m})(qx+\underline{n})$

EXAMPLE 1

Expand each product of binomials using the FOIL method.

A. $(x-2)(x+6)$

SOLUTION

$(x-2)(x+6)$

$= (x+(-2))(x+6)$

$= (x)(x) + (x)(6) + (-2)(x) + (-2)(6)$

$= x^2 + 4x - 12$

□

B. $(x+5)(x+4)$

C. $(5x+1)(3x-7)$

SOLUTION

$(5x+1)(3x-7)$

$= (5x+1)(3x+(-7))$

$= (5x)(3x) + (5x)(-7) + (1)(3x) + (1)(-7)$

$= 15x^2 + (-35x) + 3x + (-7)$

$= 15x^2 - 32x - 7$

□

D. $(5x-6)(4x-3)$

Section 34: Properties of Exponents within Quotients

Properties of Exponents

Suppose $a, b \in \mathbf{R}$ with $b \neq 0$ and $m, n \in \mathbf{N}$.
Then:

- $b^n/b^m = b^{n-m}$ if $n > m$
  $b^n/b^m = 1$ if $n = m$
  $b^n/b^m = 1/b^{m-n}$ if $n < m$
- $(a/b)^n = a^n/b^n$

Proofs

Suppose $a, b \in \mathbf{R}$ with $b \neq 0$ and $m, n \in \mathbf{N}$.

- $$\begin{aligned} n > m \Rightarrow b^n/b^m &= b^{n-m+m}/b^m \\ &= (b^{n-m}b^m)/(b^m) \\ &= b^n \cdot (b^m/b^m) \\ &= b^{n-m} \cdot 1 \\ &= b^{n-m} \end{aligned}$$

  $$\begin{aligned} n = m \Rightarrow b^n/b^m &= b^m/b^m \\ &= 1 \end{aligned}$$

  $$\begin{aligned} n < m \Rightarrow b^n/b^m &= b^n/b^{m-n+n} \\ &= b^n/(b^{m-n}b^n) \\ &= (1/b^{m-n}) \cdot (b^n/b^n) \\ &= (1/b^{m-n}) \cdot 1 \\ &= 1/b^{m-n} \end{aligned}$$

- $$\begin{aligned} (a/b)^n &= \underbrace{(a/b) \cdot (a/b) \cdot (a/b) \cdot \ldots \cdot (a/b)}_{n \text{ factors}} \\ &= (\underbrace{a \cdot a \cdot a \cdot \ldots \cdot a}_{n \text{ factors}})/(\underbrace{b \cdot b \cdot b \cdot \ldots \cdot b}_{n \text{ factors}}) \\ &= a^n/b^n \end{aligned}$$

⊡

Example 1

Simplify each expression.

A. x^7/x^2

Solution

$$\begin{aligned} x^7/x^2 &= x^{7-2} \\ &= x^5 \end{aligned}$$

□

B. x^6/x^{14}

C. $(-48m^5)/(8m^9)$

SOLUTION

$$\begin{aligned}(-48m^5)/(8m^9) &= (-48/8)(m^5/m^9)\\ &= (-6)(1/m^{9-5})\\ &= -6/m^4\end{aligned}$$

□

D. $(156x^{11})/(12x^{11})$

E. $(35x^3y^5)/(5x^3)$

SOLUTION

$$\begin{aligned}(35x^3y^5)/(5x^3) &= (35/5)\cdot(x^3/x^3)\cdot y^5\\ &= 7\cdot 1\cdot y^5\\ &= 7y^5\end{aligned}$$

□

F. $(-12a^4b^9)/(-4b^3)$

G. $\dfrac{55x^7 - 30x^{10}}{5x}$

SOLUTION

$$\begin{aligned}\frac{55x^7 - 30x^{10}}{5x} &= \frac{55x^7}{5x} - \frac{30x^{10}}{5x}\\ &= \frac{55}{5}\cdot\frac{x^7}{x} - \frac{30}{5}\cdot\frac{x^{10}}{x}\\ &= 11\cdot x^{7-1} - 6\cdot x^{10-1}\\ &= 11x^6 - 6x^9\end{aligned}$$

□

H. $\dfrac{8y^2 + 20y^5}{2y}$

I. $\dfrac{24x^5y^8 - 72x^8y^7}{-8x^4y^3}$

SOLUTION

$$\begin{aligned}&\frac{24x^5y^8 - 72x^8y^7}{-8x^4y^3}\\ &= \frac{24x^5y^8}{-8x^4y^3} - \frac{72x^8y^7}{-8x^4y^3}\\ &= \frac{24}{-8}\cdot\frac{x^5}{x^4}\cdot\frac{y^8}{y^3} - \frac{72}{-8}\cdot\frac{x^8}{x^4}\cdot\frac{y^7}{y^3}\\ &= -3\cdot x^{5-4}\cdot y^{8-3} - (-9)\cdot x^{8-4}\cdot y^{7-3}\\ &= -3\cdot x^1\cdot y^5 - (-9)\cdot x^4\cdot y^4\\ &= -3xy^5 + 9x^4y^4\end{aligned}$$

□

J. $\dfrac{-28x^5y^2 - 49x^3y + 56xy^8}{-7xy}$

K. $(n/3)^4$

SOLUTION

$$(n/3)^4 = n^4/(3)^4$$
$$= n^4/81$$

□

L. $(4/x)^2$

Section 35: Dividing Polynomials

Example 1
Divide 83,501 by 12 and describe each step of the procedure.

Solution

$\begin{array}{r} 6958 \\ 12\overline{)83501} \end{array}$	write the divisor and dividend
$\underline{72}$	write 6 on top since 12 goes into 83 six times, but not seven; $6 \cdot 12 = 72$
115	subtract 72 from 83; drop the five down
$\underline{108}$	write 9 on top since 12 goes into 115 nine times, but not ten; $9 \cdot 12 = 108$
70	subtract 108 from 115; drop the zero down
$\underline{60}$	write 5 on top since 12 goes into 70 five times, but not six; $5 \cdot 12 = 60$
101	subtract 60 from 70; drop the one down
$\underline{96}$	write 8 on top since 12 goes into 101 eight times, but not nine; $8 \cdot 12 = 96$
5	subtract 96 from 101

Thus the quotient yielded by the division is 6,958 with a remainder of 5.

□

Example 2
Divide $x^3 + 3x^2 - 5x - 10$ by $x + 2$ and describe each step of the procedure.

Solution

$\begin{array}{r} x^2 + x - 7 \\ x+2\overline{)x^3 + 3x^2 - 5x - 10} \end{array}$	write the divisor and dividend
$\underline{x^3 + 2x^2}$	write x^2 on top since $x^3/x = x^2$; $x^2 \cdot (x+2) = x^3 + 2x^2$
$x^2 - 5x$	subtract $x^3 + 2x^2$ from $x^3 + 3x^2$; bring the $-5x$ down
$\underline{x^2 + 2x}$	write x on top since $x^2/x = x$; $x \cdot (x+2) = x^2 + 2x$
$-7x - 10$	subtract $x^2 + 2x$ from $x^2 - 5x$; bring the -10 down
$\underline{-7x - 14}$	write -7 on top since $-7x/x = -7$; $-7 \cdot (x+2) = -7x - 14$
4	subtract $-7x - 14$ from $-7x - 10$

Thus the quotient yielded by the division is $x^2 + x - 7$ with a remainder of 4.

□

Formulas
The quotient and remainder yielded by dividing one polynomial (the dividend) by another (the divisor) will be related by the following equivalent equations:
- dividend/divisor = quotient + remainder/divisor
- dividend = (quotient)(divisor) + remainder

Comments

- Before we apply our division algorithm for polynomials, we should:
  - ensure that the terms of divisor and dividend are written in order of descending degree;
  - ensure that the divisor and dividend are written in a form where terms of every degree between the highest and lowest degrees of the terms in each polynomial appear; add terms with coefficients of zero as placeholders for any "missing" terms in the sequence.
- Both the quotient and remainder yielded from dividing one polynomial by another will be polynomials themselves. The degree of the remainder will be less than the degree of the divisor unless the divisor is a constant, in which case the remainder will be zero.

Example 3

Divide the polynomials as indicated and identify the quotient and remainder.

A. $(x^2 + 7x - 18) \div (x - 5)$

Solution

Division

$$\begin{array}{r|l} & x + 12 \\ \hline x - 5 & x^2 + 7x - 18 \\ & \underline{x^2 - 5x} \\ & \quad 12x - 18 \\ & \quad \underline{12x - 60} \\ & \qquad 42 \end{array}$$

Quotient and Remainder

- The quotient is $x + 12$.
- The remainder is 42.

□

B. $(30 + 10x + x^2) \div (x + 3)$

C. $(-7 - 5x + 6x^2) \div (2x - 3)$

Solution

Division

$$\begin{array}{r|l} & 3x + 2 \\ \hline 2x - 3 & 6x^2 - 5x - 7 \\ & \underline{6x^2 - 9x} \\ & \quad 4x - 7 \\ & \quad \underline{4x - 6} \\ & \qquad -1 \end{array}$$

Quotient and Remainder

- The quotient is $3x + 2$.
- The remainder is -1.

□

D. $(12x^3 + x^2 + 4x - 1) \div (3x + 1)$

E. $(v^3 - 9v - 3) \div (v + 2)$

SOLUTION
Division

$$\begin{array}{r|l} & v^2 - 2v - 5 \\ v + 2 & v^3 + 0v^2 - 9v - 3 \\ & v^3 + 2v^2 \\ & -2v^2 - 9v \\ & -2v^2 - 4v \\ & -5v - 3 \\ & -5v - 10 \\ & 7 \end{array}$$

Quotient and Remainder
- The quotient is $v^2 - 2v - 5$.
- The remainder is 7.

□

F. $(64m^3 - 1) \div (4m - 1)$

G. $(8x^4 - 6x^3 - 3x^2 + 5x + 8) \div (2x^2 - 3x + 1)$

SOLUTION
Division

$$\begin{array}{r|l} & 4x^2 + 3x + 1 \\ 2x^2 - 3x + 1 & 8x^4 - 6x^3 - 3x^2 + 5x + 8 \\ & 8x^4 - 12x^3 + 4x^2 \\ & 6x^3 - 7x^2 + 5x \\ & 6x^3 - 9x^2 + 3x \\ & 2x^2 + 2x + 8 \\ & 2x^2 - 3x + 1 \\ & 5x + 7 \end{array}$$

Quotient and Remainder
- The quotient is $4x^2 + 3x + 1$.
- The remainder is $5x + 7$.

□

H. $(5x^4 + 22x^3 - 13x^2 - 9) \div (x^2 + 5x - 3)$

Section 36: Synthetic Division

Definition

Synthetic division is a condensed method of dividing one polynomial by another polynomial of the form $x - c$.

Procedure

To divide a polynomial by a polynomial of the form $x - c$ synthetically, we create a table with three rows. Specifically, we:

1. write on the first row c and the coefficients of the dividend;
2. write on the third row the leading coefficient of the dividend directly below the same coefficient on the first row;
3. write on the second row below the second coefficient of the dividend the product of the leading coefficient and c;
4. write on the third row the sum of this product and the second coefficient;
5. write on the second row below the next coefficient the product of this sum and c;
6. write on the third row the sum of this product and the coefficient above it;
7. repeat steps (5) and (6) until the table is complete.

Comments

- We can divide sythetically only when the divisor takes the form $x - c$.
- In step (1) of our synthetic division algorithm, we should:
  - ensure that the coefficients of the dividend are written in order of descending degree;
  - use zero for the coefficients of any missing terms in the dividend.
- After step (7), the third row will show all coefficients of the quotient and the remainder.
- The quotient yielded from dividing a polynomial by a divisor of the form $x - c$ will always have a degree that is one less than the degree of the dividend, while the remainder (having degree zero since the divisor has degree one) will always be a constant.

Example 1a

Divide $x^4 + 5x^3 - 20x^2 + 8x + 3$ by $x - 2$ nonsynthetically and note redundancies.

Solution

$$
\begin{array}{rrrrrr}
 & x^3 & +\,7x^2 & -\,6x & -\,4 & \\
x-2\,\big) & x^4 & +\,5x^3 & -\,20x^2 & +\,8x & +\,3 \\
 & x^4 & -\,2x^3 & & & \\ \hline
 & & 7x^3 & -\,20x^2 & & \\
 & & 7x^3 & -\,14x^2 & & \\ \hline
 & & & -\,6x^2 & +\,8x & \\
 & & & -\,6x^2 & +\,12x & \\ \hline
 & & & & -\,4x & +\,3 \\
 & & & & -\,4x & +\,8 \\ \hline
 & & & & & -\,5
\end{array}
$$

Consider the number of times we must repeat factors within the table at left:

Variables	Coefficients
x^4 : 2	7 : 3
x^3 : 5	−20 : 2
x^2 : 6	−6 : 3
x : 6	8 : 2
	−4 : 3
	3 : 2

□

EXAMPLE 1B

Divide $x^4 + 5x^3 - 20x^2 + 8x + 3$ by $x - 2$ synthetically, showing the results at each step.

SOLUTION

Step (1):

$$\begin{array}{r|rrrrr} 2 & 1 & 5 & -20 & 8 & 3 \\ & & & & & \\ \hline & & & & & \end{array}$$

Step (2):

$$\begin{array}{r|rrrrr} 2 & 1 & 5 & -20 & 8 & 3 \\ & & & & & \\ \hline & 1 & & & & \end{array}$$

Step (3):

$$\begin{array}{r|rrrrr} 2 & 1 & 5 & -20 & 8 & 3 \\ & & 2 & & & \\ \hline & 1 & & & & \end{array}$$

Step (4):

$$\begin{array}{r|rrrrr} 2 & 1 & 5 & -20 & 8 & 3 \\ & & 2 & & & \\ \hline & 1 & 7 & & & \end{array}$$

Step (5):

$$\begin{array}{r|rrrrr} 2 & 1 & 5 & -20 & 8 & 3 \\ & & 2 & 14 & & \\ \hline & 1 & 7 & & & \end{array}$$

Step (6):

$$\begin{array}{r|rrrrr} 2 & 1 & 5 & -20 & 8 & 3 \\ & & 2 & 14 & & \\ \hline & 1 & 7 & -6 & & \end{array}$$

Step (7):

$$\begin{array}{r|rrrrr} 2 & 1 & 5 & -20 & 8 & 3 \\ & & 2 & 14 & -12 & -8 \\ \hline & 1 & 7 & -6 & -4 & \boxed{-5} \end{array}$$

The degree of the quotient must be 3. Thus after step (7), the third row of our table shows that:

- the quotient is $x^3 + 7x^2 - 6x - 4$
- the remainder is -5

□

EXAMPLE 2

Synthetically divide the polynomials as indicated and identify the quotient and remainder.

A. $(x^2 - 10x + 30) \div (x - 3)$

SOLUTION

Division

$$\begin{array}{r|rrr} 3 & 1 & -10 & 30 \\ & & 3 & -21 \\ \hline & 1 & -7 & \boxed{9} \end{array}$$

Quotient and Remainder

- The quotient is $x - 7$.
- The remainder is 9.

□

B. $(2x^2 - 13x + 7) \div (x - 5)$

C. $(2 - 13x + 9x^2 + 3x^3) \div (4 + x)$

SOLUTION

Notes

- The dividend is equivalent to
  $3x^3 + 9x^2 - 13x + 2$.
- The divisor is equivalent to $x - (-4)$.

Division

$$\begin{array}{r|rrrr} -4 & 3 & 9 & -13 & 2 \\ & & -12 & 12 & 4 \\ \hline & 3 & -3 & -1 & \boxed{6} \end{array}$$

Quotient and Remainder

- The quotient is $3x^2 - 3x - 1$.
- The remainder is 6.

□

D. $(-7 + 2x - 5x^2 + x^3) \div (x + 2)$

E. $(x^5 - 1) \div (x - 1)$

SOLUTION

Note

The dividend is equivalent to
$1x^5 + 0x^4 + 0x^3 + 0x^2 + 0x + (-1)$.

Division

$$\begin{array}{r|rrrrrr} 1 & 1 & 0 & 0 & 0 & 0 & -1 \\ & & 1 & 1 & 1 & 1 & 1 \\ \hline & 1 & 1 & 1 & 1 & 1 & \boxed{0} \end{array}$$

Quotient and Remainder

- The quotient is $x^4 + x^3 + x^2 + x + 1$.
- The remainder is zero.

□

F. $(4x^4 - 9x^2 + 2) \div (x - \frac{1}{2})$

Section 37: Nonpositive Exponents

Definitions

Suppose $b \in \mathbf{R}$ with $b \neq 0$ and $n \in \mathbf{N}$.

- $b^0 = 1$
- $b^{-n} = 1/b^n$

Comments

- The equations $b^0 = 1$ and $b^{-n} = 1/b^n$ define the meaning of nonpositive, integral exponents and require no proof.
- Our properties of exponents hold true not only for $m, n \in \mathbf{N}$ but also for $m, n \in \mathbf{Z}$.
- We can use the definitions above to condense the set of three properties given for b^n/b^m on page 85 into the single property $b^n/b^m = b^{n-m}$, since:
  - if $n > m$, then $b^n/b^m = b^{n-m}$
  - if $n = m$, then $b^n/b^m = 1 = b^0 = b^{n-n} = b^{n-m}$
  - if $n < m$, then $b^n/b^m = 1/b^{m-n} = b^{-(m-n)} = b^{-m+n} = b^{n-m}$

Example 1

Simplify each expression.

A. 3^{-4}

Solution

$3^{-4} = 1/3^4$
$\quad = 1/81$

□

B. $(4)^0$

C. $1/(3^{-1})$

Solution

$1/(3^{-1}) = 3^{-(-1)}$
$\quad = 3^1$
$\quad = 3$

□

D. $2^{-8} \cdot 2^5$

E. $(2/3)^{-2}$

Solution

$(2/3)^{-2} = (2^{-2})/(3^{-2})$
$\quad = (1/2^2)/(1/3^2)$
$\quad = (1/4)/(1/9)$
$\quad = (1/4) \cdot (9/1)$
$\quad = 9/4$

□

F. $1/[(5/2)^{-3}]$

G. $[7/(4^{-1})]^{-3}$

SOLUTION

$$\begin{aligned}[7/(4^{-1})]^{-3} &= (7^{-3})/(4^{(-1)(-3)}) \\ &= (7^{-3})/(4^{3}) \\ &= (7^{-3})\cdot[1/(4^{3})] \\ &= [1/(7^{3})]\cdot[1/(4^{3})] \\ &= 1/(7^{3}\cdot 4^{3}) \\ &= 1/(343\cdot 64) \\ &= 1/21952\end{aligned}$$

□

H. $(6^{-1}\cdot 2^{-3})^{-2}$

EXAMPLE 2

Simplify each expression. Leave your answer in a form where exponents are positive.

A. $x^{-9}x^{3}$

SOLUTION

$$\begin{aligned}x^{-9}x^{3} &= x^{-9+3} \\ &= x^{-6} \\ &= 1/x^{6}\end{aligned}$$

□

B. $m^{-5}m^{12}$

C. $(5n^{-3})(-6n^{4})$

SOLUTION

$$\begin{aligned}(5n^{-3})(-6n^{4}) &= (5)(-6)(n^{-3})(n^{4}) \\ &= -30n^{-3+4} \\ &= -30n^{1} \\ &= -30n\end{aligned}$$

□

D. $(-2x^{-5})(-9x^{-3})$

E. $\dfrac{15x^{-7}}{-6x^{-3}}$

SOLUTION

$$\begin{aligned}\frac{15x^{-7}}{-6x^{-3}} &= \frac{15}{-6}\cdot\frac{x^{-7}}{x^{-3}} \\ &= -\frac{5\cdot\cancel{3}}{2\cdot\cancel{3}}\cdot\frac{1}{x^{-3-(-7)}} \\ &= -\frac{5}{2}\cdot\frac{1}{x^{4}} \\ &= -\frac{5}{2x^{4}}\end{aligned}$$

□

F. $\dfrac{20x^{-3}}{4x^{-11}}$

G. $(8n^3)^{-2}$

SOLUTION

$$\begin{aligned}(8n^3)^{-2} &= 8^{-2} \cdot (n^3)^{-2} \\ &= (1/8^2) \cdot n^{-6} \\ &= (1/64) \cdot (1/n^6) \\ &= 1/(64n^6)\end{aligned}$$

□

H. $(2x^{-1}y^2)^7$

I. $\left(\dfrac{m^{-4}}{n^{-2}}\right)^{-3}$

SOLUTION

$$\begin{aligned}\left(\frac{m^{-4}}{n^{-2}}\right)^{-3} &= \frac{(m^{-4})^{-3}}{(n^{-2})^{-3}} \\ &= \frac{m^{(-4)(-3)}}{n^{(-2)(-3)}} \\ &= \frac{m^{12}}{n^6}\end{aligned}$$

□

J. $\left(\dfrac{3x^{-6}}{x^{-6}}\right)^{-3}$

EXAMPLE 3
Prove that $1/b^{-n} = b^n$.

PROOF

$$\begin{aligned}1/b^{-n} &= b^{-(-n)} \\ &= b^n\end{aligned}$$

⊡

EXAMPLE 4
Prove that $(1/b)^{-1} = b$.

Section 38: Factorizing by Removing a Common Factor

Definitions

- To *factorize* means to convert a sum of terms into a product of factors.
- A polynomial with integral coefficients is *factorized completely* when written as a product of nonconstant polynomials with integral coefficients and lower degree that cannot themselves be factorized into products of nonconstant polynomials with integral coefficients and lower degree.

Procedure

To factorize a sum of terms $b_1, b_2, \ldots, b_n$ with a common factor, we:

1. identify the greatest monomial factor a common to all terms in the sum;
2. write the sum $b_1 + b_2 + \cdots + b_n$ as the product $a(b_1/a + b_2/a + \cdots + b_n/a)$.

Example 1

Find the prime factorizations and greatest common factor of the numbers in each set.

A. 30, 42

Solution

Prime Factorizations

- $30 = 6 \cdot 5$
  $= \boxed{2} \cdot \boxed{3} \cdot 5$
- $42 = 6 \cdot 7$
  $= \boxed{2} \cdot \boxed{3} \cdot 7$

Greatest Common Factor

(greatest common factor) $= 2 \cdot 3 = 6$

□

B. 60, 72

C. 75, 315, 675

Solution

Prime Factorizations

- $75 = 3 \cdot 25$
  $= \boxed{3} \cdot \boxed{5} \cdot 5$
- $315 = 9 \cdot 35$
  $= \boxed{3} \cdot 3 \cdot \boxed{5} \cdot 7$
- $675 = 27 \cdot 25$
  $= \boxed{3} \cdot 3 \cdot 3 \cdot \boxed{5} \cdot 5$

Greatest Common Factor

(greatest common factor) $= 3 \cdot 5 = 15$

□

D. 56, 84, 224

EXAMPLE 2
Determine the greatest common monomial factor of the terms in each set.

A. $40n$, $48mn$

SOLUTION
Notes

- $40n = 8 \cdot 5 \cdot n$
  $= \boxed{2} \cdot \boxed{2} \cdot \boxed{2} \cdot 5 \cdot \boxed{n}$
- $48mn = 8 \cdot 6 \cdot m \cdot n$
  $= \boxed{2} \cdot \boxed{2} \cdot \boxed{2} \cdot 2 \cdot 3 \cdot m \cdot \boxed{n}$

Greatest Common Factor
(greatest common factor) $= 2 \cdot 2 \cdot 2 \cdot n$
$= 8n$

□

B. $54x^3$, $45x^5$

C. $48x^4y^5$, $80x^9y^3$, $36x^6y^4$, $16x^5y^5$

SOLUTION
Notes

- $48x^4y^5 = 8 \cdot 6 \cdot x^4 \cdot y^5$
  $= \boxed{2} \cdot \boxed{2} \cdot 2 \cdot 2 \cdot 3$
  $\cdot \boxed{x} \cdot \boxed{x} \cdot \boxed{x} \cdot \boxed{x}$
  $\cdot \boxed{y} \cdot \boxed{y} \cdot \boxed{y} \cdot y \cdot y$
- $80x^9y^3 = 8 \cdot 10 \cdot x^9 \cdot y^3$
  $= \boxed{2} \cdot \boxed{2} \cdot 2 \cdot 2 \cdot 5$
  $\cdot \boxed{x} \cdot \boxed{x} \cdot \boxed{x} \cdot \boxed{x} \cdot x \cdot x$
  $\cdot x \cdot x \cdot x \cdot \boxed{y} \cdot \boxed{y} \cdot \boxed{y}$
- $36x^6y^4 = 4 \cdot 9 \cdot x^6 \cdot y^4$
  $= \boxed{2} \cdot \boxed{2} \cdot 3 \cdot 3$
  $\cdot \boxed{x} \cdot \boxed{x} \cdot \boxed{x} \cdot \boxed{x} \cdot x \cdot x$
  $\cdot \boxed{y} \cdot \boxed{y} \cdot \boxed{y} \cdot y$
- $16x^5y^5 = 4 \cdot 4 \cdot x^5 \cdot y^5$
  $= \boxed{2} \cdot \boxed{2} \cdot 2 \cdot 2$
  $\cdot \boxed{x} \cdot \boxed{x} \cdot \boxed{x} \cdot \boxed{x} \cdot x$
  $\cdot \boxed{y} \cdot \boxed{y} \cdot \boxed{y} \cdot y \cdot y$

Greatest Common Factor
(greatest common factor)
$= 2 \cdot 2 \cdot x \cdot x \cdot x \cdot x \cdot y \cdot y \cdot y$
$= 4x^4y^3$

□

D. $55xy^4$, $25x^2y^2$, $75x^3y^2z$

Example 3
Factorize each expression completely.

A. $54x^3 + 45x^5$

Solution
$54x^3 + 45x^5$
$= 9x^3[54x^3/(9x^3) + 45x^5/(9x^3)]$
$= 9x^3(6 + 5x^2)$
□

B. $40m - 48mn$

C. $55xy^4 - 25x^2y^2$

Solution
$55xy^4 - 25x^2y^2$
$= 5xy^2[55xy^4/(5xy^2) - 25x^2y^2/(5xy^2)]$
$= 5xy^2(11y^2 - 5x)$
□

D. $48x^6y^5 + 80x^9y^3$

E. $s^5 - s^4 + 7s^3$

Solution
$s^5 - s^4 + 7s^3$
$= s^3(s^5/s^3 - s^4/s^3 + 7s^3/s^3)$
$= s^3(s^2 - s + 7)$
□

F. $12x - 21x^5 - 15x^7$

G. $m(x + 1) + 5(x + 1)$

Solution
$m(x + 1) + 5(x + 1)$
$= (x + 1)\left(\frac{m\cancel{(x+1)}}{\cancel{x+1}} + \frac{5\cancel{(x+1)}}{\cancel{x+1}}\right)$
$= (x + 1)(m + 5)$
□

H. $3n(y - 4) - 7(y - 4)$

Section 39: Factorizing by Grouping

Procedure

To factorize a sum of an even number of terms b_1, b_2, b_3, ..., b_n, where the factorizations of $b_1 + b_2$, $b_3 + b_4$, $b_5 + b_6$, ..., and $b_{n-1} + b_n$ all share a common factor, we:

1. group the pairs of terms b_1 and b_2, b_3 and b_4, b_5 and b_6, ..., and b_{n-1} and b_n together;
2. factorize the sums of the pairs of step (1) into products of the common factor;
3. factorize the sum or difference of the products yielded in step (2).

Comments

- We cannot factorize every sum of an even number of terms.
- We can factorize some expressions by first grouping terms together in sets of three or more.

Example 1

Factorize each expression.

A. $3x - 3y - mx + my$

Solution

$3x - 3y - mx + my$
$= (3x - 3y) - (mx - my)$
$= 3(\underline{x - y}) - m(\underline{x - y})$
$= (x - y)(3 - m)$

□

B. $ax + ay + 2bx + 2by$

C. $25x^2 + 10x - 20x - 8$

Solution

$25x^2 + 10x - 20x - 8$
$= (25x^2 + 10x) - (20x + 8)$
$= 5x(5x + 2) - 4(5x + 2)$
$= (5x + 2)(5x - 4)$

□

D. $12y^2 - 6y - 4y + 2$

E. $cm + cn + m + n$

Solution

$cm + cn + m + n$
$= (cm + cn) + (m + n)$
$= c(m + n) + 1(m + n)$
$= (m + n)(c + 1)$

□

F. $9s + 9t + s + t$

G. $xy - 7y - 2x + 14$

SOLUTION

$$
\begin{aligned}
&xy - 7y - 2x + 14 \\
&\quad = (xy - 7y) - (2x - 14) \\
&\quad = y(x - 7) - 2(x - 7) \\
&\quad = (x - 7)(y - 2)
\end{aligned}
$$

□

H. $-m + 5mz + 3b - 15bz$

Section 40: Factorizing the Difference of Two Squares

Property
Suppose $a, b \in \mathbf{R}$.
Then $a^2 - b^2 = (a-b)(a+b)$.

Proof
Suppose $a, b \in \mathbf{R}$.

$$\begin{aligned}(a-b)(a+b) &= (a-b)(a) + (a-b)(b)\\ &= (a)(a) - (b)(a) + (a)(b) - (b)(b)\\ &= a^2 - ab + ab - b^2\\ &= a^2 - b^2\end{aligned}$$

⊡

Procedure
We factorize a difference of squares $a^2 - b^2$ as $(a-b)(a+b)$.

Comment
We cannot factorize a sum of squares $a^2 + b^2$ over the real numbers if a and b share no common factors.

Example 1
Factorize each expression completely.

A. $x^2 - 144$

Solution

$$\begin{aligned}x^2 - 144 &= x^2 - (12)^2\\ &= (x-12)(x+12)\end{aligned}$$

□

B. $x^2 - 49$

C. $36a^2 - 25b^2$

Solution

$$\begin{aligned}36a^2 - 25b^2 &= (6a)^2 - (5b)^2\\ &= (6a-5b)(6a+5b)\end{aligned}$$

□

D. $9 - 64x^2$

E. $11x^2 - 11$

Solution

$$\begin{aligned}11x^2 - 11 &= 11(x^2 - 1)\\ &= 11(x^2 - (1)^2)\\ &= 11(x-1)(x+1)\end{aligned}$$

□

F. $5x^4 - 45x^2$

G. $28n - 63n^5$

SOLUTION

$$\begin{aligned}28n - 63n^5 &= 7n(4 - 9n^4)\\ &= 7n((2)^2 - (3n^2)^2)\\ &= 7n(2 - 3n^2)(2 + 3n^2)\end{aligned}$$

□

H. $18x^3y - 50xy^3$

I. $v^3 + 5v^2 - 4v - 20$

SOLUTION

$$\begin{aligned}&v^3 + 5v^2 - 4v - 20\\ &\quad = (v^3 + 5v^2) - (4v + 20)\\ &\quad = v^2(v + 5) - 4(v + 5)\\ &\quad = (v + 5)(v^2 - 4)\\ &\quad = (v + 5)(v^2 - (2)^2)\\ &\quad = (v + 5)(v - 2)(v + 2)\end{aligned}$$

□

J. $2w^3 - 3w^2 - 18w + 27$

Section 41: Factorizing the Difference or Sum of Two Cubes

PROPERTIES

Suppose $a, b \in \mathbf{R}$.
Then:

- $a^3 - b^3 = (a - b)(a^2 + ab + b^2)$
- $a^3 + b^3 = (a + b)(a^2 - ab + b^2)$

PROOFS

Suppose $a, b \in \mathbf{R}$.

- $$\begin{aligned}(a-b)(a^2+ab+b^2) &= (a-b)(a^2) + (a-b)(ab) + (a-b)(b^2)\\ &= (a)(a^2) - (b)(a^2) + (a)(ab) - (b)(ab) + (a)(b^2) - (b)(b^2)\\ &= a^3 - \underline{a^2b} + \underline{a^2b} - \underline{\underline{ab^2}} + \underline{\underline{ab^2}} - b^3\\ &= a^3 - b^3\end{aligned}$$

- $$\begin{aligned}(a+b)(a^2-ab+b^2) &= (a+b)(a^2) + (a+b)(-ab) + (a+b)(b^2)\\ &= (a)(a^2) + (b)(a^2) + (a)(-ab) + (b)(-ab) + (a)(b^2) + (b)(b^2)\\ &= a^3 + \underline{a^2b} - \underline{a^2b} - \underline{\underline{ab^2}} + \underline{\underline{ab^2}} - b^3\\ &= a^3 + b^3\end{aligned}$$

⊡

PROCEDURE

- We factorize a difference of cubes $a^3 - b^3$ as $(a - b)(a^2 + ab + b^2)$.
- We factorize a sum of cubes $a^3 + b^3$ as $(a + b)(a^2 - ab + b^2)$.

EXAMPLE 1

Factorize each expression.

A. $x^3 + 27$

SOLUTION

$$\begin{aligned}x^3 + 27 &= x^3 + (3)^3\\ &= (x+3)(x^2 - (x)(3) + (3)^2)\\ &= (x+3)(x^2 - 3x + 9)\end{aligned}$$

□

B. $x^3 - 1$

C. $8a^3 - 125b^3$

SOLUTION

$$\begin{aligned}&8a^3 - 125b^3\\ &= (2)^3a^3 - (5)^3b^3\\ &= (2a)^3 - (5b)^3\\ &= (2a - 5b)((2a)^2 + (2a)(5b) + (5b)^2)\\ &= (2a - 5b)(4a^2 + 10ab + 25b^2)\end{aligned}$$

□

D. $64 + m^6n^3$

Section 42: Factorizing Trinomials of the Form $x^2 + bx + c$

Property
Suppose x, m, $n \in \mathbf{R}$.
Then $x^2 + (m+n)x + mn = (x+m)(x+n)$.

Proof
Suppose x, m, $n \in \mathbf{R}$.

$$\begin{aligned}(x+m)(x+n) &= (x+m)(x) + (x+m)(n) \\ &= (x)(x) + (m)(x) + (x)(n) + (m)(n) \\ &= x^2 + mx + nx + mn \\ &= x^2 + (m+n)x + mn\end{aligned}$$

⊡

Procedure
To factorize a trinomial $x^2 + bx + c$ with integral coefficients b and c, we:
1. list all pairs (m, n) of integral factors of c;
2. identify the pair (m, n) for which $m + n = b$;
3. write $x^2 + bx + c$ as the product $(x+m)(x+n)$ for the m and n of step (2).

Comments
- We cannot factorize every trinomial of the form $x^2 + bx + c$ over the integers; possibly no integers m and n will exist where $m + n = b$ and $mn = c$.
- We may assume without loss of generality in step (1) of the procedure that $|m| \leq |n|$. Thus we should include $(1, 6)$, $(2, 3)$, $(-1, -6)$, and $(-2, -3)$ if we list the integral factors of 6, but we may omit $(3, 2)$, $(6, 1)$, $(-3, -2)$, and $(-6, -1)$.
- After some practice, we will be able to factorize mentally trinomials of the form $x^2 + bx + c$ without any need to list explicitly pairs of integral factors of the constant term.
- We can verify a factorization by comparing the expansion of that product with the original expression.

Example 1
Factorize each trinomial.

A. $x^2 + 7x + 10$

Solution
Notes
- The pairs of integral factors of 10 are $(1, 10)$, $(2, 5)$, $(-1, -10)$, and $(-2, -5)$.
- The sum of the factors 2 and 5 is 7.

Factorization
$x^2 + 7x + 10 = (x+2)(x+5)$

□

B. $x^2 + 9x + 18$

C. $z^2 - 9z + 20$

Solution

Notes

- $z^2 - 9z + 20 = z^2 + (-9)z + 20$
- The pairs of integral factors of 20 are $(1, 20)$, $(2, 10)$, $(4, 5)$, $(-1, -20)$, $(-2, -10)$, and $(-4, -5)$.
- The sum of the factors -4 and -5 is -9.

Factorization

$$\begin{aligned} z^2 - 9z + 20 &= (z + (-4))(z + (-5)) \\ &= (z - 4)(z - 5) \end{aligned}$$

□

D. $x^2 - 16x + 28$

E. $y^2 + 5y - 24$

Solution

Notes

- $y^2 + 5y - 24 = y^2 + 5y + (-24)$
- The pairs of integral factors of -24 are $(1, -24)$, $(2, -12)$, $(3, -8)$, $(4, -6)$, $(-1, 24)$, $(-2, 12)$, $(-3, 8)$, and $(-4, 6)$.
- The sum of the factors -3 and 8 is 5.

Factorization

$$\begin{aligned} y^2 + 5y - 24 &= (y + (-3))(y + 8) \\ &= (y - 3)(y + 8) \end{aligned}$$

□

F. $n^2 - n - 42$

G. $x^2 - 5xy - 14y^2$

Solution

Notes

- In order for two factors of $-14y^2$ to add up to $-5y$, both must contain a y.
- The pairs of such factors of $-14y^2$ are $(y, -14y)$, $(2y, -7y)$, $(-y, 14y)$, and $(-2y, 7y)$.
- The sum of the factors $2y$ and $-7y$ is $-5y$.

Factorization

$$\begin{aligned} x^2 - 5xy - 14y^2 &= (x + 2y)(x + (-7y)) \\ &= (x + 2y)(x - 7y) \end{aligned}$$

□

H. $a^2 - 7ab + 12b^2$

Section 43: Factorizing Trinomials of the Form $ax^2 + bx + c$

PROPERTY

Suppose x, m, n, p, $q \in \mathbf{R}$.
Then $pqx^2 + (pn + mq)x + mn = (px + m)(qx + n)$.

PROOF

Suppose x, m, n, p, $q \in \mathbf{R}$.

$$\begin{aligned}(px + m)(qx + n) &= (px + m)(qx) + (px + m)(n) \\ &= (px)(qx) + (m)(qx) + (px)(n) + (m)(n) \\ &= pqx^2 + qmx + pnx + mn \\ &= pqx^2 + (pn + qm)x + mn\end{aligned}$$

⊡

PROPERTY

Suppose a, b, c, r, $s \in \mathbf{Z}$ with $rs = ac$ and $r + s = b$.
Then $ax^2 + bx + c = (px + m)(qx + n)$,
where p is the greatest common factor of a and r, $q = a/p$, $n = r/p$, and $m = c/n$.

PROOF

Suppose a, b, c, r, $s \in \mathbf{Z}$ with $rs = ac$ and $r + s = b$.
Let p denote the greatest common factor of a and r.
Thus $a = pq$ and $r = pn$ for some $q, n \in \mathbf{Z}$ where q and n share no prime factors.
[If q and n shared some common prime factor t, then pt would divide both a and r, which would contradict our definition of p as the greatest common factor of a and r.]

But note then that:

- $ac = rs \Rightarrow (pq)(c) = (pn)(s)$
  $\Rightarrow pqc/p = pns/p$
  $\Rightarrow qc = ns$
  $\Rightarrow n$ divides c since n and q share no prime factors
  $\Rightarrow c = nm$ for some $m \in \mathbf{Z}$
- $rs = ac \Rightarrow (pn)(s) = (pq)(nm)$
  $\Rightarrow (pns)/(pn) = (pqnm)/(pn)$
  $\Rightarrow s = qm$

Therefore:

$$\begin{aligned}ax^2 + bx + c &= ax^2 + (r + s)x + c \\ &= ax^2 + rx + sx + c \\ &= pqx^2 + pnx + qmx + mn \\ &= px(qx + n) + m(qx + n) \\ &= (px + m)(qx + n),\end{aligned}$$

where p is the greatest common factor of a and r, $q = a/p$, $n = r/p$, and $m = c/n$.

⊡

PROCEDURE A (preferred)
To factorize a trinomial $ax^2 + bx + c$ with integral coefficients a, b and c, we can:
1. list all pairs (p, q) of integral factors of a;
2. list all pairs (m, n) of integral factors of c;
3. identify the pairs (p, q) and (m, n) for which $pn + qm = b$;
4. write $ax^2 + bx + c$ as the product $(px + m)(qx + n)$ for the p, q, m, and n of step (3).

PROCEDURE B (alternative)
To factorize a trinomial $ax^2 + bx + c$ with integral coefficients a, b and c, we can:
1. list all pairs (r, s) of integral factors of ac;
2. identify the pair (r, s) for which $r + s = b$;
3. write $ax^2 + bx + c$ as $ax^2 + rx + sx + c$ for the r and s of step (2);
4. factorize $ax^2 + rx + sx + c$ by grouping.

PROCEDURE C (alternative)
To factorize a trinomial $ax^2 + bx + c$ with integral coefficients a, b and c, we can:
1. multiply the trinomial by $(1/a)(a)$;
2. write the expression $(1/a)(a)(ax^2 + bx + c)$ as $(1/a)(a^2x^2 + abx + ac)$;
3. write the expression $(1/a)(a^2x^2 + abx + ac)$ as $(1/a)((ax)^2 + b(ax) + ac)$;
4. substitute u for ax;
5. factorize the trinomial $u^2 + bu + ac$;
6. substitute ax for u;
7. distribute $1/a$ to clear the product of fractions.

COMMENTS
- We cannot factorize every trinomial of the form $ax^2 + bx + c$ over the integers.
- We should remove numerical factors common to all the terms of a trinomial before we try to factorize it as a product of binomials.
- We may assume without loss of generality in step (1) of procedure A that:
  - $|p| \leq |q|$;
  - both p and q are positive if a is positive.
- We may assume without loss of generality in step (1) of procedure B that $|r| \leq |s|$.
- After some practice with procedure A, we will be able to mentally factorize trinomials of the form $ax^2 + bx + c$.

EXAMPLE 1
Factorize each trinomial completely after removing common factors.

A. $12x^2 - 24x - 36$

SOLUTION

$$\begin{aligned} 12x^2 - 24x - 36 &= 12(x^2 - 2x - 3) \\ &= 12(x - 3)(x + 1) \end{aligned}$$

□

B. $3x^2 - 18x + 24$

EXAMPLE 2

Factorize each trinomial using procedure A.

A. $2x^2 + 13x + 15$

SOLUTION

Notes

- The pair of integral factors of 2 is $(1, 2)$.
- The pairs of integral factors of 15 are $(1, 15)$, $(3, 5)$, $(5, 3)$, $(15, 1)$, $(-1, -15)$, $(-3, -5)$, $(-5, -3)$, and $(-15, -1)$.

Evaluations

- $(1)(1) + (2)(15) = 32$ X
- $(1)(3) + (2)(5) = 13$ ✓

Factorization

$$\begin{aligned} 2x^2 + 13x + 15 &= (1x + 5)(2x + 3) \\ &= (x + 5)(2x + 3) \end{aligned}$$

□

B. $6x^2 - 19x + 10$

C. $8n^2 + 6n - 9$

SOLUTION

Notes

- The pairs of integral factors of 8 are $(1, 8)$ and $(2, 4)$.
- The pairs of integral factors of -9 are $(-9, 1)$, $(-3, 3)$, $(-1, 9)$, $(1, -9)$, $(3, -3)$, and $(9, -1)$.

Evaluations

- $(1)(-9) + (8)(1) = -1$ X
- $(1)(-3) + (8)(3) = 21$ X
- $(1)(-1) + (8)(9) = 71$ X
- $(1)(1) + (8)(-9) = -71$ X
- $(1)(3) + (8)(-3) = -21$ X
- $(1)(9) + (8)(-1) = 1$ X
- $(2)(-9) + (4)(1) = -14$ X
- $(2)(-3) + (4)(3) = 6$ ✓

Factorization

$$\begin{aligned} 8n^2 + 6n - 9 &= (2n + 3)(4n + (-3)) \\ &= (2n + 3)(4n - 3) \end{aligned}$$

□

D. $3b^2 - 5b - 12$

EXAMPLE 3
Factorize each trinomial using procedure B.

A. $2x^2 + 13x + 15$

SOLUTION
Notes
- The product of 2 and 15 is 30.
- The pairs of integral factors of 30 are $(1, 30)$, $(2, 15)$, $(3, 10)$, $(5, 6)$, $(-1, -30)$, $(-2, -15)$, $(-3, -10)$, and $(-5, -6)$.
- The sum of the factors 3 and 10 is 13.

Factorization
$$\begin{aligned} 2x^2 + 13x + 15 &= 2x^2 + 3x + 10x + 15 \\ &= x(2x + 3) + 5(2x + 3) \\ &= (x + 5)(2x + 3) \end{aligned}$$
□

B. $6x^2 - 19x + 10$

C. $8n^2 + 6n - 9$

SOLUTION
Notes
- $8n^2 + 6n - 9 = 8n^2 + 6n + (-9)$
- The product of 8 and -9 is -72.
- The pairs of integral factors of -72 are $(1, -72)$, $(2, -36)$, $(3, -24)$, $(4, -18)$, $(6, -12)$, $(8, -9)$, $(-1, 72)$, $(-2, 36)$, $(-3, 24)$, $(-4, 18)$, $(-6, 12)$, and $(-8, 9)$.
- The sum of the factors -6 and 12 is 6.

Factorization
$$\begin{aligned} 8n^2 + 6n - 9 &= 8n^2 - 6n + 12n - 9 \\ &= 2n(4n - 3) + 3(4n - 3) \\ &= (2n + 3)(4n - 3) \end{aligned}$$
□

D. $3b^2 - 5b - 12$

EXAMPLE 4
Factorize each trinomial using procedure C.

A. $2x^2 + 13x + 15$

SOLUTION

$$\begin{aligned}
&2x^2 + 13x + 15\\
&\quad = (1/2)(2)(2x^2 + 13x + 15)\\
&\quad = (1/2)(4x^2 + 26x + 30)\\
&\quad = (1/2)((2x)^2 + 13(2x) + 30)\\
&\quad = (1/2)(u^2 + 13u + 30) \quad \text{where } u = 2x\\
&\quad = (1/2)(u + 3)(u + 10)\\
&\quad = (1/2)(2x + 3)(2x + 10)\\
&\quad = (2x + 3)(1/2)(2x + 10)\\
&\quad = (2x + 3)(x + 5)
\end{aligned}$$

□

B. $6x^2 - 19x + 10$

C. $8n^2 + 6n - 9$

SOLUTION

$$\begin{aligned}
&8n^2 + 6n - 9\\
&\quad = (1/8)(8)(8n^2 + 6n - 9)\\
&\quad = (1/8)(64n^2 + 48n - 72)\\
&\quad = (1/8)((8n)^2 + 6(8n) - 72)\\
&\quad = (1/8)(u^2 + 6u - 72) \quad \text{where } u = 8n\\
&\quad = (1/8)(u - 6)(u + 12)\\
&\quad = (1/8)(8n - 6)(8n + 12)\\
&\quad = (1/2)(8n - 6)(1/4)(8n + 12)\\
&\quad = (4n - 3)(2n + 3)
\end{aligned}$$

□

D. $3b^2 - 5b - 12$

Section 44: Solving Equations by Factorizing

Zero-Product Property

Suppose $a, b \in \mathbf{R}$.
Then $ab = 0$ if and only if $a = 0$ OR $b = 0$.

Proof

Suppose $a, b \in \mathbf{R}$.

($\Leftarrow$) Clearly:

- $a = 0 \Rightarrow ab = 0 \cdot b = 0$
- $b = 0 \Rightarrow ab = a \cdot 0 = 0$

Hence $a = 0$ OR $b = 0$ implies $ab = 0$.

($\Rightarrow$) Suppose $ab = 0$ but $a \neq 0$ and $b \neq 0$.
Then:

$ab = 0 \Rightarrow ab/b = 0/b$
$\Rightarrow a = 0$

Contradiction.
Hence $ab = 0$ implies $a = 0$ OR $b = 0$.

⊡

Property (proof omitted)

Suppose $a_1, a_2, a_3, \ldots, a_n \in \mathbf{R}$.
Then $a_1 \cdot a_2 \cdot a_3 \cdot \ldots \cdot a_n = 0$ if and only if $a_1 = 0$ OR $a_2 = 0$ OR $a_3 = 0$ OR $\ldots$ OR $a_n = 0$.

Procedure

To solve a nonlinear equation in one variable, we:

1. apply properties of equality to the equation until one side equals zero;
2. factorize the nonzero side of the equation completely;
3. set each nonconstant factor equal to zero and solve the resulting equations;
4. state the set of solutions to these equations.

Comments

- We cannot solve every nonlinear equation by the procedure above because we cannot factorize every expression.
- In step (3) of the procedure above, we should set only nonconstant factors equal to zero.

Example 1

Solve each equation.

A. $x^2 - 8x = 0$

Solution

$x^2 - 8x = 0$
$\Leftrightarrow x(x-8) = 0$
$\Leftrightarrow x = 0$ OR $x - 8 = 0$
$\Leftrightarrow x - 8 + 8 = 0 + 8$
$\Leftrightarrow x = 8$

Set of solutions: $\{0, 8\}$

□

B. $5x + x^2 = 0$

C. $12x^2 = -3x$

SOLUTION

$12x^2 = -3x$

$\Leftrightarrow 12x^2 + 3x = -3x + 3x$

$\Leftrightarrow 12x^2 + 3x = 0$

$\Leftrightarrow 3x(4x + 1) = 0$

$\Leftrightarrow x = 0$ OR $4x + 1 = 0$

$\Leftrightarrow 4x + 1 - 1 = 0 - 1$

$\Leftrightarrow 4x = -1$

$\Leftrightarrow 4x/4 = -1/4$

$\Leftrightarrow x = -1/4$

Set of solutions: $\{-1/4, 0\}$

□

D. $4x^2 = 8x$

E. $25x^2 = 4$

SOLUTION

$25x^2 = 4$

$\Leftrightarrow 25x^2 - 4 = 0$

$\Leftrightarrow (5x)^2 - (2)^2 = 0$

$\Leftrightarrow (5x - 2)(5x + 2) = 0$

$\Leftrightarrow 5x - 2 = 0$

$\Leftrightarrow 5x - 2 + 2 = 0 + 2$

$\Leftrightarrow 5x = 2$

$\Leftrightarrow (1/5)(5x) = (1/5)(2)$

$\Leftrightarrow x = 2/5$

OR $5x + 2 = 0$

$\Leftrightarrow 5x + 2 - 2 = 0 - 2$

$\Leftrightarrow 5x = -2$

$\Leftrightarrow (1/5)(5x) = (1/5)(-2)$

$\Leftrightarrow x = -2/5$

Set of solutions: $\{-2/5, 2/5\}$

□

F. $81 = m^2$

G. $x(3x - 9) + 11(3x - 9) = 0$

SOLUTION

$x(3x - 9) + 11(3x - 9) = 0$

$\Leftrightarrow (3x - 9)(x + 11) = 0$

$\Leftrightarrow 3(x - 3)(x + 11) = 0$

$\Leftrightarrow x - 3 = 0$ OR $x + 11 = 0$

$\Leftrightarrow x = 3$ $\quad\Leftrightarrow x = -11$

Set of solutions: $\{-11, 3\}$

□

H. $5(2x + 7) - x(2x + 7) = 0$

I. $(x-5)(x+7)=0$

SOLUTION
$(x-5)(x+7)=0$
$\Leftrightarrow x-5=0$ OR $x+7=0$
$\Leftrightarrow x=5$ $\Leftrightarrow x=-7$
Set of solutions: $\{-7,5\}$

□

J. $(x+6)(2x-8)=0$

K. $b^2-b-56=0$

SOLUTION
$b^2-b-56=0$
$\Leftrightarrow (b-8)(b+7)=0$
$\Leftrightarrow b-8=0$ OR $b+7=0$
$\Leftrightarrow b=8$ $\Leftrightarrow b=-7$
Set of solutions: $\{-7,8\}$

□

L. $x^2+7x+6=0$

M. $x(x-8)=-15$

SOLUTION
$x(x-8)=-15$
$\Leftrightarrow x(x-8)+15=-15+15$
$\Leftrightarrow x^2-8x+15=0$
$\Leftrightarrow (x-5)(x-3)=0$
$\Leftrightarrow x-5=0$ OR $x-3=0$
$\Leftrightarrow x=5$ $\Leftrightarrow x=3$
Set of solutions: $\{3,5\}$

□

N. $x^2+6x=72$

O. $5x^2-3x-14=0$

SOLUTION
$5x^2-3x-14=0$
$\Leftrightarrow 5x^2+(-3)x+(-14)=0$
$\Leftrightarrow (1x+(-2))(5x+7)=0$
$\Leftrightarrow (x-2)(5x+7)=0$
$\Leftrightarrow x-2=0$ OR $5x+7=0$
$\Leftrightarrow x=2$ $\Leftrightarrow 5x=-7$
$\Leftrightarrow 5x/5=-7/5$
$\Leftrightarrow x=-7/5$
Set of solutions: $\{-7/5,2\}$

□

P. $14c^2-43c+20=0$

Q. $12x^2 + 23x = -10$

Solution

$12x^2 + 23x = -10$
$\Leftrightarrow 12x^2 + 23x + 10 = -10 + 10$
$\Leftrightarrow 12x^2 + 23x + 10 = 0$
$\Leftrightarrow (4x + 5)(3x + 2) = 0$
$\Leftrightarrow 4x + 5 = 0$
$\Leftrightarrow 4x = -5$
$\Leftrightarrow 4x/4 = -5/4$
$\Leftrightarrow x = -5/4$
OR $3x + 2 = 0$
$\Leftrightarrow 3x = -2$
$\Leftrightarrow 3x/3 = -2/3$
$\Leftrightarrow x = -2/3$
Set of solutions: $\{-2/3, -5/4\}$

$\square$

R. $4x(x - 2) = 21$

S. $3x^3 = 12x$

Solution

$3x^3 = 12x$
$\Leftrightarrow 3x^3 - 12x = 12x - 12x$
$\Leftrightarrow 3x(x^2 - 4) = 0$
$\Leftrightarrow 3x(x - 2)(x + 2) = 0$
$\Leftrightarrow x = 0$
OR $x - 2 = 0$
$\Leftrightarrow x = 2$
OR $x + 2 = 0$
$\Leftrightarrow x = -2$
Set of solutions: $\{-2, 0, 2\}$

$\square$

T. $4x^3 = x$

U. $n^2 - 8n + 16 = 0$

Solution

$n^2 - 8n + 16 = 0$
$\Leftrightarrow (n - 4)^2 = 0$
$\Leftrightarrow n - 4 = 0$
(we need not repeat this equation as $n - 4 = 0$ OR $n - 4 = 0$)
$\Leftrightarrow n = 4$
Set of solutions: $\{4\}$

$\square$

V. $y^3 + 22y^2 + 121y = 0$

W. $(x+4)(2x+5)=-1$

SOLUTION

$(x+4)(2x+5)=-1$

$\Leftrightarrow x(2x)+x(5)+4(2x)+4(5)=-1$

$\Leftrightarrow 2x^2+5x+8x+20=-1$

$\Leftrightarrow 2x^2+5x+8x+20+1=-1+1$

$\Leftrightarrow 2x^2+13x+21=0$

$\Leftrightarrow (2x+7)(x+3)=0$

$\Leftrightarrow 2x+7=0$

$\Leftrightarrow 2x=-7$

$\Leftrightarrow 2x/2=-7/2$

$\Leftrightarrow x=-7/2$

OR $x+3=0$

$\Leftrightarrow x=-3$

Set of solutions: $\{-7/2,-3\}$

□

X. $(3x-5)(x+1)=3$

Section 45: Simplifying Fractions with Variables

Definition
A *rational expression* is a quotient of two polynomials.

Property
Suppose a, b, and c represent algebraic expressions where $a \neq 0$ and $c \neq 0$.
Then $(a \cdot b)/(a \cdot c) = b/c$.

Proof
Suppose a, b, and c represent algebraic expressions where $a \neq 0$ and $c \neq 0$.

$$\begin{aligned}(a \cdot b)/(a \cdot c) &= (a/a) \cdot (b/c)\\ &= (1) \cdot (b/c)\\ &= b/c\end{aligned}$$

⊡

Comments

- A rational expression is a kind of fraction with a variable.
- The property above suggests that if the numerator and denominator of a fraction are products with a common factor, then we can simplify the fraction by "canceling" that common factor. Note that we can only cancel common factors in products; we cannot cancel common terms in sums or differences within a numerator and denominator.

Example 1
Simplify each fraction.

A. $\dfrac{14x}{20xy}$

Solution

$$\begin{aligned}\frac{14x}{20xy} &= \frac{7 \cdot \cancel{2} \cdot \cancel{x}}{10 \cdot \cancel{2} \cdot \cancel{x} \cdot y}\\ &= \frac{7}{10y}\end{aligned}$$

□

B. $\dfrac{8m}{22n}$

C. $-\dfrac{15x^3y^4}{42x^7y^2}$

Solution

$$\begin{aligned}-\frac{15x^3y^4}{42x^7y^2} &= \frac{-5 \cdot \cancel{3} \cdot \cancel{x}^3 \cdot y^2 \cdot \cancel{y}^2}{14 \cdot \cancel{3} \cdot x^4 \cdot \cancel{x}^3 \cdot \cancel{y}^2}\\ &= -\frac{5y^2}{14x^4}\end{aligned}$$

□

D. $\dfrac{-36a^6b^2c^3}{-54a^5b^2c}$

E. $\dfrac{3x^2 - 15x}{x - 5}$

SOLUTION

$$\begin{aligned}\frac{3x^2 - 15x}{x - 5} &= \frac{(3x)(\cancel{x - 5})}{(1)(\cancel{x - 5})}\\ &= \frac{3x}{1}\\ &= 3x\end{aligned}$$

□

F. $\dfrac{8}{28a + 16b}$

G. $\dfrac{7x^2 + 19x - 6}{6x^2 + 23x + 15}$

SOLUTION

$$\begin{aligned}\frac{7x^2 + 19x - 6}{6x^2 + 23x + 15} &= \frac{(7x - 2)(\cancel{x + 3})}{(6x + 5)(\cancel{x + 3})}\\ &= \frac{7x - 2}{6x + 5}\end{aligned}$$

□

H. $\dfrac{6x^2 + 11x - 10}{9x^2 - 4}$

I. $\dfrac{x^2 - 3x}{9 - x^2}$

SOLUTION

$$\begin{aligned}\frac{x^2 - 3x}{9 - x^2} &= \frac{x(x - 3)}{(3 - x)(3 + x)}\\ &= \frac{x(x - 3)}{(-1)(-3 + x)(3 + x)}\\ &= -\frac{x(\cancel{x - 3})}{(\cancel{x - 3})(x + 3)}\\ &= -\frac{x}{x + 3}\end{aligned}$$

□

J. $\dfrac{x^2 - 25}{-x^2 - 5x}$

Section 46: Multiplying and Dividing Fractions with Variables

PROPERTY

Suppose a, b, c, and d represent algebraic expressions where $b \neq 0$ and $d \neq 0$. Then $(a/b) \cdot (c/d) = (ac)/(bd)$.

PROPERTY

Suppose a, b, c, and d represent algebraic expressions where $b \neq 0$, $c \neq 0$, and $d \neq 0$. Then $(a/b) \div (c/d) = (a/b) \cdot (d/c) = (ad)/(bc)$.

COMMENT

Rules for simplifying, multiplying, and dividing fractions that contain variables are identical to those rules for simplifying, multiplying, and dividing purely numerical fractions.

EXAMPLE 1

Perform the indicated operations and simplify the resulting fractions.

A. $\frac{5}{6} \cdot \frac{9}{10}$

SOLUTION

$$\frac{5}{6} \cdot \frac{9}{10} = \frac{\cancel{5} \cdot 3 \cdot \cancel{3}}{\cancel{3} \cdot 2 \cdot \cancel{5} \cdot 2} = \frac{3}{4}$$

□

B. $\left(-\frac{12}{7}\right) \div \left(\frac{20}{49}\right)$

C. $\frac{8xy^3}{9y} \div \left(-\frac{4xy}{15x^4}\right)$

SOLUTION

$$\begin{aligned}
\frac{8xy^3}{9y} \div \left(-\frac{4xy}{15x^4}\right) &= \frac{8xy^3}{9y} \cdot \left(-\frac{15x^4}{4xy}\right) \\
&= \frac{8xy^2}{9} \cdot \left(-\frac{15x^3}{4y}\right) \\
&= -\frac{8xy^2 \cdot 15x^3}{9 \cdot 4y} \\
&= -\frac{8 \cdot 15 \cdot x \cdot x^3 \cdot y^2}{9 \cdot 4 \cdot y} \\
&= -\frac{\cancel{4} \cdot 2 \cdot 5 \cdot \cancel{3} \cdot x^4 \cdot y}{\cancel{3} \cdot 3 \cdot \cancel{4}} \\
&= -\frac{10x^4y}{3}
\end{aligned}$$

□

D. $\frac{10x^2}{21xy} \cdot \frac{6y^3}{25x^3y^2}$

E. $\dfrac{a+b}{ab+b^3} \cdot \dfrac{b}{-a^2-ab}$

SOLUTION

$$\frac{a+b}{ab+b^3} \cdot \frac{b}{-a^2-ab}$$
$$= \frac{(a+b)(b)}{(ab+b^3)(-a^2-ab)}$$
$$= \frac{(1)\cancel{(a+b)}(\cancel{b})}{(\cancel{b})(a+b^2)(-a)\cancel{(a+b)}}$$
$$= -\frac{1}{a(a+b^2)}$$

□

F. $\dfrac{2x+10}{x^2+8x+7} \cdot \dfrac{x^2-3x-4}{x^2+5x}$

G. $\dfrac{x^2-4x+4}{6x^2+7x-5} \div \dfrac{3x^2-x-10}{2x^2+11x-6}$

SOLUTION

$$\frac{x^2-4x+4}{6x^2+7x-5} \div \frac{3x^2-x-10}{2x^2+11x-6}$$
$$= \frac{x^2-4x+4}{6x^2+7x-5} \cdot \frac{2x^2+11x-6}{3x^2-x-10}$$
$$= \frac{(x-2)\cancel{(x-2)}}{(3x+5)\cancel{(2x-1)}} \cdot \frac{\cancel{(2x-1)}(x+6)}{(3x+5)\cancel{(x-2)}}$$
$$= \frac{(x-2)(x+6)}{(3x+5)^2}$$

□

H. $\dfrac{5ab}{3ab-3b^2} \div \dfrac{a^3}{6a-6b}$

I. $\dfrac{7x^2y}{16y} \cdot \dfrac{4y^3}{21x} \div \dfrac{13x^5}{12xy}$

SOLUTION

$$\frac{7x^2y}{16y} \cdot \frac{4y^3}{21x} \div \frac{13x^5}{12xy}$$
$$= \frac{7x^2y}{16y} \cdot \frac{4y^3}{21x} \cdot \frac{12xy}{13x^5}$$
$$= \frac{7 \cdot 4 \cdot 12 \cdot x^2 \cdot x \cdot y \cdot y^3 \cdot y}{16 \cdot 21 \cdot 13 \cdot x \cdot x^5 \cdot y}$$
$$= \frac{\cancel{7} \cdot \cancel{4} \cdot \cancel{4} \cdot \cancel{3} \cdot x^3 \cdot y^5}{\cancel{4} \cdot \cancel{4} \cdot \cancel{7} \cdot \cancel{3} \cdot 13 \cdot x^6 \cdot y}$$
$$= \frac{y^4}{13x^3}$$

□

J. $\dfrac{x^2+6x+9}{x^2+4x} \div \dfrac{x}{x+4} \cdot \dfrac{3x^2-21x+30}{x^2-2x-15}$

Section 47: Adding and Subtracting Fractions with Variables

Properties

Suppose a, b, and c represent algebraic expressions where $b \neq 0$.
Then:

- $(a/b) + (c/b) = (a + c)/b$
- $(a/b) - (c/b) = (a - c)/b$

Procedure

To convert fractions to equivalent fractions with the same denominator, we:

1. factorize the denominators of the fractions;
2. identify the least common multiple of the denominators; this expression will serve as the least common denominator of the fractions;
3. for each fraction in the set:
    a. determine the quotient of the least common multiple of step (2) and the denominator of the fraction and
    b. multiply the fraction by the fraction whose numerator and denominator both equal the quotient found in step (3a).

Comments

- Rules for adding and subtracting fractions with variables are identical to those for adding and subtracting purely numerical fractions. We must convert algebraic fractions that do not have the same denominator into equivalent fractions that do have the same denominator before we can add or subtract them.
- We should factorize completely the denominators of all the fractions in a sum or difference before we attempt to find their least common multiple in step (1) of the procedure above.
- The fraction by which we multiply the fraction under consideration in step (3b) of the procedure above will be a form of 1.

Example 1

Perform the indicated operations and simplify the resulting fractions.

A. $\frac{2}{3} + \frac{11}{4}$

Solution

$$\begin{aligned} \frac{2}{3} + \frac{11}{4} &= \frac{2}{3} \cdot \frac{4}{4} + \frac{11}{4} \cdot \frac{3}{3} \\ &= \frac{8}{12} + \frac{33}{12} \\ &= \frac{8 + 33}{12} \\ &= \frac{41}{12} \end{aligned}$$

□

B. $\frac{5}{8} - \frac{3}{2}$

C. $\dfrac{5x+4}{3} - \dfrac{2x-3}{3}$

SOLUTION

$$\begin{aligned}\frac{5x+4}{3} - \frac{2x-3}{3} &= \frac{5x+4-(2x-3)}{3}\\ &= \frac{5x+4-2x+3}{3}\\ &= \frac{3x+7}{3}\end{aligned}$$

□

D. $\dfrac{5n-8}{4} + \dfrac{1-6n}{4}$

E. $\dfrac{5(m+1)}{2m} + \dfrac{3(3m-7)}{2m}$

SOLUTION

$$\begin{aligned}&\frac{5(m+1)}{2m} + \frac{3(3m-7)}{2m}\\ &= \frac{5(m+1)+3(3m-7)}{2m}\\ &= \frac{5m+5+9m-21}{2m}\\ &= \frac{14m-16}{2m}\\ &= \frac{\not{2}(7m-8)}{\not{2}m}\\ &= \frac{7m-8}{m}\end{aligned}$$

□

F. $\dfrac{4(2x-5)}{5x^3} - \dfrac{x-20}{5x^3}$

G. $\dfrac{3}{n-6} - \dfrac{4}{n}$

SOLUTION

$$\begin{aligned}\frac{3}{n-6} - \frac{4}{n} &= \frac{3}{n-6}\cdot\frac{n}{n} - \frac{4}{n}\cdot\frac{n-6}{n-6}\\ &= \frac{3n-4(n-6)}{(n-6)(n)}\\ &= \frac{3n-4n+24}{n(n-6)}\\ &= \frac{24-n}{n^2-6n}\end{aligned}$$

□

H. $\dfrac{2}{2x-5} - \dfrac{8}{3x+4}$

I. $\dfrac{13}{12x} + \dfrac{5}{4x^2}$

Solution

$$\begin{aligned}\frac{13}{12x} + \frac{5}{4x^2} &= \frac{13}{12x} \cdot \frac{x}{x} + \frac{5}{4x^2} \cdot \frac{3}{3} \\ &= \frac{13x}{12x^2} + \frac{15}{12x^2} \\ &= \frac{13x+15}{12x^2}\end{aligned}$$

□

J. $\dfrac{3}{10x} - \dfrac{5}{4x}$

K. $\dfrac{y-3}{6x} - \dfrac{x}{9y}$

Solution

$$\begin{aligned}\frac{y-3}{6x} - \frac{x}{9y} &= \frac{y-3}{6x} \cdot \frac{3y}{3y} - \frac{x}{9y} \cdot \frac{2x}{2x} \\ &= \frac{(y-3)(3y)}{18xy} - \frac{(x)(2x)}{18xy} \\ &= \frac{(y-3)(3y) - (x)(2x)}{18xy} \\ &= \frac{3y^2 - 9y - 2x^2}{18xy}\end{aligned}$$

□

L. $\dfrac{x+5}{30xy} - \dfrac{z+7}{20yz}$

M. $\dfrac{7}{x^2+5x} + \dfrac{2}{x^2-3x}$

Solution

$$\begin{aligned}&\frac{7}{x^2+5x} + \frac{2}{x^2-3x} \\ &\quad = \frac{7}{x(x+5)} + \frac{2}{x(x-3)} \\ &\quad = \frac{7}{x(x+5)} \cdot \frac{x-3}{x-3} + \frac{2}{x(x-3)} \cdot \frac{x+5}{x+5} \\ &\quad = \frac{7(x-3) + 2(x+5)}{x(x+5)(x-3)} \\ &\quad = \frac{7x - 21 + 2x + 10}{x(x+5)(x-3)} \\ &\quad = \frac{9x - 11}{x(x+5)(x-3)}\end{aligned}$$

□

N. $\dfrac{18}{x^2-3x} + \dfrac{6}{x}$

O. $\dfrac{9}{b^2+7b+12} - \dfrac{b}{2b^2+b-15}$

SOLUTION

$$\begin{aligned}
&\frac{9}{b^2+7b+12} - \frac{b}{2b^2+b-15}\\
&\quad= \frac{9}{(b+3)(b+4)} - \frac{b}{(b+3)(2b-5)}\\
&\quad= \frac{9}{(b+3)(b+4)} \cdot \frac{2b-5}{2b-5}\\
&\qquad\qquad - \frac{b}{(b+3)(2b-5)} \cdot \frac{b+4}{b+4}\\
&\quad= \frac{9(2b-5) - b(b+4)}{(b+3)(b+4)(2b-5)}\\
&\quad= \frac{18b-45-b^2-4b}{(b+3)(b+4)(2b-5)}\\
&\quad= \frac{-b^2+14b-45}{(b+3)(b+4)(2b-5)}\\
&\quad= \frac{-(b^2-14b+45)}{(b+3)(b+4)(2b-5)}\\
&\quad= -\frac{(b-5)(b-9)}{(b+3)(b+4)(2b-5)}
\end{aligned}$$

□

P. $\dfrac{x}{3x^2-7x+2} + \dfrac{2}{x^2-9x+14}$

Q. $\dfrac{3}{x^2+3x} - \dfrac{2}{x^2-3x} + \dfrac{8}{x^2-9}$

SOLUTION

$$\begin{aligned}
\frac{3}{x^2+3x} - \frac{2}{x^2-3x} + \frac{8}{x^2-9} &= \frac{3}{x(x+3)} - \frac{2}{x(x-3)} + \frac{8}{(x-3)(x+3)}\\
&= \frac{3}{x(x+3)} \cdot \frac{x-3}{x-3} - \frac{2}{x(x-3)} \cdot \frac{x+3}{x+3} + \frac{8}{(x-3)(x+3)} \cdot \frac{x}{x}\\
&= \frac{3(x-3)}{x(x+3)(x-3)} - \frac{2(x+3)}{x(x-3)(x+3)} + \frac{8(x)}{(x-3)(x+3)(x)}\\
&= \frac{3(x-3) - 2(x+3) + 8(x)}{x(x+3)(x-3)}\\
&= \frac{3x-9-2x-6+8x}{x(x+3)(x-3)}\\
&= \frac{9x-15}{x(x+3)(x-3)}
\end{aligned}$$

□

Section 48: Complex Fractions

Definition

A *complex fraction* is a fraction whose numerator or denominator contains a fraction.

Procedure A (preferred)

To simplify a complex fraction, we can:

1. simplify both the numerator and denominator of the fraction;
2. multiply the numerator by the reciprocal of the denominator;
3. simplify the product of step (2).

Procedure B (alternative)

To simplify a complex fraction, we can:

1. factorize the denominators of all fractions within the complex fraction;
2. identify the least common multiple of the denominators of step (1);
3. multiply the complex fraction by the fraction whose numerator and denominator both equal the least common multiple of step (2);
4. simplify the product of step (3).

Comments

- Procedure A is more efficient than procedure B when the denominator of the complex fraction is already simplified, but often less efficient otherwise.
- The fraction by which we multiply the complex fraction in step (3) of procedure B will be a form of 1.

Example 1

Simplify each complex fraction using procedure A.

A. $\dfrac{\dfrac{x+5}{x^2-1}}{\dfrac{x^2+4x-5}{x+1}}$

B. $\dfrac{\dfrac{x^2-6x+9}{x+7}}{\dfrac{2x-6}{x^2+5x-14}}$

Solution

$$\begin{aligned}
\frac{\dfrac{x+5}{x^2-1}}{\dfrac{x^2+4x-5}{x+1}} &= \frac{x+5}{x^2-1}\cdot\frac{x+1}{x^2+4x-5} \\
&= \frac{(x+5)(x+1)}{(x^2-1)(x^2+4x-5)} \\
&= \frac{\cancel{(x+5)(x+1)}}{(x-1)\cancel{(x+1)(x+5)}(x-1)} \\
&= \frac{1}{(x-1)^2}
\end{aligned}$$

□

C. $\dfrac{\dfrac{3-2x}{7x^2}}{\dfrac{2x}{9}-\dfrac{1}{x}}$

Solution

$$\frac{\dfrac{3-2x}{7x^2}}{\dfrac{2x}{9}-\dfrac{1}{x}} = \frac{\dfrac{3-2x}{7x^2}}{\dfrac{2x}{9}\cdot\dfrac{x}{x}-\dfrac{1}{x}\cdot\dfrac{9}{9}}$$
$$= \frac{\dfrac{3-2x}{7x^2}}{\dfrac{2x^2-9}{9x}}$$
$$= \frac{3-2x}{7x^2}\cdot\frac{9x}{2x^2-9}$$
$$= \frac{3-2x}{7x\cdot\not{x}}\cdot\frac{9\cdot\not{x}}{2x^2-9}$$
$$= \frac{27-18x}{14x^3-63x}$$

□

D. $\dfrac{\dfrac{6}{x+3}+\dfrac{1}{3}}{\dfrac{2}{x+3}-\dfrac{5}{6}}$

E. $\dfrac{\dfrac{5}{x+h}-\dfrac{5}{x}}{h}$

Solution

$$\frac{\dfrac{5}{x+h}-\dfrac{5}{x}}{h} = \left(\frac{5}{x+h}-\frac{5}{x}\right)\cdot\frac{1}{h}$$
$$= \left(\frac{5}{x+h}\cdot\frac{x}{x}-\frac{5}{x}\cdot\frac{x+h}{x+h}\right)\cdot\frac{1}{h}$$
$$= \left(\frac{5x}{x(x+h)}-\frac{5(x+h)}{x(x+h)}\right)\cdot\frac{1}{h}$$
$$= \frac{5x-5(x+h)}{x(x+h)}\cdot\frac{1}{h}$$
$$= \frac{\cancel{5x}-\cancel{5x}-5h}{x(x+h)}\cdot\frac{1}{h}$$
$$= \frac{-5\cancel{h}}{x(x+h)}\cdot\frac{1}{\cancel{h}}$$
$$= \frac{-5}{x(x+h)}$$

□

F. $\dfrac{\dfrac{1}{x+h+2}-\dfrac{1}{x+2}}{h}$

G. $\dfrac{a^{-3}+b^{-2}}{a^{-1}-b^{-2}}$

SOLUTION

$$\begin{aligned}\frac{a^{-3}+b^{-2}}{a^{-1}-b^{-2}} &= \frac{1/a^3+1/b^2}{1/a-1/b^2}\\ &= \frac{(1/a^3)(b^2/b^2)+(1/b^2)(a^3/a^3)}{(1/a)(b^2/b^2)-(1/b^2)(a/a)}\\ &= \frac{b^2/(a^3b^2)+a^3/(a^3b^2)}{b^2/(ab^2)-a/(ab^2)}\\ &= \frac{(b^2+a^3)/(a^3b^2)}{(b^2-a)/(ab^2)}\\ &= \frac{b^2+a^3}{a^3b^2}\cdot\frac{ab^2}{b^2-a}\\ &= \frac{b^2+a^3}{a^2\cdot \cancel{ab^2}}\cdot\frac{\cancel{ab^2}}{b^2-a}\\ &= \frac{b^2+a^3}{a^2b^2-a^3}\end{aligned}$$

□

H. $\dfrac{x^{-1}-y^{-1}}{x^{-1}+y^{-1}}$

I. $1+\dfrac{1}{1+\dfrac{1}{x-1}}$

SOLUTION

$$\begin{aligned}1+\frac{1}{1+\dfrac{1}{x-1}} &= 1+\frac{1}{\dfrac{x-1}{x-1}+\dfrac{1}{x-1}}\\ &= 1+\frac{1}{\dfrac{(x-\cancel{1})+\cancel{1}}{x-1}}\\ &= 1+\frac{1}{\dfrac{x}{x-1}}\\ &= 1+\frac{x-1}{x}\\ &= \frac{x}{x}+\frac{x-1}{x}\\ &= \frac{x+(x-1)}{x}\\ &= \frac{2x-1}{x}\end{aligned}$$

□

J. $1+\cfrac{6}{1+\cfrac{6}{1+\cfrac{6}{1+\cfrac{6}{1+\cfrac{6}{\vdots}}}}}$

[Hint: Let x equal the fraction above and recognize that $x=1+6/x$.]

Example 2

Simplify each complex fraction using procedure B.

A. $\dfrac{2/5 - 1/3}{1/2 + 1/4}$

Solution

Note

The LCM of 5, 3, 2 and 4 is 60.

Simplification

$$\begin{aligned}\frac{2/5 - 1/3}{1/2 + 1/4} &= \frac{2/5 - 1/3}{1/2 + 1/4} \cdot \frac{60}{60} \\ &= \frac{(2/5)(60) - (1/3)(60)}{(1/2)(60) + (1/4)(60)} \\ &= \frac{24 - 20}{30 + 15} \\ &= 4/45\end{aligned}$$

$\square$

B. $\dfrac{3/2 + 7}{-3/8 - 5/6}$

C. $\dfrac{a^{-3} + b^{-2}}{a^{-1} - b^{-2}}$

Solution

Note

The LCM of a^3, b^2, a and b^2 is a^3b^2.

Simplification

$$\begin{aligned}\frac{a^{-3} + b^{-2}}{a^{-1} - b^{-2}} &= \frac{1/a^3 + 1/b^2}{1/a - 1/b^2} \cdot \frac{a^3b^2}{a^3b^2} \\ &= \frac{\cancel{a^3}b^2/\cancel{a^3} + a^3\cancel{b^2}/\cancel{b^2}}{a^3b^2/a - a^3\cancel{b^2}/\cancel{b^2}} \\ &= \frac{b^2 + a^3}{a^2b^2 - a^3}\end{aligned}$$

$\square$

D. $\dfrac{x^{-1} - y^{-1}}{x^{-1} + y^{-1}}$

Section 49: Equations with Fractions

Property
Suppose $a, b \in \mathbf{R}$ with $b \neq 0$.
Then $a/b = 0$ if and only if $a = 0$.

Proof
Suppose $a, b \in \mathbf{R}$ with $b \neq 0$.

$(\Rightarrow)$ $a/b = 0 \Rightarrow (a/b)(b) = (0)(b)$
$\Rightarrow a = 0$

$(\Leftarrow)$ $a = 0 \Rightarrow a/b = 0/b$
$= 0$

⊡

Procedure A (preferred)
To solve an equation that contains fractions, we can:
1. apply properties of equality to the equation until one side equals zero;
2. combine expressions on the nonzero side of the equation into a single fraction;
3. set the numerator of the fraction equal to zero (refer to the property above);
4. solve the equation yielded after step (3).

Procedure B (alternative)
To solve an equation that contains fractions, we can:
1. identify the least common multiple of the factors in the denominators of the fractions;
2. eliminate the fractions by multiplying both sides of the equation by the least common denominator found in step (1) above;
3. solve the equation yielded after step (2).

Comments
- We have followed the steps of procedure B previously in solving equations and inequalities with constant denominators. We can also follow this procedure to solve equations with variable denominators; however, we cannot apply this method to solve inequalities with variable denominators unless the value of the least common multiple of the factors of those denominators is always positive or always negative. To maintain consistency between our process for solving equations and our process for solving inequalities, we will usually follow the steps of procedure A and avoid multiplying both sides of an equation by an expression that contains a variable.

- The values that we can assign to the variables in an fraction are limited to those for which its denominator will not evaluate to zero. For example, in the expression $(x^2 + 4)/(x - 3)$ we must assume that $x \neq 3$. Before we attempt to solve an equation that contains an fraction with a variable in its denominator, we should identify all restrictions on the values the variable can assume. Note that any values restricted from assignment to a variable in our original equation will remain restricted in any equivalent equation.

EXAMPLE 1
Solve each equation.

A. $x/6 + 3x/5 = 7/10$

SOLUTION
$x/6 + 3x/5 = 7/10$
$\Leftrightarrow x/6 + 3x/5 - 7/10 = 0$
$\Leftrightarrow (x/6)(5/5) + (3x/5)(6/6) - (7/10)(3/3) = 0$
$\Leftrightarrow 5x/30 + 18x/30 - 21/30 = 0$
$\Leftrightarrow (5x + 18x - 21)/30 = 0$
$\Leftrightarrow (23x - 21)/30 = 0$
$\Leftrightarrow 23x - 21 = 0$
$\Leftrightarrow 23x = 21$
$\Leftrightarrow 23x/23 = 21/23$
$\Leftrightarrow x = 21/23$
Set of solutions: $\{21/23\}$

□

B. $(x - 5)/7 - x/2 = 4/7$

C. $5/(4n) - 1 = 2/(3n)$

SOLUTION
Zeros of the Denominators
- $4n = 0 \Leftrightarrow n = 0$
- $3n = 0 \Leftrightarrow n = 0$

Set of Solutions
Assuming $n \neq 0$:
$5/(4n) - 1 = 2/(3n)$
$\Leftrightarrow 5/(4n) - 1 - 2/(3n) = 0$
$\Leftrightarrow [5/(4n)] \cdot [3/3] - 1 \cdot [12n/(12n)] - [2/(3n)] \cdot [4/4] = 0$
$\Leftrightarrow 15/(12n) - 12n/(12n) - 8/(12n) = 0$
$\Leftrightarrow (15 - 12n - 8)/(12n) = 0$
$\Leftrightarrow (-12n + 7)/(12n) = 0$
$\Leftrightarrow -12n + 7 = 0$
$\Leftrightarrow -12n + 7 - 7 = 0 - 7$
$\Leftrightarrow -12x = -7$
$\Leftrightarrow (-12x)/(-12) = (-7)/(-12)$
$\Leftrightarrow n = 7/12$
Set of solutions: $\{7/12\}$

□

D. $(t - 2)/t + 3 = 3/5$

E. $3/(x+5) = -2/(x-8)$

SOLUTION
Zeros of the Denominators
- $x+5=0 \Leftrightarrow x=-5$
- $x-8=0 \Leftrightarrow x=8$

Set of Solutions
Assuming $x \neq -5$ AND $x \neq 8$:
$3/(x+5) = -2/(x-8)$
$\Leftrightarrow 3/(x+5) + 2/(x-8) = 0$
$\Leftrightarrow \dfrac{3}{x+5}\cdot\dfrac{x-8}{x-8} + \dfrac{2}{x-8}\cdot\dfrac{x+5}{x+5} = 0$
$\Leftrightarrow \dfrac{3(x-8)+2(x+5)}{(x+5)(x-8)} = 0$
$\Leftrightarrow 3(x-8)+2(x+5) = 0$
$\Leftrightarrow 3x-24+2x+10 = 0$
$\Leftrightarrow 5x-14 = 0$
$\Leftrightarrow 5x-14+14 = 0+14$
$\Leftrightarrow 5x = 14$
$\Leftrightarrow x = 14/5$
Set of solutions: $\{14/5\}$

$\square$

F. $7/(3x+9) = 5/(x+3)+2$

G. $8x/(x-5) - x/(x+5) - 7 = 0$

SOLUTION
Zeros of the Denominators
- $x-5=0 \Leftrightarrow x=5$
- $x+5=0 \Leftrightarrow x=-5$

Set of Solutions
Assuming $x \neq -5$ AND $x \neq 5$:
$8x/(x-5) - x/(x+5) - 7 = 0$
$\Leftrightarrow \dfrac{8x}{x-5}\cdot\dfrac{x+5}{x+5} - \dfrac{x}{x+5}\cdot\dfrac{x-5}{x-5} - 7\cdot\dfrac{(x+5)(x-5)}{(x+5)(x-5)} = 0$
$\Leftrightarrow [8x(x+5) - x(x-5) - 7(x+5)(x-5)]/[(x+5)(x-5)] = 0$
$\Leftrightarrow 8x^2+40x-x^2+5x-7(x^2-25) = 0$
$\Leftrightarrow \cancel{8x^2}+40x-\cancel{x^2}+5x-\cancel{7x^2}+175 = 0$
$\Leftrightarrow 45x = -175$
$\Leftrightarrow x = -175/45$
$= -35/9$
Set of solutions: $\{-35/9\}$

$\square$

H. $1 + \dfrac{5}{x-2} = \dfrac{10}{x^2-2x}$

SOLUTION

Zeros of the Denominators

- $x - 2 = 0 \Leftrightarrow x = 2$
- $x^2 - 2x = 0 \Leftrightarrow x(x-2) = 0$

$$\Leftrightarrow x = 0 \text{ OR } x - 2 = 0$$
$$\Leftrightarrow x = 0 \text{ OR } x = 2$$

Set of Solutions

Assuming $x \neq 0$ AND $x \neq 2$:

$$1 + \frac{5}{x-2} = \frac{10}{x^2-2x}$$
$$\Leftrightarrow 1 + \frac{5}{x-2} - \frac{10}{x^2-2x} = 0$$
$$\Leftrightarrow 1 + \frac{5}{x-2} - \frac{10}{x(x-2)} = 0$$
$$\Leftrightarrow 1 \cdot \frac{x(x-2)}{x(x-2)} + \frac{5}{x-2} \cdot \frac{x}{x} - \frac{10}{x(x-2)} = 0$$
$$\Leftrightarrow \frac{x(x-2) + 5x - 10}{x(x-2)} = 0$$
$$\Leftrightarrow x(x-2) + 5x - 10 = 0$$
$$\Leftrightarrow x^2 - 2x + 5x - 10 = 0$$
$$\Leftrightarrow x^2 + 3x - 10 = 0$$
$$\Leftrightarrow (x+5)(x-2) = 0$$
$$\Leftrightarrow x + 5 = 0 \quad \text{OR } x - 2 = 0$$
$$\Leftrightarrow x = -5 \qquad \Leftrightarrow \cancel{x = 2} \quad \text{X}$$

Set of solutions: $\{-5\}$

□

I. $\dfrac{5x}{x-2} - \dfrac{10}{x+3} - \dfrac{50}{(x+3)(x-2)} = 5$

J. $x = -10/x$

SOLUTION

Assuming $x \neq 0$:

$$x = -10/x \Leftrightarrow x + 10/x = 0$$
$$\Leftrightarrow (x)(x/x) + 10/x = 0$$
$$\Leftrightarrow (x^2 + 10)/x = 0$$
$$\Leftrightarrow x^2 + 10 = 0$$
$$\Leftrightarrow \cancel{x^2 = -10} \quad \text{X}$$

The equation has no real solution (since $x^2 \geq 0$ for all $x \in \mathbf{R}$).

□

K. $\dfrac{x^2}{x-1} = \dfrac{2-x}{x-1}$

EXAMPLE 2
Solve each word problem.

A. The difference between two positive numbers is 30. When the larger number is divided by the smaller number, the quotient is 5 and the remainder is 2. Find both numbers.

SOLUTION
Let s denote the smaller number and
l denote the larger number.
Note that $l - s = 30$, so $l = 30 + s$.
$l/s = 5 + 2/s$
$\Rightarrow l/s - 5 - 2/s = 0$
$\Rightarrow (30 + s)/s - (5)(s/s) - 2/s = 0$
$\Rightarrow (30 + s)/s - 5s/s - 2/s = 0$
$\Rightarrow (30 + s - 5s - 2)/s = 0$
$\Rightarrow (-4s + 28)/s = 0$
$\Rightarrow -4s + 28 = 0$
$\Rightarrow -4s = -28$
$\Rightarrow -4s/(-4) = -28/(-4)$
$\Rightarrow s = 7$
$\Rightarrow l = 30 + 7$
$= 37$
The numbers are 7 and 37.

□

B. One number is 9 more than another number. The smaller number divided by the larger number equals 4/5. Find both numbers.

Section 50: Rates of Travel

Formula

(total distance traveled) = (speed)(total time traveled)

Example 1

Solve each word problem.

A. Two trains move in opposite directions at constant speeds after leaving a station at the same time. One train moves 7 km/hr faster than the other train. Find their speeds if they are 309 km apart 3 hours after leaving the station.

Solution

Declarations

Let d_1, r_1, and t_1 be the distance (in km), speed (in km/hr), and time (in hr) that the faster train moves and
d_2, r_2, and t_2 be the distance (in km), speed (in km/hr), and time (in hr) that the slower train moves.

Substitutions

- $d_1 = r_1 t_1$
- $d_2 = r_2 t_2$
- $r_1 = r_2 + 7$

Values

For $t_1 = t_2 = 3$:

$d_1 + d_2 = 309$

$\Rightarrow r_1 t_1 + r_2 t_2 = 309$

$\Rightarrow (r_2 + 7)(3) + (r_2)(3) = 309$

$\Rightarrow (r_2)(3) + (7)(3) + (r_2)(3) = 309$

$\Rightarrow 6r_2 + 21 = 309$

$\Rightarrow 6r_2 + \cancel{21} - \cancel{21} = 309 - 21$

$\Rightarrow 6r_2 = 288$

$\Rightarrow \cancel{6}r_2/\cancel{6} = 288/6$

$\Rightarrow r_2 = 48$

$\Rightarrow r_1 = 48 + 7$

$= 55$

The faster train moves 55 km/hr and the slower train moves 48 km/hr.

□

B. Suppose Bill runs in a straight line from his home at a rate of 600 feet per minute. Ten minutes later his wife leaves the home and runs after her husband at a rate of 750 feet per minute. How long will it take the wife to catch Bill?

C. Bob participated in a 33-mile biathalon. He swam 6 miles and then ran 27 miles more. His traveling rate on the ground was three miles per hour more than his rate in the water. Find both rates if he traveled the entire distance in 10 hours.

SOLUTION

Declarations

Let d_1, r_1, and t_1 be the distance (in mi), speed (in mi/hr), and time (in hr) that Bob traveled in the water and
d_2, r_2, and t_2 be the distance (in mi), speed (in mi/hr), and time (in hr) that Bob traveled on the ground.

Substitutions

- $d_1 = r_1t_1 \Leftrightarrow d_1/r_1 = t_1$
- $d_2 = r_2t_2 \Leftrightarrow d_2/r_2 = t_2$
- $r_2 = r_1 + 3$

Values

For $d_1 = 6$ and $d_2 = 27$:

$t_1 + t_2 = 10$

$\Rightarrow d_1/r_1 + d_2/r_2 = 10$

$\Rightarrow 6/r_1 + 27/(r_1 + 3) - 10 = 0$

$\Rightarrow 6(r_1 + 3)/(r_1(r_1 + 3)) + 27r_1/(r_1(r_1 + 3)) - 10r_1(r_1 + 3)/(r_1(r_1 + 3)) = 0$

$\Rightarrow [6(r_1 + 3) + 27r_1 - 10r_1(r_1 + 3)] / [r_1(r_1 + 3)] = 0$

$\Rightarrow 6(r_1 + 3) + 27r_1 - 10r_1(r_1 + 3) = 0$

$\Rightarrow 6r_1 + 18 + 27r_1 - 10r_1^2 - 30r_1 = 0$

$\Rightarrow -10r_1^2 + 3r_1 + 18 = 0$

$\Rightarrow -(10r_1^2 - 3r_1 - 18) = 0$

$\Rightarrow -(2r_1 - 3)(5r_1 + 6) = 0$

$\Rightarrow 2r_1 - 3 = 0$ OR $5r_1 + 6 = 0$

$\Rightarrow 2r_1 = 3$ $\qquad \Rightarrow 5r_1 = -6$

$\Rightarrow r_1 = 3/2$ $\qquad \Rightarrow \cancel{r_1 = -6/5}$

$\Rightarrow r_2 = 3/2 + 3 = 3/2 + 6/2 = 9/2$

Bob swam 3/2 (or 1.5) miles per hour and ran 9/2 (or 4.5) miles per hour. □

D. Two masked women depart the premises of a bank in opposite directions by car. One woman drives 238 km and stops. The other woman drives 2 hours longer and covers a distance of 518 km. If the women drove their cars the same average speed, then find that speed.

E. Edwin likes to run laps around a track. One day he breaks his leg and must use a wheelchair to circle the track. When he is healthy, his rate of speed is five laps per minute. In his wheelchair, he takes eighteen minutes to travel the same distance that he could run previously in three minutes. How long will it take him to roll forty-five laps around the track in the wheelchair?

SOLUTION

Declarations

Let d_1, r_1, and t_1 denote the distance (in laps), speed (in laps/min), and time (in min) that Edwin travels as a runner and

d_2, r_2, and t_2 denote the distance (in laps), speed (in laps/min), and time (in min) that Edwin travels as an invalid.

Substitutions

- $d_1 = r_1 t_1$
- $d_2 = r_2 t_2$

Value of r_2

For $r_1 = 5$, $t_1 = 3$, and $t_2 = 18$:

$$\begin{aligned} d_1 = d_2 &\Rightarrow r_1 t_1 = r_2 t_2 \\ &\Rightarrow (5)(3) = r_2(18) \\ &\Rightarrow (5)(3)/(18) = r_2(18)/(18) \\ &\Rightarrow 15/18 = r_2 \\ &\Rightarrow r_2 = 5/6 \end{aligned}$$

Value of t_2

For $r_2 = 5/6$ and $d_2 = 45$:

$$\begin{aligned} d_2 = r_2 t_2 &\Rightarrow 45 = (5/6)(t_2) \\ &\Rightarrow (6/5)(45) = (6/5)(5/6)(t_2) \\ &\Rightarrow 54 = t_2 \end{aligned}$$

Conclusion

Edwin needs 54 minutes to roll 45 laps around the track.

□

F. Fred decides to travel 28 miles to a park. He begins the trip on a bicycle, riding at a constant speed of 15 miles per hour. After some time, the bicycle breaks down and Fred is forced to walk the rest of the way at a speed of 3 miles per hour. If the entire trip takes 8 hours, then how far had Fred traveled on his bicycle before the breakdown?

Section 51: Rates of Work

Formula

(total amount of work accomplished) = (rate of work)(total time spent working)

Example 1

Solve each word problem.

A. Gary and Hal dig graves. Gary can dig a grave in 200 minutes and Hal can dig a grave in 150 minutes. If Gary and Hal work together, then how long will it take them to dig a grave?

Solution

Declarations

Let r_g and w_g be the rate (in graves/min) and amount (in graves) of work of Gary,
r_h and w_h be the rate (in graves/min) and amount (in graves) of work of Hal, and
t be the time (in min) it takes for the men to dig one grave together.

Substitutions

- $w_g = r_g t$
- $w_h = r_h t$

Values

For $r_g = 1/200$ and $r_h = 1/150$:

$$\begin{aligned} & w_g + w_h = 1 \\ & \Rightarrow (r_g)(t) + (r_h)(t) = 1 \\ & \Rightarrow (1/200)(t) + (1/150)(t) = 1 \\ & \Rightarrow t/200 + t/150 = 1 \\ & \Rightarrow (t/200)(3/3) + (t/150)(4/4) = 1 \\ & \Rightarrow 3t/600 + 4t/600 = 1 \\ & \Rightarrow 7t/600 = 1 \\ & \Rightarrow (7t/600)(600/7) = (1)(600/7) \\ & \Rightarrow t = 600/7 \\ & \quad \approx 85.71 \end{aligned}$$

It would take them about 86 minutes to dig the grave. □

B. The water from two faucets can fill a sink in 6 minutes. Water from the cold faucet alone can fill the sink in 5 minutes less time than it would take water from the hot faucet alone to fill the sink. How long would it take each faucet to fill the sink by itself?

Section 52: Sign Charts

Definition

A *sign chart* is a chart that indicates the signs of an expression on various intervals.

Procedure A (preferred)

To construct a sign chart for a product or quotient of polynomials, we can:

1. draw a number line;
2. list all factors in the product or quotient to the upper left of the number line;
3. write the entire product or quotient above the factors listed in step (2);
4. plot on the number line the points that correspond to the zeros of the factors of step (2);
5. indicate above the line segments with the endpoints of step (4) the signs (positive or negative) of the factors on the intervals that correspond to those segments;
6. indicate above the line segments with the endpoints of step (4) the signs (positive or negative) of the product or quotient on the intervals that correspond to those segments.

Procedure B (alternative)

To construct a sign chart for a product or quotient of polynomials, we can:

1. draw a number line;
2. write the entire product or quotient to the upper left of the number line;
3. plot on the number line the points that correspond to the zeros of the factors in the product or quotient;
4. evaluate the product or quotient at arbitrary values (called *test points*) within the intervals bounded by the zeros of the factors in the product or quotient;
5. indicate above the line segments with the endpoints of step (3) the signs (positive or negative) of step (4) of the product or quotient at the arbitrary values in the intervals that correspond to those segments.

Comments

- Use the signs $+$ and $-$ to indicate whether an expression is positive or negative.
- On the top line of the chart, we may write "expression", "product", or "quotient" instead of the given expression of which we want to determine the signs.
- The value of a linear factor $ax + b$ will be:
  - zero when $x = -b/a$,
  - negative when $x < -b/a$ and positive when $x > -b/a$ if $a > 0$, and
  - positive when $x < -b/a$ and negative when $x > -b/a$ if $a < 0$.
- The sign of an entire product or quotient will be:
  - negative on intervals where the number of negative factors is odd and
  - positive on intervals where the number of negative factors is even.
- As the value of a variable increases, the sign of a product or quotient of polynomials in that variable can change only at the zeros of the factors in the product or quotient.
- We will apply procedure B to construct a sign chart only in cases where we cannot find the factors of the given expression (or its numerator and denominator if a quotient) before we can find the zeros of those factors.

Example 1
Construct a sign chart for each expression.

A. $(x-3)(x+4)$

Solution
Zeros of the Factors
- $x-3=0 \Leftrightarrow x=3$
- $x+4=0 \Leftrightarrow x=-4$

Sign Chart

$(x-3)(x+4)$	$+$		$-$		$+$
$x-3$	$-$		$-$		$+$
$x+4$	$-$		$+$		$+$
		-4		3	

□

B. $\dfrac{2x+5}{(x-7)(x-1)}$

C. $\dfrac{(3x-8)(2x+10)}{(2x+6)(4x+1)}$

Solution
Zeros of the Factors
- $3x-8=0 \Leftrightarrow 3x=8 \Leftrightarrow 3x/3=8/3 \Leftrightarrow x=8/3$
- $2x+10=0 \Leftrightarrow 2x=-10 \Leftrightarrow 2x/2=-10/2 \Leftrightarrow x=-5$
- $2x+6=0 \Leftrightarrow 2x=-6 \Leftrightarrow 2x/2=8/2 \Leftrightarrow x=-3$
- $4x+1=0 \Leftrightarrow 4x=-1 \Leftrightarrow 4x/4=-1/4 \Leftrightarrow x=-1/4$

Sign Chart

quotient	$+$		$-$		$+$		$-$		$+$
$3x-8$	$-$		$-$		$-$		$-$		$+$
$2x+10$	$-$		$+$		$+$		$+$		$+$
$2x+6$	$-$		$-$		$+$		$+$		$+$
$4x+1$	$-$		$-$		$-$		$+$		$+$
		-5		-3		$-\frac{1}{4}$		$\frac{8}{3}$	

□

D. $x(x+4)(x-5)^2$

Example 2

Use procedure B to construct a sign chart for each expression given the zeros of its factors.

A. expression: $56x^3 - 166x^2 + 53x + 30$
zeros of the factors: $-2/7$, $3/4$, $5/2$

Solution

Test Points

The zeros divide the set of real numbers into four intervals. For test points:

- in $(-\infty, -2/7)$, we choose -1;
- in $(-2/7, 3/4)$, we choose 0;
- in $(3/4, 5/2)$, we choose 1;
- in $(5/2, \infty)$, we choose 3.

Table of Values

x	$56x^3 - 166x^2 + 53x + 30$
-1	$56(-1)^3 - 166(-1)^2 + 53(-1) + 30$ $= -245$
0	$56(0)^3 - 166(0)^2 + 53(0) + 30 =$ 30
1	$56(1)^3 - 166(1)^2 + 53(1) + 30 =$ -27
3	$56(3)^3 - 166(3)^2 + 53(3) + 30 =$ 207

Sign Chart

expression: $-$ | $-\frac{2}{7}$ | $+$ | $\frac{3}{4}$ | $-$ | $\frac{5}{2}$ | $+$

□

B. expression: $(x^2 - 10)/(2 - x)$
zeros of the factors: $-\sqrt{10}$, $\sqrt{10}$, 2
[Hint: $\sqrt{10} \approx 3.16$]

Section 53: Inequalities with Polynomials or Fractions

Procedure

To solve an inequality that contains polynomials or fractions, we:

1. apply properties of inequality to the inequality until one side equals zero;
2. construct a sign chart for the nonzero side of the inequality;
3. use the sign chart to identify the set of solutions of the inequality.

Comments

- The value of an inequality is greater than zero on the intervals where its sign is positive and less than zero on the intervals where its sign is negative.
- We should include a bound of an interval in the set of solutions of an inequality if the inequality is satisfied when its nonzero side equals zero, but we must exclude a bound at which the value of the denominator of a fraction within the inequality equals zero.
- The set of solutions of an inequality will usually be an interval or union of intervals.
- We must remember to change the direction of an inequality from $<$ to $>$ or vice versa when multiplying or dividing both sides of the inequality by a negative number.
- We should not multiply or divide both sides of an inequality by an expression that contains a variable unless the value of that expression is always positive or always negative.

Example 1

Solve each inequality using a sign chart.

A. $x^2 + 7x + 6 > 0$

Solution

Note

$x^2 + 7x + 6 > 0 \Leftrightarrow (x+1)(x+6) > 0$

Zeros of the Factors

- $x + 1 = 0 \Leftrightarrow x = -1$
- $x + 6 = 0 \Leftrightarrow x = -6$

Sign Chart

$(x+1)(x+6)$	$+$	$-$	$+$
$x+1$	$-$	$-$	$+$
$x+6$	$-$	$+$	$+$
	-6	-1	

Set of Solutions

Set of solutions: $(-\infty, -6) \cup (-1, \infty)$

□

B. $(3x+6)/(4x-3) < 0$

C. $\dfrac{8}{3x+7} \leq 2$

Solution

Note

$$\begin{aligned}\frac{8}{3x+7} \leq 2 &\Leftrightarrow \frac{8}{3x+7} - 2 \leq 0\\ &\Leftrightarrow \frac{8}{3x+7} - 2 \cdot \frac{3x+7}{3x+7} \leq 0\\ &\Leftrightarrow \frac{8-2(3x+7)}{3x+7} \leq 0\\ &\Leftrightarrow \frac{8-6x-14}{3x+7} \leq 0\\ &\Leftrightarrow \frac{-6x-6}{3x+7} \leq 0\end{aligned}$$

Zeros of the Factors

- $-6x-6=0 \Leftrightarrow -6x=6$
  $\Leftrightarrow -6x/(-6) = 6/(-6)$
  $\Leftrightarrow x=-1$
- $3x+7=0 \Leftrightarrow 3x=-7$
  $\Leftrightarrow 3x/3 = -7/3$
  $\Leftrightarrow x=-7/3$

Sign Chart

quotient	$-$	$+$	$-$
$-6x-6$	$+$	$+$	$-$
$3x+7$	$-$	$+$	$+$
	$-\frac{7}{3}$	-1	

Set of Solutions

Set of solutions: $(-\infty, -\frac{7}{3}) \cup [-1, \infty)$

Comments

- We include -1 in the set of solutions because $(-6x-6)/(3x+7)$ equals 0 [and so $(-6x-6)/(3x+7) \leq 0$ is satisfied] at that value of x.
- We exclude $-7/3$ from the set of solutions because the denominator $3x+7$ equals 0 at that value of x.

□

D. $x^2 - 4x \geq 12$

E. $x^3 - 4x^2 - 9x + 36 \geq 0$

SOLUTION

Note

$x^3 - 4x^2 - 9x + 36 \geq 0$
$\Leftrightarrow x^2(x-4) - 9(x-4) \geq 0$
$\Leftrightarrow (x-4)(x^2-9) \geq 0$
$\Leftrightarrow (x-4)(x-3)(x+3) \geq 0$

Zeros of the Factors

- $x - 4 = 0 \Leftrightarrow x = 4$
- $x - 3 = 0 \Leftrightarrow x = 3$
- $x + 3 = 0 \Leftrightarrow x = -3$

Sign Chart

product	$-$	$+$	$-$	$+$
$x-4$	$-$	$-$	$-$	$+$
$x-3$	$-$	$-$	$+$	$+$
$x+3$	$-$	$+$	$+$	$+$
	-3	3	4	

Set of Solutions

Set of solutions: $[-3,3] \cup [4,\infty)$

□

F. $(x^2 - 3x - 10)/(3x^2 - 4x - 4) \leq 0$

G. $\dfrac{7x^2+1}{x-9} \leq 0$

SOLUTION

Zeros of the Factors

- $7x^2 + 1 = 0$ (no solution)
- $x - 9 = 0 \Leftrightarrow x = 9$

Sign Chart

quotient	$-$	$+$
$7x^2+1$	$+$	$+$
$x-9$	$-$	$+$
	9	

Set of Solutions

Set of solutions: $(-\infty, 9)$

□

H. $\dfrac{4}{3-x} \geq \dfrac{8}{x}$

I. $|(x+2)/(3x-6)| < 1$

SOLUTION

Note

$$|(x+2)/(3x-6)| < 1 \Leftrightarrow -1 < (x+2)/(3x-6) < 1$$
$$\Leftrightarrow -1 < (x+2)/(3x-6) \text{ AND } (x+2)/(3x-6) < 1$$

First Inequality

$$-1 < \frac{x+2}{3x-6} \Leftrightarrow 0 < 1 + \frac{x+2}{3x-6}$$
$$\Leftrightarrow 0 < \frac{3x-6}{3x-6} + \frac{x+2}{3x-6}$$
$$\Leftrightarrow 0 < \frac{(3x-6)+(x+2)}{3x-6}$$
$$\Leftrightarrow 0 < \frac{4x-4}{3x-6}$$

Zeros of the Factors

- $4x - 4 = 0 \Leftrightarrow 4x = 4$
  $\Leftrightarrow 4x/4 = 4/4$
  $\Leftrightarrow x = 1$
- $3x - 6 = 0 \Leftrightarrow 3x = 6$
  $\Leftrightarrow 3x/3 = 6/3$
  $\Leftrightarrow x = 2$

Sign Chart

quotient	+	−	+
$4x-4$	−	+	+
$3x-6$	−	−	+
		1	2

Second Inequality

$$\frac{x+2}{3x-6} < 1 \Leftrightarrow \frac{x+2}{3x-6} - 1 < 0$$
$$\Leftrightarrow \frac{x+2}{3x-6} - \frac{3x-6}{3x-6} < 0$$
$$\Leftrightarrow \frac{(x+2)-(3x-6)}{3x-6} < 0$$
$$\Leftrightarrow \frac{x+2-3x+6}{3x-6} < 0$$
$$\Leftrightarrow \frac{-2x+8}{3x-6} < 0$$

Zeros of the Factors

- $-2x + 8 = 0 \Leftrightarrow -2x = -8$
  $\Leftrightarrow -2x/(-2) = (-8)/(-2)$
  $\Leftrightarrow x = 4$
- $3x - 6 = 0 \Leftrightarrow x = 2$ (see work at left)

Sign Chart

quotient	−	+	−
$-2x+8$	+	+	−
$3x-6$	−	+	+
		2	4

Set of Solutions

By our work above:

$$|(x+2)/(3x-6)| < 1 \Leftrightarrow -1 < (x+2)/(3x-6) \text{ AND } (x+2)/(3x-6) < 1$$
$$\Leftrightarrow 0 < (4x-4)/(3x-6) \text{ AND } (-2x+8)/(3x-6) < 0$$
$$\Leftrightarrow x \in (-\infty, 1) \cup (2, \infty) \text{ AND } x \in (-\infty, 2) \cup (4, \infty)$$
$$\Leftrightarrow x \in (-\infty, 1) \cup (4, \infty)$$

Set of solutions: $(-\infty, 1) \cup (4, \infty)$

□

Section 54: Roots and Radicals

Definitions

Suppose $b \in \mathbf{R}$ and $n \in \mathbf{N}$.

- An nth *root* of b is a number a for which $a^n = b$.
- The *principal* n^{th} *root* of b, denoted by $\sqrt[n]{b}$, is:
  - the nonnegative real n^{th} root of b if n is even (assuming $b \geq 0$)
  - the (only) real n^{th} root of b if n is odd
- A *radical* is an expression of the form $\sqrt[n]{b}$ that denotes a principal root.
- The *radical sign* is the symbol $\sqrt{\ }$.
- The *index* of a radical $\sqrt[n]{b}$ is the number n at the upper left of the radical sign.
- The *radicand* of a radical $\sqrt[n]{b}$ is the expression b under the radical sign.

Terminology

- A *square root* of b means a 2nd root of b.
- A *cube root* of b means a 3rd root of b.

Comments

- If $b < 0$ and n is even, then b has no real n^{th} roots.
- The radical $\sqrt{b}$, with index omitted, denotes the principal square root of b (so $\sqrt{b} = \sqrt[2]{b}$).
- The notation $a\sqrt[n]{b}$ means $a \cdot \sqrt[n]{b}$, just as ax means $a \cdot x$.
- We add or subtract terms that contain a common radical in the same way that we combine like terms. We cannot add or subtract terms that contain radicals if the indices or radicands of those radicals differ.
- The term *root* can also mean either a *zero* of a polynomial or a *solution* to an equation in one variable.

Example 1

A. 5 is a seventh root of 78,125, since $(5)^7 = 78{,}125$.
The principal seventh root of 78,125 is 5 (we write $\sqrt[7]{78{,}125} = 5$).

B. -3 and 3 are square roots of 9, since $(-3)^2 = 9$ and $3^2 = 9$.
The principal square root of 9 is 3 (we write $\sqrt{9} = 3$).

C. -4 is a cube root of -64, since $(-4)^3 = -64$.
The principal cube root of -64 is -4 (we write $\sqrt[3]{-64} = -4$).

D. No real number is a fourth root of -16, since there exists no $a \in \mathbf{R}$ such that $a^4 = -16$.

E. We cannot simplify further the expressions below:
- $\sqrt{22} + \sqrt{31}$
- $2\sqrt[4]{10} + 3\sqrt[4]{17}$
- $10\sqrt{5} - 6\sqrt[3]{5}$

□

Example 2
Evaluate each radical without a calculator.

A. $\sqrt{36}$
B. $-\sqrt{25}$
C. $\sqrt[3]{27}$
D. $\sqrt[3]{-8/125}$
E. $\sqrt[4]{81}$
F. $\sqrt{0.01}$

Solutions
A. $\sqrt{36} = 6$
B. $-\sqrt{25} = -5$
C. $\sqrt[3]{27} = 3$
D. $\sqrt[3]{-8/125} = -2/5$
E. $\sqrt[4]{81} = 3$
F. $\sqrt{0.01} = 0.1$

□

G. $\sqrt{100}$
H. $-\sqrt{81}$
I. $\sqrt[3]{8}$
J. $-\sqrt[3]{1/64}$
K. $\sqrt[5]{-32}$
L. $\sqrt{-4}$

Example 3
Simplify each expression.

A. $11\sqrt[4]{6} - 2\sqrt[4]{6}$

Solution
$$\begin{aligned} 11\sqrt[4]{6} - 2\sqrt[4]{6} &= (11-2)\sqrt[4]{6} \\ &= 9\sqrt[4]{6} \end{aligned}$$

□

B. $\sqrt{5} + 12\sqrt{5}$

C. $4\sqrt{3} - 2\sqrt{3} + 9\sqrt{3}$

Solution
$$\begin{aligned} 4\sqrt{3} - 2\sqrt{3} + 9\sqrt{3} &= (4-2+9)\sqrt{3} \\ &= 11\sqrt{3} \end{aligned}$$

□

D. $-5\sqrt[3]{4} - 7\sqrt[3]{4} - 2\sqrt[3]{4}$

E. $8\sqrt{2} - 2\sqrt{3} + 5\sqrt{2} + 17\sqrt{3}$

Solution
$$\begin{aligned} &\underline{8\sqrt{2}} - \underline{\underline{2\sqrt{3}}} + \underline{5\sqrt{2}} + \underline{\underline{17\sqrt{3}}} \\ &= (8+5)\sqrt{2} + (-2+17)\sqrt{3} \\ &= 13\sqrt{2} + 15\sqrt{3} \end{aligned}$$

□

F. $\sqrt{7} + 5\sqrt[3]{7} - 7\sqrt{7} - \sqrt[3]{7} - 4\sqrt[3]{7}$

Section 55: Properties of Radicals

Properties of Radicals (proofs omitted)
Suppose $a, b \in \mathbf{R}$ and $n \in \mathbf{N}$.
Then:

- $\sqrt[n]{a^n} = a$ if n is odd
  $\sqrt[n]{a^n} = |a|$ if n is even
- $\sqrt[n]{ab} = \sqrt[n]{a}\sqrt[n]{b}$ provided $a \geq 0$ and $b \geq 0$ if n is even
- $\sqrt[n]{a/b} = \sqrt[n]{a}/\sqrt[n]{b}$ provided $a \geq 0$ and $b > 0$ if n is even, $b \neq 0$ if n is odd

Example 1
Simplify each expression.

A. $\sqrt[5]{(8)^5}$

Solution
$\sqrt[5]{(8)^5} = 8$

□

B. $\sqrt[3]{(-4)^3}$

C. $\sqrt[4]{(7)^4}$

Solution
$\sqrt[4]{(7)^4} = 7$

□

D. $\sqrt{(8)^2}$

E. $\sqrt[4]{(-7)^4}$

Solution
$$\begin{aligned}\sqrt[4]{(-7)^4} &= |-7| \\ &= 7\end{aligned}$$

□

F. $\sqrt{(-8)^2}$

G. $\sqrt{4 \cdot 3}$

Solution
$$\begin{aligned}\sqrt{4 \cdot 3} &= \sqrt{4}\sqrt{3} \\ &= 2\sqrt{3}\end{aligned}$$

□

H. $\sqrt{9 \cdot 2}$

I. $\sqrt{11/9}$

Solution
$$\begin{aligned}\sqrt{11/9} &= \sqrt{11}/\sqrt{9} \\ &= \sqrt{11}/3\end{aligned}$$

□

J. $\sqrt[3]{5/8}$

Section 56: Simplest Radical Form

Definition

An algebraic expression that contains a radical is in *simplest radical form* if the following three conditions are satisfied:

- no fraction appears under a radical sign;
- no radical appears in the denominator of a fraction;
- no factor of a radicand is an mth power of an integer or variable where m is greater than or equal to the index of the radical within which the radicand lies.

Procedure A

To simplify an expression of the form $\sqrt[n]{a}$ where the mth power of some term is a factor of a and $m \geq n$ (in violation of the third condition above), we:

1. identify the product c of the greatest integer and the greatest powers of variables whose n^{th} powers are factors of a;
2. rewrite a as $c^n \cdot b$, where $b = a/c^n$;
3. convert the expression $\sqrt[n]{a}$ into the expression $c \cdot \sqrt[n]{b}$ as follows:

$$\begin{aligned}\sqrt[n]{a} &= \sqrt[n]{c^n \cdot b} \\ &= \sqrt[n]{c^n} \cdot \sqrt[n]{b} \\ &= c \cdot \sqrt[n]{b}\end{aligned}$$

Procedure B

To simplify an expression of the form $a/\sqrt[n]{b}$ (in violation of the second condition above), we:

1. multiply the expression $a/\sqrt[n]{b}$ by $\sqrt[n]{b^{n-1}}/\sqrt[n]{b^{n-1}}$ and simplify as follows:

$$\begin{aligned}\frac{a}{\sqrt[n]{b}} &= \frac{a}{\sqrt[n]{b}} \cdot \frac{\sqrt[n]{b^{n-1}}}{\sqrt[n]{b^{n-1}}} \\ &= \frac{a \cdot \sqrt[n]{b^{n-1}}}{\sqrt[n]{b \cdot b^{n-1}}} \\ &= \frac{a\sqrt[n]{b^{n-1}}}{\sqrt[n]{b^n}} \\ &= \frac{a\sqrt[n]{b^{n-1}}}{b}\end{aligned}$$

2. apply procedure A to the radical in the numerator of step (1) if such violates the third condition.

Procedure C

To simplify an expression of the form $\sqrt[n]{a/b}$ (in violation of the first condition above), we:

1. rewrite the expression as $\sqrt[n]{a}/\sqrt[n]{b}$;
2. apply procedure A to the radicals in the numerator and denominator;
3. apply procedure B if necessary.

Comments

- We have assumed that all variables (except a in procedure B) in the procedures on the previous page represent positive real numbers when n is even. The variables can represent negative numbers when n is odd.
- The expression $\sqrt[n]{b^{n-1}}/\sqrt[n]{b^{n-1}}$ in procedure B is a form of 1.
- If the term c in step (1) of procedure A is not the product of the greatest integer and greatest powers of variables whose n^{th} powers are factors of a, then the expression $\sqrt[n]{b}$ in step (3) will not be in simplest radical form and we must apply the procedure again to this radical. But we need not let such inefficiencies disturb us excessively. In some cases where we cannot identify easily the greatest integer whose n^{th} power is a factor of the radicand, we can simplify a radical more quickly in the long run by deliberately identifying smaller factors of the radicand and repeating procedure A.

Example 1

Convert each radical into simplest radical form using procedure A. Assume that all variables represent positive real numbers.

A. $\sqrt{63}$

Solution

$$\begin{aligned}\sqrt{63} &= \sqrt{9 \cdot 7}\\ &= \sqrt{9} \cdot \sqrt{7}\\ &= 3\sqrt{7}\end{aligned}$$

□

B. $\sqrt{200}$

C. $\sqrt[3]{m^5}$

Solution

$$\begin{aligned}\sqrt[3]{m^5} &= \sqrt[3]{m^3 \cdot m^2}\\ &= \sqrt[3]{m^3} \cdot \sqrt[3]{m^2}\\ &= m\sqrt[3]{m^2}\end{aligned}$$

□

D. $\sqrt[7]{x^{10}}$

E. $\sqrt{50x^7}$

Solution

$$\begin{aligned}\sqrt{50x^7} &= \sqrt{25 \cdot 2 \cdot x^6 \cdot x}\\ &= \sqrt{25 \cdot x^6 \cdot 2 \cdot x}\\ &= \sqrt{25} \cdot \sqrt{x^6} \cdot \sqrt{2x}\\ &= 5\sqrt{(x^3)^2}\sqrt{2x}\\ &= 5x^3\sqrt{2x}\end{aligned}$$

□

F. $\sqrt{490c^3}$

G. $\sqrt[3]{-32x^7y^2z^3}$

Solution

$$\begin{aligned}\sqrt[3]{-32x^7y^2z^3} &= \sqrt[3]{-8x^6z^3\cdot 4xy^2}\\ &= \sqrt[3]{-8x^6z^3}\cdot\sqrt[3]{4xy^2}\\ &= \sqrt[3]{(-2x^2z)^3}\cdot\sqrt[3]{4xy^2}\\ &= -2x^2z\sqrt[3]{4xy^2}\end{aligned}$$

□

H. $\frac{2}{3}\sqrt[4]{324a^{17}b^7}$

I. $\sqrt{768}$

Solution

We will apply procedure A repeatedly using small perfect square (2nd power) factors of the radicand.

$$\begin{aligned}\sqrt{768} &= \sqrt{4\cdot 192}\\ &= \sqrt{4\cdot 4\cdot 48}\\ &= \sqrt{4\cdot 4\cdot 4\cdot 12}\\ &= \sqrt{4\cdot 4\cdot 4\cdot 4\cdot 3}\\ &= \sqrt{4}\sqrt{4}\sqrt{4}\sqrt{4}\sqrt{3}\\ &= 2\cdot 2\cdot 2\cdot 2\cdot\sqrt{3}\\ &= 16\sqrt{3}\end{aligned}$$

□

J. $\sqrt{3645}$
[Hint: 9 is a factor of 3645.]

Example 2

Convert each expression into simplest radical form using procedure B. Assume that all variables represent positive real numbers.

A. $\dfrac{6}{\sqrt{7}}$

Solution

$$\begin{aligned}\frac{6}{\sqrt{7}} &= \frac{6}{\sqrt{7}}\cdot\frac{\sqrt{7}}{\sqrt{7}}\\ &= \frac{6\cdot\sqrt{7}}{\sqrt{7\cdot 7}}\\ &= \frac{6\sqrt{7}}{\sqrt{7^2}}\\ &= \frac{6\sqrt{7}}{7}\end{aligned}$$

□

B. $\dfrac{8}{\sqrt{6}}$

C. $\dfrac{11y}{\sqrt[4]{x}}$

Solution

$$\begin{aligned}\frac{11y}{\sqrt[4]{x}} &= \frac{11y}{\sqrt[4]{x}} \cdot \frac{\sqrt[4]{x^3}}{\sqrt[4]{x^3}} \\ &= \frac{11y\sqrt[4]{x^3}}{\sqrt[4]{x \cdot x^3}} \\ &= \frac{11y\sqrt[4]{x^3}}{\sqrt[4]{x^4}} \\ &= \frac{11y\sqrt[4]{x^3}}{x}\end{aligned}$$

□

D. $\dfrac{x}{4\sqrt[5]{z^2}}$

Example 3

Convert each radical into simplest radical form using procedure C. Assume that all variables represent positive real numbers.

A. $-\sqrt{23/4}$

Solution

$$\begin{aligned}-\sqrt{23/4} &= -\sqrt{23}/\sqrt{4} \\ &= -\sqrt{23}/2\end{aligned}$$

□

B. $\sqrt[3]{15/64}$

C. $\sqrt[3]{1/x^9}$

Solution

$$\begin{aligned}\sqrt[3]{1/x^9} &= \sqrt[3]{1}/\sqrt[3]{x^9} \\ &= \sqrt[3]{1}/\sqrt[3]{(x^3)^3} \\ &= 1/x^3\end{aligned}$$

□

D. $\sqrt{4/n^{10}}$

Example 4

Simplify each expression.

A. $3\sqrt{80} - 7\sqrt{45}$

Solution

$$\begin{aligned}3\sqrt{80} - 7\sqrt{45} &= 3\sqrt{16 \cdot 5} - 7\sqrt{9 \cdot 5} \\ &= 3(4)\sqrt{5} - 7(3)\sqrt{5} \\ &= 12\sqrt{5} - 21\sqrt{5} \\ &= -9\sqrt{5}\end{aligned}$$

□

B. $11\sqrt{32} + 5\sqrt{98} - 2\sqrt{50}$

C. $5\sqrt{7}/\sqrt{28}$

Solution

$$\begin{aligned} 5\sqrt{7}/\sqrt{28} &= (5\sqrt{7})/(\sqrt{4 \cdot 7}) \\ &= (5\sqrt{7})/(\sqrt{4}\sqrt{7}) \\ &= (5\cancel{\sqrt{7}})/(2\cancel{\sqrt{7}}) \\ &= 5/2 \end{aligned}$$

□

D. $(3\sqrt{8})/(4\sqrt{18})$

Section 57: Products of Expressions with Radicals

Procedure

To simplify a product of expressions that contain radicals with the same indices, we:

1. expand the product completely;
2. apply the property $\sqrt[n]{a}\sqrt[n]{b} = \sqrt[n]{ab}$ to the products of radicals;
3. convert all radicals to simplest radical form;
4. combine like terms.

Example 1

Simplify each product. Assume that all variables represent positive real numbers.

A. $(-4\sqrt{7})(9\sqrt{7})$

Solution

$$\begin{aligned}(-4\sqrt{7})(9\sqrt{7}) &= (-4)(9)(\sqrt{7})(\sqrt{7})\\ &= (-36)(\sqrt{(7)^2})\\ &= (-36)(7)\\ &= -252\end{aligned}$$

□

B. $(5\sqrt{2})(3\sqrt{18})$

C. $\sqrt{3}\sqrt{6}$

Solution

$$\begin{aligned}\sqrt{3}\sqrt{6} &= \sqrt{3\cdot 6}\\ &= \sqrt{18}\\ &= \sqrt{9\cdot 2}\\ &= \sqrt{9}\sqrt{2}\\ &= 3\sqrt{2}\end{aligned}$$

□

D. $\sqrt{5}\sqrt{15}$

E. $\sqrt{10}(\sqrt{2}-\sqrt{5})$

Solution

$$\begin{aligned}\sqrt{10}(\sqrt{2}-\sqrt{5}) &= \sqrt{10}(\sqrt{2}) - \sqrt{10}(\sqrt{5})\\ &= \sqrt{10\cdot 2} - \sqrt{10\cdot 5}\\ &= \sqrt{20} - \sqrt{50}\\ &= \sqrt{4\cdot 5} - \sqrt{25\cdot 2}\\ &= \sqrt{4}\sqrt{5} - \sqrt{25}\sqrt{2}\\ &= 2\sqrt{5} - 5\sqrt{2}\end{aligned}$$

□

F. $\sqrt{12}(4+\sqrt{18})$

G. $\sqrt{2}\sqrt{13}$

SOLUTION
$$\begin{aligned}\sqrt{2}\sqrt{13} &= \sqrt{2\cdot 13}\\ &= \sqrt{26}\end{aligned}$$
□

H. $\sqrt{5}\sqrt{11}$

I. $(\sqrt{14}-\sqrt{2})(\sqrt{7}+3)$

SOLUTION
$$\begin{aligned}&(\sqrt{14}-\sqrt{2})(\sqrt{7}+3)\\ &= (\sqrt{14}-\sqrt{2})(\sqrt{7})+(\sqrt{14}-\sqrt{2})(3)\\ &= \sqrt{14}\sqrt{7}-\sqrt{2}\sqrt{7}+\sqrt{14}\cdot 3-\sqrt{2}\cdot 3\\ &= \sqrt{14\cdot 7}-\sqrt{2\cdot 7}+3\sqrt{14}-3\sqrt{2}\\ &= \sqrt{98}-\sqrt{14}+3\sqrt{14}-3\sqrt{2}\\ &= \sqrt{49\cdot 2}-\sqrt{14}+3\sqrt{14}-3\sqrt{2}\\ &= \sqrt{49}\sqrt{2}+2\sqrt{14}-3\sqrt{2}\\ &= 7\sqrt{2}-3\sqrt{2}+2\sqrt{14}\\ &= 4\sqrt{2}+2\sqrt{14}\end{aligned}$$
□

J. $(5+\sqrt{7})(5-\sqrt{7})$

K. $\sqrt{ab^2}\sqrt{ab}$

SOLUTION
$$\begin{aligned}\sqrt{ab^2}\sqrt{ab} &= \sqrt{ab^2\cdot ab}\\ &= \sqrt{a^2b^3}\\ &= \sqrt{a^2\cdot b^2\cdot b}\\ &= \sqrt{a^2}\sqrt{b^2}\sqrt{b}\\ &= ab\sqrt{b}\end{aligned}$$
□

L. $(\sqrt{x^3})(3\sqrt{xy})$

M. $\sqrt[3]{4n}(\sqrt{5}+\sqrt[3]{16n^7})$

SOLUTION
$$\begin{aligned}&\sqrt[3]{4n}(\sqrt{5}+\sqrt[3]{16n^7})\\ &= \sqrt[3]{4n}(\sqrt{5})+\sqrt[3]{4n}(\sqrt[3]{16n^7})\\ &= \sqrt{5}(\sqrt[3]{4n})+\sqrt[3]{4n\cdot 16n^7}\\ &= \sqrt{5}(\sqrt[3]{4n})+\sqrt[3]{64n^8}\\ &= \sqrt{5}(\sqrt[3]{4n})+(\sqrt[3]{64n^6\cdot n^2})\\ &= \sqrt{5}(\sqrt[3]{4n})+(\sqrt[3]{64n^6})(\sqrt[3]{n^2})\\ &= \sqrt{5}(\sqrt[3]{4n})+(\sqrt[3]{(4n^2)^3})(\sqrt[3]{n^2})\\ &= \sqrt{5}(\sqrt[3]{4n})+4n^2(\sqrt[3]{n^2})\end{aligned}$$
□

N. $(\sqrt{z}-8)(\sqrt{z}+6)$

o. $(10\sqrt{3}+2\sqrt{15})(10\sqrt{3}-2\sqrt{15})$

Solution

$(10\sqrt{3}+2\sqrt{15})(10\sqrt{3}-2\sqrt{15})$
$= (10\sqrt{3}+2\sqrt{15})(10\sqrt{3}) - (10\sqrt{3}+2\sqrt{15})(2\sqrt{15})$
$= (10\sqrt{3})(10\sqrt{3}) + (2\sqrt{15})(10\sqrt{3}) - [(10\sqrt{3})(2\sqrt{15}) + (2\sqrt{15})(2\sqrt{15})]$
$= (10)(10)\sqrt{3}\sqrt{3} + (2)(10)\sqrt{15}\sqrt{3} - (10)(2)\sqrt{3}\sqrt{15} - (2)(2)\sqrt{15}\sqrt{15}$
$= 100\sqrt{3\cdot 3} + 20\sqrt{15\cdot 3} - 20\sqrt{3\cdot 15} - 4\sqrt{15\cdot 15}$
$= 100\sqrt{9} + \cancel{20\sqrt{45}} - \cancel{20\sqrt{45}} - 4\sqrt{225}$
$= 100(3) - 4(15)$
$= 300 - 60$
$= 240$

□

Section 58: Quotients of Expressions with Radicals

Definition
The *conjugate* of an expression of the form $a + \sqrt{b}$ is $a - \sqrt{b}$.

Terminology
To "rationalize a denominator" of a fraction means to convert the fraction into a form in which its denominator does not contain a radical.

Property
Suppose $a, b \in \mathbf{R}$ with $b \geq 0$.
Then the product of $a + \sqrt{b}$ and $a - \sqrt{b}$ does not contain a radical.

Proof
Suppose $a, b \in \mathbf{R}$ with $b \geq 0$.

$$\begin{aligned}(a + \sqrt{b})(a - \sqrt{b}) &= (a)(a) + (a)(-\sqrt{b}) + (\sqrt{b})(a) + (\sqrt{b})(-\sqrt{b}) \\ &= a^2 - \cancel{a\sqrt{b}} + \cancel{a\sqrt{b}} - (\sqrt{b})^2 \\ &= a^2 - b\end{aligned}$$

Thus the product of $a + \sqrt{b}$ and $a - \sqrt{b}$ does not contain a radical.

⊡

Property
The conjugate of an expression of the form $a - \sqrt{b}$ is $a + \sqrt{b}$.

Proof

$$\begin{aligned}(\text{conjugate of } a - \sqrt{b}) &= (\text{conjugate of } a + (-\sqrt{b})) \\ &= a - (-\sqrt{b}) \\ &= a + \sqrt{b}\end{aligned}$$

⊡

Procedure
To rationalize a denominator of the form $\sqrt[n]{b}$, we follow procedure B on page 148.

Procedure
To rationalize a denominator of the form $a + \sqrt{b}$ or $a - \sqrt{b}$, we:

1. multiply the given fraction by the fraction whose numerator and denominator both equal the conjugate of the denominator of the given fraction;
2. simplify the fraction that results from step (1).

Comments

- In both of the procedures above, we multiply the given fraction by a form of 1.
- In some situations we may want to rationalize the numerator of an fraction.

Quotients of Expressions with Radicals

Example 1

Simplify each expression. Assume that all variables represent positive real numbers.

A. $\sqrt{3/7}$

Solution

$$\begin{aligned}\sqrt{3/7} &= (\sqrt{3}/\sqrt{7})(\sqrt{7}\sqrt{7})\\ &= \sqrt{3\cdot 7}/\sqrt{7\cdot 7}\\ &= \sqrt{21}/7\end{aligned}$$

□

B. $\sqrt{2/3}$

C. $\sqrt{2}/(4\sqrt{10})$

Solution

$$\begin{aligned}\sqrt{2}/(4\sqrt{10}) &= [\sqrt{2}/(4\sqrt{10})]\cdot(\sqrt{10}/\sqrt{10})\\ &= \sqrt{2\cdot 10}/(4\sqrt{10\cdot 10})\\ &= \sqrt{20}/(4\sqrt{10^2})\\ &= \sqrt{4\cdot 5}/(4\cdot 10)\\ &= (\sqrt{4}\sqrt{5})/40\\ &= (\cancel{2}\sqrt{5})/(\cancel{2}\cdot 20)\\ &= \sqrt{5}/20\end{aligned}$$

□

D. $(8\sqrt{5})/\sqrt{6}$

E. $\sqrt{3x}/\sqrt{12y}$

Solution

$$\begin{aligned}\sqrt{3x}/\sqrt{12y} &= \sqrt{3x}/\sqrt{4\cdot 3y}\\ &= \sqrt{3x}/(\sqrt{4}\sqrt{3y})\cdot(\sqrt{3y}/\sqrt{3y})\\ &= \sqrt{3x\cdot 3y}/(2\sqrt{3y\cdot 3y})\\ &= (\sqrt{3^2}\cdot\sqrt{xy})/(2\sqrt{(3y)^2})\\ &= (\cancel{3}\cdot\sqrt{xy})/(2\cdot\cancel{3}y)\\ &= \sqrt{xy}/(2y)\end{aligned}$$

□

F. $(10\sqrt{m})/\sqrt{2mn}$

G. $2/\sqrt{y^3}$

Solution

$$\begin{aligned}2/\sqrt{y^3} &= (2/\sqrt{y^3})\cdot(\sqrt{y}/\sqrt{y})\\ &= 2\sqrt{y}/\sqrt{y^3\cdot y}\\ &= 2\sqrt{y}/\sqrt{y^4}\\ &= 2\sqrt{y}/\sqrt{(y^2)^2}\\ &= 2\sqrt{y}/y^2\end{aligned}$$

□

H. $\sqrt{32/(5x^6)}$

I. $\sqrt[5]{7}/\sqrt[5]{(2)^{11}}$

Solution

$\sqrt[5]{7}/\sqrt[5]{(2)^{11}}$
$= (\sqrt[5]{7}/\sqrt[5]{(2)^{11}}) \cdot (\sqrt[5]{(2)^4}/\sqrt[5]{(2)^4})$
$= \sqrt[5]{7 \cdot (2)^4}/\sqrt[5]{(2)^{11} \cdot (2)^4}$
$= \sqrt[5]{7 \cdot 16}/\sqrt[5]{(2)^{15}}$
$= \sqrt[5]{112}/\sqrt[5]{((2)^3)^5}$
$= \sqrt[5]{112}/(2)^3$
$= \sqrt[5]{112}/8$

□

J. $\sqrt[3]{4/5}$

K. $2\sqrt{21} + 4\sqrt{3/7}$

Solution

$2\sqrt{21} + 4\sqrt{3/7}$
$= 2\sqrt{21} + (4\sqrt{3}/\sqrt{7}) \cdot (\sqrt{7}/\sqrt{7})$
$= 2\sqrt{21} + 4\sqrt{3 \cdot 7}/\sqrt{7 \cdot 7}$
$= 2\sqrt{21} + 4\sqrt{21}/7$
$= (2\sqrt{21})(7/7) + 4\sqrt{21}/7$
$= 14\sqrt{21}/7 + 4\sqrt{21}/7$
$= (14\sqrt{21} + 4\sqrt{21})/7$
$= 18\sqrt{21}/7$

□

L. $4\sqrt{3} + 11\sqrt{12} - 2/\sqrt{3}$

M. $\dfrac{26}{4 - \sqrt{3}}$

Solution

$$\frac{26}{4-\sqrt{3}} = \frac{26}{4-\sqrt{3}} \cdot \frac{4+\sqrt{3}}{4+\sqrt{3}}$$
$$= \frac{26(4+\sqrt{3})}{(4-\sqrt{3})(4+\sqrt{3})}$$
$$= \frac{26(4+\sqrt{3})}{16 + \cancel{4\sqrt{3}} - \cancel{4\sqrt{3}} - \sqrt{3}\sqrt{3}}$$
$$= \frac{26(4+\sqrt{3})}{16-3}$$
$$= \frac{(\cancel{13})(2)(4+\sqrt{3})}{\cancel{13}}$$
$$= 2(4+\sqrt{3})$$
$$= 8 + 2\sqrt{3}$$

□

N. $\dfrac{15}{\sqrt{5} + \sqrt{2}}$

O. $\dfrac{\sqrt{11}}{7+2\sqrt{11}}$

SOLUTION

$$\begin{aligned}
&\frac{\sqrt{11}}{7+2\sqrt{11}}\\
&=\frac{\sqrt{11}}{7+2\sqrt{11}}\cdot\frac{7-2\sqrt{11}}{7-2\sqrt{11}}\\
&=\frac{\sqrt{11}(7)-\sqrt{11}(2\sqrt{11})}{7^2-\cancel{7(2\sqrt{11})}+\cancel{7(2\sqrt{11})}-(2\sqrt{11})^2}\\
&=\frac{7\sqrt{11}-2(11)}{49-(2)^2(\sqrt{11})^2}\\
&=\frac{7\sqrt{11}-22}{49-4(11)}\\
&=\frac{-22+7\sqrt{11}}{5}
\end{aligned}$$

□

P. $\dfrac{\sqrt{x}+5}{\sqrt{x}-6}$

EXAMPLE 2

Rationalize the numerator of each expression.

A. $\dfrac{\sqrt{10}-2}{3}$

SOLUTION

$$\begin{aligned}
&\frac{\sqrt{10}-2}{3}\\
&=\frac{\sqrt{10}-2}{3}\cdot\frac{\sqrt{10}+2}{\sqrt{10}+2}\\
&=\frac{\sqrt{10}\sqrt{10}+\cancel{2\sqrt{10}}-\cancel{2\sqrt{10}}-2(2)}{3(\sqrt{10}+2)}\\
&=\frac{10-4}{3(\sqrt{10}+2)}\\
&=\frac{6}{3(\sqrt{10}+2)}\\
&=\frac{\cancel{3}\cdot 2}{\cancel{3}(\sqrt{10}+2)}\\
&=\frac{2}{2+\sqrt{10}}
\end{aligned}$$

□

B. $\dfrac{7+3\sqrt{13}}{17}$

Section 59: Equations with Radicals

Properties
Suppose $a, b \in \mathbf{R}$ and $n \in \mathbf{N}$.
Then:

- $a = b \Rightarrow a^n = b^n$
- $(\sqrt[n]{a})^n = a$ provided $a \geq 0$ if n is even

Proofs (second property only)
Suppose $a \in \mathbf{R}$ and $n \in \mathbf{N}$.

$$\begin{aligned}(\sqrt[n]{a})^n &= \underbrace{\sqrt[n]{a} \cdot \sqrt[n]{a} \cdot \sqrt[n]{a} \cdot \ldots \cdot \sqrt[n]{a}}_{n \text{ factors}} \\ &= \sqrt[n]{\underbrace{a \cdot a \cdot a \cdot \ldots \cdot a}_{n \text{ factors}}} \\ &= \sqrt[n]{a^n} \\ &= a\end{aligned}$$

⊡

Procedure
To solve an equation that contains a radical, we:

1. isolate the term that contains the radical on one side of the equation;
2. raise both sides of the equation to an exponent equal to the index of the radical;
3. solve the equation that results from step (2);
4. verify the potential solutions determined in step (3);
5. state the set of solutions verified in step (4).

Comments

- We can use the property $|x|^2 = x^2$ to eliminate an absolute value symbol from an equation.
- Step (4) of the procedure is imperative when the index n of the radical is even because the converse of the property $a = b \Rightarrow a^n = b^n$ is not true then. If $a^n = b^n \not\Rightarrow a = b$, solutions of an equation that results from step (2) of the procedure may not be solutions of the original equation.
- To solve some equations, we may need to raise both sides to exponents multiple times.

Example 1
The following examples demonstrate that $a^n = b^n \not\Rightarrow a = b$.

A. Observe that $9 = 9$, but $-3 \neq 3$.
Thus $(-3)^2 = (3)^2 \not\Rightarrow -3 = 3$

B. If $x^2 = 36$, then x could equal either -6 or 6.
Thus $x^2 = (6)^2 \not\Rightarrow x = 6$.

□

EXAMPLE 2
Solve each equation.

A. $\sqrt{5x} = -10$

SOLUTION
Potential Solutions

$$\begin{aligned}\sqrt{5x} = -10 &\Rightarrow (\sqrt{5x})^2 = (-10)^2\\ &\Rightarrow 5x = 100\\ &\Rightarrow 5x/5 = 100/5\\ &\Rightarrow x = 20\end{aligned}$$

Check

20: $\sqrt{5 \cdot 20} \stackrel{?}{=} -10$

$\Leftrightarrow \sqrt{100} \stackrel{?}{=} -10$ X

Set of Solutions
Set of solutions: $\emptyset$

□

B. $\sqrt{3x - 11} = 2$

C. $|x/3| = 4$

SOLUTION
Potential Solutions

$$\begin{aligned}&|x/3| = 4\\ &\Rightarrow |x/3|^2 = (4)^2\\ &\Rightarrow (x/3)^2 = 16\\ &\Rightarrow x^2/9 = 16\\ &\Rightarrow 9(x^2/9) = 9(16)\\ &\Rightarrow x^2 = 144\\ &\Rightarrow x^2 - 144 = 0\\ &\Rightarrow (x-12)(x+12) = 0\\ &\Rightarrow x - 12 = 0 \quad \text{OR } x + 12 = 0\\ &\quad \Rightarrow x = 12 \qquad\quad \Rightarrow x = -12\end{aligned}$$

Checks

- 12: $|12/3| \stackrel{?}{=} 4$

  $\Leftrightarrow |4| \stackrel{?}{=} 4$ ✓

- −12: $|-12/3| \stackrel{?}{=} 4$

  $\Leftrightarrow |-4| \stackrel{?}{=} 4$ ✓

Set of Solutions
Set of solutions: $\{-12, 12\}$

□

D. $|m + 6| = 4$

E. $2\sqrt{a+7} = 9$

Solution

Potential Solutions

$$\begin{aligned}2\sqrt{a+7} = 9 &\Rightarrow \sqrt{a+7} = 9/2\\ &\Rightarrow (\sqrt{a+7})^2 = (9/2)^2\\ &\Rightarrow a+7 = 81/4\\ &\Rightarrow a = 81/4 - 7\\ &\quad\;\; = 81/4 - 28/4\\ &\quad\;\; = 53/4\end{aligned}$$

Check

$$\begin{aligned}53/4: \quad & 2\sqrt{53/4+7} \stackrel{?}{=} 9\\ \Leftrightarrow \quad & 2\sqrt{81/4} \stackrel{?}{=} 9\\ \Leftrightarrow \quad & 2(9/2) \stackrel{?}{=} 9 \quad \checkmark\end{aligned}$$

Set of Solutions

Set of solutions: $\{53/4\}$

□

F. $\sqrt{13b+6} = \sqrt{9b-26}$

G. $-3\sqrt{x} = 4 - x$

Solution

Potential Solutions

$$\begin{aligned}&-3\sqrt{x} = 4 - x\\ &\Rightarrow (-3\sqrt{x})^2 = (4-x)^2\\ &\Rightarrow (-3)^2(\sqrt{x})^2 = (4)^2 - 2(4)(x) + (x)^2\\ &\Rightarrow 9x = 16 - 8x + x^2\\ &\Rightarrow 0 = x^2 - 8x + 16 - 9x\\ &\qquad = x^2 - 17x + 16\\ &\qquad = (x-16)(x-1)\\ &\Rightarrow x - 16 = 0 \quad \text{OR } x - 1 = 0\\ &\qquad \Rightarrow x = 16 \qquad\quad \Rightarrow x = 1\end{aligned}$$

Checks

- 16: $-3\sqrt{16} \stackrel{?}{=} 4 - 16$
  $\Leftrightarrow \; -3(4) \stackrel{?}{=} -12 \quad \checkmark$
- 1: $-3\sqrt{1} \stackrel{?}{=} 4 - 1$
  $\Leftrightarrow \; -3(1) \stackrel{?}{=} 3 \quad$ X

Set of Solutions

Set of solutions: $\{16\}$

□

H. $\sqrt{c^2 - 2x + 14} = c + 12$

I. $\sqrt[6]{x} = 2$

SOLUTION

Potential Solutions

$$\sqrt[6]{x} = 2 \Rightarrow (\sqrt[6]{x})^6 = (2)^6$$
$$\Rightarrow x = 64$$

Check

64: $\sqrt[6]{64} \stackrel{?}{=} 2$ ✓

Set of Solutions

Set of solutions: $\{64\}$

□

J. $\sqrt[4]{9x} = -3$

K. $\sqrt[3]{x-5} = 2$

SOLUTION

$$\sqrt[3]{x-5} = 2 \Leftrightarrow (\sqrt[3]{x-5})^3 = (2)^3$$
$$\Leftrightarrow x - 5 = 8$$
$$\Leftrightarrow x = 8 + 5$$
$$= 13$$

Set of solutions: $\{13\}$

□

L. $\sqrt[3]{3x-1} = -4$

M. $\sqrt[3]{x^3-19} = x - 1$

SOLUTION

$$\sqrt[3]{x^3-19} = x - 1$$
$$\Leftrightarrow (\sqrt[3]{x^3-19})^3 = (x-1)^3$$
$$\Leftrightarrow x^3 - 19$$
$$= (x-1)(x-1)^2$$
$$= (x-1)(x^2 - 2(1)(x) + (-1)^2)$$
$$= (x-1)(x^2 - 2x + 1)$$
$$= x(x^2) + x(-2x) + x(1)$$
$$- 1(x^2) - 1(-2x) - 1(1)$$
$$= x^3 - 2x^2 + x - x^2 + 2x - 1$$
$$= x^3 - 3x^2 + 3x - 1$$
$$\Leftrightarrow \cancel{x^3} - 19 - \cancel{x^3} + 3x^2 - 3x + 1 = 0$$
$$\Leftrightarrow 3x^2 - 3x - 18 = 0$$
$$\Leftrightarrow 3(x^2 - x - 6) = 0$$
$$\Leftrightarrow 3(x-3)(x+2) = 0$$
$$\Leftrightarrow x - 3 = 0 \quad \text{OR } x + 2 = 0$$
$$\Leftrightarrow x = 3 \qquad \Leftrightarrow x = -2$$

Set of solutions: $\{-2, 3\}$

□

N. $m = \sqrt[3]{7m^2 + 4m - 28}$

O. $\sqrt{2x-3} = 8 - \sqrt{4x+1}$

Solution

Potential Solutions

$$\begin{aligned}
&\sqrt{2x-3} = 8 - \sqrt{4x+1} \\
&\Rightarrow (\sqrt{2x-3})^2 = (8-\sqrt{4x+1})^2 \\
&\Rightarrow 2x-3 = (8)^2 - 2(8)\sqrt{4x+1} + (\sqrt{4x+1})^2 \\
&\qquad = 64 - 16\sqrt{4x+1} + (4x+1) \\
&\qquad = 4x + 65 - 16\sqrt{4x+1} \\
&\Rightarrow 2x - 3 - 4x - 65 = -16\sqrt{4x+1} \\
&\Rightarrow -2x - 68 = -16\sqrt{4x+1} \\
&\Rightarrow (-2x-68)^2 = (-16\sqrt{4x+1})^2 \\
&\Rightarrow (-2x)^2 + 2(-2x)(-68) + (-68)^2 = (-16)^2(\sqrt{4x+1})^2 \\
&\Rightarrow 4x^2 + 272x + 4624 = 256(4x+1) \\
&\qquad = 1024x + 256 \\
&\Rightarrow 4x^2 + 272x + 4624 - 1024x - 256 = 0 \\
&\Rightarrow 4x^2 - 752x + 4368 = 0 \\
&\Rightarrow 4(x^2 - 188x + 1092) = 0 \\
&\Rightarrow 4(x-6)(x-182) = 0 \\
&\Rightarrow x - 6 = 0 \quad \text{OR } x - 182 = 0 \\
&\quad \Rightarrow x = 6 \qquad \Rightarrow x = 182
\end{aligned}$$

Checks

- 6: $\sqrt{2(6)-3} \stackrel{?}{=} 8 - \sqrt{4(6)+1}$
  $\Leftrightarrow \sqrt{9} \stackrel{?}{=} 8 - \sqrt{25}$
  $\Leftrightarrow 3 \stackrel{?}{=} 8 - 5$ ✓

- 182: $\sqrt{2(182)-3} \stackrel{?}{=} 8 - \sqrt{4(182)+1}$
  $\Leftrightarrow \sqrt{361} \stackrel{?}{=} 8 - \sqrt{729}$
  $\Leftrightarrow 19 \stackrel{?}{=} 8 - 27$ X

Set of Solutions

Set of solutions: $\{6\}$

□

P. $\sqrt{x+4} + \sqrt{6x-5} = 8$

Section 60: Equations with Constant Exponents

PROPERTIES (proofs omitted)
Suppose $x, b \in \mathbf{R}$ and $n \in \mathbf{N}$.
Then:

- $x^n = b \Leftrightarrow x = \sqrt[n]{b}$ if n is odd
- $x^n = b \Leftrightarrow x = \sqrt[n]{b}$ OR $x = -\sqrt[n]{b}$ if n is even, provided $b \geq 0$

NOTATION
"$x = \pm m$" means "$x = m$ OR $x = -m$"

COMMENTS

- The properties above imply that:
  - if n is odd, then one real n^{th} root of b exists;
  - if n is even and $b > 0$, then two real n^{th} roots of b exist (one positive and one negative);
  - if n is even and $b < 0$, then no real n^{th} roots of b exist.
- We can condense the second property above to "$x^n = b \Leftrightarrow x = \pm\sqrt[n]{b}$ if n is even", though we will usually employ the longer form of this property when we must apply some operation to both the $\sqrt[n]{b}$ and the $-\sqrt[n]{b}$ terms.

EXAMPLE 1
Solve each equation.

A. $x^4 = 16$

SOLUTION
$x^4 = 16 \Leftrightarrow x = \pm\sqrt[4]{16}$
$= \pm 2$
Set of solutions: $\{-2, 2\}$
□

B. $x^3 = 216$

C. $9x^2 + 6 = 31$

SOLUTION
$9x^2 + 6 = 31$
$\Leftrightarrow 9x^2 = 31 - 6$
$= 25$
$\Leftrightarrow 9x^2/9 = 25/9$
$\Leftrightarrow x^2 = 25/9$
$\Leftrightarrow x = \pm\sqrt{25/9}$
$= \pm\sqrt{25}/\sqrt{9}$
$= \pm 5/3$
Set of solutions: $\{-5/3, 5/3\}$
□

D. $5x^2 - 60 = 0$

E. $(x+4)^2 = 100$

Solution

$(x+4)^2 = 100$

$\Leftrightarrow x+4 = \sqrt{100}$

$\Leftrightarrow x+4 = 10$

$\Leftrightarrow x = 10-4$

$= 6$

or $x+4 = -\sqrt{100}$

$\Leftrightarrow x+4 = -10$

$\Leftrightarrow x = -10-4$

$= -14$

Set of solutions: $\{-14, 6\}$

□

F. $(2x-5)^2 = 9$

G. $4(3x+1)^2 + 7 = 56$

Solution

$4(3x+1)^2 + 7 = 56$

$\Leftrightarrow 4(3x+1)^2 = 56-7$

$\Leftrightarrow 4(3x+1)^2 = 49$

$\Leftrightarrow 4(3x+1)^2/4 = 49/4$

$\Leftrightarrow (3x+1)^2 = 49/4$

$\Leftrightarrow 3x+1 = \sqrt{49/4}$

$\Leftrightarrow 3x+1 = 7/2$

$\Leftrightarrow 3x = 7/2-1$

$= 7/2-2/2$

$= 5/2$

$\Leftrightarrow (1/3)(3x) = (1/3)(5/2)$

$\Leftrightarrow x = 5/6$

or $3x+1 = -\sqrt{49/4}$

$\Leftrightarrow 3x+1 = -7/2$

$\Leftrightarrow 3x = -7/2-1$

$= -7/2-2/2$

$= -9/2$

$\Leftrightarrow (1/3)(3x) = (1/3)(-9/2)$

$\Leftrightarrow x = -3/2$

Set of solutions: $\{-3/2, 5/6\}$

□

H. $2(4m)^2 - 13 = 51$

Section 61: The Pythagorean Theorem

Definitions

- The *hypotenuse* of a right triangle is the longest side of the triangle.
- A *leg* of a right triangle is one of the shorter two sides of the triangle.

Pythagorean Theorem

In a right triangle, the square of the length of the hypotenuse is equal to the sum of the squares of the lengths of the legs.

Proof

Suppose a right triangle has a hypotenuse of length c and legs of lengths a and b. We can arrange four copies of the triangle in the configuration shown at lower right.

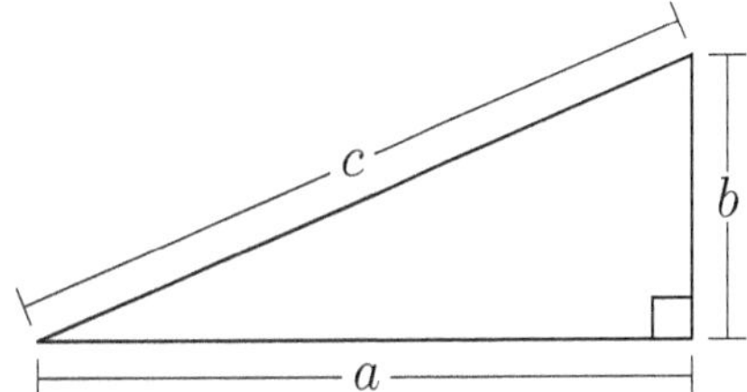

Observe that:

- (area of the larger square) = (area of the smaller square) + $4 \cdot$ (area of each triangle)
- (area of the larger square) = $(a+b)^2$
- (area of the smaller square) = c^2
- (area of each triangle) = $(1/2)(ab)$

Therefore:

$$(a+b)^2 = c^2 + 4 \cdot (1/2)(ab)$$
$$\Rightarrow a^2 + 2ab + b^2 = c^2 + 2ab$$
$$\Rightarrow a^2 + 2ab + b^2 - 2ab = c^2 + \cancel{2ab} - \cancel{2ab}$$
$$\Rightarrow a^2 + b^2 = c^2$$

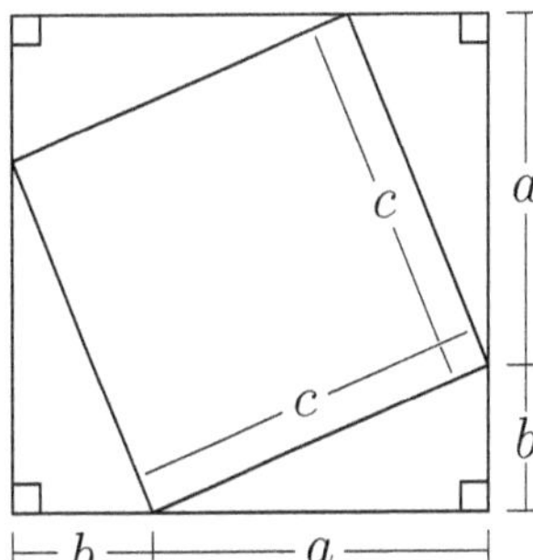

⊡

Example 1

Determine the value of the indicated variable. Assume that a and b represent the lengths of the legs of a right triangle and c represents the length of its hypotenuse.

A. Find a if $b = 24$ and $c = 25$.

Solution

For $b = 24$ and $c = 25$:

$$a^2 + b^2 = c^2$$
$$\Rightarrow a^2 + (24)^2 = (25)^2$$
$$\Rightarrow a^2 + 576 = 625$$
$$\Rightarrow a^2 = 625 - 576$$
$$= 49$$
$$\Rightarrow a = \sqrt{49} \quad \text{OR } a = -\sqrt{49}$$
$$\Rightarrow \boxed{a = 7} \qquad \Rightarrow \cancel{a = -7} \ \text{X}$$

□

B. Find c if $a = 5$ and $b = 8$.

EXAMPLE 2
Solve each word problem.

A. One leg of a right triangle is 7 feet longer than the other leg. The hypotenuse of the triangle is 8 feet longer than the shorter leg. Find the lengths of all three sides of the triangle.

SOLUTION
Declarations
Let a be the length of the shorter leg,
b be the length of the longer leg, and
c be the length of the hypotenuse.

Substitutions
- $b = a + 7$
- $c = a + 8$

Values
$a^2 + b^2 = c^2$
$\Rightarrow a^2 + (a+7)^2 = (a+8)^2$
$\Rightarrow a^2 + a^2 + 14a + 49 = a^2 + 16a + 64$
$\Rightarrow 2a^2 + 14a + 49 = a^2 + 16a + 64$
$\Rightarrow 2a^2 + 14a + 49 - a^2 - 16a - 64 = 0$
$\Rightarrow a^2 - 2a - 15 = 0$
$\Rightarrow (a-5)(a+3) = 0$
$\Rightarrow a - 5 = 0$ OR $a + 3 = 0$
$\Rightarrow a = 5$ $\Rightarrow$ ~~$a = -3$~~ X
$\Rightarrow b = 5 + 7$
$= 12$
AND $c = 5 + 8$
$= 13$
The lengths of the sides of the triangle are 5 feet, 12 feet, and 13 feet.

□

B. Suppose one leg of a right triangle is 3 feet shorter than the other leg. The hypotenuse is 3 feet longer than the longer leg. Find the lengths of all three sides of the triangle.

Section 62: Completing the Square

Terminology
To "complete the square" means to convert an expression from the form $ax^2 + bx + c$ to the form $a(x-h)^2 + k$.

Lemma
Suppose $x, a, b \in \mathbf{R}$.
Then $x^2 + (b/a)x + (b/2a)^2 = (x + b/2a)^2$.

Proof
Suppose $x, a, b \in \mathbf{R}$.

$$\begin{aligned}(x+b/2a)^2 &= x^2 + 2(x)(b/2a) + (b/2a)^2 \\ &= x^2 + (b/a)x + (b/2a)^2\end{aligned}$$

⊡

Procedure
To complete the square on a trinomial of the form $ax^2 + bx + c$, we follow the steps below:

$$\begin{aligned} ax^2 + bx + c &= a[x^2 + (b/a)x] + c \\ &= a[x^2 + (b/a)x + ((b/a)/2)^2 - ((b/a)/2)^2] + c \\ &= a[x^2 + (b/a)x + (b/2a)^2 - (b/2a)^2] + c \\ &= a[(x + b/2a)^2 - (b/2a)^2] + c \\ &= a(x + b/2a)^2 - a(b/2a)^2 + c \end{aligned}$$

Comment
We add a form of 0 (namely $(b/2a)^2 - (b/2a)^2$) to a trinomial in the process of completing the square. This operation is legitimate because, like multiplying by 1 (or a form thereof), adding 0 to an expression does not change its value.

Example 1
Complete the square on each expression.

A. $x^2 + 6x - 2$

Solution

$$\begin{aligned} &x^2 + 6x - 2 \\ &\quad = x^2 + 6x + [(6/2)^2 - (6/2)^2] - 2 \\ &\quad = x^2 + 6x + (3)^2 - (3)^2 - 2 \\ &\quad = (x^2 + 6x + (3)^2) - (3)^2 - 2 \\ &\quad = (x+3)^2 - 9 - 2 \\ &\quad = (x+3)^2 - 11 \end{aligned}$$

□

B. $x^2 - 8x + 7$

C. $x^2 - 7x + 20$

D. $x^2 + 5x - 6$

SOLUTION

$x^2 - 7x + 20$
$= x^2 + (-7)x + 20$
$= x^2 + (-7)x + [(-7/2)^2 - (-7/2)^2] + 20$
$= (x^2 + (-7)x + (-7/2)^2) - 49/4 + 20$
$= (x + (-7/2))^2 - 49/4 + 80/4$
$= (x - 7/2)^2 + 31/4$

□

E. $2x^2 - 8x - 1$

F. $3x^2 + 10x + 5$

SOLUTION

$2x^2 - 8x - 1$
$= 2[x^2 - 4x] - 1$
$= 2[x^2 - 4x + ((-4/2)^2 - (-4/2)^2)] - 1$
$= 2[x^2 - 4x + (-2)^2 - (-2)^2] - 1$
$= 2[(x^2 - 4x + (-2)^2) - (-2)^2] - 1$
$= 2[(x - 2)^2 - 4] - 1$
$= 2(x - 2)^2 - 2(4) - 1$
$= 2(x - 2)^2 - 8 - 1$
$= 2(x - 2)^2 - 9$

□

EXAMPLE 2

Solve each equation by completing the square.

A. $x^2 - 4x - 6 = 0$

B. $x^2 + 10x + 9 = 0$

SOLUTION

$x^2 - 4x - 6 = 0$
$\Leftrightarrow x^2 - 4x + (-4/2)^2 - (-4/2)^2 - 6 = 0$
$\Leftrightarrow x^2 - 4x + (-2)^2 - (-2)^2 - 6 = 0$
$\Leftrightarrow (x + (-2))^2 - 4 - 6 = 0$
$\Leftrightarrow (x - 2)^2 - 10 = 0$
$\Leftrightarrow (x - 2)^2 - 10 + 10 = 0 + 10$
$\Leftrightarrow (x - 2)^2 = 10$
$\Leftrightarrow x - 2 = \pm\sqrt{10}$
$\Leftrightarrow x = 2 \pm \sqrt{10}$

Set of solutions: $\{2 - \sqrt{10},\ 2 + \sqrt{10}\}$

□

C. $x(x+12)=13$

SOLUTION

$x(x+12)=13$

$\Leftrightarrow x^2+12x=13$

$\Leftrightarrow x^2+12x+(12/2)^2-(12/2)^2=13$

$\Leftrightarrow x^2+12x+(6)^2-(6)^2=13$

$\Leftrightarrow (x+6)^2-36=13$

$\Leftrightarrow (x+6)^2=13+36$

$=49$

$\Leftrightarrow x+6=\pm\sqrt{49}$

$=\pm 7$

$\Leftrightarrow x=-6\pm 7$

$\Leftrightarrow x=-6+7$ OR $x=-6-7$

$=1$ $\quad$ $=-13$

Set of solutions: $\{-13,1\}$

□

D. $-x^2-3x=-4$

E. $3x^2+6x-17=0$

SOLUTION

$3x^2+6x-17=0$

$\Leftrightarrow 3[x^2+2x]-17=0$

$\Leftrightarrow 3[x^2+2x+(2/2)^2-(2/2)^2]-17$

$=0$

$\Leftrightarrow 3[(x^2+2x+(1)^2)-(1)^2]-17=0$

$\Leftrightarrow 3[(x+1)^2-1]-17=0$

$\Leftrightarrow 3(x+1)^2-3(1)-17=0$

$\Leftrightarrow 3(x+1)^2-3-17=0$

$\Leftrightarrow 3(x+1)^2-20=0$

$\Leftrightarrow 3(x+1)^2=20$

$\Leftrightarrow 3(x+1)^2/3=20/3$

$\Leftrightarrow (x+1)^2=20/3$

$\Leftrightarrow x+1=\pm\sqrt{20/3}$

$=\pm(\sqrt{20}/\sqrt{3})(\sqrt{3}/\sqrt{3})$

$=\pm\sqrt{20\cdot 3}/\sqrt{3\cdot 3}$

$=\pm\sqrt{60}/\sqrt{3^2}$

$=\pm\sqrt{4\cdot 15}/3$

$=\pm\sqrt{4}\sqrt{15}/3$

$=\pm 2\sqrt{15}/3$

$\Leftrightarrow x=-1\pm 2\sqrt{15}/3$

Set of solutions:

$\{-1-2\sqrt{15}/3,\ -1+2\sqrt{15}/3\}$

□

F. $2x^2-12x+5=0$

G. $5x^2 - 5x + 2 = 0$

Solution

$5x^2 - 5x + 2 = 0$

$\Leftrightarrow 5[x^2 - x] + 2 = 0$

$\Leftrightarrow 5[x^2 - x + (-1/2)^2 - (-1/2)^2] + 2 = 0$

$\Leftrightarrow 5[(x + (-1/2))^2 - 1/4] + 2 = 0$

$\Leftrightarrow 5(x - 1/2)^2 - 5(1/4) + 2 = 0$

$\Leftrightarrow 5(x - 1/2)^2 - 5/4 + 8/4 = 0$

$\Leftrightarrow 5(x - 1/2)^2 + 3/4 = 0$

$\Leftrightarrow 5(x - 1/2)^2 = -3/4$

$\Leftrightarrow (1/5)(5(x - 1/2)^2) = (1/5)(-3/4)$

$\Leftrightarrow (x - 1/2)^2 = -3/20$

$\Leftrightarrow x - 1/2 = \pm\sqrt{-3/20}$

$\Leftrightarrow \cancel{x = 1/2 \pm \sqrt{-3/20}}$ X

Set of solutions: $\emptyset$

$\square$

H. $4x^2 + 36x - 3 = 0$

Section 63: The Quadratic Formula

Property (proof omitted)
Suppose $a, b, c \in \mathbf{R}$ with $a \neq 0$.
Then $ax^2 + bx + c = 0$ if and only if $x = (-b \pm \sqrt{b^2 - 4ac}\,)/(2a)$.

Definition
The *quadratic formula* is the formula $x = (-b \pm \sqrt{b^2 - 4ac}\,)/(2a)$.

Example 1
Use the quadratic formula to solve each equation.

A. $x^2 - 5x - 14 = 0$

Solution
$x^2 - 5x - 14 = 0$
$\Leftrightarrow (1)x^2 + (-5)x + (-14) = 0$
$\Leftrightarrow x = \dfrac{-(-5) \pm \sqrt{(-5)^2 - 4(1)(-14)}}{2(1)}$
$= (5 \pm \sqrt{25 + 56}\,)/2$
$= (5 \pm \sqrt{81}\,)/2$
$= (5 \pm 9)/2$
$\Leftrightarrow x = (5 + 9)/2$ OR $x = (5 - 9)/2$
$= 14/2$ $\quad = -4/2$
$= 7$ $\quad = -2$
Set of solutions: $\{-2, 7\}$

□

B. $x^2 + 7x = -6$

C. $2n^2 + 6n = -3$

Solution
$2n^2 + 6n = -3$
$\Leftrightarrow 2n^2 + 6n + 3 = 0$
$\Leftrightarrow n = (-6 \pm \sqrt{(6)^2 - 4(2)(3)}\,)/(2(2))$
$= (-6 \pm \sqrt{36 - 24}\,)/4$
$= (-6 \pm \sqrt{12}\,)/4$
$= (-6 \pm \sqrt{4 \cdot 3}\,)/4$
$= (-6 \pm 2\sqrt{3}\,)/4$
$= \not{2}(-3 \pm \sqrt{3}\,)/(\not{2} \cdot 2)$
$= (-3 \pm \sqrt{3}\,)/2$
Set of solutions:
$\{(-3 - \sqrt{3})/2, (-3 + \sqrt{3})/2\}$

□

D. $-3b^2 - 5b + 8 = 0$

E. $5x^2 - 8x = 0$

SOLUTION

$$
\begin{aligned}
&5x^2 - 8x = 0\\
&\quad \Leftrightarrow 5x^2 + (-8)x + 0 = 0\\
&\quad \Leftrightarrow x = \frac{-(-8) \pm \sqrt{(-8)^2 - 4(5)(0)}}{2(5)}\\
&\qquad\quad = (8 \pm \sqrt{64 - 0})/10\\
&\qquad\quad = (8 \pm \sqrt{64})/10\\
&\qquad\quad = (8 \pm 8)/10\\
&\quad \Leftrightarrow x = (8 + 8)/10 \text{ OR } x = (8 - 8)/10\\
&\qquad\quad = 16/10 \qquad\qquad\quad = 0/10\\
&\qquad\quad = 8/5 \qquad\qquad\qquad = 0
\end{aligned}
$$

Set of solutions: $\{0, 8/5\}$

□

F. $2x^2 - x + 4 = 0$

Section 64: Quadratic Equations in One Variable

Definition

A *quadratic* (or *second-degree*) *equation in one variable* x is an equation that can be written equivalently in the form $ax^2 + bx + c = 0$ for some constants $a, b, c \in \mathbf{R}$ with $a \neq 0$.

Procedures

We have four techniques for solving a quadratic equation in the form $ax^2 + bx + c = 0$, each of which has limitations or drawbacks.

A. Factorization
If we can factorize the trinomial on the nonzero side of the equation, then we can solve the equation by performing that factorization, equating all factors to zero, and solving the equations that result; however, we may have difficulty with factorizing the trinomial unless the solutions to the equation are rational numbers.

B. Application of the Property $x^2 = k \Leftrightarrow x = \pm\sqrt{k}$
If the coefficient b in the trinomial on the nonzero side of the equation equals zero (that is, if the quadratic equation takes the form $ax^2 + c = 0$), then after a few algebraic steps we can apply the property $x^2 = k \Leftrightarrow x = \pm\sqrt{k}$; however, such is usually not the case.

C. Application of the Quadratic Formula
We can use the quadratic formula to solve any quadratic equation; however, the formula will produce complicated expressions that require simplification. We should always consider procedures A and B first.

D. Completing the Square
We can also use the process of completing the square to solve any quadratic equation. This process will yield the same expressions that require simplification as the process of applying the quadratic formula, but has more steps.

Example 1

Solve each equation.

A. $x^2 + 6x = 16$

Solution

$x^2 + 6x = 16$

$\Leftrightarrow x^2 + 6x - 16 = 0$

$\Leftrightarrow (x+8)(x-2) = 0$

$\Leftrightarrow x + 8 = 0$ OR $x - 2 = 0$

$\Leftrightarrow x = -8$ $\Leftrightarrow x = 2$

Set of solutions: $\{-8, 2\}$

□

B. $2x^2 - 12 = 0$

C. $x^2 + \sqrt{12}x - 1 = 0$

SOLUTION

$$
\begin{aligned}
& x^2 + \sqrt{12}x - 1 = 0 \\
& \Leftrightarrow x = \frac{-\sqrt{12} \pm \sqrt{(\sqrt{12})^2 - 4(1)(-1)}}{2(1)} \\
& \quad = (-\sqrt{12} \pm \sqrt{12+4}\,)/2 \\
& \quad = (-\sqrt{4 \cdot 3} \pm \sqrt{16}\,)/2 \\
& \quad = (-2\sqrt{3} \pm 4)/2 \\
& \quad = \not{2}(-\sqrt{3} \pm 2)/\not{2} \\
& \quad = -\sqrt{3} \pm 2 \\
& \quad = \pm 2 - \sqrt{3}
\end{aligned}
$$

Set of solutions: $\{-2-\sqrt{3},\ 2-\sqrt{3}\}$ □

D. $\sqrt{5}x = 3x^2$

E. $-\dfrac{3}{x+1} - \dfrac{4}{x} + 7 = 0$

SOLUTION

Assuming $x \neq -1$ AND $x \neq 0$:

$$
\begin{aligned}
& -\frac{3}{x+1} - \frac{4}{x} + 7 = 0 \\
& \Leftrightarrow \frac{-3}{x+1} \cdot \frac{x}{x} + \frac{-4}{x} \cdot \frac{x+1}{x+1} \\
& \qquad\qquad + 7 \cdot \frac{x(x+1)}{x(x+1)} = 0 \\
& \Leftrightarrow \frac{-3x - 4(x+1) + 7x(x+1)}{x(x+1)} = 0 \\
& \Leftrightarrow -3x - 4(x+1) + 7x(x+1) = 0 \\
& \Leftrightarrow -3x - 4x - 4 + 7x^2 + 7x = 0 \\
& \Leftrightarrow 7x^2 - 4 = 0 \\
& \Leftrightarrow 7x^2 = 4 \\
& \Leftrightarrow 7x^2/7 = 4/7 \\
& \Leftrightarrow x^2 = 4/7 \\
& \Leftrightarrow x = \pm\sqrt{4/7} \\
& \quad = (\pm\sqrt{4}/\sqrt{7}\,) \cdot (\sqrt{7}/\sqrt{7}\,) \\
& \quad = (\pm\sqrt{4} \cdot \sqrt{7}\,)/\sqrt{7^2} \\
& \quad = \pm 2\sqrt{7}/7
\end{aligned}
$$

Set of solutions: $\{-2\sqrt{7}/7,\ 2\sqrt{7}/7\}$ □

F. $3x - \dfrac{1}{x} + 1 = 0$

Section 65: Word Problems: Quadratic Applications

Example 1
Solve each word problem.

A. Three times the square of a positive number equals fifteen times the number. Find the number.

Solution
Let n represent the number in question.
$3 \cdot n^2 = 15 \cdot n$
$\Leftrightarrow 3n^2 - 15n = 15n - 15n$
$\Leftrightarrow 3n(n-5) = 0$
$\Leftrightarrow \cancel{n = 0}$ X OR $n - 5 = 0$
$\Leftrightarrow n = 5$
The number is 5 (we reject 0 because n must be positive). □

B. The cube of a positive number equals 16 times the number. Find the number.

C. The width of a rectangle is two-thirds of its length. Its area is twenty-four square feet. Find the length and width of the rectangle.

Solution
Let l denote the length and
w denote the width of the rectangle.
Note that $w = (2/3)(l)$.
(area of rectangle) $= 24$
$\Rightarrow l \cdot w = 24$
$\Rightarrow l \cdot (2/3)(l) = 24$
$\Rightarrow (2/3)(l^2) = 24$
$\Rightarrow (3/2)(2/3)(l^2) = (3/2)(24)$
$\Rightarrow l^2 = 36$
$\Rightarrow l^2 - 36 = 36 - 36$
$\Rightarrow (l-6)(l+6) = 0$
$\Rightarrow l - 6 = 0$ OR $l + 6 = 0$
$\Rightarrow l = 6$ $\Rightarrow \cancel{l = -6}$ X
$\Rightarrow w = (2/3)(6)$
$= 4$
The rectangle has a length of 6 feet and a width of 4 feet. □

D. The area of a triangle is 72 cm^2. One side of the triangle has the same length as the altitude to that side. Find the length of the side.

E. The product of two consecutive odd whole numbers is 195. Identify them.

SOLUTION
Let m denote the lower odd number and
n denote the higher odd number.
Note that $n = m + 2$.
$mn = 195$
$\Rightarrow m(m + 2) = 195$
$\Rightarrow m(m) + m(2) = 195$
$\Rightarrow m^2 + 2m - 195 = 195 - 195$
$\Rightarrow (m - 13)(m + 15) = 0$
$\Rightarrow m - 13 = 0$ OR $m + 15 = 0$
$\Rightarrow m = 13$ $\Rightarrow$ ~~$m = -15$~~ X
$\Rightarrow n = 13 + 2$
$= 15$
The numbers are 13 and 15.

□

F. A certain number is 4 units larger than another number. The sum of their squares is 80. Find the numbers.

G. A farmer of tobacco grows 130 plants in even rows. The number of plants in each row is 17 less than three times the number of rows. Find the number of rows and the number of plants per row.

SOLUTION
Let r be the number of rows and
n be the number of plants per row.
Note that $n = 3r - 17$.
(total number of plants) $= 130$
$\Rightarrow r \cdot n = 130$
$\Rightarrow r(3r - 17) = 130$
$\Rightarrow r(3r - 17) - 130 = 130 - 130$
$\Rightarrow 3r^2 - 17r - 130 = 0$
$\Rightarrow (3r + 13)(r - 10) = 0$
$\Rightarrow 3r + 13 = 0$ OR $r - 10 = 0$
$\Rightarrow 3r = -13$ $\Rightarrow r = 10$
$\Rightarrow 3r/3 = -13/3$
$\Rightarrow$ ~~$r = -13/3$~~ X
$\Rightarrow n = 3(10) - 17$
$= 13$
There are 10 rows with 13 plants each.

□

H. The area of a large rectangular carpet is 120 square feet. Find the dimensions of the carpet if its length is two feet more than seven times its width.

I. The sum of the areas of the 64 squares on a chessboard is 144 square inches. Find the length of the side of each square.

Solution

Let s denote the length of the side of one of the squares on the board.

$64 \cdot (\text{area of a square}) = 144$
$\Rightarrow 64s^2 = 144$
$\Rightarrow 64s^2/64 = 144/64$
$\Rightarrow s^2 = 144/64$
$\Rightarrow s = \sqrt{144/64}$
$\Rightarrow s = 12/8$
$= 3/2$
$= 1.5$

The length of a side is 1.5 inches.

□

J. The area of a circle is 90π square units. Find the radius of the circle.

K. The sum of the areas of two circles is 97π square inches. The radius of one circle is 5 inches larger than the radius of the other. Find the radii.

Solution

Let r_1 represent the smaller radius and r_2 represent the larger radius.

Note that $r_2 = r_1 + 5$.

$(\text{sum of areas}) = 97\pi$
$\Rightarrow \pi r_1^2 + \pi r_2^2 = 97\pi$
$\Rightarrow \pi r_1^2 + \pi r_2^2 - 97\pi = 97\pi - 97\pi$
$\Rightarrow \pi r_1^2 + \pi r_2^2 - 97\pi = 0$
$\Rightarrow \pi(r_1^2 + r_2^2 - 97) = 0$
$\Rightarrow \pi(r_1^2 + (r_1+5)^2 - 97) = 0$
$\Rightarrow \pi(r_1^2 + r_1^2 + 10r_1 + 25 - 97) = 0$
$\Rightarrow \pi(2r_1^2 + 10r_1 - 72) = 0$
$\Rightarrow \pi(2[r_1^2 + 5r_1 - 36]) = 0$
$\Rightarrow 2\pi(r_1^2 + 5r_1 - 36) = 0$
$\Rightarrow 2\pi(r_1 - 4)(r_1 + 9) = 0$
$\Rightarrow r_1 - 4 = 0$ or $r_1 + 9 = 0$
$\Rightarrow r_1 = 4$ $\quad \Rightarrow \cancel{r_1 = -9}$ X
$\Rightarrow r_2 = 4 + 5$
$= 9$

The radii are 4 inches and 9 inches.

□

L. Alex is three years older than Bill. Carl is four years younger than Bill. The product of the ages of Alex and Carl equals 78. Find the ages of Alex, Bill, and Carl.

M. Mona runs 4 km every morning. One day she feels sluggish and runs 2 km per hour slower than usual. As a result, she takes 10 minutes longer than usual to complete her route. Find the length of time that Mona usually takes to run the route and her usual speed.

Solution

Declarations

Let t_1 and r_1 denote Mona's usual time (in hr) and speed (in km/hr) and
t_2 and r_2 denote Mona's time (in hr) and speed (in km/hr) on the day she feels sluggish.

Substitutions

- $r_2 = r_1 - 2$
- $t_2 = t_1 + 1/6$ (since 10 min = 1/6 hr)
- $r_1 t_1 = 4 \Leftrightarrow t_1 = 4/r_1$

Values

$r_2 t_2 = 4$

$\Rightarrow (r_1 - 2)(t_1 + 1/6) = 4$

$\Rightarrow (r_1 - 2)(4/r_1 + 1/6) = 4$

$\Rightarrow (r_1 - 2)((4/r_1)(6/6) + (1/6)(r_1/r_1)) = 4$

$\Rightarrow (r_1 - 2)(24/(6r_1) + r_1/(6r_1)) = 4$

$\Rightarrow (r_1 - 2)((24 + r_1)/(6r_1)) - 4 = 0$

$\Rightarrow (24r_1 + r_1^2 - 2(24) - 2r_1)/(6r_1) - 4 \cdot (6r_1/6r_1) = 0$

$\Rightarrow (22r_1 + r_1^2 - 48)/(6r_1) - (24r_1)/6r_1) = 0$

$\Rightarrow [(22r_1 + r_1^2 - 48) - 24r_1]/(6r_1) = 0$

$\Rightarrow (r_1^2 - 2r_1 - 48)/(6r_1) = 0$

$\Rightarrow r_1^2 - 2r_1 - 48 = 0$

$\Rightarrow (r_1 - 8)(r_1 + 6) = 0$

$\Rightarrow r_1 - 8 = 0$ OR $r_1 + 6 = 0$

$\Rightarrow r_1 = 8$ $\qquad \Rightarrow \cancel{r_1 = -6}$ X

$\Rightarrow t_1 = 4/8$

$= 1/2$

Mona usually runs for 1/2 hour (or 30 minutes) at a speed of 8 km/hr.

□

N. The rate at which a capitalist on a riding lawnmower cuts grass is 1250 m^2/hr more than the rate at which a socialist with a sickle cuts grass. Find the time it takes the capitalist to cut the grass in a field with area 4675 m^2 if that time is 34 hours less than the time it takes the socialist to cut the field.

o. The product of two numbers is 8. The sum of the same numbers is 7. Identify the numbers.

Solution
Let x and y represent the numbers.
Note that $x + y = 7$, so $y = 7 - x$.
$xy = 8$
$\Rightarrow x(7 - x) = 8$
$\Rightarrow 7x - x^2 = 8$
$\Rightarrow 7x - x^2 - 8 = 0$
$\Rightarrow (-1)x^2 + 7x + (-8) = 0$
$\Rightarrow x = (-7 \pm \sqrt{(7)^2 - 4(-1)(-8)}\,)/(2(-1))$
$= (-7 \pm \sqrt{49 - 32}\,)/(-2)$
$= (-7 \pm \sqrt{17}\,)/(-2)$
$= [(-1)(7 \mp \sqrt{17}\,)]/[(-1)(2)]$
$= (7 \mp \sqrt{17})/2$

For $x = (7 - \sqrt{17})/2$:
$y = 7 - (7 - \sqrt{17})/2$
$= 14/2 - (7 - \sqrt{17})/2$
$= (14 - (7 - \sqrt{17}))/2$
$= (14 - 7 + \sqrt{17})/2$
$= (7 + \sqrt{17})/2$

For $x = (7 + \sqrt{17})/2$:
$y = 7 - (7 + \sqrt{17})/2$
$= 14/2 - (7 + \sqrt{17})/2$
$= (14 - (7 + \sqrt{17}))/2$
$= (14 - 7 - \sqrt{17})/2$
$= (7 - \sqrt{17})/2$

The numbers are $(7 - \sqrt{17})/2$ and $(7 + \sqrt{17})/2$ (in both cases).

□

p. Given two numbers, suppose the ratio of the larger to the smaller equals the ratio of their sum to the larger. Find the value of this ratio (called the "golden ratio").

Section 66: Rational Exponents

Definitions

Suppose $b \in \mathbf{R}$, $r \in \mathbf{Z}$, and $s \in \mathbf{N}$, where r and s have no common integral factors other than 1 and $b \geq 0$ if s is even. Then:

- $b^{1/s} = \sqrt[s]{b}$
- $b^{r/s} = \sqrt[s]{b^r} = (\sqrt[s]{b})^r$

Comments

- Bases must be nonnegative under rational exponents with even denominators just as radicands must be nonnegative within radicals with even indices.
- The properties of exponents we saw on pages 77 and 85 hold true not only for $m, n \in \mathbf{N}$, but also for $m, n \in \mathbf{Q}$.
- We can rewrite the properties on page 165 in terms of rational exponents:
  - $x^n = b \Leftrightarrow x = b^{1/n}$ if n is odd
  - $x^n = b \Leftrightarrow x = b^{1/n}$ OR $x = -b^{1/n}$ if n is even, provided $b \geq 0$

Example 1

Evaluate each numerical expression.

A. $(-64)^{1/3}$

Solution

$$(-64)^{1/3} = \sqrt[3]{-64} = -4$$

□

B. $-25^{1/2}$

C. $-81^{3/4}$

Solution

$$-81^{3/4} = -(81^{3/4}) = -(\sqrt[4]{81})^3 = -(3)^3 = -27$$

□

D. $27^{4/3}$

E. $5^{5/2} \cdot 5^{1/2}$

Solution

$$5^{5/2} \cdot 5^{1/2} = 5^{(5/2+1/2)} = 5^{6/2} = 5^3 = 125$$

□

F. $7^{4/5}/7^{-1/5}$

G. $(1/4)^{-5/2}$

SOLUTION

$$\begin{aligned}(1/4)^{-5/2} &= ((1/4)^{-1})^{5/2}\\ &= (4)^{5/2}\\ &= (\sqrt{4})^5\\ &= (2)^5\\ &= 32\end{aligned}$$

□

H. $243^{-1/5}$

EXAMPLE 2

Simplify each expression and leave your answers with positive exponents. Assume that all variables represent positive real numbers.

A. $(8x^{5/4}y^{-3/2})^2$

SOLUTION

$$\begin{aligned}(8x^{5/4}y^{-3/2})^2 &= (8)^2(x^{5/4})^2(y^{-3/2})^2\\ &= 64(x^{5/2})(y^{-3})\\ &= 64x^{5/2}/y^3\end{aligned}$$

□

B. $(5x^{3/4})(7x^{-1/3})$

C. $(9c^{1/2})/(3c^{5/7})$

SOLUTION

$$\begin{aligned}(9c^{1/2})/(3c^{5/7}) &= (9/3)(c^{(1/2-5/7)})\\ &= 3c^{(7/14-10/14)}\\ &= 3c^{-3/14}\\ &= 3/c^{3/14}\end{aligned}$$

□

D. $[(2m^{5/6})/(6m^{1/4})]^3$

EXAMPLE 3

Solve each equation.

A. $x^{2/3}+1=5$

SOLUTION

$$\begin{aligned}x^{2/3}+1=5 &\Leftrightarrow x^{2/3}+1-1=5-1\\ &\Leftrightarrow x^{2/3}=4\\ &\Leftrightarrow (x^{2/3})^3=(4)^3\\ &\Leftrightarrow x^2=64\\ &\Leftrightarrow x=\pm\sqrt{64}\\ &=\pm 8\end{aligned}$$

Set of solutions: $\{-8, 8\}$

□

B. $2x^{5/2}=-64$

C. $2t^{3/4} - 8 = 46$

Solution

Potential Solutions

$$\begin{aligned}2t^{3/4} - 8 = 46 &\Rightarrow 2t^{3/4} - 8 + 8 = 46 + 8\\ &\Rightarrow 2t^{3/4} = 54\\ &\Rightarrow \not{2}t^{3/4}/\not{2} = 54/2\\ &\Rightarrow (t^{1/4})^3 = 27\\ &\Rightarrow t^{1/4} = \sqrt[3]{27}\\ &\qquad = 3\\ &\Rightarrow (t^{1/4})^4 = (3)^4\\ &\Rightarrow t = 81\end{aligned}$$

Check

$$\begin{aligned}81:\quad & 2(81)^{3/4} - 8 \stackrel{?}{=} 46\\ \Leftrightarrow\ & 2(\sqrt[4]{81})^3 - 8 \stackrel{?}{=} 46\\ \Leftrightarrow\ & 2(3)^3 - 8 \stackrel{?}{=} 46\\ \Leftrightarrow\ & 2(27) - 8 \stackrel{?}{=} 46\\ \Leftrightarrow\ & 54 - 8 \stackrel{?}{=} 46 \quad \checkmark\end{aligned}$$

Set of Solutions

Set of solutions: $\{81\}$

□

D. $\frac{1}{5}a^{4/3} + 17 = 142$

Example 4

Use rational exponents to simplify each expression. Leave each answer as a radical.

A. $\sqrt[9]{x^3}$

Solution

$$\begin{aligned}\sqrt[9]{x^3} &= x^{3/9}\\ &= x^{1/3}\\ &= \sqrt[3]{x}\end{aligned}$$

□

B. $\sqrt[20]{x^2}$

C. $\sqrt[4]{25}$

Solution

$$\begin{aligned}\sqrt[4]{25} &= (25)^{1/4}\\ &= (5^2)^{1/4}\\ &= (5)^{2/4}\\ &= (5)^{1/2}\\ &= \sqrt{5}\end{aligned}$$

□

D. $\sqrt[6]{27}$

E. $\sqrt[3]{x} \cdot \sqrt[4]{x}$

SOLUTION

$$\begin{aligned}\sqrt[3]{x} \cdot \sqrt[4]{x} &= x^{1/3} \cdot x^{1/4} \\ &= x^{4/12} \cdot x^{3/12} \\ &= x^{(4/12+3/12)} \\ &= x^{7/12} \\ &= \sqrt[12]{x^7}\end{aligned}$$

□

F. $\sqrt[5]{x^2} \cdot \sqrt[6]{x^5}$

G. $\sqrt[8]{x^7} / \sqrt[5]{x^4}$

SOLUTION

$$\begin{aligned}\sqrt[8]{x^7} / \sqrt[5]{x^4} &= x^{7/8} / x^{4/5} \\ &= x^{35/40} / x^{32/40} \\ &= x^{(35/40-32/40)} \\ &= x^{3/40} \\ &= \sqrt[40]{x^3}\end{aligned}$$

□

H. $\sqrt[3]{x^2} / \sqrt[7]{x^6}$

Section 67: Factorizing by Substitution

Procedure

To facilitate the factorization of an expression, we can:

1. substitute a variable for a term within that expression;
2. factorize the resulting expression;
3. replace the variable with the original term for which we had substituted it.

Example 1

Factorize each expression completely.

A. $x^{2/7} + 2x^{1/7} - 15$

Solution

$x^{2/7} + 2x^{1/7} - 15$
$= (x^{1/7})^2 + 2x^{1/7} - 15$
$= u^2 + 2u - 15 \quad \text{where } u = x^{1/7}$
$= (u + 5)(u - 3)$
$= (x^{1/7} + 5)(x^{1/7} - 3)$

□

B. $x^{4/3} - x^{1/3}$

C. $(11n + 5)^2 - 9$

Solution

$(11n + 5)^2 - 9$
$= u^2 - 9 \quad \text{where } u = 11n + 5$
$= u^2 - (3)^2$
$= (u + 3)(u - 3)$
$= ((11n + 5) + 3)((11n + 5) - 3)$
$= (11n + 8)(11n + 2)$

□

D. $2(3t - 1)^2 + 7(3t - 1) + 6$

E. $x^{-10} + 8x^{-4}$

Solution

$x^{-10} + 8x^{-4}$
$= (x^{-2})^5 + 8(x^{-2})^2$
$= u^5 + 8u^2 \quad \text{where } u = x^{-2}$
$= u^2(u^3 + 8)$
$= u^2(u^3 + (2)^3)$
$= u^2(u + 2)(u^2 - (u)(2) + (2)^2)$
$= u^2(u + 2)(u^2 - 2u + 4)$
$= (x^{-2})^2(x^{-2} + 2)((x^{-2})^2 - 2x^{-2} + 4)$
$= (x^{-4})(x^{-2} + 2)(x^{-4} - 2x^{-2} + 4)$

□

F. $c^{-3} - c^{-3/2} - 2$

EXAMPLE 2
Solve each equation.

A. $20x = 12 - 8\sqrt{x}$

SOLUTION
Potential Solutions
$20x = 12 - 8\sqrt{x}$
$\Rightarrow 20x - 12 + 8\sqrt{x} = 0$
$\Rightarrow 4(5x - 3 + 2\sqrt{x}) = 0$
$\Rightarrow 4(5x + 2\sqrt{x} - 3) = 0$
$\Rightarrow 4(5(\sqrt{x})^2 + 2\sqrt{x} - 3) = 0$
$\Rightarrow 4(5u^2 + 2u - 3) = 0$ where $u = \sqrt{x}$
$\Rightarrow 4(5u - 3)(u + 1) = 0$
$\Rightarrow 4(5\sqrt{x} - 3)(\sqrt{x} + 1) = 0$
$\Rightarrow 5\sqrt{x} - 3 = 0$
$\Rightarrow 5\sqrt{x} - 3 + 3 = 0 + 3$
$\Rightarrow 5\sqrt{x} = 3$
$\Rightarrow \not{5}\sqrt{x}/\not{5} = 3/5$
$\Rightarrow \sqrt{x} = 3/5$
$\Rightarrow (\sqrt{x})^2 = (3/5)^2$
$\Rightarrow x = 9/25$
OR $\sqrt{x} + 1 = 0$
$\Rightarrow \sqrt{x} = -1$
$\Rightarrow (\sqrt{x})^2 = (-1)^2$
$\Rightarrow x = 1$

Checks
- 9/25: $20(9/25) \stackrel{?}{=} 12 - 8\sqrt{9/25}$
  $\Leftrightarrow 4(9/5) \stackrel{?}{=} 12 - 8(\sqrt{9}/\sqrt{25})$
  $\Leftrightarrow 36/5 \stackrel{?}{=} 12 - 8(3/5)$
  $\Leftrightarrow 36/5 \stackrel{?}{=} 60/5 - 24/5$
  $\Leftrightarrow 36/5 \stackrel{?}{=} (60 - 24)/5$ ✓
- 1: $20(1) \stackrel{?}{=} 12 - 8\sqrt{1}$
  $\Leftrightarrow 20 \stackrel{?}{=} 12 - 8(1)$
  $\Leftrightarrow 20 \stackrel{?}{=} 12 - 8$ X

Set of Solutions
Set of solutions: $\{9/25\}$

□

B. $x^{2/3} - x^{1/3} = 12$

C. $t^{-2} - 6t^{-1} + 8 = 0$

SOLUTION

$t^{-2} - 6t^{-1} + 8 = 0$

$\Leftrightarrow (t^{-1})^2 - 6t^{-1} + 8 = 0$

$\Leftrightarrow u^2 - 6u + 8 = 0 \quad \text{where } u = t^{-1}$

$\Leftrightarrow (u-2)(u-4) = 0$

$\Leftrightarrow (t^{-1} - 2)(t^{-1} - 4) = 0$

$\Leftrightarrow t^{-1} - 2 = 0$ OR $t^{-1} - 4 = 0$

$\Leftrightarrow t^{-1} = 2$	$\Leftrightarrow t^{-1} = 4$
$\Leftrightarrow 1/t = 2$	$\Leftrightarrow 1/t = 4$
$\Leftrightarrow 1 = 2t$	$\Leftrightarrow 1 = 4t$
$\Leftrightarrow 1/2 = t$	$\Leftrightarrow 1/4 = t$

Set of solutions: $\{1/4, 1/2\}$

□

D. $x^4 + 13x^2 + 36 = 0$

Section 68: The Cartesian Plane

Definitions

- The *Cartesian plane* is a plane in which each point represents an ordered pair of numbers.
- The *axes* of the Cartesian plane are two perpendicular number lines that intersect at a point called the *origin* and divide the plane into four regions called *quadrants*.
- The *x-coordinate* (or *abscissa*) of the point that represents the ordered pair (a, b) is a.
- The *y-coordinate* (or *ordinate*) of the point that represents the ordered pair (a, b) is b.

Procedure

To construct a portion of the Cartesian plane, we draw a horizontal line segment to serve as a portion of one axis of the plane and draw a vertical line segment to serve as a portion of the other. By convention:

- we number quadrants as I, II, III, and IV, starting with the upper right quadrant and continuing in a counterclockwise direction;
- we call the horizontal axis the x-axis;
- we call the vertical axis the y-axis;
- we label the origin as O.

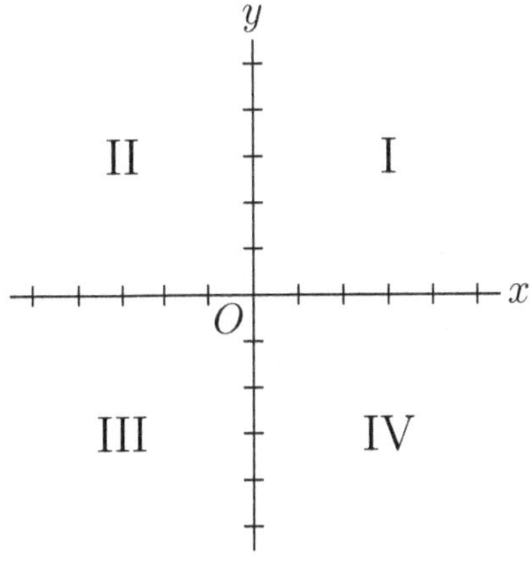

Procedure

To depict points with given coordinates in the Cartesian plane, we:

1. construct a portion of the Cartesian plane;
2. place tick marks along the x-axis at regular intervals according to a scale where:
   - the value at the leftmost mark is slightly less than the lowest x-coordinate and
   - the value at the rightmost mark is slightly greater than the highest x-coordinate;
3. place tick marks along the y-axis at regular intervals according to a scale where:
   - the value at the lowermost mark is slightly less than the lowest y-coordinate and
   - the value at the uppermost mark is slightly greater than the highest y-coordinate;
4. identify for each point with coordinates (a, b) its proper location in the plane, where the point should lie:
   - $|a|$ units to the right of the y-axis if $a \geq 0$ or
   - $|a|$ units to the left of the y-axis if $a < 0$, and
   - $|b|$ units above the x-axis if $b \geq 0$ or
   - $|b|$ units below the x-axis if $b < 0$;
5. plot each point in the plane in reference to the tick marks on the axes.

Notation

- The notation $P(a, b)$ means that the point P has the coordinates (a, b).
- We denote the set of all ordered pairs of real numbers by $\mathbf{R}\times\mathbf{R}$ or $\mathbf{R}^2$. Symbolically, we write that $\mathbf{R}^2 = \mathbf{R}\times\mathbf{R} = \{\,(x, y) \mid x, y \in \mathbf{R}\,\}$.

COMMENTS

- Both axes of the Cartesian plane extend without termination in two directions.
- We may omit indications of scale on tick marks when we space them one unit apart.
- The origin represents the ordered pair $(0, 0)$.
- Instead of saying "the point with coordinates (a, b)", we may simply say "the point (a, b)".
- We will refer to the portion of the x-axis that lies to the right of the y-axis as its "positive side" and the portion that lies to the left of the y-axis as its "negative side".

EXAMPLE 1

Identify the x- and y-coordinates of the point that corresponds to the given ordered pair.

A. $(5, -2)$

SOLUTION

- The x-coordinate is 5.
- The y-coordinate is -2.

□

B. $(0, 8)$

EXAMPLE 2

Construct a portion of the Cartesian plane that will contain the points below and then plot those points in the plane.

A. $A(4, 2)$
B. $B(-2, -5)$
C. $C(-3, 1)$
D. $D(3, -3)$
E. $E(0, -2)$
F. $F(1, 0)$

SOLUTION

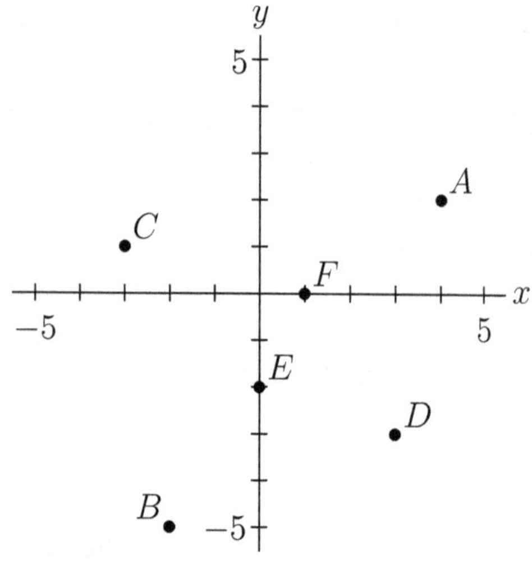

□

EXAMPLE 3

Construct a portion of the Cartesian plane that will contain the points below and then plot those points in the plane.

G. $G(-6, -2)$
H. $H(-1, 0)$
I. $I(3, -1)$
J. $J(0, 5)$
K. $K(-4, 4)$
L. $L(2, 7)$

Section 69: Distance and Midpoint Formulas, Two Dimensions

Distance and Midpoint Formulas

Suppose points P and Q have the coordinates (x_1, y_1) and (x_2, y_2), respectively. Then:

- $d(P, Q) = \sqrt{(x_2 - x_1)^2 + (y_2 - y_1)^2}$
- the midpoint of the line segment PQ has the coordinates $((x_1 + x_2)/2,\ (y_1 + y_2)/2)$

Comments

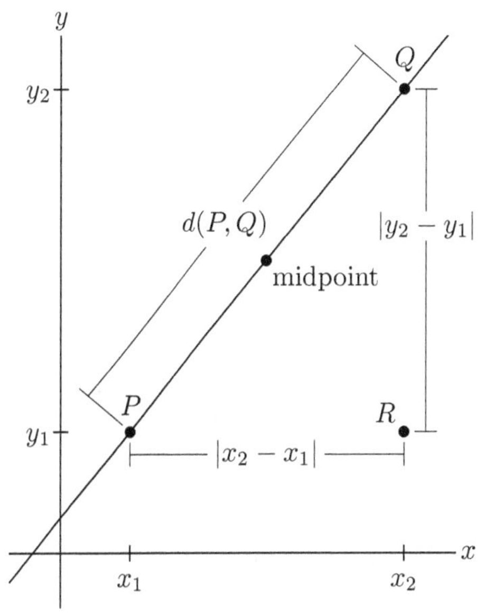

- The coordinates of the midpoint of PQ are the arithmetic means (the averages) of the respective coordinates of P and Q.
- If the points $P(x_1, y_1)$ and $Q(x_2, y_2)$ lie on the:
  - same horizontal line, then $d(P, Q) = |x_2 - x_1|$;
  - same vertical line, then $d(P, Q) = |y_2 - y_1|$.
- We derive the distance formula above by applying the Pythagorean Theorem to the right triangle with vertices $P(x_1, y_1)$, $Q(x_2, y_2)$, and $R(x_2, y_1)$:

$$\begin{aligned}(d(P, Q))^2 &= (d(P, R))^2 + (d(Q, R))^2 \\ &= |x_2 - x_1|^2 + |y_2 - y_1|^2 \\ &= (x_2 - x_1)^2 + (y_2 - y_1)^2 \\ \Rightarrow d(P, Q) &= \sqrt{(x_2 - x_1)^2 + (y_2 - y_1)^2}\end{aligned}$$

Example 1

Given the coordinates of points P and Q, determine:

- the distance $d(P, Q)$ between the points and
- the coordinates of the midpoint of the line segment PQ.

A. $P(2, 7)$, $Q(-6, 1)$

Solution

- $d(P, Q) = \sqrt{(-6 - 2)^2 + (1 - 7)^2}$
  $= \sqrt{(-8)^2 + (-6)^2}$
  $= \sqrt{64 + 36}$
  $= \sqrt{100}$
  $= 10$

- coordinates of midpoint
  $= ((2 + (-6))/2,\ (7 + 1)/2)$
  $= (-4/2,\ 8/2)$
  $= (-2,\ 4)$

□

B. $P(-9, -2)$, $Q(-8, 4)$

EXAMPLE 2
Determine the area of the triangle whose vertices lie at the points $A(1,2)$, $B(6,5)$, and $C(8,2)$.

SOLUTION
Graph
We draw the triangle for visual aid:

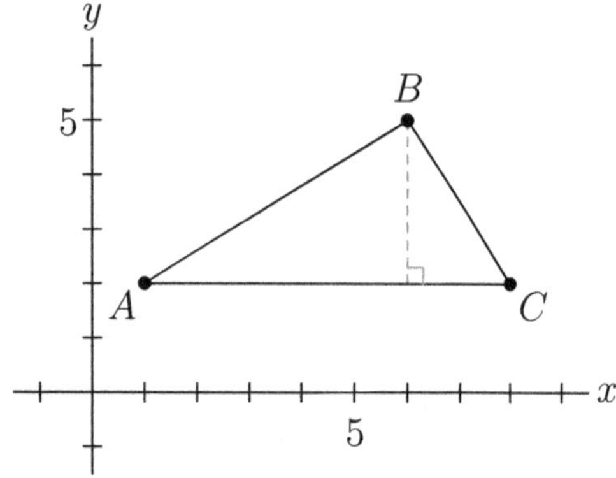

Declarations
Let h denote the height,
b denote the length of the base, and
D denote the point on the base of $\triangle ABC$ that lies directly below B.

Values of Variables
- D has coordinates $(6,2)$
- $b = d(A,C)$
$= \sqrt{(8-1)^2 + (2-2)^2}$
$= \sqrt{(7)^2 + (0)^2}$
$= \sqrt{49+0}$
$= 7$
- $h = d(B,D)$
$= \sqrt{(6-6)^2 + (5-2)^2}$
$= \sqrt{(0)^2 + (3)^2}$
$= \sqrt{0+9}$
$= 3$

Solution
$$\begin{aligned}(\text{area of } \triangle ABC) &= (1/2)bh \\ &= (1/2)(7)(3) \\ &= 21/2\end{aligned}$$

□

EXAMPLE 3
Which lies closer to the point $P(4,3)$: the point $A(11,2)$ or the point $B(0,-3)$?

Section 70: Graphs of Equations in Two Variables

Definitions

- The *graph of an algebraic statement* is the locus of all points that correspond to solutions of that statement.
- An *x-intercept* of a graph in the Cartesian plane is the x-coordinate of a point at which the graph crosses the x-axis.
- A *y-intercept* of a graph in the Cartesian plane is the y-coordinate of a point at which the graph crosses the y-axis.

Procedure

To find solutions of an equation in two variables x and y, we:

1. solve the equation for y in terms of x;
2. substitute an arbitrary value for the variable x;
3. calculate the corresponding value for y that makes the equation true;
4. identify the ordered pair (x, y) as a solution to the equation;
5. repeat steps (2), (3), and (4) as many times as desired.

Procedure

To find the x-intercepts of the graph of an equation, we:

1. substitute 0 for y in the given equation;
2. solve the equation resulting from step (1) for x.

Procedure

To find the y-intercepts of the graph of an equation, we:

1. substitute 0 for x in the given equation;
2. solve the equation resulting from step (1) for y.

Procedure

To draw the graph of an equation in two variables in the Cartesian plane, we:

1. identify a sufficient number of ordered pairs that satisfy the equation;
2. construct a portion of the Cartesian plane that will contain the points that correspond to the pairs identified in step (1);
3. plot the points in the plane;
4. draw a smooth curve through the points plotted in step (3).

Comments

- A solution to an equation in two variables is an ordered pair of numbers.
- We can state multiple solutions of an equation concisely within the framework of a table.
- We can find multiple solutions to some equations more easily after isolating x rather than y.
- The number of points we need to plot in order to draw accurately the graph of an equation varies.
- If a graph has an x-intercept of a, then the point $(a, 0)$ lies on the graph.
- If a graph has a y-intercept of b, then the point $(0, b)$ lies on the graph.

Example 1
Solve each equation for the specified variable.

A. $4x + 3y = 11$ for y

Solution
$4x + 3y = 11$
$\Leftrightarrow 3y = -4x + 11$
$\Leftrightarrow (1/3)(3y) = (1/3)(-4x + 11)$
$\Leftrightarrow y = (1/3)(-4x) + (1/3)(11)$
$\Leftrightarrow y = -(4/3)x + 11/3$

□

B. $-x + 2y = 8$ for x

Example 2
Draw the graph of each equation.

A. $x + y = 6$

Solution
Note
$x + y = 6 \Leftrightarrow y = -x + 6$

Table of Values

x	y	solution
0	$-0 + 6 = 6$	$(0, 6)$
2	$-2 + 6 = 4$	$(2, 4)$
4	$-4 + 6 = 2$	$(4, 2)$
6	$-6 + 6 = 0$	$(6, 0)$

Graph

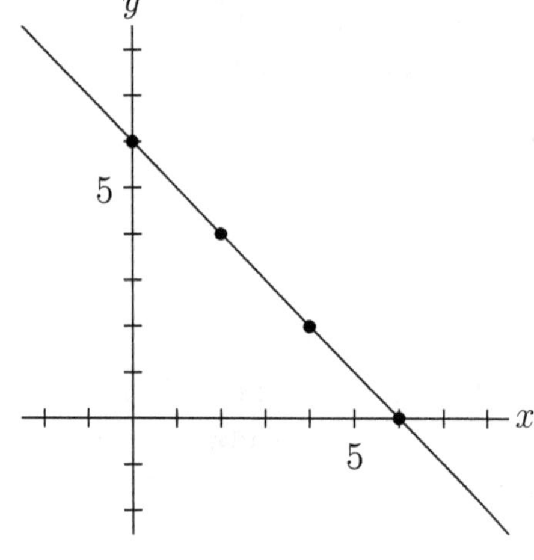

□

B. $-2x + y = -9$

C. $y - 1 = (x+3)^2$

SOLUTION

Note

$y - 1 = (x+3)^2 \Leftrightarrow y = (x+3)^2 + 1$

Table of Values

x	y	solution
-6	$(-6+3)^2+1 = 10$	$(-6, 10)$
-5	$(-5+3)^2+1 = 5$	$(-5, 5)$
-4	$(-4+3)^2+1 = 2$	$(-4, 2)$
-3	$(-3+3)^2+1 = 1$	$(-3, 1)$
-2	$(-2+3)^2+1 = 2$	$(-2, 2)$
-1	$(-1+3)^2+1 = 5$	$(-1, 5)$
0	$(0+3)^2+1 = 10$	$(0, 10)$

Graph

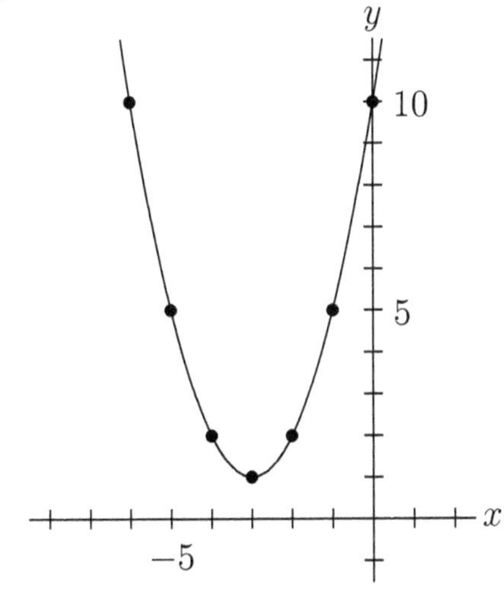

□

D. $x = (y-2)^2$

E. $x = -2y^3$

SOLUTION

Table of Values

y	x	solution
-3	$-2(-3)^3 = 54$	$(54, -3)$
-2	$-2(-2)^3 = 16$	$(16, -2)$
-1	$-2(-1)^3 = 2$	$(2, -1)$
0	$-2(0)^3 = 0$	$(0, 0)$
1	$-2(1)^3 = -2$	$(-2, 1)$
2	$-2(2)^3 = -16$	$(-16, 2)$
3	$-2(3)^3 = -54$	$(-54, 3)$

Graph

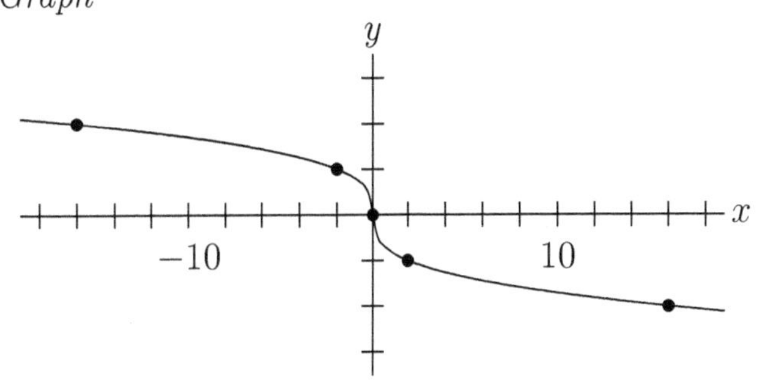

□

F. $y = 7 - |3x - 6|$

SOLUTION

Note

$$\begin{aligned} y &= 7 - |3x-6| \\ &= \begin{cases} 7-(3x-6) & \text{if } 3x-6 \geq 0 \\ 7-(-(3x-6)) & \text{if } 3x-6 < 0 \end{cases} \\ &= \begin{cases} 7-3x+6 & \text{if } 3x \geq 6 \\ 7+3x-6 & \text{if } 3x < 6 \end{cases} \\ &= \begin{cases} 13-3x & \text{if } x \geq 2 \\ 1+3x & \text{if } x < 2 \end{cases} \end{aligned}$$

Table of Values

x	y	solution
-1	$1+3(-1) = -2$	$(-1,-2)$
0	$1+3(0) = 1$	$(0,1)$
1	$1+3(1) = 4$	$(1,4)$
2	$13-3(2) = 7$	$(2,7)$
3	$13-3(3) = 4$	$(3,4)$
4	$13-3(4) = 1$	$(4,1)$
5	$13-3(5) = -2$	$(5,-2)$

Graph

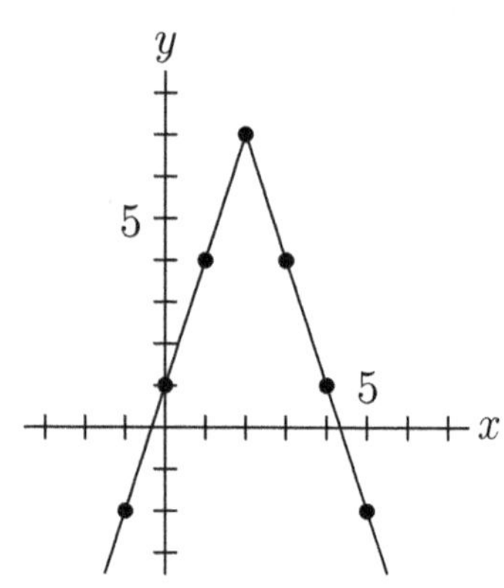

□

G. $x = |12 + 4y|$

H. $|2x| = 8$

SOLUTION

Note

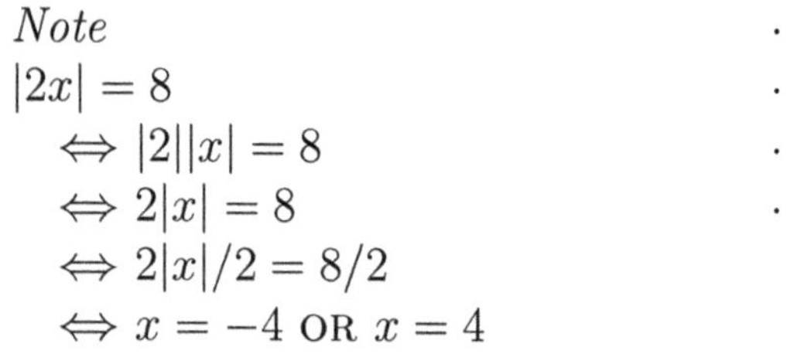

$|2x| = 8$

$\Leftrightarrow |2||x| = 8$

$\Leftrightarrow 2|x| = 8$

$\Leftrightarrow 2|x|/2 = 8/2$

$\Leftrightarrow x = -4$ OR $x = 4$

Graph

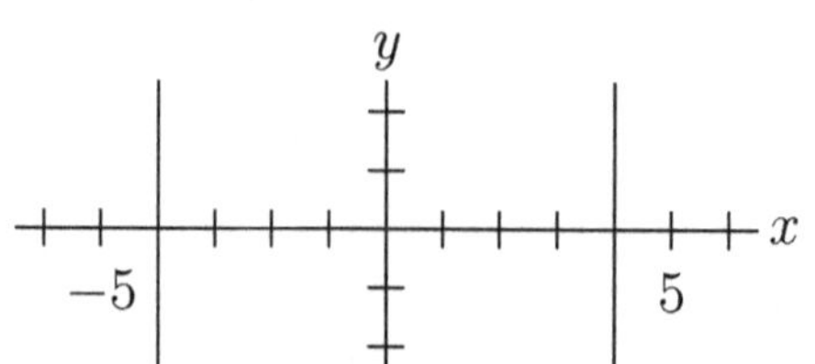

□

EXAMPLE 3

Identify the intercepts of the graph of each equation.

A. $x + 9 = y^2$

SOLUTION

- $x = 0 \Rightarrow 0 + 9 = y^2$
  $\Rightarrow y^2 = 9$
  $\Rightarrow y = \pm\sqrt{9}$
  $= \pm 3$

  The y-intercepts are -3 and 3.

- $y = 0 \Rightarrow x + 9 = 0^2$
  $\Rightarrow x = 0 - 9$
  $= -9$

  The x-intercept is -9.

□

B. $4x + y = 17$

Section 71: Lines

Definitions

- A *line* is a locus of points equidistant from two fixed points.
- The *slope* of a line is a measure of the steepness of the line.

Equation

The *standard form* of the equation of a line is $Ax + By = C$, where A, B, and C are real constants with $A \neq 0$ OR $B \neq 0$.

Derivation

Consider the line that comprises the points equidistant from the points (x_1, y_1) and (x_2, y_2), where $x_1 \neq x_2$ or $y_1 \neq y_2$.
Let (x, y) be the coordinates of a point on the line.
By the definition of line:

(the distance from (x, y) to (x_1, y_1)) = (the distance from (x, y) to (x_2, y_2))

$$\Rightarrow \sqrt{(x - x_1)^2 + (y - y_1)^2} = \sqrt{(x - x_2)^2 + (y - y_2)^2}$$
$$\Rightarrow \left(\sqrt{(x - x_1)^2 + (y - y_1)^2}\right)^2 = \left(\sqrt{(x - x_2)^2 + (y - y_2)^2}\right)^2$$
$$\Rightarrow (x - x_1)^2 + (y - y_1)^2 = (x - x_2)^2 + (y - y_2)^2$$
$$\Rightarrow x^2 - 2xx_1 + x_1^2 + y^2 - 2yy_1 + y_1^2 = x^2 - 2xx_2 + x_2^2 + y^2 - 2yy_2 + y_2^2$$
$$\Rightarrow -2xx_1 - 2yy_1 + 2xx_2 + 2yy_2 = \cancel{x^2} + x_2^2 + \cancel{y^2} + y_2^2 - \cancel{x^2} - x_1^2 - \cancel{y^2} - y_1^2$$
$$\Rightarrow -2xx_1 + 2xx_2 - 2yy_1 + 2yy_2 = x_2^2 + y_2^2 - x_1^2 - y_1^2$$
$$\Rightarrow (-2x_1 + 2x_2)x + (-2y_1 + 2y_2)y = -x_1^2 + x_2^2 - y_1^2 + y_2^2$$
$$\Rightarrow Ax + By = C, \text{ where } A = -2x_1 + 2x_2,\ B = -2y_1 + 2y_2,\ C = -x_1^2 + x_2^2 - y_1^2 + y_2^2$$

⊡

Formula

Suppose (x_1, y_1) and (x_2, y_2) denote the coordinates of two distinct points on a nonvertical line and m denotes the slope of the line.
Then $m = (y_2 - y_1)/(x_2 - x_1)$.

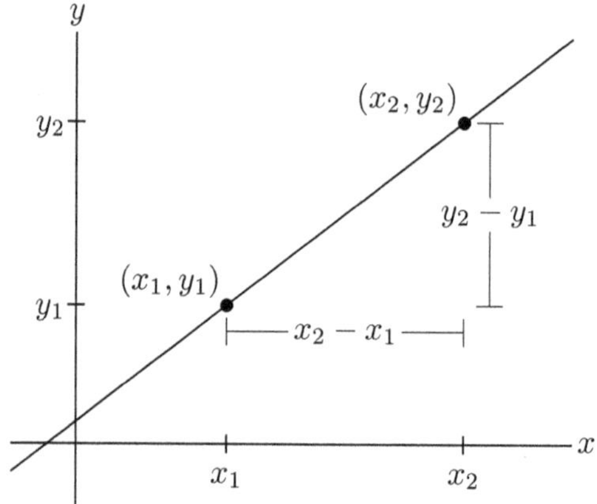

Notation

The Greek letter Δ ("delta") symbolizes change.
Thus:

- the notation Δq means "the change in q";
- if the value of q changes from q_1 to q_2, then $\Delta q = q_2 - q_1$.

Procedure

Suppose the point with coordinates (a, b) lies on a nonvertical line with slope m. To find a second point on a this line, we:

1. identify values Δx and Δy for which $m = \Delta y/\Delta x$;
2. choose an arbitrary real number c;
3. calculate the coordinates of a second point on the line as $(a + c \cdot \Delta x, b + c \cdot \Delta y)$.

Comments

- For a vertical line, our formula for slope will yield a fraction with a denominator of zero because all points on the line will have the same x-coordinate. Hence the slope of a vertical line is undefined.
- Given two points on a line, our formula for slope will yield the same result regardless of which pair of coordinates we label as (x_1, y_1) and which we label as (x_2, y_2).
- Geometrically:
  - a line that slants upward as we trace it from left to right will have positive slope;
  - a line that slants downward as we trace it from left to right will have negative slope;
  - a horizontal line will have a slope of zero.
- The slope m of a line equals the ratio of the amount of vertical change to the amount of horizontal change in a trace of the line from the point (x_1, y_1) to the point (x_2, y_2). We may thus rewrite the formula for the slope as $m = \Delta y/\Delta x$, since:
  - the change in x-coordinate is given by Δx, which equals $x_2 - x_1$, and
  - the change in y-coordinate is given by Δy, which equals $y_2 - y_1$.
- Informally, we say slope means "rise over run".
- We may repeat the procedure on the previous page with different values of c to determine the coordinates of as many points on the line as we want.

Example 1

Find the slope m of the line that passes through the points with the given coordinates.

A. $(-1, 2)$ and $(3, -6)$

Solution

$$\begin{aligned} m &= (-6-2)/(3-(-1)) \\ &= -8/4 \\ &= -2 \end{aligned}$$

□

B. $(2, -2)$ and $(5, 8)$

Example 2

Suppose the line through the points $(4, 8)$ and $(10, y)$ has slope $2/3$. Find y.

Solution

$$\begin{aligned} &(y-8)/(10-4) = 2/3 \\ &\quad\Rightarrow (y-8)/6 = 2/3 \\ &\quad\Rightarrow ((y-8)/6)(6) = (2/3)(6) \\ &\quad\Rightarrow y - 8 = 4 \\ &\quad\Rightarrow y = 4 + 8 \\ &\qquad\quad = 12 \end{aligned}$$

□

Example 3

Suppose the line through the points $(x, 1)$ and $(-4, 7)$ has slope $-5/2$. Find x.

EXAMPLE 4
Given a point (a, b) on a line with slope m, find three other points on the line.

A. $(a, b) = (6, 8)$, $m = -1/4$

SOLUTION
For $\Delta x = 4$ and $\Delta y = -1$:

- $(a - 1 \cdot \Delta x,\ b - 1 \cdot \Delta y)$
  $= (6 - 1(4),\ 8 - 1(-1))$
  $= (2, 9)$
- $(a + 1 \cdot \Delta x,\ b + 1 \cdot \Delta y)$
  $= (6 + 1(4),\ 8 + 1(-1))$
  $= (10, 7)$
- $(a + 2 \cdot \Delta x,\ b + 2 \cdot \Delta y)$
  $= (6 + 2(4),\ 8 + 2(-1))$
  $= (14, 6)$

The coordinates of three other points on the line are $(2, 9)$, $(10, 7)$, and $(14, 6)$. □

B. $(a, b) = (-7, -10)$, $m = 2$

EXAMPLE 5
Find the coordinates of two points on the line with the given equation and then use those points to find the slope m of the line.

A. $4x + y = 7$

SOLUTION
Notes

- $x = 0 \Rightarrow 4(0) + y = 7$
  $\Rightarrow 0 + y = 7$
  $\Rightarrow y = 7$
- $x = 1 \Rightarrow 4(1) + y = 7$
  $\Rightarrow 4 + y = 7$
  $\Rightarrow -4 + 4 + y = -4 + 7$
  $\Rightarrow y = 3$

Points
Points $(0, 7)$ and $(1, 3)$ lie on the line.

Slope
$m = (7 - 3)/(0 - 1)$
$= 4/(-1)$
$= -4$ □

B. $-2x - 5y = 12$

Section 72: Linear Equations in Two Variables

Definition

A *linear equation in two variables* x and y is an equation that can be written equivalently in the form $Ax + By = C$ for some constants $A, B, C \in \mathbf{R}$ with $A \neq 0$ or $B \neq 0$.

Forms of the Equation of a Line

- *standard form*: $Ax + By = C$, where A, B, and C are real constants with $A \neq 0$ or $B \neq 0$
- *point-slope form*: $y - y_1 = m(x - x_1)$, where:
  - m represents the slope of the line
  - (x_1, y_1) represent the coordinates of a point on the line
- *slope-intercept form*: $y = mx + b$, where:
  - m represents the slope of the line
  - b represents the y-intercept of the line

Comments

- The graph of a linear equation in two variables is a line in a two-dimensional plane.
- We can determine the equation of a line with only two points.
- Given its equation, the points on a line that we can identify most easily are those where it crosses the axes, since either the x- or the y-coordinate of these points will equal zero.
- The graph of an equation of the form:
  - $x = a$ is a vertical line that passes through all points with an x-coordinate of a;
  - $y = b$ is a horizontal line that passes through all points with an y-coordinate of b.
- Equations of vertical lines cannot take point-slope form or slope-intercept form.
- When we leave a linear equation in standard form $Ax + By = C$, we should ensure (when possible) that:
  - A, B, and C are integers with no common factors and
  - A is nonnegative.

  We cannot always leave linear equations with irrational coefficients in this ideal form.
- All equations that represent the same line are algebraically equivalent.
- We have described above the most conventional forms of linear equations, though we could conceive innumerable forms of less use or simplicity. Slope-intercept form is especially useful because it allows us to immediately identify the slope and y-intercept of a line.

Procedures

To convert a linear equation in two variables:

- from standard form to slope-intercept form, we solve the equation for y in terms of x;
- from point-slope form to slope-intercept form, we solve the equation for y in terms of x;
- from slope-intercept form to standard form, we subtract the mx term from both sides;
- from standard form to the ideal standard form, we:
  1. clear the equation of fractions (if necessary);
  2. multiply both sides by -1 to yield a positive coefficient on the x-term (if necessary);
  3. simplify the equation by eliminating common factors of the coefficients (if necessary).

EXAMPLE 1
Determine the equation of the line that contains the given point and has the given slope. Leave the equation in standard form.

A. $(6, -2)$; $m = 4/3$

SOLUTION
$y - (-2) = (4/3)(x - 6)$
$\Leftrightarrow y + 2 = (4/3)(x) - (4/3)(6)$
$= (4/3)x - 8$
$\Leftrightarrow y = (4/3)x - 8 - 2$
$= (4/3)x - 10$
$\Leftrightarrow -(4/3)x + y = -10$
$\Leftrightarrow (-3)(-(4/3)x + y) = (-3)(-10)$
$\Leftrightarrow (-3)(-(4/3)x) + (-3)(y) = 30$
$\Leftrightarrow 4x - 3y = 30$

□

B. $(4, 1)$; $m = 2/7$

EXAMPLE 2
Determine the equation of the line on which the given points lie. Leave the equation in standard form.

A. $(-7, 10)$ and $(-4, 1)$

SOLUTION
Slope
Let m denote the slope of the line.
$m = (1 - 10)/(-4 - (-7))$
$= -9/3$
$= -3$

Equation
$y - 1 = -3(x - (-4))$
$\Leftrightarrow y = -3(x + 4) + 1$
$= -3x - 12 + 1$
$= -3x - 11$
$\Leftrightarrow 3x + y = -11$

[Note that our choice of the point $(-4, 1)$ instead of $(-7, 10)$ for substitution into the point-slope form of the equation was somewhat arbitrary. Each point will yield the same result.]

□

B. $(-2, 6)$ and $(6, 10)$

EXAMPLE 3
Determine the equation of the line with the given slope m and y-intercept b. Leave the equation in standard form.

A. $m = 2/5;\ b = -7$

SOLUTION
$y = (2/5)x + (-7)$
$\Leftrightarrow -(2/5)x + y = -7$
$\Leftrightarrow (-5)((-2/5)x + y) = (-5)(-7)$
$\Leftrightarrow (-5)((-2/5)x) + (-5)(y) = 35$
$\Leftrightarrow 2x - 5y = 35$

□

B. $m = 7/3;\ b = 4$

EXAMPLE 4
Determine the intercepts of the graph of each equation and then draw that graph.

A. $-3x + 2y = 12$

SOLUTION
Intercepts

- $x = 0 \Rightarrow -3(0) + 2y = 12$
  $\Rightarrow 2y/2 = 12/2$
  $\Rightarrow y = 6$

  The y-intercept is 6.

- $y = 0 \Rightarrow -3x + 2(0) = 12$
  $\Rightarrow -3x/(-3) = 12/(-3)$
  $\Rightarrow x = -4$

  The x-intercept is -4.

Graph

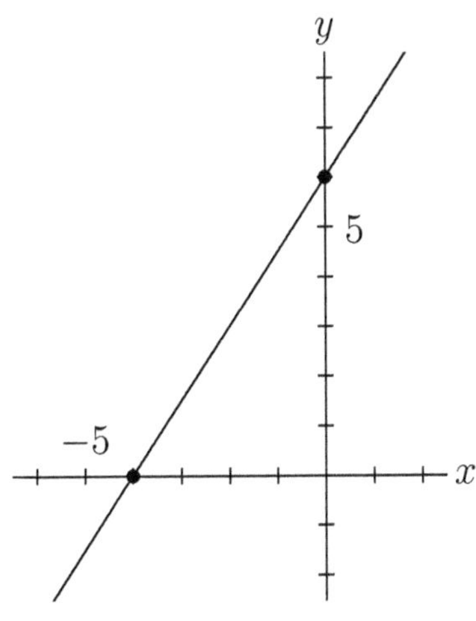

□

B. $y = -\frac{1}{2}x + 10$

C. $y = 5$

SOLUTION

Intercepts

- The y-intercept is 5.

- The graph has no x-intercept.
(The y-coordinate of every point on the line must be 5, so no point on the line can have a y-coordinate of 0.)

Graph

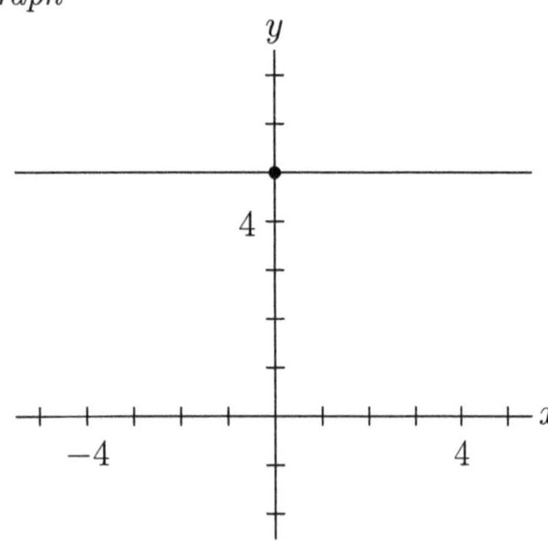

□

D. $x + 2 = 0$

EXAMPLE 5

Convert each linear equation to slope-intercept form (if necessary) and then determine the slope m and y-intercept b of the line with that equation.

A. $6x + 11y = 2$

SOLUTION

Note

$6x + 11y = 2$

$\Leftrightarrow 11y = -6x + 2$

$\Leftrightarrow (1/11)(11y) = (1/11)(-6x + 2)$

$\Leftrightarrow y = (1/11)(-6x) + (1/11)(2)$

$\Leftrightarrow y = (-6/11)x + 2/11$

Values

- $m = -6/11$
- $b = 2/11$

□

B. $7x - 2y = -10$

C. $y = 8x - 3$

SOLUTION

Note

$y = 8x - 3 \Leftrightarrow y = 8x + (-3)$

Values

- $m = 8$
- $b = -3$

$\square$

D. $y = (-1/6)x + 10$

Section 73: Parallel and Perpendicular Lines

Definitions

- Two lines in the Cartesian plane are *parallel* if they do not intersect.
- Two lines in the Cartesian plane are *perpendicular* if they intersect to form right angles.

Notation

- The symbol $||$ means "parallel".
- The symbol $\perp$ means "perpendicular".

Graphical Relationships

Suppose two nonvertical lines have slopes m_1 and m_2.
Then:

- the lines are parallel if and only if $m_1 = m_2$
- the lines are perpendicular if and only if $m_1 \cdot m_2 = -1$

Example 1

A. Lines l_1 and l_2 below are parallel.

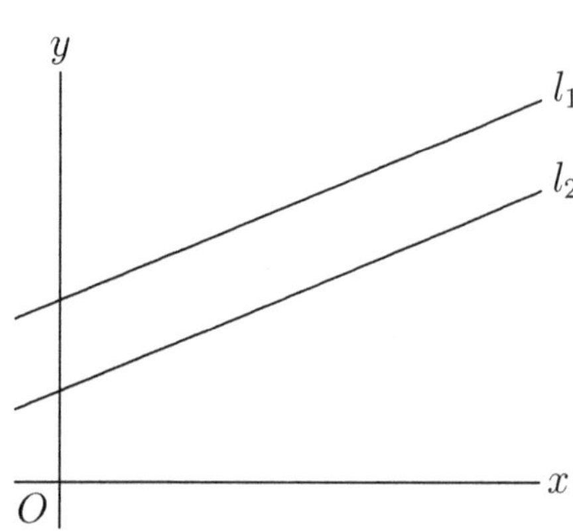

□

B. Lines l_3 and l_4 below are perpendicular.

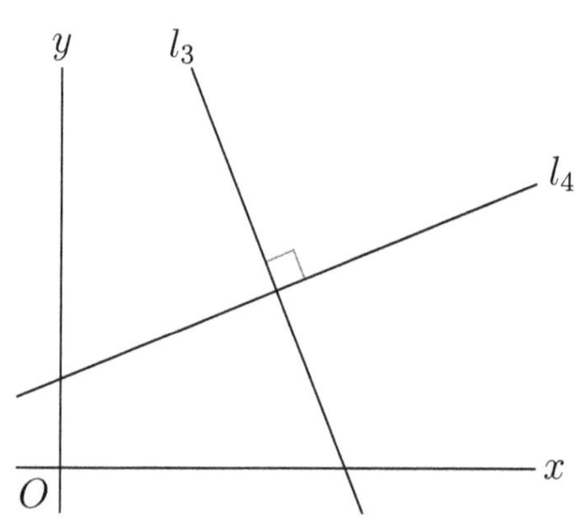

□

Example 2

Given the slope m of a line, find:

- the slope $m_{||}$ of any line parallel to the given line and
- the slope $m_{\perp}$ of any line perpendicular to the given line.

A. $m = 2/3$

Solution

- $m_{||} = m \Rightarrow m_{||} = 2/3$
- $m \cdot m_{\perp} = -1$

  $\Rightarrow (2/3) \cdot m_{\perp} = -1$

  $\Rightarrow (3/2)(2/3)(m_{\perp}) = (3/2)(-1)$

  $\Rightarrow m_{\perp} = -3/2$

□

B. $m = 5$

EXAMPLE 3

Determine the equation of the line that satisfies the given conditions. Leave the equation in slope-intercept form.

A. passes through the point $(-10, 4)$; runs perpendicular to the line with equation $6x + 3y = 7$

SOLUTION

Declarations

Let m_1 denote the slope of the line with equation $6x + 3y = 7$ and
m_2 denote the slope of the line in question.

Values

- $6x + 3y = 7 \Rightarrow 3y = -6x + 7$
  $\Rightarrow 3y/3 = (-6x + 7)/3$
  $\Rightarrow y = -6x/3 + 7/3$
  $= -2x + 7/3$
  $\Rightarrow m_1 = -2$
- $m_1 \cdot m_2 = -1 \Rightarrow -2 \cdot m_2 = -1$
  $\Rightarrow m_2 = -1/(-2)$
  $\Rightarrow m_2 = 1/2$

Equation

$y - 4 = (1/2)(x - (-10))$
$\Leftrightarrow y = (1/2)(x + 10) + 4$
$\Leftrightarrow y = (1/2)(x) + 5 + 4$
$\Leftrightarrow y = (1/2)x + 9$

□

B. passes through the point $(1, 4)$; runs parallel to the line with equation $2x - 5y = -11$

C. passes through the point $(-12, 7)$; runs perpendicular to the line with equation $y = 2$

SOLUTION

the line with equation $y = 2$ is horizontal
$\Rightarrow$ the line in question is vertical
$\Rightarrow$ the equation of the line in question takes the form $x = a$, where a is the x-coordinate of all points on the line
$\Rightarrow$ the equation of the line in question is $x = -12$

□

D. passes through the point $(4, 9)$; runs parallel to the x-axis

E. passes through the point $(7,1)$; runs parallel to the line that passes through the points $(-3,-4)$ and $(-1,2)$

Solution

Declarations

Let m_1 denote the slope of the line through $(-3,-4)$ and $(-1,2)$ and
m_2 denote the slope of the line in question.

Values

$$\begin{aligned} m_2 &= m_1 \\ &= (2-(-4))/(-1-(-3)) \\ &= 6/2 \\ &= 3 \end{aligned}$$

Equation

$$\begin{aligned} y-1=3(x-7) &\Leftrightarrow y=3x-21+1 \\ &\Leftrightarrow y=3x-20 \end{aligned}$$

□

F. passes through the point $(-3,-2)$; runs perpendicular to the line that passes through the points $(7,-8)$ and $(5,2)$

Example 4

Show that the points $A(-4,-4)$, $B(3,7)$, and $C(6,2)$ serve as the vertices of a right triangle.

Solution

After plotting A, B, and C in the Cartesian plane for visual aid (see below), we see that the points serve as the vertices of a triangle with longest side AB and shorter sides AC and BC. We must verify that the shorter two sides of $\triangle ABC$ are perpendicular.

Note that:

- $(\text{slope of } AC) = (2-(-4))/(6-(-4)) = 6/10 = 3/5$
- $(\text{slope of } BC) = (2-7)/(6-3) = -5/3$

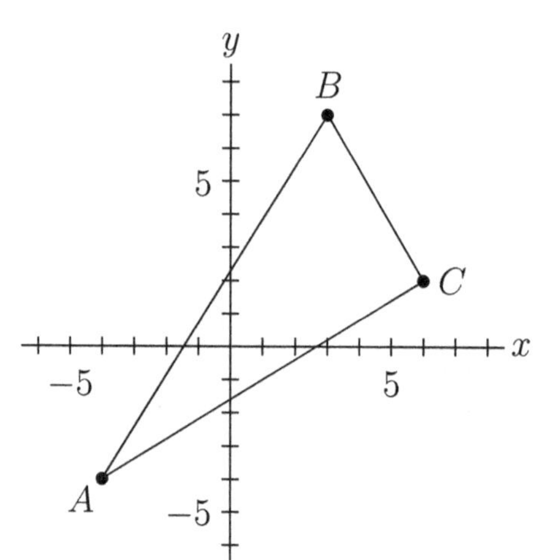

So:

$$\begin{aligned} (\text{slope of } AC)\cdot(\text{slope of } BC) &= (3/5)(-5/3) \\ &= -1 \end{aligned}$$

$\Rightarrow AC$ and BC are perpendicular
$\Rightarrow \angle ACB$ is a right angle
$\Rightarrow \triangle ABC$ is a right triangle

□

Section 74: Solving Systems of Linear Equations by Graphing

Definitions

- A *solution of a system of equations* is a solution of every equation in the system.
- A system of equations is *consistent* if it has at least one solution.
- A system of equations is *inconsistent* if it has no solution.
- The equations of a system are *dependent* if the system has infinitely many solutions.

Graphical Relationship

The solutions of a system of equations correspond to the points where the graphs of those equations intersect because a solution satisfies every equation in a system if and only if the point corresponding to the solution lies on the graph of every equation in the system.

Procedure

To solve a system of two linear equations in the variables x and y with graphs, we:

1. draw the graphs of both equations in the system;
2. identify the coordinates of the point where the graphs intersect;
3. state the set of solutions as a set of ordered pairs of the form (x, y).

Comments

- The graphs of two linear equations in two variables:
  - intersect if the system of equations is consistent;
  - run parallel and do not intersect if the system of equations is inconsistent;
  - coincide as the same line if the equations are dependent.
- The technique of using graphs to solve equations has limited value because we cannot usually ascertain by sight the exact values of the coordinates of a point at which two graphs intersect unless we happen to know that those coordinates are integers.

Example 1

Determine whether the given ordered pair is a solution of the given system of equations.

A. system: $\begin{cases} 2x + y = 2 \\ x - 3y = 1 \end{cases}$; pair: $(1, 0)$

Solution

Checks

- $2x + y = 2$: $\quad 2(1) + 0 \stackrel{?}{=} 2 \iff 2 + 0 \stackrel{?}{=} 2$ ✓
- $x - 3y = 1$: $\quad 1 - 3(0) \stackrel{?}{=} 1 \iff 1 - 0 \stackrel{?}{=} 1$ ✓

Conclusion

Yes, $(1, 0)$ is a solution of the system.

□

B. system: $\begin{cases} 3x - 2y = 0 \\ 2x + 3y = -13 \end{cases}$; pair: $(2, 3)$

EXAMPLE 2

A. The graphs of the equations in the system

$$\begin{cases} y = x - 1 \\ y = -3x + 15 \end{cases}$$

intersect in only one point, namely $(4, 3)$.
Hence:

- the system has exactly one solution, namely $(4, 3)$;
- the system is consistent.

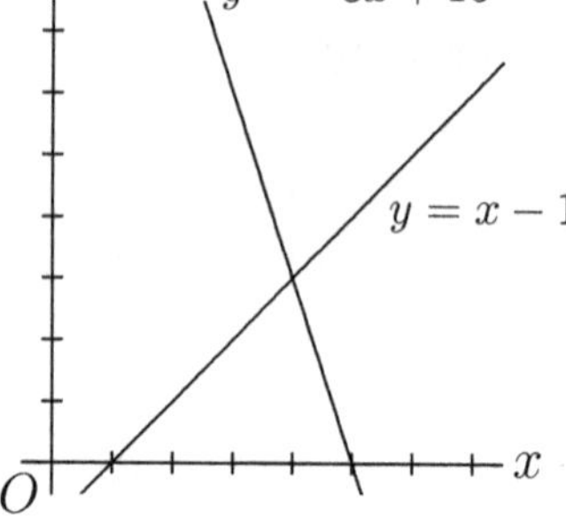

□

B. The graphs of the equations in the system

$$\begin{cases} x - 2y = -6 \\ x - 2y = -2 \end{cases}$$

do not intersect.
Hence:

- the system has no solution;
- the system is inconsistent.

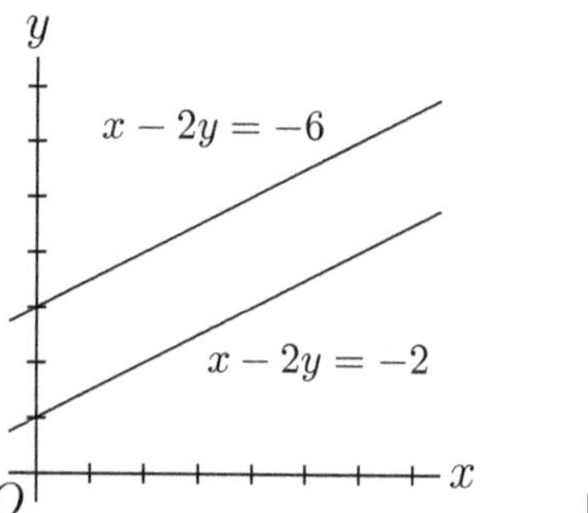

□

C. The graphs of the equations in the system

$$\begin{cases} x + y = 6 \\ 2x + 2y = 12 \end{cases}$$

coincide as the same line.
Hence:

- the system has infinitely many solutions, namely the ordered pairs in the set
$$\{\,(x, y) \in \mathbf{R}^2 \mid y = 6 - x\,\};$$
- the system is consistent;
- the equations in the system are dependent.

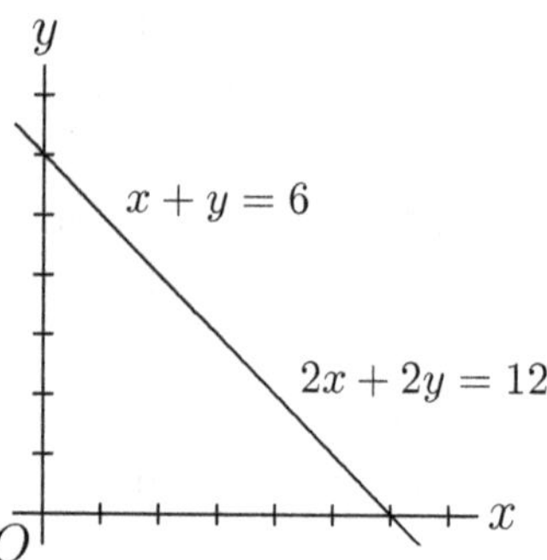

□

D. The graphs of the equations in the system

$$\begin{cases} y = -2x + 14 \\ y = x + 1 \\ y = 2 \end{cases}$$

do not all intersect at any point.
Hence:

- the system has no solution;
- the system is inconsistent.

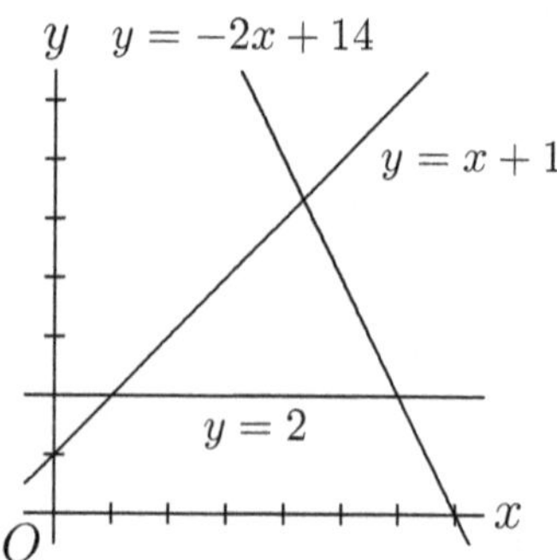

□

Example 3

Use graphs to solve each system of equations. Assume that each solution is an ordered pair of integers.

A. $\begin{cases} 2x + y = 2 \\ x - 3y = 1 \end{cases}$

Solution

Points on Lines

For $2x + y = 2$:

- $x = 0 \Rightarrow 2(0) + y = 2$
  $\Rightarrow y = 2$
- $x = 3 \Rightarrow 2(3) + y = 2$
  $\Rightarrow 6 + y - 6 = 2 - 6$
  $\Rightarrow y = -4$

So $(0, 2)$ and $(3, -4)$ satisfy $2x + y = 2$.

For $x - 3y = 1$:

- $x = 4 \Rightarrow 4 - 3y = 1$
  $\Rightarrow -3y = 1 - 4$
  $\Rightarrow (-3y)/3 = (1 - 4)/3$
  $\Rightarrow y = 1$
- $x = -2 \Rightarrow -2 - 3y = 1$
  $\Rightarrow -3y = 1 + 2$
  $\Rightarrow (-3y)/3 = (1 + 2)/3$
  $\Rightarrow y = -1$

So $(4, 1)$ and $(-2, -1)$ satisfy $x - 3y = 1$.

Graphs

Set of Solutions

Set of solutions: $\{(1, 0)\}$

□

B. $\begin{cases} 3x - 2y = 0 \\ 2x + 3y = -13 \end{cases}$

Section 75: Solving Systems of Linear Equations by Elimination

Terminology
To "eliminate a variable" from a set of two equations means to derive algebraically from those equations a third equation that does not contain the variable.

Property
Suppose $A = B$ and $C = D$, where A, B, C, and D represent algebraic expressions. Then $A + C = B + D$.

Proof
Suppose $A = B$ and $C = D$.

$$\begin{aligned} A = B &\Rightarrow A + C = B + C \\ &\Rightarrow A + C = B + D \end{aligned}$$

⊡

Procedure A
To eliminate a variable from a set of two linear equations, we:

1. write both equations in standard form;
2. identify the least common multiple of the coefficients of the variable in the equations;
3. multiply both sides of one of the equations by the quotient of the least common multiple of step (2) and the coefficient of the variable in that equation;
4. multiply both sides of the other equation by the quotient of the additive inverse of the least common multiple of step (2) and the coefficient of the variable in that equation;
5. equate the sum of the left sides of the equations resulting from steps (3) and (4) to the sum of the right sides of those equations;
6. simplify the equation resulting from step (5).

Procedure B
To solve a system of two linear equations in two variables x and y by elimination, we:

1. eliminate x from the equations;
2. solve the equation resulting from step (1) for y;
3. substitute the solution of step (2) for y in one of the equations of the original system;
4. solve the equation resulting from step (3) for x;
5. state the set of solutions as a set of ordered pairs of the form (x, y).

Comments

- The equation that results from step (1) of procedure B may contain no variables. This equation will be:
  - false if the system is inconsistent;
  - true if both equations in the system define the same line.
- We can solve some systems of equations by elimination more easily by interchanging the roles of x and y in the first four steps of procedure B.

EXAMPLE 1

Determine constants to multiply to the sides of the equations in the given system in order to eliminate x from that system.

A. $\begin{cases} 10x - 3y = -2 \\ -8x + 4y = 1 \end{cases}$

SOLUTION

Notes

- The LCM of 10 and −8 is 40.
- $40/10 = 4$
- $-40/(-8) = 5$

Conclusion

We should multiply both sides of:

- the first equation by 4 and
- the second equation by 5.

□

B. $\begin{cases} 8x + 5y = 11 \\ 12x - 7y = 2 \end{cases}$

EXAMPLE 2

Solve each system of equations by elimination.

A. $\begin{cases} 3x - 2y = 0 \\ 2x + 3y = -13 \end{cases}$

SOLUTION

$$\begin{cases} 3x - 2y = 0 \\ 2x + 3y = -13 \end{cases}$$

$$\Rightarrow \begin{cases} 2(3x - 2y) = 2(0) \\ -3(2x + 3y) = -3(-13) \end{cases}$$

$$\Rightarrow \begin{cases} 6x - 4y = 0 \\ -6x - 9y = 39 \end{cases}$$

$$-13y = 39$$

$\Rightarrow -13y/(-13) = 39/(-13)$

$\Rightarrow y = -3$

$\Rightarrow 3x - 2(-3) = 0$ (since $3x - 2y = 0$)

$\Rightarrow 3x + 6 = 0$

$\Rightarrow 3x = -6$

$\Rightarrow 3x/3 = -6/3$

$\Rightarrow x = -2$

Set of solutions: $\{(-2, -3)\}$

□

B. $\begin{cases} 2x + y = 2 \\ x - 3y = 1 \end{cases}$

C. $\begin{cases} 5x + 3y = -2 \\ 15x + 9y = -6 \end{cases}$

SOLUTION

Values

$$\begin{cases} 5x + 3y = -2 \\ 15x + 9y = -6 \end{cases}$$

$$\Rightarrow \begin{cases} -3(5x + 3y) = -3(-2) \\ 15x + 9y = -6 \end{cases}$$

$$\Rightarrow \begin{cases} -15x - 9y = 6 \\ 15x + 9y = -6 \end{cases}$$

$$0 = 0$$

Implications

- The equation $0 = 0$ is always true, so all solutions to the first equation are also solutions to the second equation.
- $5x + 3y = -2 \Leftrightarrow 3y = -2 - 5x$
  $\Leftrightarrow y = (-2 - 5x)/3$

Set of Solutions

Set of solutions:

$\{ (x, y) \in \mathbf{R}^2 \mid y = (-2 - 5x)/3 \}$.

□

D. $\begin{cases} 4x - 2y = 1 \\ 6x - 3y = 5 \end{cases}$

E. $\begin{cases} -3x + 4y = 5 \\ 6x - 8y = 7 \end{cases}$

SOLUTION

Values

$$\begin{cases} -3x + 4y = 5 \\ 6x - 8y = 7 \end{cases}$$

$$\Rightarrow \begin{cases} 2(-3x + 4y) = 2(5) \\ 6x - 8y = 7 \end{cases}$$

$$\Rightarrow \begin{cases} -6x + 8y = 10 \\ 6x - 8y = 7 \end{cases}$$

$$0 = 17$$

Set of Solutions

The set of solutions is $\emptyset$ (since the equation $0 = 17$ is false for every pair of values (x, y)).

□

F. $\begin{cases} -9x - 15y = 6 \\ 3x + 5y = -2 \end{cases}$

EXAMPLE 3

Solve each problem.

A. Henry buys 37 stamps for \$12.97. Some stamps cost \$0.73 each and the rest cost \$0.19 each. How many stamps of each kind did he buy?

SOLUTION

Declarations

Let x denote the number of stamps that Henry bought for \$0.73 and
y denote the number of stamps that that Henry bought for \$0.19.

Values

$$\begin{cases} x + y = 37 \\ 0.73x + 0.19y = 12.97 \end{cases}$$

$$\Rightarrow \begin{cases} -0.73(x + y) = -0.73(37) \\ 0.73x + 0.19y = 12.97 \end{cases}$$

$$\Rightarrow \begin{cases} -0.73x - 0.73y = -27.01 \\ 0.73x + 0.19y = 12.97 \end{cases}$$

$$-0.54y = -14.04$$

$$\Rightarrow -0.54y/(-0.54) = -14.04/(-0.54)$$

$$\Rightarrow y = 26$$

$$\Rightarrow x + 26 = 37 \quad (\text{since } x + y = 37)$$

$$\Rightarrow x = 37 - 26$$

$$= 11$$

Conclusion

Henry bought:

- 11 stamps that cost \$0.73 each and
- 26 stamps that cost \$0.19 each.

□

B. The mass of a blue ball is four times the mass of a yellow ball and the total mass of three yellow balls and eight blue balls is forty-nine slugs. How much mass does a single ball of each color have?

Section 76: Solving Systems of Linear Equations by Substitution

PROCEDURE

To solve a system of two linear equations in two variables x and y by substitution, we:

1. solve one of the given equations for y in terms of x;
2. write the other equation in terms only of x using the substitutive property of equality;
3. solve the equation resulting from step (2) for x;
4. substitute the solution of step (3) for x in the result of step (1) and evaluate for y;
5. state the set of solutions as a set of ordered pairs of the form (x, y).

COMMENTS

- The equation that results from step (2) of the procedure above may contain no variables after simplification. This equation will be:
  - false if the system is inconsistent;
  - true if both equations in the system define the same line.
- We can solve some systems of equations by substitution more easily by interchanging the roles of x and y in the first four steps of the procedure above.
- We can also use the method of substitution to solve systems of nonlinear equations.

EXAMPLE 1

Solve each system of equations by substitution.

A. $\begin{cases} 3x - 2y = 0 \\ 2x + 3y = -13 \end{cases}$

SOLUTION

Substitution

$3x - 2y = 0$

$\Rightarrow 3x = 2y$

$\Rightarrow 3x/3 = 2y/3$

$\Rightarrow x = (2/3)y$

Solution

$2x + 3y = -13$

$\Rightarrow 2((2/3)y) + 3y = -13$

$\Rightarrow (4/3)y + 3y = -13$

$\Rightarrow (4/3)y + (9/3)y = -13$

$\Rightarrow (13/3)y = -13$

$\Rightarrow (3/13)(13/3)y = (3/13)(-13)$

$\Rightarrow y = -3$

$\Rightarrow x = (2/3)(-3)$

$= -2$

Set of solutions: $\{(-2, -3)\}$

□

B. $\begin{cases} 5x - 7y = 5 \\ 3x - 2y = -8 \end{cases}$

C. $\begin{cases} 6x + 5y = 23 \\ -3x + y = -8 \end{cases}$

SOLUTION

Substitution

$-3x + y = -8$

$\Rightarrow 3x + (-3x) + y = 3x + (-8)$

$\Rightarrow y = 3x - 8$

Solution

$6x + 5y = 23$

$\Rightarrow 6x + 5(3x - 8) = 23$

$\Rightarrow 6x + 5(3x) - 5(8) = 23$

$\Rightarrow 6x + 15x - 40 = 23$

$\Rightarrow 21x - 40 = 23$

$\Rightarrow 21x = 23 + 40$

$\Rightarrow 21x = 63$

$\Rightarrow 21x/21 = 63/21$

$\Rightarrow x = 3$

$\Rightarrow y = 3(3) - 8$

$= 1$

Set of solutions: $\{(3, 1)\}$

$\square$

D. $\begin{cases} 2x + y = 2 \\ x - 3y = 1 \end{cases}$

E. $\begin{cases} 7x - 7y = 4 \\ -x + y = 5 \end{cases}$

SOLUTION

Substitution

$-x + y = 5$

$\Rightarrow x + (-x) + y = x + 5$

$\Rightarrow y = x + 5$

Solution

$7x - 7y = 4$

$\Rightarrow 7x - 7(x + 5) = 4$

$\Rightarrow 7x - 7x - 35 = 4$

$\Rightarrow -35 = 4$

The set of solutions is $\emptyset$ (since the equation $-35 = 4$ is false for every pair of values (x, y)).

$\square$

F. $\begin{cases} x + 4y = 6 \\ -2x - 8y = -12 \end{cases}$

Example 2
Solve each word problem.

A. The sum of the digits of a two-digit number is 11. The number yielded by reversing the digits is 27 less than the original number. Find the original number.

Solution
Declarations
Let x represent the tens digit and
y represent the ones digit of the original number.

System
Observe that:

- the original number equals $10x + y$;
- the number with digits reversed equals $10y + x$.

Thus the given information yields the system:

$$\begin{cases} x + y = 11 \\ 10y + x = (10x + y) - 27 \end{cases}$$

Substitution
$x + y = 11 \Rightarrow y = 11 - x$

Values

$$\begin{aligned} 10y + x &= (10x + y) - 27 \\ &\Rightarrow 10(11 - x) + x = (10x + (11 - x)) - 27 \\ &\Rightarrow 110 - 10x + x = 9x - 16 \\ &\Rightarrow -9x + 110 = 9x - 16 \\ &\Rightarrow -9x - 9x + 110 = -9x + 9x - 16 \\ &\Rightarrow -18x + 110 = -16 \\ &\Rightarrow -18x + 110 - 110 = -16 - 110 \\ &\Rightarrow -18x = -126 \\ &\Rightarrow -18x/(-18) = -126/(-18) \\ &\Rightarrow x = 7 \\ &\Rightarrow y = 11 - 7 \\ &\quad = 4 \end{aligned}$$

The original number is 74 (and the number with digits reversed is 47).

□

B. Eric divides $1000 dollars into two bank accounts. He earns simple interest on the accounts at the annual rates of 5% and 9%, respectively. After one year, the total amount of interest accrued in the accounts is $63.72. How much money did Eric place in each account?

Section 77: Linear Inequalities in Two Variables

Definition
A *linear inequality in two variables* x and y is an inequality that can be written equivalently in the form $Ax + By < C$ for some constants $A, B, C \in \mathbf{R}$ with $A \neq 0$ OR $B \neq 0$.

Procedure
To draw the graph of the inequality $Ax + By < C$, we:

1. draw as a dashed line the graph of the equation $Ax + By = C$; this line will separate the plane into two regions;
2. choose any point (called the "test point") that does not lie on the line of step (1);
3. determine whether the coordinates of the test point satisfy the inequality and:
    - if so, shade the region of the plane that contains the test point, but
    - if not, shade the region of the plane that does not contain the test point.

Comments

- We also follow the procedure above to draw graphs of statements of the form $Ax + By \leq C$, but we use a solid (not dashed) line in step (1).
- Linear inequalities define shaded regions in the same way that linear equations define lines; graphs of both kinds contain the points whose coordinates satisfy a algebraic statement.
- The graph of the set of solutions of a system of inequalities is the intersection of the graphs of all the inequalities in the system.

Example 1
Draw the graph of each inequality.

A. $2x + y > 4$

Solution
Notes

- if $2x + y = 4$, then:
    - $y = 4 - 2x$
    - $x = 0 \Rightarrow y = 4 - 2(0) = 4$
    - $x = 2 \Rightarrow y = 4 - 2(2) = 0$
- $(0, 0)$ does not satisfy $2x + y > 4$.

Procedure

- We plot holes at $(0, 4)$ and $(2, 0)$.
- We draw a dashed line through the holes, dividing the plane into two regions.
- We shade the region of the plane that does not contain the point $(0, 0)$.

Graph

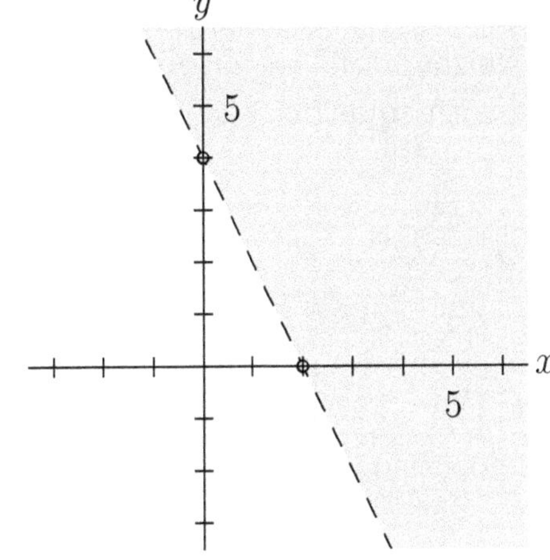

□

B. $y \le x/3$

SOLUTION

Notes

- if $y = x/3$, then:
  - $x = 0 \Rightarrow y = 0/3 = 0$
  - $x = 6 \Rightarrow y = 6/3 = 2$
- $(1, 0)$ satisfies $y \le x/3$.

Procedure

- We plot points at $(0, 0)$ and $(6, 2)$.
- We draw a solid line through the points.
- We shade the region that contains $(1, 0)$.

Graph

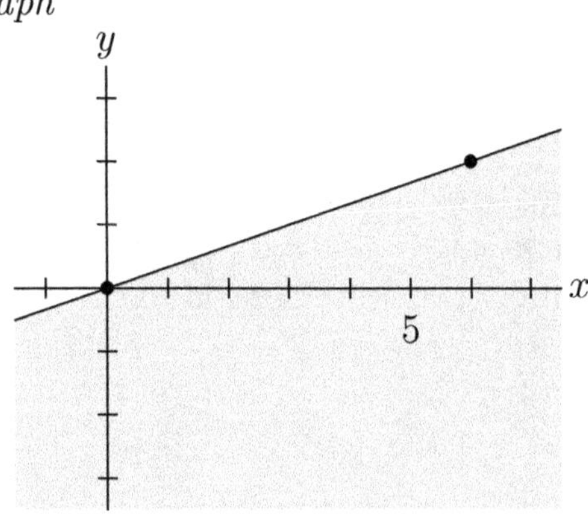

□

C. $3x - 2y > -6$

D. $x < 2$

SOLUTION

Note that $(0, 0)$ satisfies $x < 2$.
Hence we shade the region to the left of the line with equation $x = 2$:

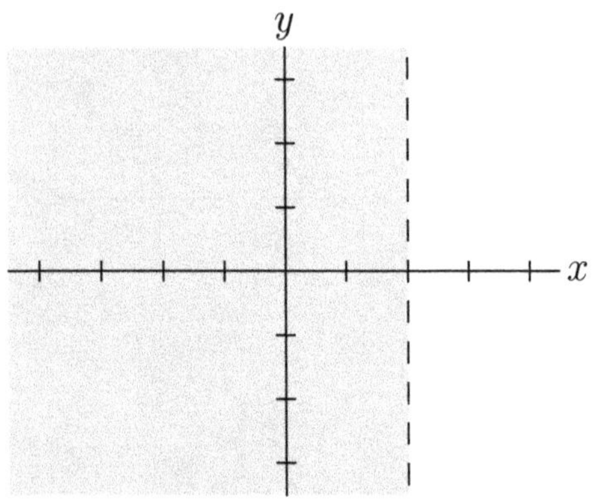

□

E. $y \ge 3$

F. $|x| + |2y| \geq 6$

SOLUTION

Notes

- $|x| + |2y| = 6$

$$\begin{aligned}
&\Leftrightarrow |x| = 6 - |2y| \\
&\qquad = 6 - |2||y| \\
&\Leftrightarrow x = \begin{cases} 6 - 2|y| & \text{if } x \geq 0 \\ -(6 - 2|y|) & \text{if } x < 0 \end{cases} \\
&\qquad = \begin{cases} 6 - 2y & \text{if } x \geq 0,\ y \geq 0 \\ 6 - 2(-y) & \text{if } x \geq 0,\ y < 0 \\ -(6 - 2y) & \text{if } x < 0,\ y \geq 0 \\ -(6 - 2(-y)) & \text{if } x < 0,\ y < 0 \end{cases} \\
&\qquad = \begin{cases} 6 - 2y & \text{if } x \geq 0,\ y \geq 0 \\ 6 + 2y & \text{if } x \geq 0,\ y < 0 \\ -6 + 2y & \text{if } x < 0,\ y \geq 0 \\ -6 - 2y & \text{if } x < 0,\ y < 0 \end{cases}
\end{aligned}$$

- The graph consists of four line segments with endpoints on lines $x = 0$ and $y = 0$.
- $(2, 2)$ and $(4, 1)$ satisfy $x = 6 - 2y$.
- $(2, -2)$ and $(4, -1)$ satisfy $x = 6 + 2y$.
- $(-2, 2)$ and $(-4, 1)$ satisfy $x = -6 + 2y$.
- $(-2, -2)$ and $(-4, -1)$ satisfy $x = -6 - 2y$.
- $(0, 0)$ does not satisfy $|x| + |2y| \geq 6$.

Procedure

- We plot the four pairs of points listed above that satisfy $|x| + |2y| = 6$.
- We draw a solid line segment bounded by the axes through each pair of points.
- We shade the region that does not contain the point $(0, 0)$.

Graph

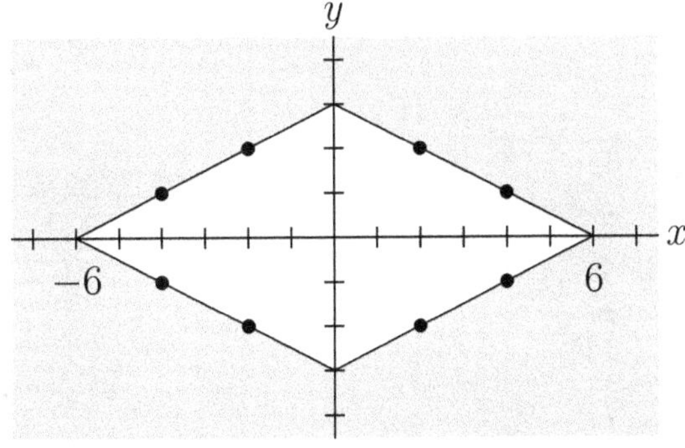

□

G. $x < |y/2| - 5$

EXAMPLE 2
Draw the graph of the set of solutions of each system of linear inequalities.

A. $\begin{cases} y \geq -x+1 \\ y < \ 3x+3 \end{cases}$

SOLUTION
Notes
- if $y = -x + 1$, then:
  - $x = 0 \Rightarrow y = -0 + 1 = 1$
  - $x = 4 \Rightarrow y = -4 + 1 = -3$
- if $y = 3x + 3$, then:
  - $x = 0 \Rightarrow y = 3(0) + 3 = 3$
  - $x = -2 \Rightarrow y = 3(-2) + 3 = -3$
- $(0, 0)$ does not satisfy $y \geq -x + 1$.
- $(0, 0)$ does satisfy $y < 3x + 3$.

Procedure
- We plot points at $(0, 1)$ and $(4, -3)$ and draw a solid line through the points.
- We plot holes at $(0, 3)$ and $(-2, -3)$ and draw a dashed line through the holes.
- We shade the region that lies both:
  - above the graph of $y = -x + 1$ (and does not contain $(0, 0)$);
  - below the graph of $y = 3x + 3$ (and does contain $(0, 0)$).

Graph

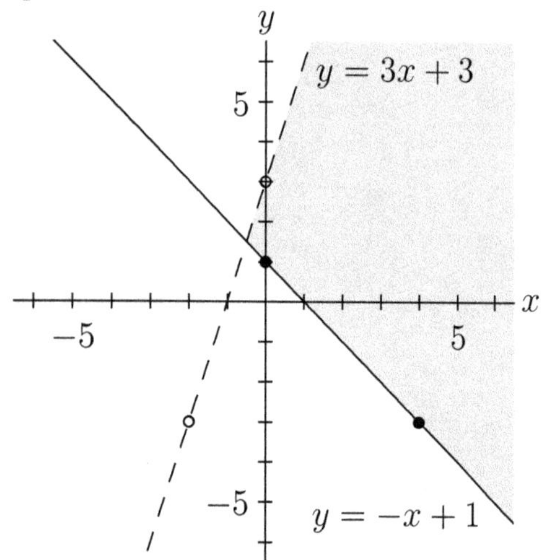

□

B. $\begin{cases} 2x + 4y > -6 \\ x + 2y \leq 4 \end{cases}$

Section 78: Linear Equations in Three Variables

DEFINITION
A *linear equation in three variables* x, y, and z is an equation that can be written equivalently in the form $Ax + By + Cz = D$ for some constants $A, B, C, D \in \mathbf{R}$ with $A \neq 0$ OR $B \neq 0$ OR $C \neq 0$.

PROCEDURE A (preferred)
To solve a system of three linear equations in three variables x, y, and z, we can:

1. eliminate x from the first and second equations and replace the second equation with the equation that results;
2. eliminate x from the first and third equations and replace the third equation with the equation that results;
3. eliminate y from the new second and third equations and replace the new third equation with the equation that results; at this point, the system will assume the "triangular" form:

$$\begin{cases} Ax + By + Cz = D & \leftarrow \text{original first equation} \\ \phantom{Ax + {}} Ey + Fz = G & \leftarrow \text{new second equation} \\ \phantom{Ax + By + {}} Hz = I & \leftarrow \text{newest third equation} \end{cases}$$

4. multiply both sides of the third equation by the multiplicative inverse of the coefficient of the term that contains z and replace the third equation with the equation that results;
5. eliminate z from the second and third equations and replace the second equation with the equation that results;
6. eliminate z from the first and third equations and replace the first equation with the equation that results;
7. multiply both sides of the second equation by the multiplicative inverse of the coefficient of the term that contains y and replace the second equation with the equation that results;
8. eliminate y from the first and second equations and replace the first equation with the equation that results;
9. multiply both sides of the first equation by the multiplicative inverse of the coefficient of the term that contains x and replace the first equation with the equation that results; at this point, the system will assume the "reduced" form:

$$\begin{cases} x \phantom{{}+y+z} = J \\ \phantom{x+{}} y \phantom{{}+z} = K \\ \phantom{x+y+{}} z = L \end{cases}$$

10. state the set of solutions as a set of ordered triples of the form (x, y, z).

PROCEDURE B (alternative)

To solve a system of three linear equations in three variables x, y, and z, we can follow the first three steps of procedure A and then the steps below:

4. solve the third equation for z;
5. substitute the solution of step (4) for z in the second equation;
6. solve the equation resulting from step (5) for y;
7. substitute the solutions of steps (4) and (6) for z and y, respectively, in the first equation;
8. solve the equation resulting from step (7) for x;
9. state the set of solutions as a set of ordered triples of the form (x, y, z).

NOTATION

- In procedures A and B, we replace one equation with another multiple times. We will use the notation "$aR_m + bR_n$" next to a new equation in a system to indicate its source. For example, suppose within a given system that we add four times each side of the first equation to three times the corresponding side of the second equation, equate these sums, and replace the second equation in the system with this new equation. Then next to the new equation, we will write $4R_1 + 3R_2$, as shown below:

$$\begin{cases} 3x + 5y - z = 21 \\ -4x - 3y + 7z = -5 \\ 5x - y - 3z = 10 \end{cases} \Rightarrow \begin{cases} 3x + 5y - z = 21 \\ 11y + 17z = 69 \quad 4R_1 + 3R_2 \\ 5x - y - 3z = 10 \end{cases}$$

since

$$\begin{array}{rcrcr} 3x + 5y - z = 21 & \Rightarrow & 4(3x + 5y - z) = 4(21) & \Rightarrow & 12x + 20y - 4z = 84 \\ -4x - 3y + 7z = -5 & \Rightarrow & 3(-4x - 3y + 7z) = 3(-5) & \Rightarrow & -12x - 9y + 21z = -15 \\ \hline & & & & 11y + 17z = 69 \end{array}$$

- We denote the set of all ordered triples of real numbers by $\mathbf{R}\times\mathbf{R}\times\mathbf{R}$ or $\mathbf{R}^3$. Symbolically, we write that $\mathbf{R}^3 = \mathbf{R}\times\mathbf{R}\times\mathbf{R} = \{\, (x, y, z) \mid x, y, z \in \mathbf{R} \,\}$.

COMMENTS

- A solution to an equation in three variables is an ordered triple of numbers.
- The graph of a linear equation in three variables is a plane in three-dimensional space.
- To convert a system into triangular form most expediently, we may wish to rearrange the order of the equations. In general, we should immediately:
  - place an equation without both x and y in third position;
  - place an equation without an x in second or third position;
  - place the equation with the lowest nonzero coefficient of x in first position.
- We can use extended versions of the procedures in this section to solve systems of n linear equations in n variables where $n \geq 4$.
- The equations that result from steps (1), (2), or (3) of the procedures may contain no variables. These equations will be:
  - false if the system is inconsistent;
  - true if the equations in the system are dependent.

EXAMPLE 1

Solve each system of equations.

A. $\begin{cases} 4x + 5y - z = 21 \\ -2x - 3y + 7z = -5 \\ 5x - y - 3z = 10 \end{cases}$

B. $\begin{cases} 3x + 3y - 2z = -13 \\ 8x + 7y + 3z = 7 \\ -4x - 5y - z = -1 \end{cases}$

SOLUTION

Solution

$$\begin{cases} 4x + 5y - z = 21 \\ -2x - 3y + 7z = -5 \\ 5x - y - 3z = 10 \end{cases}$$

$$\Rightarrow \begin{cases} 4x + 5y - z = 21 & R_1 \\ -y + 13z = 11 & R_1 + 2R_2 \\ 29y + 7z = 65 & 5R_1 - 4R_3 \end{cases}$$

$$\Rightarrow \begin{cases} 4x + 5y - z = 21 & R_1 \\ -y + 13z = 11 & R_2 \\ 384z = 384 & 29R_2 + R_3 \end{cases}$$

$$\Rightarrow \begin{cases} 4x + 5y - z = 21 & R_1 \\ -y + 13z = 11 & R_2 \\ z = 1 & R_3/384 \end{cases}$$

$$\Rightarrow \begin{cases} 4x + 5y = 22 & R_1 + R_2 \\ -y = -2 & R_2 - 13R_3 \\ z = 1 & R_3 \end{cases}$$

$$\Rightarrow \begin{cases} 4x + 5y = 22 & R_1 \\ y = 2 & R_2/(-1) \\ z = 1 & R_3 \end{cases}$$

$$\Rightarrow \begin{cases} 4x = 12 & R_1 - 5R_2 \\ y = 2 & R_2 \\ z = 1 & R_3 \end{cases}$$

$$\Rightarrow \begin{cases} x = 3 & R_1/4 \\ y = 2 & R_2 \\ z = 1 & R_3 \end{cases}$$

Set of solutions: $\{(3, 2, 1)\}$

Checks

- $4(3) + 5(2) - 1 \stackrel{?}{=} 21$ ✓
- $-2(3) - 3(2) + 7(1) \stackrel{?}{=} -5$ ✓
- $5(3) - 2 - 3(1) \stackrel{?}{=} 10$ ✓

□

C. $\begin{cases} 3x - 8y - z = -26 \\ 3x - 8y + 6z = 16 \\ 4x + 2y + 7z = 28 \end{cases}$

SOLUTION

Solution

$$\begin{cases} 3x - 8y - z = -26 \\ 3x - 8y + 6z = 16 \\ 4x + 2y + 7z = 28 \end{cases}$$

$$\Rightarrow \begin{cases} 3x - 8y - z = -26 & R_1 \\ 7z = 42 & R_2 - R_1 \\ 38y + 25z = 188 & 3R_3 - 4R_1 \end{cases}$$

$$\Rightarrow \begin{cases} 3x - 8y - z = -26 & R_1 \\ 38y + 25z = 188 & R_3 \\ 7z = 42 & R_2 \end{cases}$$

$$\Rightarrow \begin{cases} 3x - 8y - z = -26 & R_1 \\ 38y + 25z = 188 & R_2 \\ z = 6 & R_3/7 \end{cases}$$

$$\Rightarrow \begin{cases} 3x - 8y = -20 & R_1 + R_2 \\ 38y = 38 & R_2 - 25R_3 \\ z = 6 & R_3 \end{cases}$$

$$\Rightarrow \begin{cases} 3x - 8y = -20 & R_1 \\ y = 1 & R_2/38 \\ z = 6 & R_3 \end{cases}$$

$$\Rightarrow \begin{cases} 3x = -12 & R_1 + 8R_2 \\ y = 1 & R_2 \\ z = 6 & R_3 \end{cases}$$

$$\Rightarrow \begin{cases} x = -4 & R_1/3 \\ y = 1 & R_2 \\ z = 6 & R_3 \end{cases}$$

Set of solutions: $\{(-4, 1, 6)\}$

Comment

We interchanged equations R_2 and R_3 in the third step by necessity; we could not convert to triangular form otherwise.

Checks

- $3(-4) - 8(1) - 6 \stackrel{?}{=} -26$ ✓
- $3(-4) - 8(1) + 6(6) \stackrel{?}{=} 16$ ✓
- $4(-4) + 2(1) + 7(6) \stackrel{?}{=} 28$ ✓

□

D. $\begin{cases} 2x - 3y + 7z = -23 \\ x + 5y - 2z = 32 \\ 3x + y + 5z = 4 \end{cases}$

[Hint: Interchange the first and second equations before proceeding further.]

EXAMPLE 2

Set up a system of equations for each word problem and then solve it.

A. Find the measures of all three angles of a triangle if:
- the sum of the first and three times the second is equal to the third and
- the sum of two times the first and four times the second is equal to the sum of the third and 80 degrees.

SOLUTION

Declarations

Let a, b, and c denote the measures of the first, second, and third angles, respectively.

System

$$\begin{cases} a + b + c = 180 \\ a + 3b = c \\ 2a + 4b = c + 80 \end{cases}$$

$$\Rightarrow \begin{cases} a + b + c = 180 \\ a + 3b - c = 0 \\ 2a + 4b - c = 80 \end{cases}$$

$$\Rightarrow \begin{cases} a + b + c = 180 & R_1 \\ 2b - 2c = -180 & R_2 - R_1 \\ 2b - 3c = -280 & R_3 - 2R_1 \end{cases}$$

$$\Rightarrow \begin{cases} a + b + c = 180 & R_1 \\ 2b - 2c = -180 & R_2 \\ c = 100 & R_2 - R_3 \end{cases}$$

$$\Rightarrow \begin{cases} a + b = 80 & R_1 - R_3 \\ 2b = 20 & R_2 + 2R_3 \\ c = 100 & R_2 - R_3 \end{cases}$$

$$\Rightarrow \begin{cases} a + b = 80 & R_1 \\ b = 10 & R_2/2 \\ c = 100 & R_3 \end{cases}$$

$$\Rightarrow \begin{cases} a = 70 & R_1 - R_2 \\ b = 10 & R_2 \\ c = 100 & R_3 \end{cases}$$

Conclusion

The measures are 70°, 10°, and 100°.

□

B. A slave transports water from a stream to a pool using buckets of various sizes. The total amounts (in gallons) of water he transports on three consecutive days are 106, 65, and 29. The numbers of trips he makes those days with buckets A, B, and C are as follows:

first day — A: 6 B: 7 C: 6
second day — A: 3 B: 4 C: 5
third day — A: 2 B: 2 C: 1

Find the capacity of each bucket.

EXAMPLE 3

State the set of solutions for each system of dependent equations using set builder notation.

A. $$\begin{cases} x - 3y + 2z = 10 \\ 2x - 5y + 7z = 25 \\ 5x - 14y + 13z = 55 \end{cases}$$

SOLUTION

Values

$$\begin{cases} x - 3y + 2z = 10 \\ 2x - 5y + 7z = 25 \\ 5x - 14y + 13z = 55 \end{cases}$$

$$\Rightarrow \begin{cases} x - 3y + 2z = 10 & R_1 \\ y + 3z = 5 & R_2 - 2R_1 \\ y + 3z = 5 & R_3 - 5R_1 \end{cases}$$

$$\Rightarrow \begin{cases} x - 3y + 2z = 10 & R_1 \\ y + 3z = 5 & R_2 \\ 0 = 0 & R_3 - R_2 \end{cases}$$

$$\Rightarrow \begin{cases} x + 11z = 25 & R_1 + 3R_2 \\ y + 3z = 5 & R_2 \\ 0 = 0 & R_3 \end{cases}$$

Notes

- $x + 11z = 25 \Leftrightarrow x = 25 - 11z$
- $y + 3z = 5 \Leftrightarrow y = 5 - 3z$

Set of Solutions

Set of solutions:

$\{ (x, y, z) \in \mathbf{R}^3 \mid x = 25 - 11z,\ y = 5 - 3z \}$

□

B. $$\begin{cases} x - 3y + 2z = 10 \\ 2x - 6y + 4z = 20 \\ 5x - 15y + 10z = 50 \end{cases}$$

SOLUTION

Values

$$\begin{cases} x - 3y + 2z = 10 \\ 2x - 6y + 4z = 20 \\ 5x - 15y + 10z = 50 \end{cases}$$

$$\Rightarrow \begin{cases} x - 3y + 2z = 10 & R_1 \\ 0 = 0 & R_2 - 2R_1 \\ 0 = 0 & R_3 - 5R_1 \end{cases}$$

Note

$x - 3y + 2z = 10 \Leftrightarrow x = 10 + 3y - 2z$

Set of Solutions

The set of solutions is

$\{ (x, y, z) \in \mathbf{R}^3 \mid x = 10 + 3y - 2z \}$.

□

Section 79: Symmetry

DEFINITIONS

- A graph is *symmetric with respect to the line* l if for every point Q on the graph but not on l there exists a point R on the graph such that l is the perpendicular bisector of QR.
- A graph is *symmetric with respect to the point* P if for every point Q on the graph but not on P there exists a point R on the graph such that P is the midpoint of QR.

GRAPHICAL RELATIONSHIPS

A graph is symmetric with respect to:

- the x-axis if it comprises the same points as its reflection through the x-axis;
- the y-axis if it comprises the same points as its reflection through the y-axis;
- the origin if it comprises the same points as:
  - the reflection through the y-axis of its reflection through the x-axis or
  - the reflection through the x-axis of its reflection through the y-axis.

GRAPHICAL RELATIONSHIPS

The following statements regarding the graph of an equation are equivalent:

- The graph is symmetric with respect to the x-axis.
- (a, b) satisfies the equation if and only if $(a, -b)$ satisfies the equation.
- The point (a, b) lies on the graph if and only if the point $(a, -b)$ lies on the graph.
- The portion of the graph that lies above the x-axis is the mirror image of the portion of the graph that lies below the x-axis.

GRAPHICAL RELATIONSHIPS

The following statements regarding the graph of an equation are equivalent:

- The graph is symmetric with respect to the y-axis.
- (a, b) satisfies the equation if and only if $(-a, b)$ satisfies the equation.
- The point (a, b) lies on the graph if and only if the point $(-a, b)$ lies on the graph.
- The portion of the graph that lies to the right of the y-axis is the mirror image of the portion of the graph that lies to the left of the y-axis.

GRAPHICAL RELATIONSHIPS

The following statements regarding the graph of an equation are equivalent:

- The graph is symmetric with respect to the origin.
- (a, b) satisfies the equation if and only if $(-a, -b)$ satisfies the equation.
- The point (a, b) lies on the graph if and only if the point $(-a, -b)$ lies on the graph.

GRAPHICAL RELATIONSHIPS

The following statements regarding the graph of an equation are equivalent:

- The graph is symmetric with respect to the line $y = x$.
- (a, b) satisfies the equation if and only if (b, a) satisfies the equation.
- The point (a, b) lies on the graph if and only if the point (b, a) lies on the graph.

COMMENTS

- A graph can have more than one kind of symmetry or no kind of symmetry.
- In many equations of graphs that are symmetric with respect to the x-axis, the variable y serves as the base for an even exponent or as the argument of an absolute value.
- In many equations of graphs that are symmetric with respect to the y-axis, the variable x serves as the base for an even exponent or as the argument of an absolute value.
- We may speak loosely of two points having a symmetry when we mean technically that the graph that comprises those two points has that symmetry. For example, we might say that the points with coordinates (c, d) and $(c, -d)$ are symmetric with respect to the x-axis.

EXAMPLE 1

The graphs below exemplify symmetries with respect to the x-axis, y-axis, and origin.

symmetry with respect to the x-axis

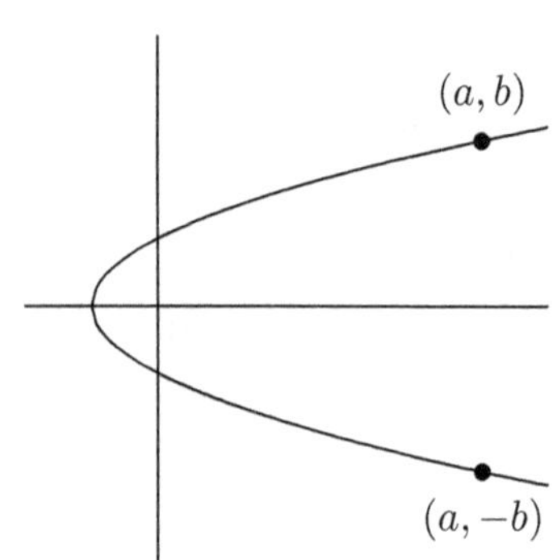

symmetry with respect to the y-axis

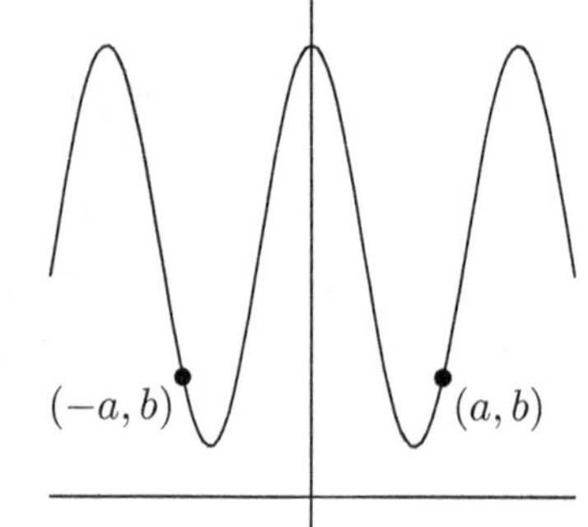

symmetry with respect to the origin

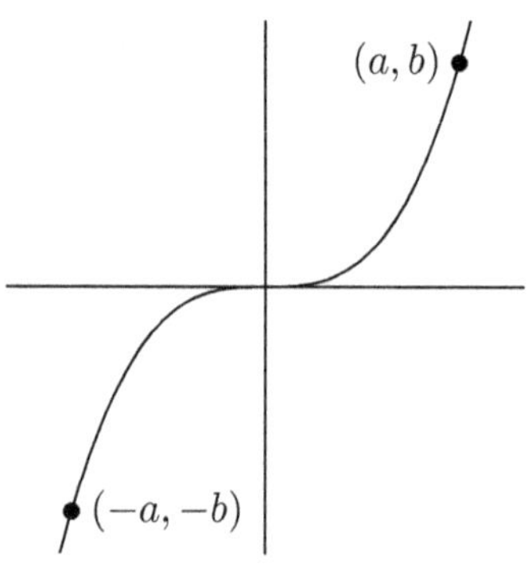

□

EXAMPLE 2

Prove that the graph of $x^2 + y^2 = 9$ is symmetric with respect to the origin.

PROOF

Suppose $(a, b) \in \mathbf{R}^2$.

$(-a, -b)$ satisfies $x^2 + y^2 = 9$

$\Leftrightarrow (-a)^2 + (-b)^2 = 9$

$\Leftrightarrow (-1 \cdot a)^2 + (-1 \cdot b)^2 = 9$

$\Leftrightarrow (-1)^2(a)^2 + (-1)^2(b)^2 = 9$

$\Leftrightarrow (1)(a^2) + (1)(b^2) = 9$

$\Leftrightarrow a^2 + b^2 = 9$

$\Leftrightarrow (a, b)$ satisfies $x^2 + y^2 = 9$

Therefore, the graph of $x^2 + y^2 = 9$ is symmetric with respect to the origin.

⊡

EXAMPLE 3

Prove that the graph of $y = x^4 + 7$ is symmetric with respect to the y-axis.

Example 4
Provide a counterexample to show that the graph of $y = x^3 - 1$ is not symmetric with respect to the y-axis.

Counterexample
Clearly $(2, 7)$ satisfies $y = x^3 - 1$
but $(-2, 7)$ does not.
Therefore, the graph of $y = x^3 - 1$ is not symmetric with respect to the y-axis.

□

Example 5
Provide a counterexample to show that the graph of $y = \sqrt{x+5}$ is not symmetric with respect to the origin.

Example 6
Draw the graph of each equation and determine whether the graph has any symmetries with respect to the x-axis, y-axis, or origin.

A. $y = -5x^2$

Solution
Table of Values

x	y	solution
-2	$-5(-2)^2 = -20$	$(-2, -20)$
-1	$-5(-1)^2 = -5$	$(-1, -5)$
0	$-5(0)^2 = 0$	$(0, 0)$
1	$-5(1)^2 = -5$	$(1, -5)$
2	$-5(2)^2 = -20$	$(2, -20)$

Graph

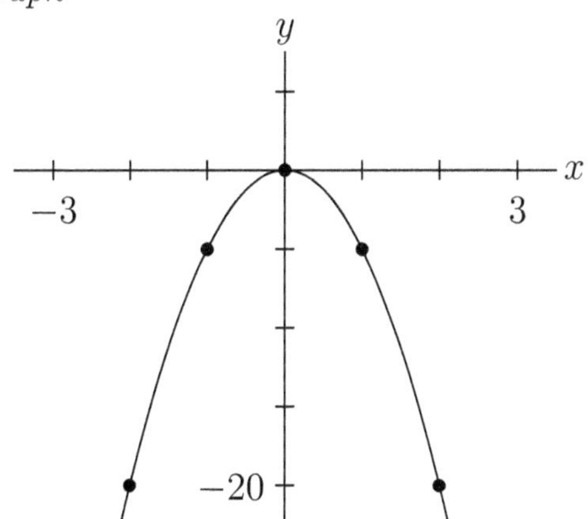

Symmetries
- not symmetric w.r.t. the x-axis
- symmetric w.r.t. the y-axis
- not symmetric w.r.t. the origin

□

B. $xy = 8$

C. $x = |y|$

Solution

Table of Values

y	x	solution
-3	$\|-3\| = 3$	$(3, -3)$
-2	$\|-2\| = 2$	$(2, -2)$
-1	$\|-1\| = 1$	$(1, -1)$
0	$\|0\| = 0$	$(0, 0)$
1	$\|1\| = 1$	$(1, 1)$
2	$\|2\| = 2$	$(2, 2)$
3	$\|3\| = 3$	$(3, 3)$

Graph

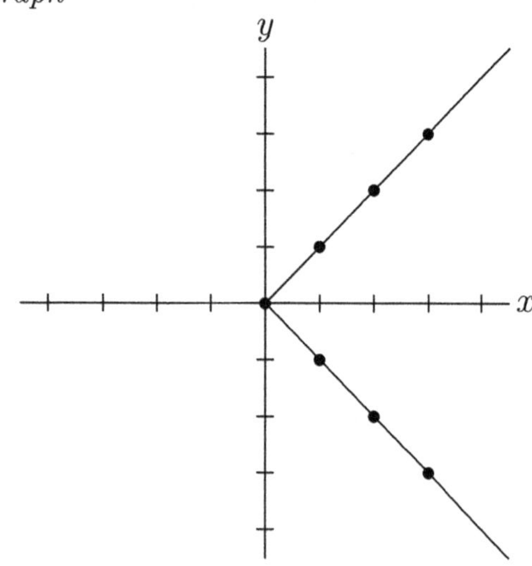

Symmetries

- symmetric w.r.t. the x-axis
- not symmetric w.r.t. the y-axis
- not symmetric w.r.t. the origin

□

D. $x = 7 - y$

Part Two

Section 80: Functions

Definitions

- A *function* is a correspondence between two sets that assigns to each element in one set (the *domain*) exactly one element in the other set (the *codomain*).
- The *image* of an element in the domain of a function is the element in the codomain of the function that the function assigns to it.
- The *range* of a function is the set of all images of elements in its domain.
- A *parameter* is a variable that represents an element in the domain of a function.
- A *piecewise-defined function* is a function in which the formula used to determine the image of an element in its domain depends upon the value of that element.

Notation

- If f is a function with domain X and codomain Y, then we write "$f : X \to Y$".
- We denote the image that a function f assigns to an element x in its domain by "$f(x)$".

Terminology

Suppose the function f is defined by formula through the parameter x.
If the formula equates $f(x)$ to a:

- constant, then we say "$f(x)$ is a constant";
- polynomial, then we say "$f(x)$ is a polynomial";
- (kind of expression), then we say "$f(x)$ is a (kind of expression)".

Comments

- We read the expression "$f(x)$" aloud as "f of x".
- The range of a function is a subset of its codomain. Not every element in the codomain of a function need be assigned to an element in its domain.
- We can define a function:
  - with explicit assignments of elements in its codomain to elements in its domain,
  - with ordered pairs,
  - with a figure, or
  - with a formula.
- To determine $f(c)$ where f is defined by a formula, we substitute c for x in the expression provided for $f(x)$ and simplify the result.
- The notation "$x(3)$" means "x times 3" if x is a variable, but the notation "$f(3)$" does **not** mean "f times 3" if f is a function; rather, "$f(3)$" is an indivisible term that denotes the image of 3 under the correspondence f.

Example 1

A. The function C defined by the formula $C(r) = 2\pi r$ has the parameter r.
B. The function f defined by the formula $f(n) = 3n^2$ has the parameter n.
C. The function h defined by the formula $h(m) = m + 1$ has the parameter m.
D. The function P defined by the formula $P(x, y) = 2x + 2y$ has the parameters x and y.

□

EXAMPLE 2

The figure at right defines a function $f : A \to B$, where $A = \{2, 3, 4, 8\}$ and $B = \{1, 2, 7, 8, 9\}$. Here:

- the domain of f is A
- the codomain of f is B
- the range of f is $\{1, 7, 9\}$

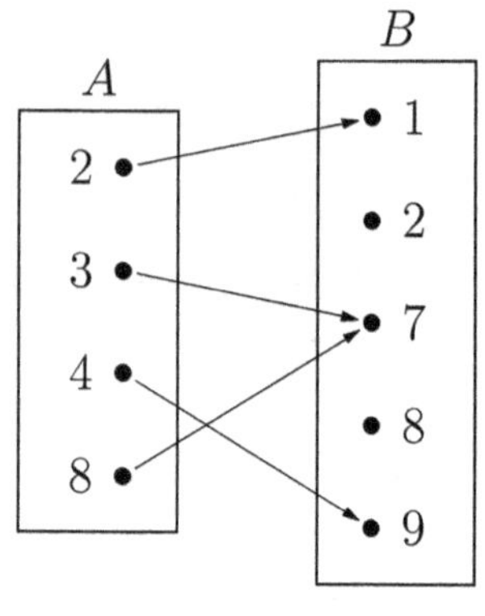

We could also have defined f by providing:

- the assignments $f(2) = 1$, $f(3) = 7$, $f(4) = 9$, $f(8) = 7$
- the set of ordered pairs

$$\{\,(x, f(x)) \mid x \in A\,\} = \{(2,1), (3,7), (4,9), (8,7)\}$$

□

EXAMPLE 3

The function $g : \mathbf{R} \to \mathbf{R}$ is defined by:

$$g(x) = 3x + 5$$

Find $g(-1)$, $g(\pi)$, and $g(c + 2)$.

SOLUTIONS

- $g(-1) = 3(-1) + 5$
  $= 2$
- $g(\pi) = 3\pi + 5$
- $g(c + 2) = 3(c + 2) + 5$
  $= 3c + 6 + 5$
  $= 3c + 11$

□

EXAMPLE 4

The function $h : \mathbf{R} \to \mathbf{R}$ is defined by:

$$h(x) = x^2 + 3x - 1$$

Find $h(7)$, $h(1/2)$, and $h(t)$.

EXAMPLE 5

The function $f : \mathbf{R} \to \mathbf{R}$ is defined piecewise by:

$$f(x) = \begin{cases} x + 4 & \text{if } x \leq 5 \\ 3x & \text{if } x > 5 \end{cases}$$

Find $f(8)$, $f(-3)$, and $f(5)$.

SOLUTIONS

- $f(8) = 3(8)$ (since $8 > 5$)
  $= 24$
- $f(-3) = (-3) + 4$ (since $-3 \leq 5$)
  $= 1$
- $f(5) = 5 + 4$ (since $5 \leq 5$)
  $= 9$

□

EXAMPLE 6

The function $p : \mathbf{R} \to \mathbf{R}$ is defined piecewise by:

$$p(x) = \begin{cases} 5 & \text{if } x \leq -1 \\ x - 2 & \text{if } -1 < x \leq 1 \\ x^2 & \text{if } x > 1 \end{cases}$$

Find $p(0)$, $p(-10)$, $p(7)$, and $p(1)$.

EXAMPLE 7
Find $[f(a+h) - f(a)]/h$ for the given definition of f.

A. $f(x) = 1/x$

SOLUTION

$$
\begin{aligned}
&[f(a+h) - f(a)]/h \\
&\quad = (1/h) \cdot [f(a+h) - f(a)] \\
&\quad = \frac{1}{h}\left[\frac{1}{a+h} - \frac{1}{a}\right] \\
&\quad = \frac{1}{h}\left[\frac{1}{a+h} \cdot \frac{a}{a} - \frac{1}{a} \cdot \frac{a+h}{a+h}\right] \\
&\quad = \frac{1}{h}\left[\frac{a}{a(a+h)} - \frac{a+h}{a(a+h)}\right] \\
&\quad = \frac{1}{h}\left[\frac{a-(a+h)}{a(a+h)}\right] \\
&\quad = \frac{1}{h}\left[\frac{\cancel{a} - \cancel{a} - h}{a(a+h)}\right] \\
&\quad = \frac{1}{\cancel{h}}\left[\frac{-\cancel{h}}{a(a+h)}\right] \\
&\quad = \frac{-1}{a(a+h)}
\end{aligned}
$$

□

B. $f(x) = x^2 + 4$

Section 81: Relations

Definitions

- A *relation* is a set of ordered pairs.
- The *domain* of a relation is the set of the first elements of the pairs.
- The *range* of a relation is the set of the second elements of the pairs.

Comment

Recall that we can define functions in terms of ordered pairs; thus a function is a kind of relation. Specifically, a function is a relation in which no two ordered pairs in the relation have the same first element.

Example 1

Let r denote the set of ordered pairs $\{(3,8),(4,11),(5,6),(6,13),(4,2),(8,10),(15,15)\}$. Then r constitutes a relation with domain $\{3,4,5,6,8,15\}$ and range $\{2,6,8,10,11,13,15\}$. This relation r is not a function since the element 4 of the domain is assigned two elements (namely 11 and 2) of the range.

□

Section 82: Domains and Ranges of Functions

Procedure A

To determine the domain of a function defined by formula, we:

1. identify all real numbers that, upon substitution into the formula, produce either:
    - a denominator of zero in a fraction or
    - a negative radicand under a radical with even index;
2. state the domain as all real numbers except those values found in step (1).

Procedure B

To determine the range of a function defined by formula that has domain **R**, we:

1. equate the variable y to the image of the parameter x;
2. solve the equation of step (1) for x in terms of y;
3. identify all real numbers that, upon substitution into the equation of step (2) for y, produce either:
    - a denominator of zero in a fraction or
    - a negative radicand within a radical with even index;
4. state the range as all real numbers except those values found in step (3).

Notation

We will usually denote the domain and range of a function f by $\mathbb{D}$ and $\mathbb{R}$, respectively, or by $\mathbb{D}_f$ and $\mathbb{R}_f$ when necessary to avoid ambiguity.

Comments

- The assumption underlying procedure A is that unless the elements of a domain are explicitly identified, the domain is the set of all real numbers with a real image.
- We assume the codomain of a function is all real numbers unless stated otherwise.
- Excluding from a domain all real numbers that produce a negative radicand is equivalent to including in a domain only real numbers that do not produce a negative radicand.
- If $f(x)$ is a polynomial, then the domain of f is all real numbers, since polynomials contain neither fractions with variables in denominators nor radicals with variables in radicands.
- We can find a range effectively through procedure B because by solving for x in terms of y, we guarantee for every value that we can substitute legitimately for y the existence of an element x in the domain to which the function assigns that value. Since every real number that we can substitute for y in the equation of step (2) must then be the image of some element in the domain and every real number that we cannot substitute for y cannot be the image of any element in the domain, all real numbers that we can substitute must constitute the range of the function.
- We should not use procedure B if the domain of the function is not all real numbers or if we cannot complete step (2). In such cases, we may try to find the range by observing that:
    - the lower bound of powers with even exponents and radicals with even indices is 0;
    - the bounds of the range of a function that is monotonic (see page 250) on its domain will be the the values of the function at the bounds of its domain.

EXAMPLE 1

Determine the domain $\mathbb{D}$ of the function with the given definition.

A. $f(x) = 1/(5x - 2)$

SOLUTION

Condition for Exclusion from Domain

$$5x - 2 = 0 \Leftrightarrow 5x = 2$$
$$\Leftrightarrow 5x/5 = 2/5$$
$$\Leftrightarrow x = 2/5$$

Domain

$$\mathbb{D} = \{\, x \in \mathbf{R} \mid x \neq 2/5 \,\}$$
$$= (-\infty, 2/5) \cup (2/5, \infty)$$

□

B. $f(x) = 7/[(x+4)(x-1)]$

C. $g(x) = \sqrt{10 - 2x}$

SOLUTION

Condition for Inclusion in Domain

$$10 - 2x \geq 0 \Leftrightarrow -2x \geq -10$$
$$\Leftrightarrow -2x/(-2) \leq -10/(-2)$$
$$\Leftrightarrow x \leq 5$$

Domain

$$\mathbb{D} = \{\, x \in \mathbf{R} \mid x \leq 5 \,\}$$
$$= (-\infty, 5]$$

□

D. $P(r) = \sqrt[4]{4r + 8}$

E. $g(x) = \sqrt[6]{3m + 7}$

SOLUTION

Condition for Exclusion from Domain

(We apply the condition for exclusion rather than inclusion only to demonstrate another approach to finding a domain.)

$$3m + 7 < 0 \Leftrightarrow 3m < -7$$
$$\Leftrightarrow 3m/3 < -7/3$$
$$\Leftrightarrow m < -7/3$$

Domain

$$\mathbb{D} = \{\, m \in \mathbf{R} \mid m \not< -7/3 \,\}$$
$$= \{\, m \in \mathbf{R} \mid m \geq -7/3 \,\}$$
$$= [-7/3, \infty)$$

□

F. $P(r) = \sqrt{-5x - 11}$

G. $h(x) = \sqrt[8]{(x-2)(x-4)}$

SOLUTION

Zeros of the Factors

- $x - 2 = 0 \Leftrightarrow x = 2$
- $x - 4 = 0 \Leftrightarrow x = 4$

Sign Chart

$(x-2)(x-4)$	$+$	$-$	$+$
$x-2$	$-$	$+$	$+$
$x-4$	$-$	$-$	$+$
	2	4	

Domain

$$\mathbb{D} = \{\, x \in \mathbf{R} \mid (x-2)(x-4) \geq 0 \,\} = (-\infty, 2] \cup [4, \infty)$$

□

H. $g(x) = \sqrt{3x^2 + 2x - 5}$

I. $h(x) = \sqrt{(x^2 - 8x + 7)/(x+2)}$

SOLUTION

Note

$h(x) = \sqrt{(x-1)(x-7)/(x+2)}$

Zeros of the Factors

- $x - 1 = 0 \Leftrightarrow x = 1$
- $x - 7 = 0 \Leftrightarrow x = 7$
- $x + 2 = 0 \Leftrightarrow x = -2$

Sign Chart

product	$-$	$+$	$-$	$+$
$x-1$	$-$	$-$	$+$	$+$
$x-7$	$-$	$-$	$-$	$+$
$x+2$	$-$	$+$	$+$	$+$
	-2	1	7	

Domain

$$\mathbb{D} = \{x \in \mathbf{R} \mid (x-1)(x-7)/(x+2) \geq 0 \text{ AND } x \neq -2\} = (-2, 1] \cup [7, \infty)$$

□

J. $f(x) = \sqrt{3x+6} + 5\sqrt{7-x}$

EXAMPLE 2
Determine the range ℝ of the function with the given definition.

A. $F(x) = x^3$

SOLUTION
Note
$F(x) = x^3 \Leftrightarrow y = x^3$ where $y = F(x)$
$\Leftrightarrow x = \sqrt[3]{y}$

Range
$\mathbb{R} = \mathbf{R}$

□

B. $g(x) = 1/x$

C. $f(x) = x^2 + 5$

SOLUTION
Note
$f(x) = x^2 + 5 \Leftrightarrow y = x^2 + 5$
where $y = f(x)$
$\Leftrightarrow y - 5 = x^2$
$\Leftrightarrow x = \pm\sqrt{y-5}$

Condition for Inclusion in Range
$y - 5 \geq 0 \Leftrightarrow y \geq 5$

Range
$\mathbb{R} = \{\, y \in \mathbf{R} \mid y \geq 5 \,\}$
$= [5, \infty)$

□

D. $h(x) = (x + 7)/(4x - 5)$

E. $g(x) = x + 2$ where $g : (5, 10] \to \mathbf{R}$

SOLUTION
Implication
$5 < x \leq 10$
$\Rightarrow 5 + 2 < x + 2 \leq 10 + 2$
$\Rightarrow 7 < g(x) \leq 12$

Range
$\mathbb{R} = (7, 12]$

□

F. $f(x) = \sqrt{x - 3}$

G. $h(x) = -2x^2 - 5$ where $h : [-7, 4] \to \mathbf{R}$

SOLUTION

Notes

- $-7 \le x \le 0$
  $\Rightarrow 0 \le x^2 \le (-7)^2$
  $\Rightarrow 0 \le x^2 \le 49$
  $\Rightarrow -2(0) \ge -2(x^2) \ge -2(49)$
  $\Rightarrow 0 \ge -2x^2 \ge -98$
  $\Rightarrow 0 - 5 \ge -2x^2 - 5 \ge -98 - 5$
  $\Rightarrow -5 \ge h(x) \ge -103$

- $0 \le x \le 4$
  $\Rightarrow 0 \le x^2 \le (4)^2$
  $\Rightarrow 0 \le x^2 \le 16$
  $\Rightarrow -2(0) \ge -2(x^2) \ge -2(16)$
  $\Rightarrow 0 \ge -2x^2 \ge -32$
  $\Rightarrow 0 - 5 \ge -2x^2 - 5 \ge -32 - 5$
  $\Rightarrow -5 \ge h(x) \ge -37$

Range

$\mathbb{R} = [-103, -5] \cup [-37, -5]$
$= [-103, -5]$

□

H. $g(x) = 6x + 13$ where $f : [-10, -1] \to \mathbf{R}$

I. $f(x) = x^2 - 6x + 1$

SOLUTION

Note

$x^2 - 6x + 1$
$= x^2 - 6x + ((-6)/2)^2 - ((-6)/2)^2 + 1$
$= x^2 - 6x + (-3)^2 - (-3)^2 + 1$
$= (x^2 - 6x + (-3)^2) - (-3)^2 + 1$
$= (x - 3)^2 - 9 + 1$
$= (x - 3)^2 - 8$

Implication

$0 \le (x - 3)^2$
$\Rightarrow 0 - 8 \le (x - 3)^2 - 8$
$\Rightarrow -8 \le x^2 - 6x + 1$
$\Rightarrow -8 \le g(x)$

Range

$\mathbb{R} = [-8, \infty)$

□

J. $h(x) = -2x^2 - 4x + 11$

Section 83: Formulations of Functions

Definitions

- An *independent* variable is a variable that can assume an arbitrary value.
- A *dependent* variable is a variable whose value depends on the value of another variable.

Comments

- If the the equation $y = f(x)$ relates the variables x and y, then:
  - x is the *independent* variable and
  - y is the *dependent* variable.
- We should not refer to $f(x)$ as a function; the symbol f denotes the function, whereas the symbol $f(x)$ denotes the image of x under the correspondence f.

Example 1

For each equation that relates x and y, determine the function f for which $y = f(x)$.

A. $x + y = 5$

Solution

$x + y = 5 \Leftrightarrow y = -x + 5$

So $f(x) = -x + 5$.

□

B. $x/y = 10$

Example 2

Define a function A such that $A(r)$ represents the area of a circle of radius r and state the domain $\mathbb{D}$ of that function.

Solution

- We define A by $A(r) = \pi r^2$.
- $\mathbb{D} = [0, \infty)$
  (since radii must be nonnegative)

□

Example 3

Define a function P such that $P(x)$ represents the perimeter of a square with sides of length x and state the domain $\mathbb{D}$ of that function.

Example 4

Define a function P such that $P(w)$ represents the perimeter of a rectangle of length 4 and width w.

Solution

$$\begin{aligned} \text{perimeter} &= 2 \cdot 4 + 2 \cdot w \\ &= 8 + 2w \end{aligned}$$

We define P by $P(w) = 8 + 2w$.

□

Example 5

Define a function V such that $V(h)$ represents the volume of a cone of radius 5 and height h.

EXAMPLE 6
Define a function h such that $h(r)$ represents the height of a cylinder of radius r and volume 8.

SOLUTION
Let y denote the height of the cylinder.
volume $= \pi(\text{radius})^2(\text{height})$
$\Rightarrow 8 = \pi r^2 y$
$\Rightarrow 8/(\pi r^2) = (\pi r^2 y)/(\pi r^2)$
$\Rightarrow 8/(\pi r^2) = y$
We define h by $h(r) = 8/(\pi r^2)$.

□

EXAMPLE 7
Define a function b such that $b(h)$ represents the length of the base of a triangle of height h and area 10.

EXAMPLE 8
A man six-feet tall stands near a light at the top of a nine-foot pole. Define a function S where $S(d)$ denotes the length of the man's shadow if he stands a distance d from the pole (see figure).

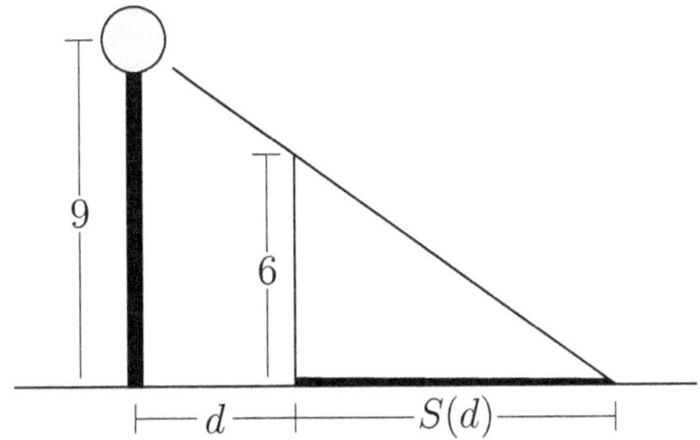

SOLUTION
Let y denote the length of the shadow.
By properties of similar triangles:
$y/6 = (d+y)/9$
$\Rightarrow 54(y/6) = 54((d+y)/9)$
$\Rightarrow 9y = 6(d+y)$
$= 6d + 6y$
$\Rightarrow 9y - 6y = 6d + 6y - 6y$
$\Rightarrow 3y = 6d$
$\Rightarrow 3y/3 = 6d/3$
$\Rightarrow y = 2d$
We define S by $S(d) = 2d$.

□

EXAMPLE 9
Suppose a circle circumscribes a square. Define a function A where $A(r)$ denotes the area of the square if the radius of the circle is r (see figure).

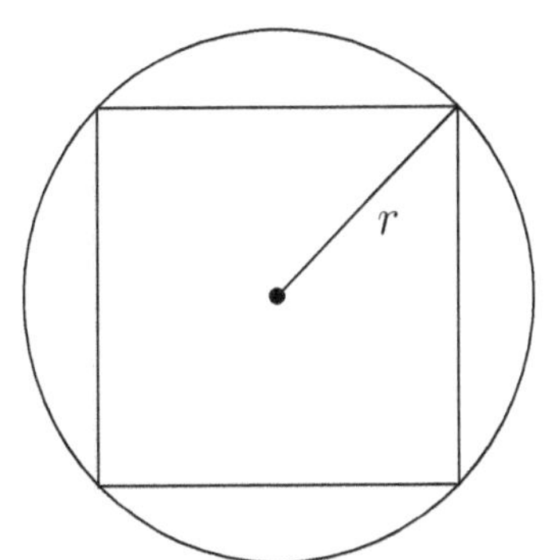

Example 10
Define a function V such that $V(S)$ represents the volume of a cube with surface area S.

Solution
Declarations
Let x represent the length of the side of the cube.

Formulas
- (surface area) = $6 \cdot$ (area of a side)
- (volume) = (length of a side)^3

Implications
- $S = 6x^2$
  $\Rightarrow S/6 = \not{6}x^2/\not{6}$
  $\Rightarrow S/6 = x^2$
  $\Rightarrow x = \sqrt{S/6}$

- (volume) $= x^3$
  $= (\sqrt{S/6})^3$
  $= ((S/6)^{1/2})^3$
  $= (S/6)^{3/2}$

Definition
We define V by $V(S) = (S/6)^{3/2}$.

□

Example 11
Define a function C such that $C(A)$ represents the circumference of a circle with area A.

Section 84: Graphs of Functions

Definition
The *graph of a function* f with domain $\mathbb{D}$ is the locus of all points in $\{(x, f(x)) \mid x \in \mathbb{D}\}$.

Vertical Line Test
A curve is the graph of a function if and only if no vertical line intersects the curve in more than one point.

Comments
- The definition above implies that the graph of f is the graph of the equation $y = f(x)$.
- No two points on the graph of a function f can lie on the same vertical line because f cannot assign to any element in its domain more than one element in its codomain. If the graph were to contain the points (a, b_1) and (a, b_2) where $b_1 \neq b_2$, then both $f(a) = b_1$ and $f(a) = b_2$, and such would contradict the premise that f is a function.

Example 1
Determine whether each curve is the graph of a function.

A. No

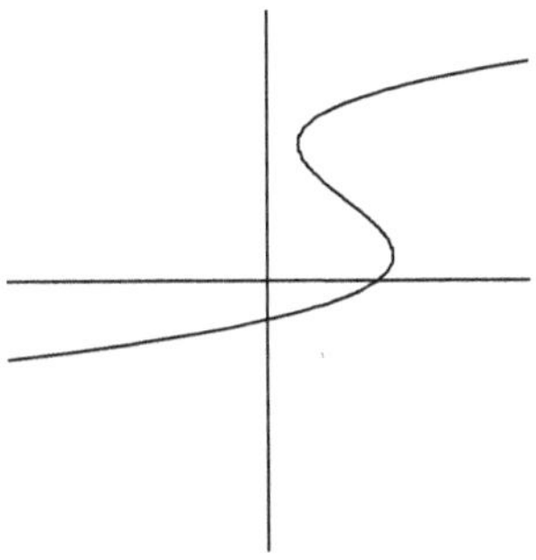

B. Yes

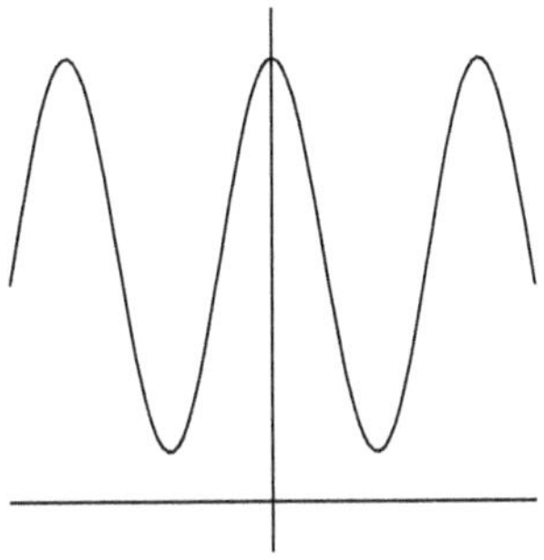

C. No

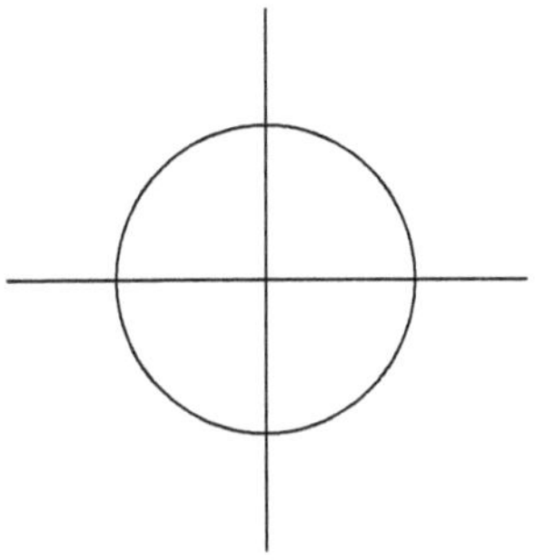

D. ____

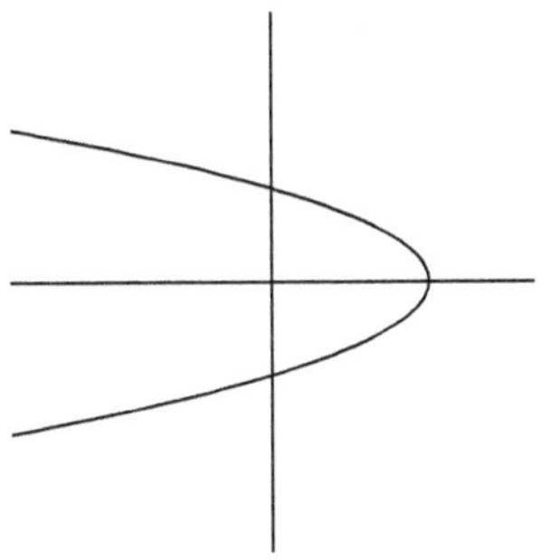

E. ____

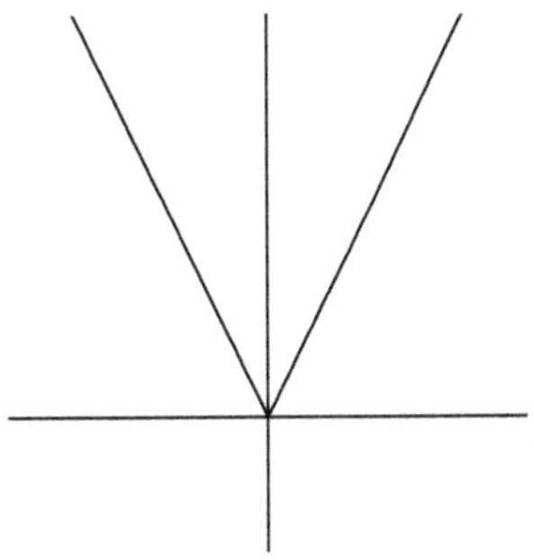

F. ____

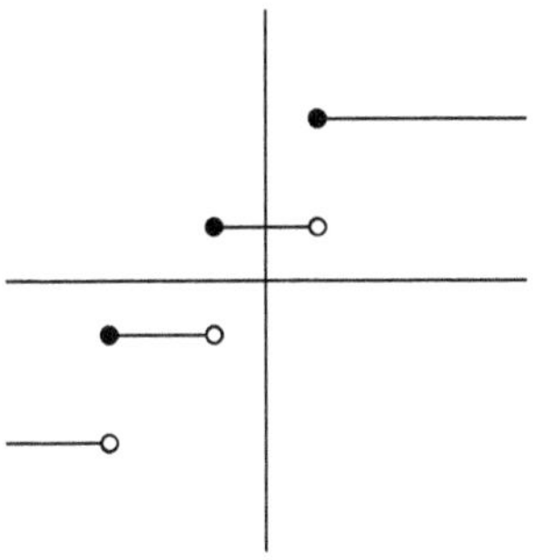

□

EXAMPLE 2
Draw the graph of the function with the given definition.

A. $f(t) = -2t/3 + 1$

SOLUTION
Table of Values

t	$f(t)$	solution
0	$-2(0)/3 + 1 = 1$	$(0, 1)$
3	$-2(3)/3 + 1 = -1$	$(3, -1)$

Graph

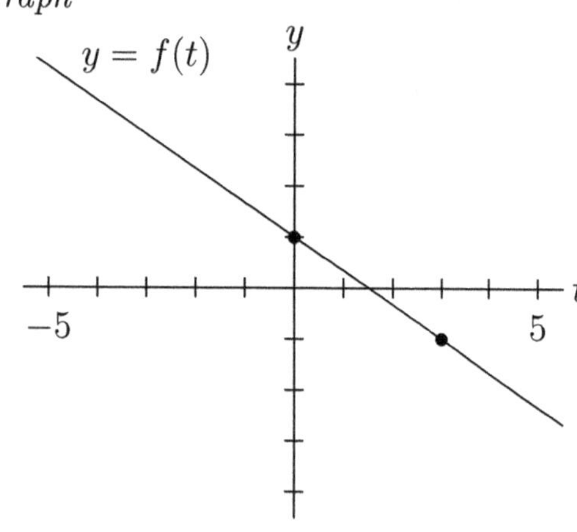

☐

B. $g(x) = 4$

C. $\Phi(x) = \sqrt{x + 3}$
[Hint: The domain of Φ is $[-3, \infty)$.]

SOLUTION
Table of Values

x	$\Phi(x)$	solution
-3	$\sqrt{-3+3} = 0$	$(-3, 0)$
-2	$\sqrt{-2+3} = 1$	$(-2, 1)$
1	$\sqrt{1+3} = 2$	$(1, 2)$
6	$\sqrt{6+3} = 3$	$(6, 3)$

Graph

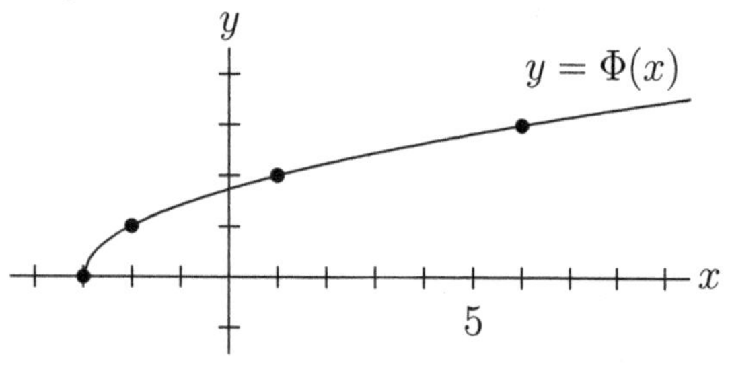

☐

D. $F(t) = |t|/t$

E. $f(x) = \begin{cases} 2 & \text{if } x \leq 1 \\ 3x - 7 & \text{if } x > 1 \end{cases}$

SOLUTION

Notes

- The graph of f will have two pieces.
- The graph of $y = 2$ on $(-\infty, 1]$ is a ray with right terminus at $x = 1$; we plot this terminus as a dot since $1 \in (-\infty, 1]$.
- The graph of $y = 3x - 7$ on $(1, \infty)$ is a ray with left terminus at $x = 1$; we plot this terminus as a hole since $1 \notin (1, \infty)$.

Tables of Values

-

x	2	solution
-4	2	$(-4, 2)$
1	2	$(1, 2)$

-

x	$3x - 7$	solution
1	$3(1) - 7 = -4$	$(1, -4)$
2	$3(2) - 7 = -1$	$(2, -1)$
4	$3(4) - 7 = 5$	$(4, 5)$

Graph

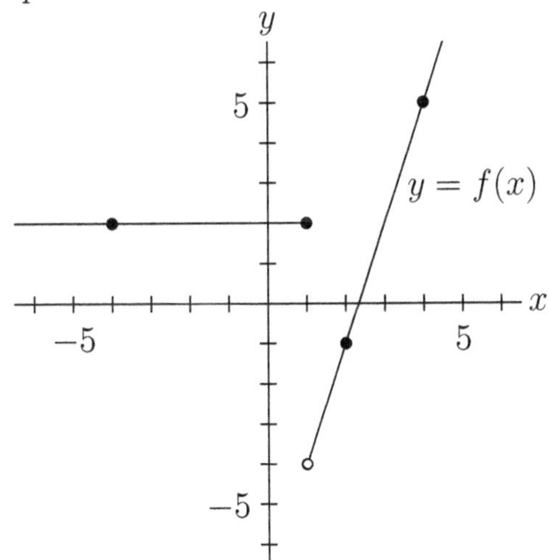

□

F. $g(x) = \begin{cases} x/2 + 4 & \text{if } x \leq -2 \\ -x + 1 & \text{if } -2 < x \leq 5 \\ -4 & \text{if } 5 < x \end{cases}$

Section 85: Characteristics of Functions

Definitions

- A *value of a function* is a value of an image of an element in its domain.
- An *argument of a function* is an expression that serves as an element in its domain.
- A *turning point* in the graph of a function is a point at which a particle that traces the graph from left to right ceases to rise and begins to fall, or vice versa.

Definitions

Suppose f is a function with domain $\mathbb{D}$ where $c \in \mathbb{D}$.

- The *absolute maximum* of f is the value $f(c)$ for which $f(c) \geq f(x)$ for all $x \in \mathbb{D}$.
- The *absolute minimum* of f is the value $f(c)$ for which $f(c) \leq f(x)$ for all $x \in \mathbb{D}$.
- A *relative maximum* of f is a value $f(c)$ for which $f(c) \geq f(x)$ for all $x \in \mathbb{D}$ near c.
- A *relative minimum* of f is a value $f(c)$ for which $f(c) \leq f(x)$ for all $x \in \mathbb{D}$ near c.
- An *extremum* of f is a maximum or minimum of f.

Definitions

Suppose f is a function defined on an interval I with $x_1, x_2 \in I$.

- f is *increasing* on I if $f(x_1) < f(x_2)$ whenever $x_1 < x_2$.
- f is *decreasing* on I if $f(x_1) > f(x_2)$ whenever $x_1 < x_2$.
- f is *monotonic* on I if it is either increasing or decreasing on I.

Formula

Suppose f is a function with domain $\mathbb{D}$ where $a, b \in \mathbb{D}$.
The *average rate of change of f*, denoted $\Delta f/\Delta x$, on the interval $[a, b]$ is given by:

$$\frac{\Delta f}{\Delta x} = \frac{f(b) - f(a)}{b - a}$$

Comments

- Not every function has absolute or relative extrema.
- We can use the graph of a function to identify some of its characteristics:
    - the maximum of a function is the y-coordinate of the highest point on its graph;
    - the minimum of a function is the y-coordinate of the lowest point on its graph;
    - a function is increasing where a particle that traces its graph from left to right rises;
    - a function is decreasing where a particle that traces its graph from left to right falls.
- Absolute extrema will also be relative extrema.
- The average rate of change of f on the interval $[a, b]$ is equal to the slope of the line that passes through the points $(a, f(a))$ and $(b, f(b))$.

Example 1

Suppose the function $g : \mathbf{R} \to \mathbf{R}$ is defined by $g(x) = 3x + 5$.

A. The value of g at 2 is 11, since $g(2) = 3(2) + 5 = 11$.
B. The value of g at π is $3\pi + 5$, since $g(\pi) = 3\pi + 5$.

□

Example 2

A. The argument of the function f in the expression $f(2)$ is 2.
B. The argument of the function g in the expression $g(\pi)$ is π.
C. The argument of the function h in the expression $h(a)$ is a.
D. The argument of the function F in the expression $F(c+5)$ is $c+5$.

□

Example 3

Given the graph of a function, identify:

- the turning points in the graph,
- the intervals on which the function is increasing and decreasing, and
- the extrema of the function and the points in its domain where it attains those values.

A.

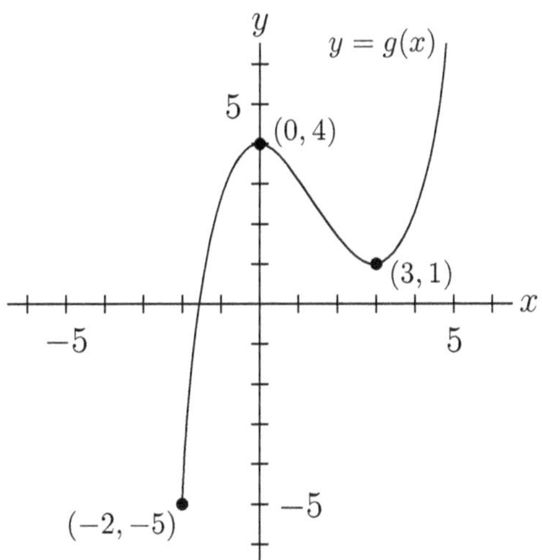

B.

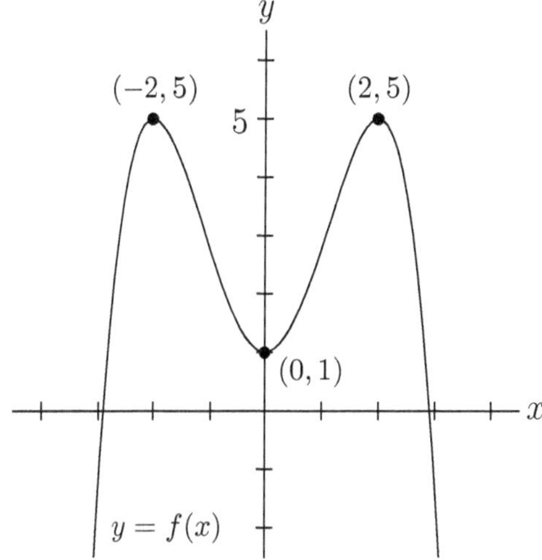

Solution

- The turning points are $(0,4)$ and $(3,1)$.

- Function g is:
  - decreasing on $[0,3]$ and
  - increasing on $[-2,0]$ and $[3,\infty)$.

- Function g has:
  - no absolute maximum;
  - an absolute (and relative) minimum of -5, which it attains at $x=-2$;
  - a relative maximum of 4, which it attains at $x=0$;
  - a relative minimum of 1, which it attains at $x=3$.

□

EXAMPLE 4
Determine the rate of change of the given function on the given interval.

A. $f(x) = 3x - 8$; $[-2, 5]$

SOLUTION
On $[-2, 5]$:

$$\begin{aligned}\Delta f/\Delta x &= (f(5) - f(-2))/(5 - (-2)) \\ &= ((3(5) - 8) - (3(-2) - 8))/7 \\ &= ((15 - 8) - (-6 - 8))/7 \\ &= (7 - (-14))/7 \\ &= (7 + 14)/7 \\ &= 21/7 \\ &= 3\end{aligned}$$

□

B. $f(x) = 4 - 7x$; $[1, 10]$

C. $g(x) = x^2 + 11$; $[0, 6]$

SOLUTION
On $[0, 6]$:

$$\begin{aligned}\Delta g/\Delta x &= (g(6) - g(0))/(6 - 0) \\ &= (((6)^2 + 11) - ((0)^2 + 11))/6 \\ &= ((36 + 11) - (0 + 11))/6 \\ &= (47 - 11)/6 \\ &= 36/6 \\ &= 6\end{aligned}$$

□

D. $h(x) = 5 + \sqrt{x}$; $[16, 49]$

E. $h(x) = |x|$; $[-9, 9]$

SOLUTION
On $[-9, 9]$:

$$\begin{aligned}\Delta h/\Delta x &= (h(9) - h(-9))/(9 - (-9)) \\ &= (|9| - |-9|)/(9 + 9) \\ &= (9 - 9)/18 \\ &= 0/18 \\ &= 0\end{aligned}$$

□

F. $g(x) = 4$; $[-10, -3]$

Section 86: Transformations of Graphs of Functions

Graphical Relationship

Suppose f and g are functions and A, B, C, and D are real constants.
If $g(x) = A \cdot f(B(x - C)) + D$ for all x in the domain of g and the point (a, b) lies on the graph of f, then the point $(a/B + C, Ab + D)$ lies on the graph of g.

Proof

Suppose the premises of the property.

$$\begin{aligned} g(a/B + C) &= A \cdot f(B((a/B + C) - C)) + D \\ &= A \cdot f(B(a/B)) + D \\ &= A \cdot f(a) + D \\ &= Ab + D \end{aligned}$$

Thus $(a/B + C, Ab + D)$ lies on the graph of g.

⊡

Graphical Relationships

Suppose f and g are functions and A, B, C, and D are real constants.
Let $\mathbb{F}$ and $\mathbb{G}$ denote the graphs of f and g, respectively.

- If $g(x) = f(x) + D$, then the point (a, b) on $\mathbb{F}$ translates to the point $(a, b + D)$ on $\mathbb{G}$, so:
  - $\mathbb{G}$ is the translation of $\mathbb{F}$ up $|D|$ units if $D > 0$
  - $\mathbb{G}$ is the translation of $\mathbb{F}$ down $|D|$ units if $D < 0$

- If $g(x) = f(x - C)$, then the point (a, b) on $\mathbb{F}$ translates to the point $(a + C, b)$ on $\mathbb{G}$, so:
  - $\mathbb{G}$ is the translation of $\mathbb{F}$ right $|C|$ units if $C > 0$
  - $\mathbb{G}$ is the translation of $\mathbb{F}$ left $|C|$ units if $C < 0$

- If $g(x) = f(Bx)$, then the point (a, b) on $\mathbb{F}$ translates to the point $(a/B, b)$ on $\mathbb{G}$, so:
  - $\mathbb{G}$ is a horizontal contraction of $\mathbb{F}$ if $1 < B$
  - $\mathbb{G}$ is a horizontal expansion of $\mathbb{F}$ if $0 < B < 1$
  - $\mathbb{G}$ is a horizontal contraction of the reflection of $\mathbb{F}$ through the y-axis if $B < -1$
  - $\mathbb{G}$ is the reflection of $\mathbb{F}$ through the y-axis if $B = -1$
  - $\mathbb{G}$ is a horizontal expansion of the reflection of $\mathbb{F}$ through the y-axis if $-1 < B < 0$

- If $g(x) = A \cdot f(x)$, then the point (a, b) on $\mathbb{F}$ translates to the point (a, Ab) on $\mathbb{G}$, so:
  - $\mathbb{G}$ is a vertical contraction of $\mathbb{F}$ if $0 < A < 1$
  - $\mathbb{G}$ is a vertical expansion of $\mathbb{F}$ if $1 < A$
  - $\mathbb{G}$ is a vertical contraction of the reflection of $\mathbb{F}$ through the x-axis if $-1 < A < 0$
  - $\mathbb{G}$ is the reflection of $\mathbb{F}$ through the x-axis if $A = -1$
  - $\mathbb{G}$ is a vertical expansion of the reflection of $\mathbb{F}$ through the x-axis if $A < -1$

Notation

To indicate a translation of a point (a, b) on one graph to a point (c, d) on another graph, we will write "$(a, b) \mapsto (c, d)$".

PROCEDURE

Suppose f and g are functions and A, B, C, and D are real constants.
To transform the graph of f into the graph of g where $g(x) = A \cdot f(B(x - C)) + D$, we:

1. translate salient points on the graph of f to points on the graph of g;
2. draw the graph of g through the translated points with the general (though possibly shifted, distorted, and/or reflected) shape of the graph of f.

COMMENT

Where functions f and g are related by an equation of the form $g(x) = \Gamma(f(\Phi(x))$ for some functions Γ and Φ, assume (unless informed otherwise) that:

- the equation holds true for all x in the domain of g;
- x lies in the domain of g if and only $\Phi(x)$ lies in the domain of f.

EXAMPLE 1

Suppose $g(x) = 5 \cdot f(2(x - 3)) + (-9)$.
Translate points $(-5, 0)$ and $(8, -4)$ on the graph of f to the graph of g.

SOLUTION

- $(-5\,,0) \mapsto (-5/2 + 3\,, 5 \cdot 0 + (-9))$
  $= (-5/2 + 6/2\,, 0 - 9)$
  $= (1/2\,, -9)$
- $(8\,, -4) \mapsto (8/2 + 3\,, 5 \cdot (-4) + (-9))$
  $= (4 + 3\,, -20 - 9)$
  $= (7\,, -29)$

□

EXAMPLE 2

Suppose $g(x) = -f(3x - 3) + 2$.
Translate points $(6, -2)$ and $(0, 10)$ on the graph of f to the graph of g.

EXAMPLE 3

Determine the point on the graph of the equation $y = (-1/2)(x + 5)^3 - 2$ that corresponds to the point $(2, 8)$ on the graph of the equation $y = x^3$.

SOLUTION

Relationship between Functions

Let $f(x) = x^3$ and
$g(x) = (-1/2)(x + 5)^3 - 2$.
So $g(x) = (-1/2) \cdot f(x + 5) - 2$
$= (-1/2) \cdot f(x - (-5)) + (-2)$.

Translation

$(2\,, 8) \mapsto (2 + (-5)\,, (-1/2) \cdot 8 + (-2))$
$= (-3\,, -6)$
The point on $y = g(x)$ is $(-3\,, -6)$.

□

EXAMPLE 4

Determine the point on the graph of the equation $y = |x/3 - 5|$ that corresponds to the point $(-4, 4)$ on the graph of the equation $y = |x|$.

EXAMPLE 5

Consider the following translations of the graph of $y = x^2$ shown at upper left.

$y = x^2$

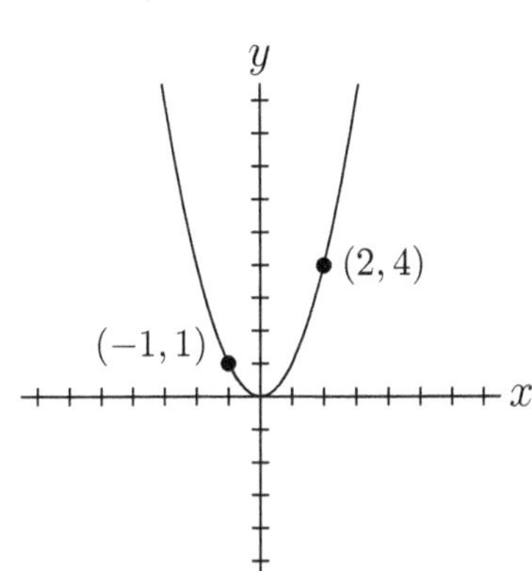

$y = x^2 + 3$

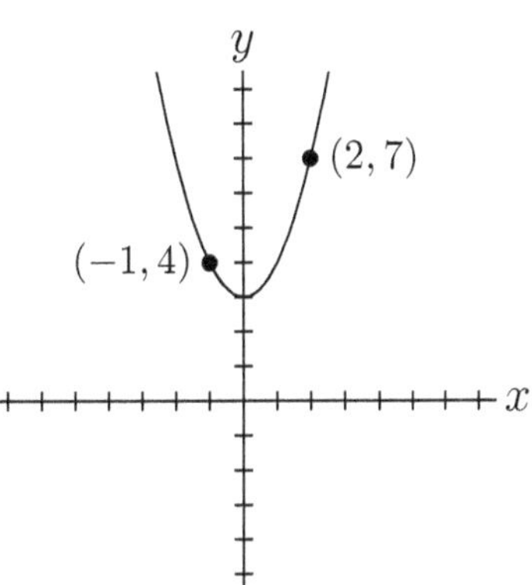

$y = x^2 - 3$

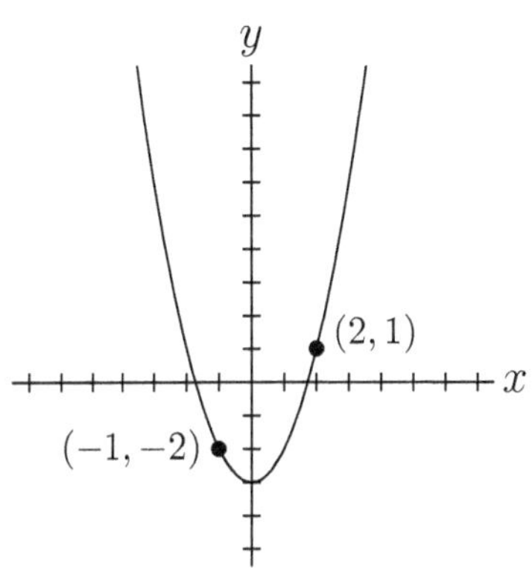

$y = (x + 3)^2$

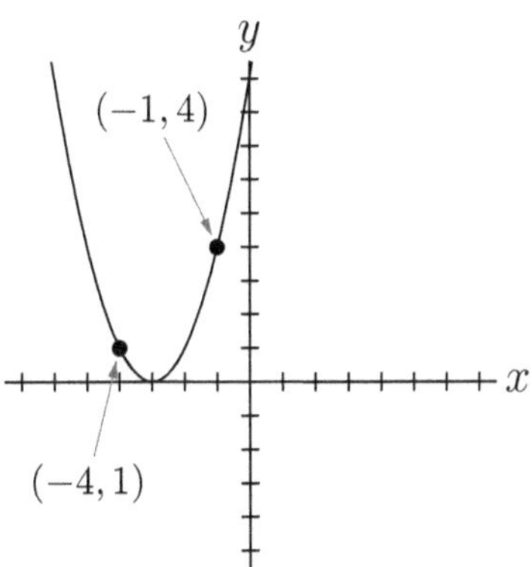

$y = (x - 3)^2$

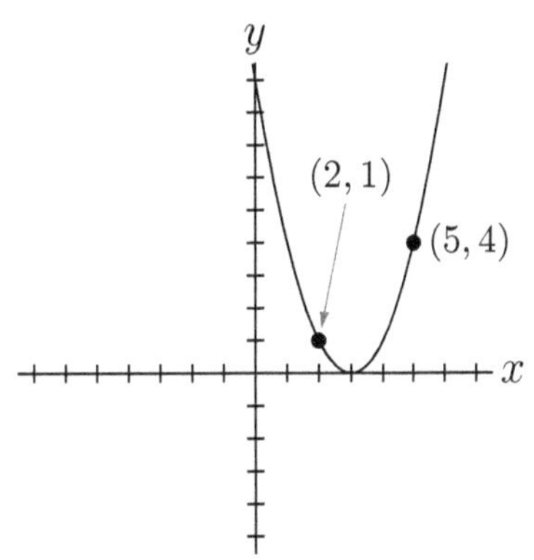

$y = -x^2$

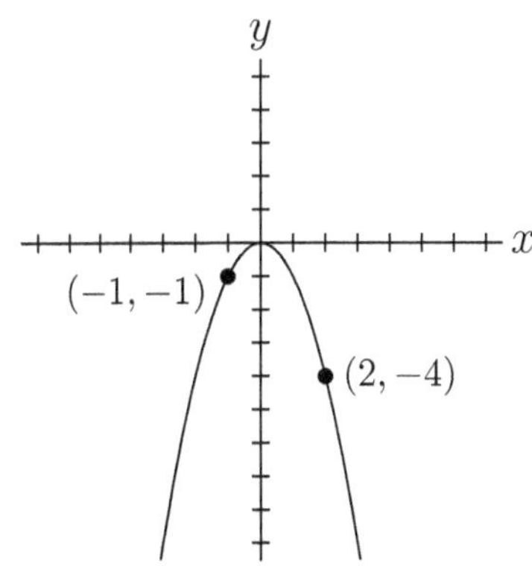

$y = (-x)^2$

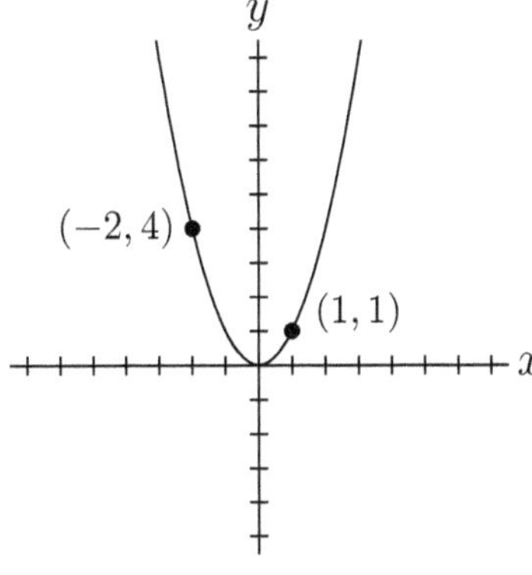

$y = 2x^2$

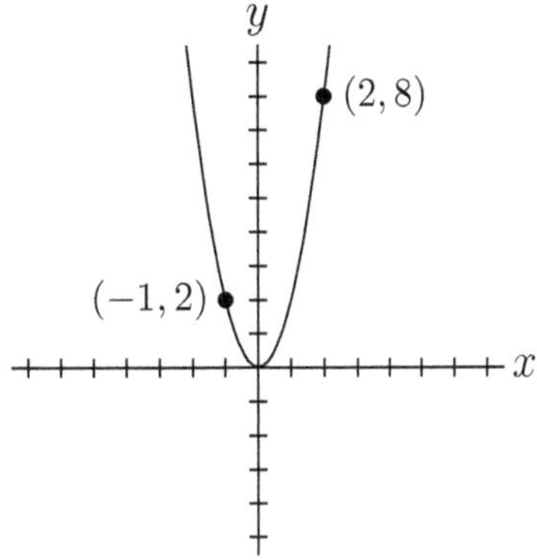

$y = (\frac{1}{2}x)^2$

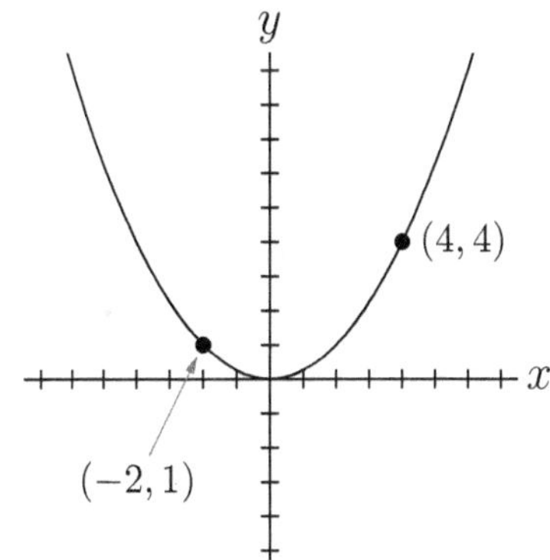

□

EXAMPLE 6
Suppose $g(x) = -f(x/2 + 1) + 5$.
Draw the graph of function g given the graph of function f shown below:

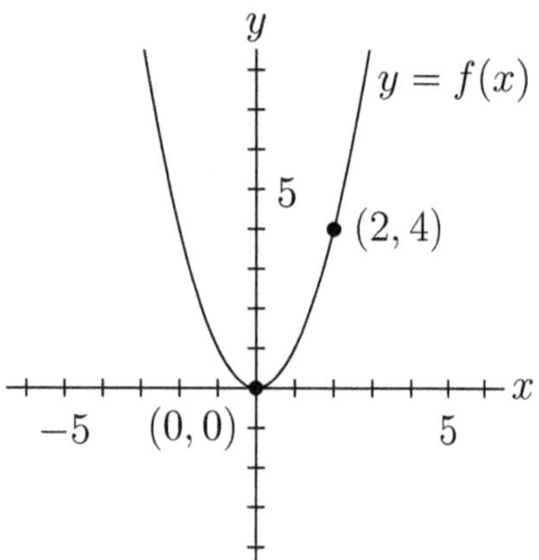

SOLUTION
Note

$$\begin{aligned} g(x) &= (-1) \cdot f((1/2)(x+2)) + 5 \\ &= (-1) \cdot f((1/2)(x-(-2))) + 5 \end{aligned}$$

Translations

- $(0\,,0) \mapsto (0/(1/2) + (-2)\,, (-1)\cdot 0 + 5)$
  $= (-2\,,5)$
- $(2\,,4) \mapsto (2/(1/2) + (-2)\,, (-1)\cdot 4 + 5)$
  $= (2\cdot(2/1) - 2\,, -4 + 5)$
  $= (2\,,1)$

Graph

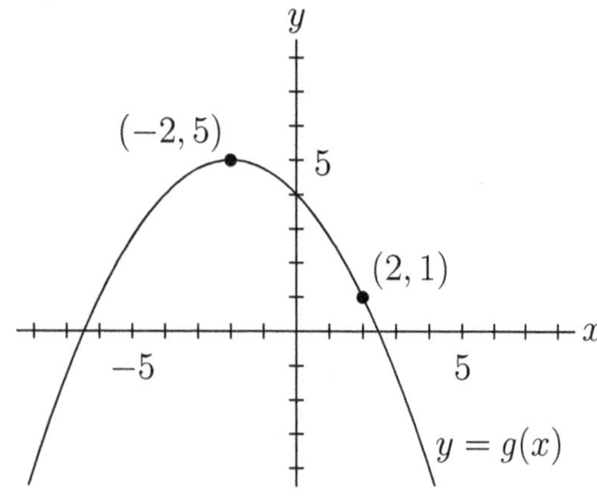

□

EXAMPLE 7
Suppose $v(x) = 2 \cdot h(x-4) - 3$.
Draw the graph of function v given the graph of function h shown below:

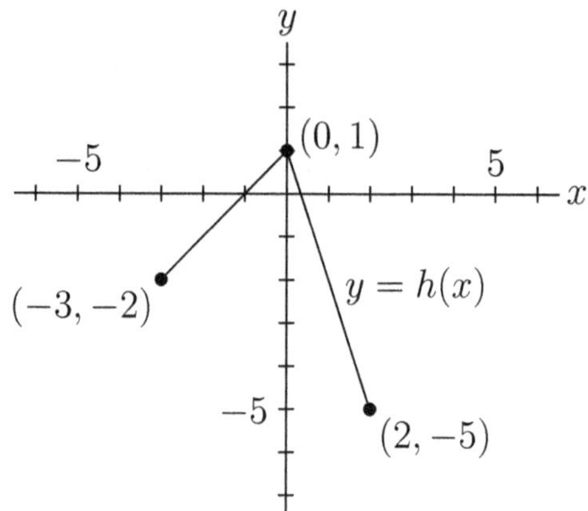

Section 87: Translations of Graphs of Equations

Property (proof omitted)
Suppose E_1 and E_2 are equations in the variables x and y where E_2 is equivalent to the equation formed by replacing in E_1 every x with $x-h$ and every y with $y-k$ for some constants $h, k \in \mathbf{R}$. Then $(a+h, b+k)$ satisfies E_2 if and only if (a, b) satisfies E_1.

Graphical Relationships
Suppose E_1 and E_2 are equations.

- If E_2 is equivalent to the equation formed by replacing in E_1 every x with $x-h$, then each point (a, b) on the graph of E_1 translates to the point $(a+h, b)$ on the graph of E_2, so:
  - the graph of E_2 is the translation of the graph of E_1 right $|h|$ units if $h > 0$
  - the graph of E_2 is the translation of the graph of E_1 left $|h|$ units if $h < 0$

- If E_2 is equivalent to the equation formed by replacing in E_1 every y with $y-k$, then each point (a, b) on the graph of E_1 translates to the point $(a, b+k)$ on the graph of E_2, so:
  - the graph of E_2 is the translation of the graph of E_1 up $|k|$ units if $k > 0$
  - the graph of E_2 is the translation of the graph of E_1 down $|k|$ units if $k < 0$

Procedure
Suppose E_1 and E_2 are equations that meet the conditions described in the property above. To translate the graph of E_1 into the graph of E_2, we:
1. translate salient points on the graph of E_1 to points on the graph of E_2;
2. draw the graph of E_2 through the translated points with the shape of the graph of E_1.

Example 1
Translate the points $(2\sqrt{2}, 2\sqrt{2})$, $(0, -4)$, and $(4, 0)$ on the graph of the equation $x^2 + y^2 = 16$ to points on the graph of the equation $(x-2)^2 + (y+3)^2 = 16$.

Solution
Note
$(x-2)^2 + (y+3)^2 = 16$
$\quad \Leftrightarrow (x-2)^2 + (y-(-3))^2 = 16$

Translations
- $(2\sqrt{2}, 2\sqrt{2}) \mapsto (2\sqrt{2}+2, 2\sqrt{2}+(-3))$
  $= (2\sqrt{2}+2, 2\sqrt{2}-3)$
- $(0, -4) \mapsto (0+2, -4+(-3))$
  $= (2, -7)$
- $(4, 0) \mapsto (4+2, 0+(-3))$
  $= (6, -3)$

□

Example 2
Suppose the focus of the parabola with the equation $8x = y^2$ is the point $(2, 0)$. Determine the focus of the parabola with the equation $8(x+5) = (y+1)^2$.

Example 3
Use the graph below to draw the graph of the equation $x + 5 = (y - 2)^2$.

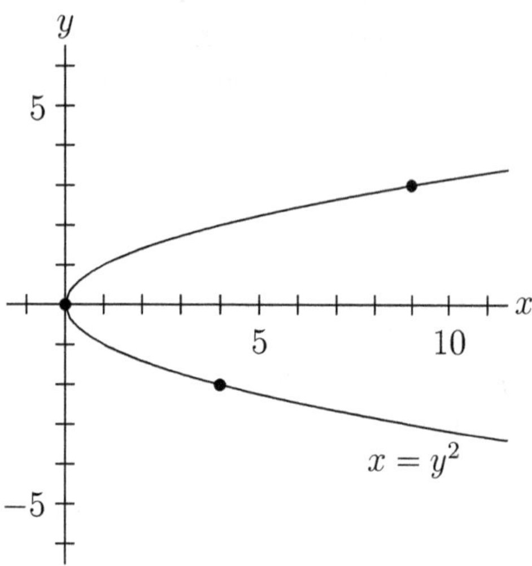

Solution
Translations

- $(9\,,3) \mapsto (9-5\,,3+2)$
  $= (4\,,5)$
- $(0\,,0) \mapsto (0-5\,,0+2)$
  $= (-5\,,2)$
- $(4\,,-2) \mapsto (4-5\,,-2+2)$
  $= (-1\,,0)$

Graph

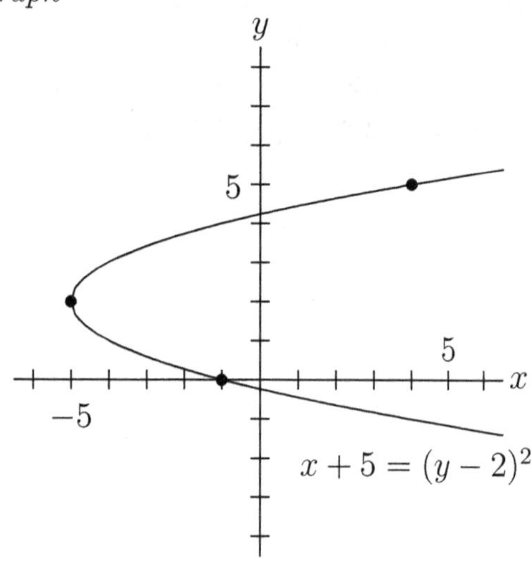

□

Example 4
Use the graph below to draw the graph of the equation $(x - 1)^2/4 + (y + 3)^2/16 = 1$.

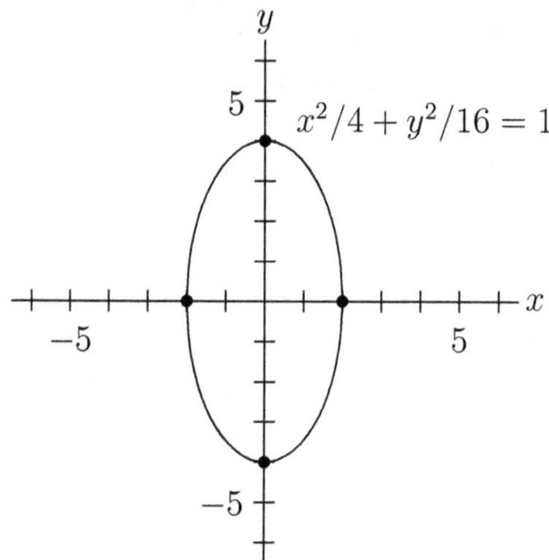

Section 88: Even and Odd Functions

Definitions

Suppose f is a function.

- f is an *even function* if $f(-x) = f(x)$ for all x in the domain of f.
- f is an *odd function* if $f(-x) = -f(x)$ for all x in the domain of f.

Graphical Relationships

- The graph of an even function is symmetric with respect to the y-axis.
- The graph of an odd function is symmetric with respect to the origin.

Proofs

- Suppose f is an even function and $(a, b) \in \mathbf{R}^2$.

$$\begin{aligned} (-a, b) \text{ satisfies } y = f(x) &\Leftrightarrow b = f(-a) \\ &\Leftrightarrow b = f(a) \\ &\Leftrightarrow (a, b) \text{ satisfies } y = f(x) \end{aligned}$$

So the graph of f is symmetric with respect to the y-axis.

- Suppose f is an odd function and $(a, b) \in \mathbf{R}^2$.

$$\begin{aligned} (-a, -b) \text{ satisfies } y = f(x) &\Leftrightarrow -b = f(-a) \\ &\Leftrightarrow -b = -f(a) \\ &\Leftrightarrow (-1)(-b) = (-1)(-f(a)) \\ &\Leftrightarrow b = f(a) \\ &\Leftrightarrow (a, b) \text{ satisfies } y = f(x) \end{aligned}$$

So the graph of f is symmetric with respect to the origin.

⊡

Comments

- A function can be even, odd, or neither even nor odd.
- If a function f is both even and odd, then $f(x) = 0$ for all x in the domain of f.

Example 1

Suppose $f(x) = x^3$.
Prove that the function f is odd.

Proof

Suppose $x \in \mathbf{R}$.

$$\begin{aligned} f(-x) &= (-x)^3 \\ &= (-1 \cdot x)^3 \\ &= (-1)^3(x)^3 \\ &= (-1)x^3 \\ &= -f(x) \end{aligned}$$

So $f(-x) = -f(x)$ for all $x \in \mathbf{R}$.
Therefore, f is odd.

⊡

Example 2

Suppose $h(x) = x^4 + 7$.
Prove that the function h is even.

Example 3
Suppose $g(x) = |x^2 - 5x - 4|$.
Provide a counterexample to show that the function g is not even.

Counterexample
Observe that:

- $g(-1) = |(-1)^2 - 5(-1) - 4|$
  $= |1 + 5 - 4|$
  $= |2|$
  $= 2$
- $g(1) = |(1)^2 - 5(1) - 4|$
  $= |1 - 5 - 4|$
  $= |-8|$
  $= 8$

Clearly $g(-1) \neq g(1)$.
So there exists an $x \in \mathbf{R}$ (namely 1) such that $g(-x) \neq g(x)$.
Therefore, g is not even.

$\square$

Example 4
Suppose $f(x) = \sqrt{x+9}$.
Provide a counterexample to show that the function f is not odd.

Section 89: One-to-One Functions

Definition
A function is *one-to-one* if each element of its range is assigned to one and only one element of its domain.

Property
Suppose f is a function with domain $\mathbb{D}$. Then the following statements are equivalent:

- f is one-to-one
- $x_1 \neq x_2 \Rightarrow f(x_1) \neq f(x_2)$ for all $x_1, x_2 \in \mathbb{D}$
- $f(x_1) = f(x_2) \Rightarrow x_1 = x_2$ for all $x_1, x_2 \in \mathbb{D}$

Horizontal Line Test
A function is one-to-one if and only if no horizontal line intersects its graph in more than one point.

Comments

- The term "injective" means "one-to-one".
- A function defined by formula may be one-to-one on an explicitly restricted domain though not one-to-one on an implicitly assumed domain of all real numbers. For example:
  - the function $f : \mathbf{R} \to \mathbf{R}$ defined by $f(x) = x^2$ is not one-to-one;
  - the function $f : [0, \infty) \to \mathbf{R}$ defined by $f(x) = x^2$ is one-to-one.
- No two points on the graph of a one-to-one function f can lie on the same horizontal line because f cannot assign any element in its range to more than one element in its domain. If the graph were to contain the points (a_1, b) and (a_2, b) where $a_1 \neq a_2$, then $f(a_1) = f(a_2)$ since both $f(a_1) = b$ and $f(a_2) = b$. Such would contradict the premise that f is one-to-one.
- A curve that no horizontal line intersects in more than one point may not be the graph of a one-to-one function; if it fails the vertical line test, then it is not the graph of a function in the first place.

Example 1
Determine whether each curve is the graph of a one-to-one function.

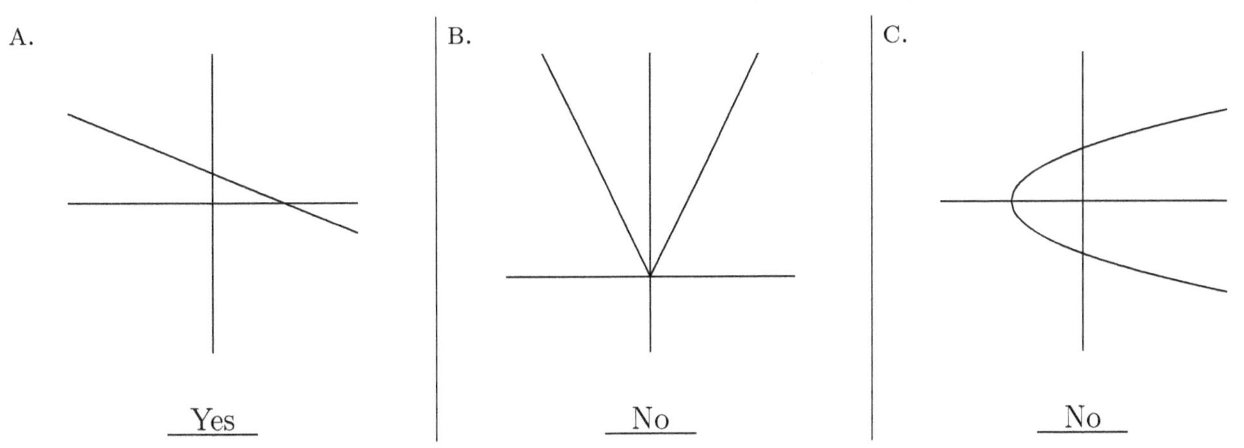

□

EXAMPLE 2
Suppose $g(x) = 5x + 2$.
Prove that the function g is one-to-one.

PROOF
Suppose $x_1, x_2 \in \mathbf{R}$.
$g(x_1) = g(x_2)$
$\Rightarrow 5x_1 + 2 = 5x_2 + 2$
$\Rightarrow 5x_1 + 2 - 2 = 5x_2 + 2 - 2$
$\Rightarrow 5x_1 = 5x_2$
$\Rightarrow 5x_1/5 = 5x_2/5$
$\Rightarrow x_1 = x_2$
So $g(x_1) = g(x_2) \Rightarrow x_1 = x_2 \;\forall\; x_1, x_2 \in \mathbf{R}$.
Therefore, g is one-to-one.

$\boxdot$

EXAMPLE 3
Suppose $f(x) = 7x^5$.
Prove that the function f is one-to-one.

EXAMPLE 4
Suppose $f(x) = (x - 5)^4$.
Provide a counterexample to show that the function f is not one-to-one.

COUNTEREXAMPLE
Notes

- $f(3) = (3-5)^4 = (-2)^4 = 16$
- $f(7) = (7-5)^4 = (2)^4 = 16$

Conclusion
Clearly $f(3) = f(7)$ but $3 \neq 7$.
So there exist $x_1, x_2 \in \mathbf{R}$ (namely 3 and 7) such that $f(x_1) = f(x_2) \not\Rightarrow x_1 = x_2$.
Therefore, f is not one-to-one.

$\square$

EXAMPLE 5
Suppose $h(x) = x^2 + 2x + 5$.
Provide a counterexample to show that the function h is not one-to-one.

Section 90: Combinations of Functions

Definitions

Suppose f and g are functions.

- The *sum* of f and g, denoted $f+g$, is defined by $(f+g)(x) = f(x)+g(x)$.
- The *difference* of f and g, denoted $f-g$, is defined by $(f-g)(x) = f(x)-g(x)$.
- The *product* of f and g, denoted fg, is defined by $(fg)(x) = (f(x))(g(x))$.
- The *quotient* of f and g, denoted f/g, is defined by $(f/g)(x) = f(x)/g(x)$.
- The *composition* of f and g, denoted $f \circ g$, is defined by $(f \circ g)(x) = f(g(x))$.

Comments

- A combination of two functions is itself a function.
- The difference, quotient, and composition of f and g are not the same as the difference, quotient, and composition of g and f, respectively.
- We will denote the domains of the combinations above by $\mathbb{D}_{f+g}$, $\mathbb{D}_{f-g}$, $\mathbb{D}_{fg}$, $\mathbb{D}_{f/g}$, and $\mathbb{D}_{f\circ g}$. These domains relate to the domains $\mathbb{D}_f$ and $\mathbb{D}_g$ as follows:
  - $\mathbb{D}_{f+g} = \mathbb{D}_f \cap \mathbb{D}_g$
  - $\mathbb{D}_{f-g} = \mathbb{D}_f \cap \mathbb{D}_g$
  - $\mathbb{D}_{fg} = \mathbb{D}_f \cap \mathbb{D}_g$
  - $\mathbb{D}_{f/g} = \{\, x \in \mathbb{D}_f \cap \mathbb{D}_g \mid g(x) \neq 0 \,\}$
  - $\mathbb{D}_{f\circ g} = \{\, x \in \mathbb{D}_g \mid g(x) \in \mathbb{D}_f \,\}$

Example 1

Suppose f and g are functions where:

$$\begin{array}{llll} f(2)=7 & f(4)=2 & f(7)=6 & f(8)=4 \\ g(2)=4 & g(5)=7 & g(6)=2 & g(8)=3 \end{array}$$

A. Evaluate $(f+g)(8)$, $(fg)(2)$, $(f \circ g)(5)$, and $(f \circ f)(8)$.

Solution

- $$\begin{aligned}(f+g)(8) &= f(8)+g(8) \\ &= 4+3 \\ &= 7\end{aligned}$$
- $$\begin{aligned}(fg)(2) &= (f(2))(g(2)) \\ &= (7)(4) \\ &= 28\end{aligned}$$
- $$\begin{aligned}(f \circ g)(5) &= f(g(5)) \\ &= f(7) \\ &= 6\end{aligned}$$
- $$\begin{aligned}(f \circ f)(8) &= f(f(8)) \\ &= f(4) \\ &= 2\end{aligned}$$

□

B. Evaluate $(f-g)(2)$, $(g/f)(8)$, $(g \circ f)(7)$, and $(g \circ g)(6)$.

Example 2

Given the definitions of functions f and g:

- determine formulas that define $f+g$, $f-g$, fg, and f/g,
- determine the domains of $f+g$ and f/g, and
- evaluate $(f+g)(9)$, $(f-g)(4)$, $(fg)(1)$, and $(f/g)(2)$.

A. $f(x) = x+2$, $g(x) = x^2 - 3x - 4$

Solution

Formulas

- $(f+g)(x) = (x+2)+(x^2-3x-4)$
  $= x^2 - 2x - 2$
- $(f-g)(x) = (x+2)-(x^2-3x-4)$
  $= -x^2 + 4x + 6$
- $(fg)(x) = (x+2)(x^2-3x-4)$
- $(f/g)(x) = (x+2)/(x^2-3x-4)$

Domains

- $\mathbb{D}_{f+g} = \mathbb{D}_f \cap \mathbb{D}_g$
  $= \mathbf{R} \cap \mathbf{R}$
  $= \mathbf{R}$
- $g(x) = 0 \Leftrightarrow x^2 - 3x - 4 = 0$
  $\Leftrightarrow (x-4)(x+1) = 0$
  $\Leftrightarrow x - 4 = 0 \text{ OR } x + 1 = 0$
  $\Leftrightarrow x = 4 \text{ OR } x = -1$

$\therefore \mathbb{D}_{f/g} = \{\, x \in \mathbb{D}_f \cap \mathbb{D}_g \mid g(x) \neq 0 \,\}$
$= \{\, x \in \mathbf{R} \cap \mathbf{R} \mid g(x) \neq 0 \,\}$
$= \{\, x \in \mathbf{R} \mid x \neq -1 \text{ AND } x \neq 4 \,\}$
$= (-\infty, -1) \cup (-1, 4) \cup (4, \infty)$

Evaluations

- $(f+g)(9) = (9)^2 - 2(9) - 2$
  $= 61$
- $(f-g)(4) = -(4)^2 + 4(4) + 6$
  $= 6$
- $(fg)(1) = (1+2)((1)^2 - 3(1) - 4)$
  $= (3)(-6)$
  $= -18$
- $(f/g)(2) = (2+2)/((2)^2 - 3(2) - 4)$
  $= 4/(-6)$
  $= -2/3$

□

B. $f(x) = \sqrt{x}$, $g(x) = 3/x$

EXAMPLE 3

Given the definitions of functions f and g:

- determine formulas that define $f \circ g$, $g \circ f$, and $f \circ f$, and
- determine the domains of $f \circ g$, $g \circ f$, and $f \circ f$.

A. $f(x) = 5/x$, $g(x) = x + 2$

SOLUTION

Formulas

- $(f \circ g)(x) = f(g(x))$
  $= f(x+2)$
  $= 5/(x+2)$
- $(g \circ f)(x) = g(f(x))$
  $= g(5/x)$
  $= (5/x) + 2$
  $= (5 + 2x)/x$
- $(f \circ f)(x) = f(f(x))$
  $= f(5/x)$
  $= 5/(5/x)$
  $= \cancel{5} \cdot (x/\cancel{5})$
  $= x$

Domains

- $\mathbb{D}_{f \circ g} = \{ x \in \mathbb{D}_g \mid g(x) \in \mathbb{D}_f \}$
  $= \{ x \in \mathbf{R} \mid g(x) \neq 0 \}$
  $= \{ x \in \mathbf{R} \mid x + 2 \neq 0 \}$
  $= \{ x \in \mathbf{R} \mid x \neq -2 \}$
  $= (-\infty, -2) \cup (-2, \infty)$
- $\mathbb{D}_{g \circ f} = \{ x \in \mathbb{D}_f \mid f(x) \in \mathbb{D}_g \}$
  $= \{ x \in \mathbb{D}_f \mid f(x) \in \mathbf{R} \}$
  $= \mathbb{D}_f$
  $= \{ x \in \mathbf{R} \mid x \neq 0 \}$
  $= (-\infty, 0) \cup (0, \infty)$
- $\mathbb{D}_{f \circ f} = \{ x \in \mathbb{D}_f \mid f(x) \in \mathbb{D}_f \}$
  $= \{ x \in \mathbb{D}_f \mid f(x) \neq 0 \}$
  $= \{ x \in \mathbb{D}_f \mid 5/x \neq 0 \}$
  $= \mathbb{D}_f$
  $= \{ x \in \mathbf{R} \mid x \neq 0 \}$
  $= (-\infty, 0) \cup (0, \infty)$

Comment

$(f \circ f)(x) = x$, but $\mathbb{D}_{f \circ f} \neq \mathbf{R}$.

□

B. $f(x) = 7 - x$, $g(x) = \sqrt{x - 5}$

EXAMPLE 4

Given the definitions of functions f and g:

- determine formulas that define $f \circ g$, $g \circ f$, and $g \circ g$,
- evaluate $(f \circ g)(4)$ and $(g \circ f)(4)$ using the formulas determined for $f \circ g$ and $g \circ f$, and
- evaluate $(f \circ g)(4)$ and $(g \circ f)(4)$ using the definition of composition directly.

A. $f(x) = 5x - 2$, $g(x) = 2x + 1$

SOLUTION

Formulas

- $$\begin{aligned} (f \circ g)(x) &= f(g(x)) \\ &= f(2x+1) \\ &= 5(2x+1) - 2 \\ &= 10x + 5 - 2 \\ &= 10x + 3 \end{aligned}$$
- $$\begin{aligned} (g \circ f)(x) &= g(f(x)) \\ &= g(5x-2) \\ &= 2(5x-2) + 1 \\ &= 10x - 4 + 1 \\ &= 10x - 3 \end{aligned}$$
- $$\begin{aligned} (g \circ g)(x) &= g(g(x)) \\ &= g(2x+1) \\ &= 2(2x+1) + 1 \\ &= 4x + 2 + 1 \\ &= 4x + 3 \end{aligned}$$

Evaluations by Formula

- $$\begin{aligned} (f \circ g)(4) &= 10(4) + 3 \\ &= 43 \end{aligned}$$
- $$\begin{aligned} (g \circ f)(4) &= 10(4) - 3 \\ &= 37 \end{aligned}$$

Evaluations by Definition

- $$\begin{aligned} (f \circ g)(4) &= f(g(4)) \\ &= f(2(4)+1) \\ &= f(9) \\ &= 5(9) - 2 \\ &= 43 \end{aligned}$$
- $$\begin{aligned} (g \circ f)(4) &= g(f(4)) \\ &= g(5(4)-2) \\ &= f(18) \\ &= 2(18) + 1 \\ &= 37 \end{aligned}$$

□

B. $f(x) = 3x - 1$, $g(x) = x^2 - 2x$

EXAMPLE 5

Given the definition of function F, define functions f and g so that $F = f \circ g$.

A. $F(x) = (x+3)^4$

SOLUTION

Let $f(x) = x^4$ and
$g(x) = x + 3$.

□

B. $F(x) = 5/(x^2 - 3)$

C. $F(x) = \sqrt{7 - x}$

SOLUTION

Let $f(x) = \sqrt{x}$ and
$g(x) = 7 - x$.

□

D. $F(x) = |x| - \pi$

EXAMPLE 6

Given the definitions of functions f, g, and h, determine a formula that defines $f \circ g \circ h$.

A. $f(x) = \sqrt{x-2}$
$g(x) = x^2 - 2x + 8$
$h(x) = x - 1$

SOLUTION

$$
\begin{aligned}
&(f \circ g \circ h)(x) \\
&\quad = f((g \circ h)(x)) \\
&\quad = f(g(h(x)) \\
&\quad = f(g(x-1) \\
&\quad = f((x-1)^2 - 2(x-1) + 8) \\
&\quad = f(x^2 - 2x + 1 - 2x + 2 + 8) \\
&\quad = f(x^2 - 4x + 11) \\
&\quad = \sqrt{(x^2 - 4x + 11) - 2} \\
&\quad = \sqrt{x^2 - 4x + 9}
\end{aligned}
$$

□

B. $f(x) = x + 9$
$g(x) = x^6$
$h(x) = \sqrt[3]{x}$

Section 91: Inverses of Functions

Definition

Suppose f is a one-to-one function with domain $\mathbb{D}$ and range $\mathbb{R}$.
The *inverse* of f, denoted by f^{-1}, is the function with domain $\mathbb{R}$ and range $\mathbb{D}$ defined by:

$$f^{-1}(y) = x \Leftrightarrow f(x) = y \quad \text{for all } x \in \mathbb{D}$$

Definition

The *identity function*, denoted by I, is the function defined by $I(x) = x$ for all x.

Properties

Suppose f and g are one-to-one functions with domains $\mathbb{D}_f$ and $\mathbb{D}_g$, respectively.
Then:

- $g = f^{-1} \Leftrightarrow (f \circ g)(x) = x \ \forall\ x \in \mathbb{D}_g$ AND $(g \circ f)(x) = x \ \forall\ x \in \mathbb{D}_f$
- $g = f^{-1} \Leftrightarrow f \circ g = I$ AND $g \circ f = I$
- $g = f^{-1} \Leftrightarrow f = g^{-1}$

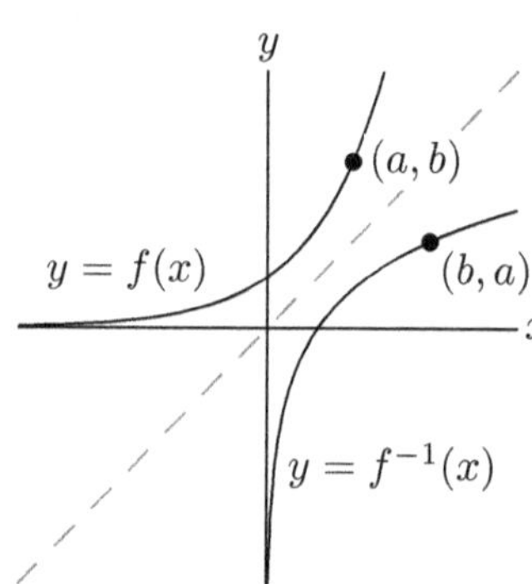

Graphical Relationship

The graph of the inverse of a function f is the reflection of the graph of f through the line $y = x$ because for each point (a, b) that lies on the graph of $y = f(x)$, the point (b, a) lies on the graph of $y = f^{-1}(x)$.

Procedure

Suppose f is a one-to-one function defined by formula.
To determine the formula that defines its inverse f^{-1}, we:

1. let $y = f^{-1}(x)$;
2. solve the equation $f(y) = x$ for y in terms of x;
3. state that $f^{-1}(x)$ equals the expression that y equals in step (2).

Comments

- We read the expression "f^{-1}" aloud as "f inverse."
- If $g = f^{-1}$ and $f = g^{-1}$, then we say that f and g are inverses of each other.
- In step (2) of the procedure above, $f(y) = x$ because $f(f^{-1}(x)) = x$.
- A complete definition for f^{-1} will include its domain as well as its formula.
- The expression $f^{-1}(x)$ represents the image of x under the function f^{-1}; $f^{-1}(x)$ does not represent the quotient $1/f(x)$, which is equivalent to $[f(x)]^{-1}$.
- Technically we use I to represent "an" identity function rather than "the" identity function; the domain of a function I will depend on context and can be any subset of $\mathbf{R}$.
- Although we will limit our discussion of inverses to those of one-to-one functions, we could define more broadly the inverse R^{-1} of any relation R by $R^{-1} = \{\,(b, a) \mid (a, b) \in R\}$.

EXAMPLE 1
Suppose f and g are one-to-one functions where:

$$\begin{array}{lll} f(1) = 6 & f(4) = -9 & f(8) = 2 \\ g(-2) = -9 & g(0) = 3 & g(4) = 7 \end{array}$$

A. Find $f^{-1}(2)$, $g^{-1}(-9)$, and $(f \circ g^{-1})(7)$.

SOLUTION
- $f^{-1}(2) = 8$
- $g^{-1}(-9) = -2$
- $(f \circ g^{-1})(7) = f(g^{-1}(7))$
  $= f(4)$
  $= -9$

□

B. Find $f^{-1}(6)$, $g^{-1}(3)$, and $(f^{-1} \circ g)(-2)$.

EXAMPLE 2
Verify that the given functions are inverses of each other.

A. $f(x) = (x-6)/2$, $g(x) = 2x+6$

SOLUTION
First Test
Suppose $x \in \mathbb{D}_g$.

$$\begin{aligned}(f \circ g)(x) &= f(g(x)) \\ &= f(2x+6) \\ &= ((2x+6)-6)/2 \\ &= (2x)/2 \\ &= x\end{aligned}$$

Second Test
Suppose $x \in \mathbb{D}_f$.

$$\begin{aligned}(g \circ f)(x) &= g(f(x)) \\ &= g((x-6)/2) \\ &= 2((x-6)/2)+6 \\ &= (x-6)+6 \\ &= x\end{aligned}$$

Conclusion
Since $(f \circ g)(x) = x$ for all $x \in \mathbb{D}_g$
and $(g \circ f)(x) = x$ for all $x \in \mathbb{D}_f$,
the functions f and g are inverses of each other.

□

B. $f(x) = x^3 + 5$, $g(x) = \sqrt[3]{x-5}$

Example 3

Fill in the blanks.

A. If f is a one-to-one function with range $\mathbf{R}$, then $(f \circ f^{-1})(5) =$ _____ .

B. If g is a one-to-one function and m lies in the domain of g, then $g^{-1}(g(m)) =$ _____ .

□

Example 4

Provide a counterexample to show that given functions are not inverses of each other.

A. $f(x) = 5x$, $g(x) = x - 3$

Counterexample

Notes

- $f(2) = 5(2)$
  $= 10$
- $g(10) = 10 - 3$
  $= 7$

Conclusion

$f(2) = 10$ but $g(10) \neq 2$, so there exists an $x \in \mathbf{R}$ (viz. 2) such that $g(f(x)) \neq x$. Therefore, $g \neq f^{-1}$.

□

B. $f(x) = x^2$, $g(x) = \sqrt{x}$

Example 5

Determine the formula and domain that define the inverse of the given function.

A. $g(x) = 4 - 3x$

Solution

Formula

Let $y = g^{-1}(x)$.

$g(g^{-1}(x)) = x$

$\Rightarrow g(y) = x$

$\Rightarrow 4 - 3y = x$

$\Rightarrow -3y = -4 + x$

$\Rightarrow (-3y)/(-3) = (-4 + x)/(-3)$

$\Rightarrow y = (4 - x)/3$

$\Rightarrow g^{-1}(x) = (4 - x)/3$

Domain

(domain of g^{-1}) $= \mathbf{R}$

□

B. $h(x) = 7x + 1$

C. $f(x) = \sqrt{-x-8}$

Solution

Formula

Let $y = f^{-1}(x)$.

$f(f^{-1}(x)) = x$
$\Rightarrow f(y) = x$
$\Rightarrow \sqrt{-y-8} = x$
$\Rightarrow (\sqrt{-y-8})^2 = (x)^2$
$\Rightarrow -y-8 = x^2$
$\Rightarrow -y-8+8 = x^2+8$
$\Rightarrow -y = x^2+8$
$\Rightarrow (-1)(-y) = (-1)(x^2+8)$
$\Rightarrow y = -x^2-8$
$\Rightarrow f^{-1}(x) = -x^2-8$

Domain

(range of f) $= [0, \infty)$
$\Rightarrow$ (domain of f^{-1}) $= [0, \infty)$

$\square$

D. $f(x) = (2x-1)/(4x+7)$

Section 92: Linear Functions

Definitions

- A function f is *constant* if defined by the formula $f(x) = b$ for some constant $b \in \mathbf{R}$.
- A function f is *linear* if defined by the formula $f(x) = ax + b$ for some constants $a, b \in \mathbf{R}$.

Comments

- The definition above implies that:
  - f is a constant function if $f(x)$ is a polynomial of degree zero;
  - f is a linear function if $f(x)$ is a polynomial of degree one or zero.
- A constant function is a kind of linear function.
- The graph of any constant function is a horizontal line.
- The graph of any linear function is a nonvertical line.
- The graph of the function defined by $f(x) = ax + b$ has slope a and y-intercept b.
- The domain of any linear function is all real numbers.

Example 1

Draw the graph of each linear function.

A. $g(x) = 2x - 5$

Solution

Table of Values

x	$g(x)$	solution
0	$2(0) - 5 = -5$	$(0, -5)$
5	$2(5) - 5 = 5$	$(5, 5)$

Graph

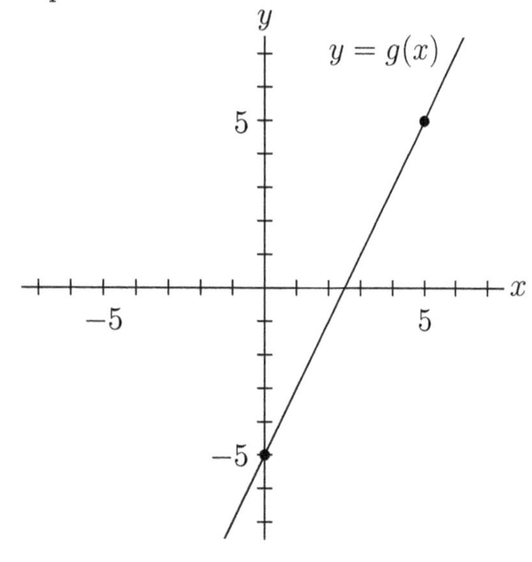

□

B. $\Gamma(x) = (-1/3)x - 3$

C. $f(x) = -6$

SOLUTION
We draw the graph of $y = -6$.

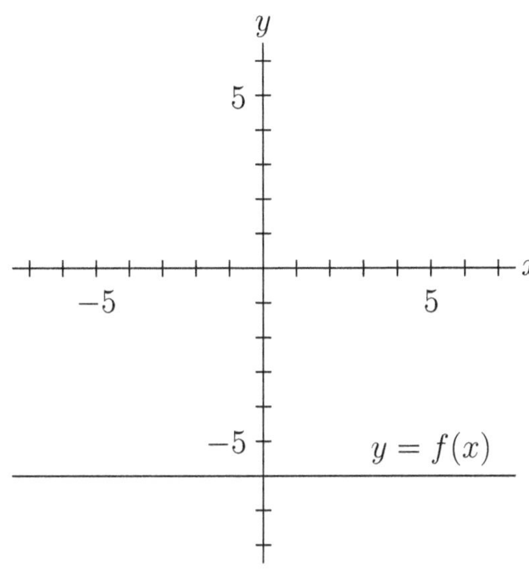

□

D. $f(x) = 2$

EXAMPLE 2
Define a linear function f that satisfies the given conditions.

A. the graph passes through the points $(2, 8)$ and $(7, 23)$

SOLUTION
Note
$f(x) = ax + b$ for some $a, b \in \mathbf{R}$.

Values of the Coefficients
- $a = (23 - 8)/(7 - 2)$
  $= 15/5$
  $= 3$
  Thus $f(x) = 3x + b$.

- $f(2) = 8 \Leftrightarrow 3(2) + b = 8$
  $\Leftrightarrow 6 + b = 8$
  $\Leftrightarrow 6 + b - 6 = 8 - 6$
  $\Leftrightarrow b = 2$

Definition
$f(x) = 3x + 2$

□

B. the graph passes through the points $(1, -5)$ and $(-3, 9)$

C. $f(-5) = 4$ and $f(-9) = 10$

SOLUTION

Notes

- $f(x) = ax + b$ for some $a, b \in \mathbf{R}$
- the graph of f passes through the points $(-5, 4)$ and $(-9, 10)$

Values of the Coefficients

- $a = (10 - 4)/(-9 - (-5))$
  $= 6/(-4)$
  $= -3/2$
  Thus $f(x) = (-3/2)x + b$.

- $f(-5) = 4 \Leftrightarrow (-3/2)(-5) + b = 4$
  $\Leftrightarrow 15/2 + b = 4$
  $\Leftrightarrow b = 4 - 15/2$
  $= 8/2 - 15/2$
  $= -7/2$

Definition

$f(x) = (-3/2)x - 7/2$

□

D. $f(-11) = -24$ and $f(-9) = -6$

Section 93: Quadratic Functions

Definition

A function f is *quadratic* if defined by the formula $f(x) = ax^2 + bx + c$ for some constants $a, b, c \in \mathbf{R}$ with $a \neq 0$.

Properties

Suppose f is a quadratic function defined by $f(x) = a(x-h)^2 + k$. Then:

- if $a > 0$, then f attains a minimum of k at $x = h$
- if $a < 0$, then f attains a maximum of k at $x = h$

Proofs

- Suppose $f(x) = a(x-h)^2 + k$ and $a > 0$.
  Note that:
  - for all $x \in \mathbf{R}$:
    $$\begin{aligned} &(x-h)^2 \geq 0 \\ &\quad \Rightarrow a \cdot (x-h)^2 \geq a \cdot 0 \\ &\quad \Rightarrow a(x-h)^2 + k \geq a \cdot 0 + k \\ &\quad \Rightarrow f(x) \geq k \end{aligned}$$
  - $$\begin{aligned} f(h) &= a(h-h)^2 + k \\ &= a(0)^2 + k \\ &= k \end{aligned}$$

  Therefore, k is the minimum value of f and f attains that minimum at $x = h$.

- Suppose $f(x) = a(x-h)^2 + k$ and $a < 0$.
  Note that:
  - for all $x \in \mathbf{R}$:
    $$\begin{aligned} &(x-h)^2 \geq 0 \\ &\quad \Rightarrow a \cdot (x-h)^2 \leq a \cdot 0 \\ &\quad \Rightarrow a(x-h)^2 + k \leq a \cdot 0 + k \\ &\quad \Rightarrow f(x) \leq k \end{aligned}$$
  - $$\begin{aligned} f(h) &= a(h-h)^2 + k \\ &= a(0)^2 + k \\ &= k \end{aligned}$$

  Therefore, k is the maximum value of f and f attains that maximum at $x = h$.

$\boxdot$

Procedure

To find the extremum of a quadratic function f defined by $f(x) = ax^2 + bx + c$, we:

1. complete the square on the expression $ax^2 + bx + c$ and rewrite the formula for f in the form $f(x) = a(x-h)^2 + k$;
2. identify the value k in the formula of step (1) as the extremum of f;
3. identify the value h of step (1) as the value of x at which f attains its extremum.

Graphs

The graph of a function f defined by $f(x) = a(x-h)^2 + k$ will:

- resemble one of the curves shown at right,
- open up if $a > 0$ and open down if $a < 0$,
- have an extreme point with coordinates (h, k), and
- be symmetric with respect to the line with equation $x = h$.

Terminology

- The graph of a quadratic function is called a *parabola*.
- The extreme point on a parabola is called its *vertex*.
- The line with respect to which a parabola is symmetric is called its *axis of symmetry*.

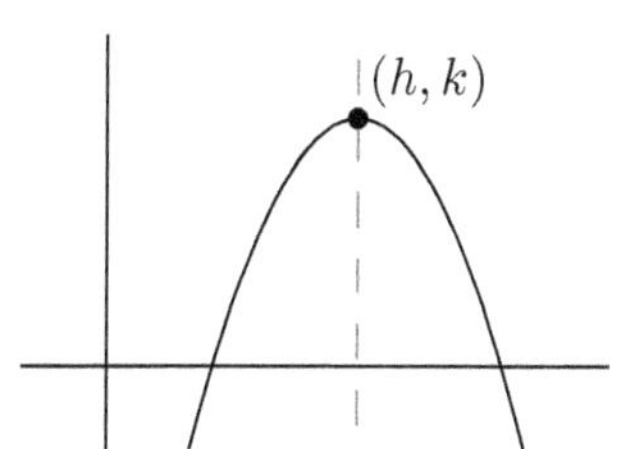

COMMENTS

- The definition on the previous page implies that f is a quadratic function if $f(x)$ is a polynomial of degree two.
- We define formally the terms parabola, vertex, and axis of symmetry on page 415.
- The domain of a quadratic function is all real numbers.

EXAMPLE 1

Identify the vertex and draw the graph of the quadratic function with the given definition.

A. $f(x) = 3x^2 + 18x + 29$

SOLUTION

Vertex

$$\begin{aligned} f(x) &= 3x^2 + 18x + 29 \\ &= 3(x^2 + 6x) + 29 \\ &= 3(x^2 + 6x + 9 + (-9)) + 29 \\ &= 3(x^2 + 6x + 9) + 3(-9) + 29 \\ &= 3(x+3)^2 + (-27) + 29 \\ &= 3(x+3)^2 + 2 \\ &= 3(x - (-3))^2 + 2 \end{aligned}$$

The vertex is the point $(-3, 2)$.

Table of Values

x	$f(x)$	solution
-5	$3(-5+3)^2+2 = 14$	$(-5, 14)$
-4	$3(-4+3)^2+2 = 5$	$(-4, 5)$
-2	$3(-2+3)^2+2 = 5$	$(-2, 5)$
-1	$3(-1+3)^2+2 = 14$	$(-1, 14)$

Graph

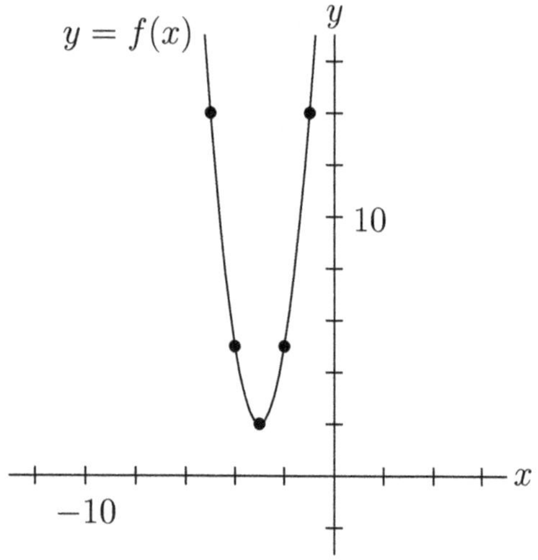

$y = f(x)$

y

10

x

-10

□

B. $f(x) = -x^2 + 8x - 19$

C. $H(x) = (-1/3)(x-4)^2 + 6$

SOLUTION

Vertex

The vertex is the point $(4, 6)$.

Table of Values

x	$H(x)$
-2	$(-1/3)(-2-4)^2 + 6 = -6$
1	$(-1/3)(1-4)^2 + 6 = 3$
7	$(-1/3)(7-4)^2 + 6 = 3$
10	$(-1/3)(10-4)^2 + 6 = -6$

Graph

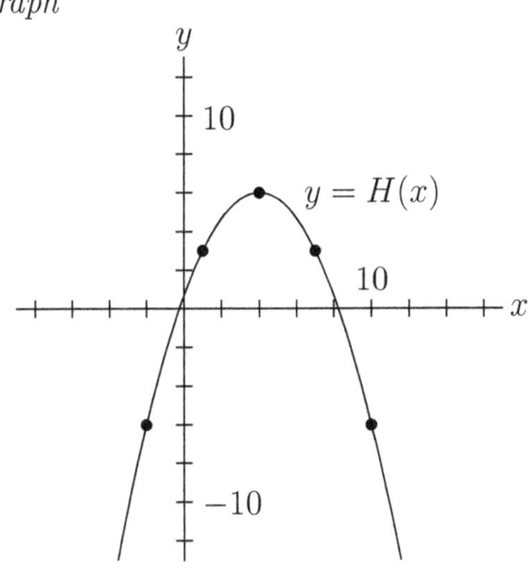

□

D. $g(x) = 2(x+10)^2 - 11$

EXAMPLE 2

Find the vertex and axis of symmetry of the graph of the function with the given definition.

A. $q(x) = x^2 - 8x + 19$

SOLUTION

Note

$$\begin{aligned} q(x) &= x^2 - 8x + 19 \\ &= x^2 - 8x + (-4)^2 - (-4)^2 + 19 \\ &= (x^2 - 8x + (-4)^2) - 16 + 19 \\ &= (x-4)^2 + 3 \end{aligned}$$

Vertex and Axis

- The vertex is the point $(4, 3)$.
- The axis of symmetry is the line $x = 4$.

□

B. $F(x) = x^2 + 6x + 11$

Example 3
Identify the extremum of the quadratic function with the given definition.

A. $f(x) = -3(x-4)^2 + 7$

Solution
The maximum of f is 7.

□

B. $f(x) = (1/2)(x+3)^2 + 8$

C. $g(x) = 2x^2 + 6x + 1$

Solution
Note

$$\begin{aligned} g(x) &= 2x^2 + 6x + 1 \\ &= 2(x^2 + 3x) + 1 \\ &= 2(x^2 + 3x + (3/2)^2 - (3/2)^2) + 1 \\ &= 2((x+3/2)^2 - 9/4) + 1 \\ &= 2(x+3/2)^2 - 9/2 + 2/2 \\ &= 2(x-(-3/2))^2 - 7/2 \end{aligned}$$

Extremum
The minimum of g is $-7/2$.

□

D. $g(x) = -3x^2 + 12x - 5$

Example 4
Suppose the formula $s(t) = 64t - 16t^2$ gives the height (in feet) above the ground of a rocket t seconds after launch. Find the maximum height that the rocket attains and the time at which it attains this height.

Solution
Note

$$\begin{aligned} s(t) &= 64t - 16t^2 \\ &= -16t^2 + 64t \\ &= -16(t^2 - 4t) \\ &= -16(t^2 - 4t + (-2)^2 - (-2)^2) \\ &= -16((t-2)^2 - 4) \\ &= -16(t-2)^2 + 64 \end{aligned}$$

Conclusion
The rocket attains a maximum height of 64 feet at time 2 seconds after launch.

□

Example 5
Suppose the average cost (in francs) of producing x slingshots at a factory is given by $A(x) = 2x^2 - 1128x + 159055$. Find the level x of production at which the average cost is minimized.

Example 6

Suppose a jailer has 100 yards of razor wire with which to build a fence. Determine the dimensions of the largest (by area) rectangular pen that he can enclose with a fence built from that material.

Solution

Declarations

Let w represent the width,
l represent the length, and
A represent the area of the pen.

Substitution

(perimeter of pen) $= 100$

$$\begin{aligned} &\Leftrightarrow 2w + 2l = 100 \\ &\Leftrightarrow 2l = -2w + 100 \\ &\Leftrightarrow (1/2)(2l) = (1/2)(-2w + 100) \\ &\Leftrightarrow l = -w + 50 \end{aligned}$$

Completion of Square

$$\begin{aligned} A &= lw \\ &= (-w + 50)(w) \\ &= -w^2 + 50w \\ &= -(w^2 - 50w) \\ &= -(w^2 - 50w + (-25)^2 - (-25)^2) \\ &= -(w^2 - 50w + (-25)^2) + (-25)^2 \\ &= -(w - 25)^2 + 625 \end{aligned}$$

Extremum

A attains a maximum of 625 at:

- $w = 25$
- $l = -25 + 50$
  $= 25$

Conclusion

The largest pen will have:

- a width of 25 yards and
- a length of 25 yards.

□

Example 7

There are many pairs of numbers that differ by 10. Find the pair where the product of the numbers is the least.

Section 94: The Remainder and Factor Theorems

REMAINDER THEOREM

Suppose $f(x)$ is a polynomial and c is a constant.
Then the remainder yielded by dividing $f(x)$ by $x - c$ equals $f(c)$.

PROOF

Suppose $f(x)$ is a polynomial and c is a constant.
Let $q(x)$ represent the quotient yielded by dividing $f(x)$ by $x - c$ and
R represent the remainder yielded by dividing $f(x)$ by $x - c$.

$$\begin{aligned} f(x) = q(x)(x-c) + R \Rightarrow f(c) &= q(c)(c-c) + R \\ &= q(c)(0) + R \\ &= R \\ \Rightarrow R &= f(c) \end{aligned}$$

□

FACTOR THEOREM

Suppose $f(x)$ is a polynomial and c is a constant.
Then $x - c$ is a factor of $f(x)$ if and only if $f(c) = 0$.

PROOF

Suppose $f(x)$ is a polynomial and c is a constant.

($\Rightarrow$) $x - c$ is a factor of $f(x)$

$$\begin{aligned} &\Rightarrow f(x) = q(x)(x-c) \text{ for some } q(x) \\ &\Rightarrow f(c) = q(c)(c-c) \\ &\qquad\quad = q(c)(0) \\ &\qquad\quad = 0 \end{aligned}$$

($\Leftarrow$) Suppose $f(c) = 0$.

$$\begin{aligned} &f(x) = q(x)(x-c) + f(c) \text{ for some } q(x) \\ &\Rightarrow f(x) = q(x)(x-c) + 0 \\ &\Rightarrow f(x) = q(x)(x-c) \\ &\Rightarrow x - c \text{ is a factor of } f(x) \end{aligned}$$

□

EXAMPLE 1

Use the Remainder Theorem to find $f(-4)$ where $f(x) = -x^6 - 7x^5 - 8x^4 - 4x^2 - 10x - 3$.

SOLUTION

Division

We divide $f(x)$ by $x - (-4)$ synthetically:

$$\begin{array}{r|rrrrrrr} -4 & -1 & -7 & -8 & 0 & -4 & -10 & -3 \\ & & 4 & 12 & -16 & 64 & -240 & 1000 \\ \hline & -1 & -3 & 4 & -16 & 60 & -250 & \boxed{997} \end{array}$$

Value

The remainder yielded by the division of $f(x)$ by $x - (-4)$ is 997.
Therefore, $f(-4) = 997$.

□

EXAMPLE 2

For each polynomial $f(x)$ and constant c:

- determine $f(c)$ using the Remainder Theorem and
- evaluate $f(c)$ directly.

A. $f(x) = 2x^3 - 3x^2 - 7x + 1$; $c = 2$

SOLUTION

Evaluation by Remainder Theorem

We divide $f(x)$ by $x - 2$ synthetically:

$$\begin{array}{r|rrrr} 2 & 2 & -3 & -7 & 1 \\ & & 4 & 2 & -10 \\ \hline & 2 & 1 & -5 & \boxed{-9} \end{array}$$

Thus $f(2) = -9$.

Evaluation by Substitution

$$\begin{aligned} f(2) &= 2(2)^3 - 3(2)^2 - 7(2) + 1 \\ &= 2(8) - 3(4) - 14 + 1 \\ &= 16 - 12 - 14 + 1 \\ &= -9 \end{aligned}$$

□

B. $f(x) = 3x^4 + 2x^3 - x - 9$; $c = -2$

EXAMPLE 3

Use the Factor Theorem to determine whether $g(x)$ is a factor of $f(x)$.

A. $f(x) = 2x^4 + 4x^3 - 6x^2 - 7x + 5$;
$g(x) = x + 1$

SOLUTION

Notes

- $g(x) = x - (-1)$
- $f(-1) = 2(-1)^4 + 4(-1)^3 - 6(-1)^2 - 7(-1) + 5$
  $= 2(1) + 4(-1) - 6(1) + 7 + 5$
  $= 2 - 4 - 6 + 7 + 5$
  $= 4$

Implications

$f(-1) \neq 0$
$\Rightarrow x - (-1)$ is not a factor of $f(x)$
$\Rightarrow g(x)$ is not a factor of $f(x)$

□

B. $f(x) = x^3 - 4x^2 + x + 6$;
$g(x) = x - 3$

EXAMPLE 4

Factorize each polynomial completely.

A. $g(x) = 2x^3 + x^2 - 23x + 20$
[Hint: $g(1) = 0$]

SOLUTION

First Factor

$g(1) = 0$ (by the hint)
$\Rightarrow x - 1$ is a factor of $g(x)$
$\Rightarrow g(x) = (q(x))(x - 1)$ for some $q(x)$

Division

$$\begin{array}{r|rrrr} 1 & 2 & 1 & -23 & 20 \\ & & 2 & 3 & -20 \\ \hline & 2 & 3 & -20 & \boxed{0} \end{array}$$

Factorization

$$\begin{aligned} g(x) &= (2x^2 + 3x - 20)(x - 1) \\ &= (2x - 5)(x + 4)(x - 1) \end{aligned}$$

□

B. $f(x) = 7x^3 + x^2 - 28x - 4$
[Hint: $f(-2) = 0$]

C. $h(x) = 2x^2 - 8x + 2$

SOLUTION

Notes

- $h(x)$ has exactly two linear factors since the degree of $h(x)$ is 2.
- The leading coefficient of $h(x)$ is 2.

Zeros of the Polynomial

$h(x) = 0$

$$\begin{aligned} &\Rightarrow 2x^2 - 8x + 2 = 0 \\ &\Rightarrow x = \left(8 \pm \sqrt{(-8)^2 - 4(2)(2)}\,\right)/(2(2)) \\ &\qquad = (8 \pm \sqrt{48})/4 \\ &\qquad = (8 \pm \sqrt{16 \cdot 3})/4 \\ &\qquad = (8 \pm 4\sqrt{3})/4 \\ &\qquad = \cancel{4}(2 \pm \sqrt{3})/\cancel{4} \\ &\qquad = 2 \pm \sqrt{3} \end{aligned}$$

Factorization

$$\begin{aligned} h(x) &= 2(x^2 - 4x + 1) \\ &= 2(x - (2 - \sqrt{3}))(x - (2 + \sqrt{3})) \\ &= 2(x - 2 + \sqrt{3})(x - 2 - \sqrt{3}) \end{aligned}$$

□

D. $v(x) = x^2 - 2x - 6$

Section 95: The Rational Zeros Theorem

Rational Zeros Theorem (proof omitted)
Suppose that $f(x) = a_n x^n + a_{n-1}x^{n-1} + \ldots + a_1 x + a_0$ for constants $a_0, a_1, a_2, \ldots, a_n \in \mathbf{Z}$ and $n \in \mathbf{N}$ with $a_n \neq 0$
and that p and q are integers that share no common factors.
If p/q is a zero of $f(x)$, then p is a factor of a_0 and q is a factor of a_n.

Example 1
Determine the set of all possible rational zeros of the given polynomial.

A. $2x^3 + 4x^2 - 7x - 6$

Solution
Factors of Coefficients
- constant term (-6): $\pm 6, \pm 3, \pm 2, \pm 1$
- leading coefficient (2): $\pm 2, \pm 1$

Possible Rational Zeros
The set A of the possible rational zeros of $2x^3 + 4x^2 - 7x - 6$ is given by:

$$\begin{aligned} A &= \{\pm\tfrac{6}{2}, \pm\tfrac{3}{2}, \pm\tfrac{2}{2}, \pm\tfrac{1}{2}, \pm\tfrac{6}{1}, \pm\tfrac{3}{1}, \pm\tfrac{2}{1}, \pm\tfrac{1}{1}\} \\ &= \{\pm 3, \pm\tfrac{3}{2}, \pm 1, \pm\tfrac{1}{2}, \pm 6, \pm 3, \pm 2, \pm 1\} \\ &= \{\pm 3, \pm\tfrac{3}{2}, \pm 1, \pm\tfrac{1}{2}, \pm 6, \pm 2\} \\ &= \{\pm\tfrac{1}{2}, \pm 1, \pm\tfrac{3}{2}, \pm 2, \pm 3, \pm 6\} \end{aligned}$$

□

B. $14x^5 + x^2 + 7x + 5$

Example 2
Show that the polynomial $x^3 - 7x - 1$ has no rational zeros.

Solution
Possible Rational Zeros
Let $f(x) = x^3 - 7x + 1$.
The factors of:
- the constant term -1 are ± 1;
- the leading coefficient 1 are ± 1.

Hence the only possible rational zeros of the polynomial $f(x)$ are $\pm 1/1$, or ± 1.

Values of the Polynomial
- $f(-1) = (-1)^3 - 7(-1) - 1$
  $= -1 + 7 - 1$
  $= 5$
- $f(1) = (1)^3 - 7(1) - 1$
  $= 1 - 7 - 1$
  $= -7$

Conclusion
Both $f(-1) \neq 0$ and $f(1) \neq 0$, so neither of the possible rational zeros -1 and 1 actually is a zero of the polynomial $f(x)$. Therefore, the polynomial $x^3 - 7x - 1$ has no rational zeros.

□

EXAMPLE 3
Identify one rational zero of the given polynomial.

A. $3x^4 - x^3 + 9x - 3$

SOLUTION
Possible Rational Zeros
Let $f(x) = 3x^4 - x^3 + 9x - 3$.
The factors of:

- the constant term (-3) are ± 3, ± 1;
- the leading coefficient (3) are ± 3, ± 1.

So the set A of the possible rational zeros of $f(x)$ is given by:

$$\begin{aligned} A &= \{\pm 3/3, \pm 3/1, \pm 1/3, \pm 1/1\} \\ &= \{\pm 1/3, \pm 1, \pm 3\} \end{aligned}$$

Test of Possibilities
We attempt through direct evaluation to find an $a \in A$ for which $f(a) = 0$:

- $$\begin{aligned} f(1) &= 3(1)^4 - (1)^3 + 9(1) - 3 \\ &= 3 - 1 + 9 - 3 \\ &= 8 \\ &\neq 0 \end{aligned}$$

- $$\begin{aligned} f(-1) &= 3(-1)^4 - (-1)^3 + 9(-1) - 3 \\ &= 3 + 1 - 9 - 3 \\ &= -8 \\ &\neq 0 \end{aligned}$$

(Since $f(-1) < 0$ and $f(1) > 0$, the graph of f must cross the x-axis at some point between $x = -1$ and $x = 1$. Hence $f(c) = 0$ for some $c \in (-1, 1)$. Therefore, we check next the possible zeros between -1 and 1, namely 1/3 and $-1/3$.)

- $$\begin{aligned} f(1/3) &= 3(1/3)^4 - (1/3)^3 + 9(1/3) - 3 \\ &= 3(1/81) - 1/27 + \cancel{3} - \cancel{3} \\ &= 1/27 - 1/27 \\ &= 0 \end{aligned}$$

Rational Zero
1/3 is a rational zero of $f(x)$.

□

B. $x^3 - 2x^2 - 33x + 10$

EXAMPLE 4

Factorize the polynomial $x^3 - 13x^2 + 44x - 14$ completely.

SOLUTION

Possible Rational Zeros

Let $f(x)$ equal $x^3 - 13x^2 + 44x - 14$.

The factors of:

- the constant term (-14) are ±14, ±7, ±2, ±1;
- the leading coefficient (1) are ±1.

So the set A of the possible rational zeros of $f(x)$ is given by:

$$\begin{aligned} A &= \{\pm14/1, \pm7/1, \pm2/1, \pm1/1\} \\ &= \{\pm1, \pm2, \pm7, \pm14\} \end{aligned}$$

Identification of Rational Zero and Partial Factorization of $f(x)$

We now use synthetic division to try to find an $a \in A$ for which the remainder yielded by dividing $f(x)$ by $x - a$ is zero (that is, for which $f(a) = 0$):

$$\begin{array}{r|rrrr} 1 & 1 & -13 & 44 & -14 \\ & & 1 & -12 & 32 \\ \hline & 1 & -12 & 32 & \boxed{18} \end{array} \qquad \begin{array}{r|rrrr} -1 & 1 & -13 & 44 & -14 \\ & & -1 & 14 & -58 \\ \hline & 1 & -14 & 58 & \boxed{-72} \end{array} \qquad \begin{array}{r|rrrr} 2 & 1 & -13 & 44 & -14 \\ & & 2 & -22 & 44 \\ \hline & 1 & -11 & 22 & \boxed{30} \end{array}$$

$$\begin{array}{r|rrrr} -2 & 1 & -13 & 44 & -14 \\ & & -2 & 30 & -148 \\ \hline & 1 & -15 & 74 & \boxed{-162} \end{array} \qquad \begin{array}{r|rrrr} 7 & 1 & -13 & 44 & -14 \\ & & 7 & -42 & 14 \\ \hline & 1 & -6 & 2 & \boxed{0} \end{array}$$

Thus $f(7) = 0$ and $f(x) = (x - 7)(x^2 - 6x + 2)$.

Factorization of $x^2 - 6x + 2$

Note that:

- $x^2 - 6x + 2$ has exactly two linear factors since the degree of $x^2 - 6x + 2$ is 2
- the leading coefficient of $x^2 - 6x + 2$ is 1
- $\begin{aligned}[t] x^2 - 6x + 2 = 0 \iff x &= (-(-6) \pm \sqrt{(-6)^2 - 4(1)(2)}\,)/(2(1)) \\ &= (6 \pm \sqrt{36 - 8})/2 \\ &= (6 \pm \sqrt{28})/2 \\ &= (6 \pm 2\sqrt{7})/2 \\ &= \cancel{2}(3 \pm \sqrt{7})/\cancel{2} \\ &= 3 \pm \sqrt{7} \end{aligned}$

Thus $x^2 - 6x + 2 = 1 \cdot (x - (3 - \sqrt{7})(x - (3 + \sqrt{7}))$.

Complete Factorization of $f(x)$

$$\begin{aligned} x^3 - 13x^2 + 44x - 14 &= (x - 7)(x^2 - 6x + 2) \\ &= (x - 7) \cdot [1 \cdot (x - (3 - \sqrt{7})(x - (3 + \sqrt{7}))] \\ &= (x - 7)(x - 3 + \sqrt{7})(x - 3 - \sqrt{7}) \end{aligned}$$

□

Section 96: Descartes' Rule of Signs

Definition

A *variation in sign* is a difference in the signs of coefficients of successive terms of a polynomial with real coefficients, assuming the polynomial takes the form where terms appear in descending order of degree and only if nonzero.

Descartes' Rule of Signs

Suppose $f(x)$ is a polynomial with real coefficients.
Then:

- the number of positive real zeros of $f(x)$ is either:
  - equal to the number of variations in sign within $f(x)$, or
  - less than the number of variations in sign within $f(x)$ by a even whole number;
- the number of negative real zeros of $f(x)$ is either:
  - equal to the number of variations in sign within $f(-x)$, or
  - less than the number of variations in sign within $f(-x)$ by a even whole number.

Example 1

A. The polynomial $x^5 - 7x^4 + 9x^3 - 2x^2 + 8x - 11$ contains 5 variations in sign.
B. The polynomial $-3x^4 + 5x^3 - x^2 - 6x - 3$ contains 2 variations in sign.
C. The polynomial $10x^6 + x^5 + 2x^4 + x^3 + 5x^2 + 3x + 7$ contains 0 variations in sign.

□

Example 2

Determine the possible numbers of positive and negative real zeros of the polynomial.

A. $x^4 - 7x^3 + 9x^2 - 2x + 8$

Solution

Declarations

Let $f(x) = x^4 - 7x^3 + 9x^2 - 2x + 8$.
Thus
$$\begin{aligned} f(-x) &= (-x)^4 - 7(-x)^3 + 9(-x)^2 \\ &\qquad - 2(-x) + 8 \\ &= x^4 + 7x^3 + 9x^2 + 2x + 8 \end{aligned}$$

Variations in Sign

- $f(x)$ contains 4 variations in sign
- $f(-x)$ contains 0 variations in sign

Numbers of Zeros

- $f(x)$ has 4, 2, or 0 positive real zeros
- $f(x)$ has 0 negative real zeros

□

B. $5x^3 - x^2 - 6x - 3$

Section 97: Dominant Terms of Polynomials

Definition

The *dominant term* of a polynomial is the term of the polynomial of highest degree.

Notation

- "$x \to \infty$" means that the value of x increases without (upper) bound.
- "$x \to -\infty$" means that the value of x decreases without (lower) bound.

Properties

Suppose $f(x) = a_n x^n + a_{n-1}x^{n-1} + \ldots + a_1 x + a_0$ for some constants $a_0, a_1, a_2, \ldots, a_n \in \mathbf{R}$. The table below describes the behavior of $f(x)$ as the value of x increases or decreases without bound.

n	Sign of a_n	Behavior of x	Behavior of $f(x)$
even	$+$	$x \to \infty$	$f(x) \to \infty$
even	$+$	$x \to -\infty$	$f(x) \to \infty$
even	$-$	$x \to \infty$	$f(x) \to -\infty$
even	$-$	$x \to -\infty$	$f(x) \to -\infty$
odd	$+$	$x \to \infty$	$f(x) \to \infty$
odd	$+$	$x \to -\infty$	$f(x) \to -\infty$
odd	$-$	$x \to \infty$	$f(x) \to -\infty$
odd	$-$	$x \to -\infty$	$f(x) \to \infty$

Comments

- When the value of x is large (either positive or negative), the value of a polynomial $f(x)$ will depend primarily upon the value of its dominant term $a_n x^n$ because the magnitude of that dominant term will exceed the magnitudes of its other terms. Therefore, as the value of x increases or decreases without bound:
  - the behavior of $f(x)$ will depend primarily upon the sign of a_n and the value of n;
  - the ratio of $a_n x^n$ to $f(x)$ will approach 1.
- The dominant term of a polynomial is equivalent to the product of the dominant terms of its factors.

Example 1

Suppose $f(x) = -5x^3 - 7x^2 + 20x - 80$. The table below suggests that as $x \to -\infty$:
- the magnitude of $-5x^3$ will exceed the magnitude of the other terms of $f(x)$;
- the value of $f(x)$ will depend primarily on the value of the term $-5x^3$.

x	$-5x^3$	$-7x^2$	$20x$	-80	$f(x)$	$-5x^3/f(x)$
-1	5	-7	-20	-80	-102	-0.049
-10	5,000	-700	-200	-80	4,020	1.244
-100	5,000,000	$-70,000$	$-2,000$	-80	4,927,920	1.015
-1000	5,000,000,000	$-7,000,000$	$-20,000$	-80	4,992,979,920	1.001

□

EXAMPLE 2

State the dominant term and determine the behavior as $x \to -\infty$ and $x \to \infty$ of each polynomial.

A. $g(x) = -2x^8 + 10x^3 - 7x$

SOLUTION

Dominant Term

(dominant term of $g(x)$) $= -2x^8$

Notes

- $-2x^8 < 0$ when $x < 0$
- $-2x^8 < 0$ when $x > 0$

Behavior

- $g(x) \to -\infty$ as $x \to -\infty$
- $g(x) \to -\infty$ as $x \to \infty$

□

B. $f(x) = 800 - x + 6x - 4x^3$

C. $f(x) = 3x^3(x-1)(x-8)$

SOLUTION

Dominant Term

$$\begin{aligned}(\text{dominant term of } f(x)) &= 3x^3(x)(x) \\ &= 3x^5\end{aligned}$$

Notes

- $3x^5 < 0$ when $x < 0$
- $3x^5 > 0$ when $x > 0$

Behavior

- $f(x) \to -\infty$ as $x \to -\infty$
- $f(x) \to \infty$ as $x \to \infty$

□

D. $h(x) = 7(3x-5)(x-1)(x+4)(x+9)$

Section 98: Polynomial Functions

Definition

A function f is *polynomial* if defined by $f(x) = a_n x^n + a_{n-1}x^{n-1} + \ldots + a_1 x + a_0$ for some constants $n \in \mathbf{N} \cup \{0\}$ and $a_0, a_1, a_2, \ldots, a_n \in \mathbf{R}$.

Procedure A

To draw the graph of a polynomial function f where $f(x)$ has degree 0, 1, or 2, we follow the procedures described previously for drawing the graphs of constant, linear, and quadratic functions. Recall that:

- if $f(x)$ has degree 0, then the graph of f will be a horizontal line;
- if $f(x)$ has degree 1, then the graph of f will be a nonvertical line;
- if $f(x)$ has degree 2, then the graph of f will be a parabola.

Procedure B

To sketch the graph of a polynomial function f where $f(x)$ has degree 3 or more, we:

1. identify the intercepts of the graph of f by finding:
   - the real zeros of $f(x)$ (the x-intercepts) and
   - the value of $f(0)$ (the y-intercept);
2. identify the signs of the values of $f(x)$ between its zeros;
3. identify the behavior of $f(x)$ as $x \to \infty$ and as $x \to -\infty$;
4. sketch the graph with the characteristics from steps (1) through (3) in recognition that:
   - the graph of f will lie above the x-axis on intervals where $f(x)$ is positive,
   - the graph of f will lie below the x-axis on intervals where $f(x)$ is negative,
   - beyond some point, a particle that traces the graph of f from right to left will:
     - rise steeply without bound or downturn if $f(x) \to \infty$ as $x \to -\infty$ and
     - fall steeply without bound or upturn if $f(x) \to -\infty$ as $x \to -\infty$, and
   - beyond some point, a particle that traces the graph of f from left to right will:
     - rise steeply without bound or downturn if $f(x) \to \infty$ as $x \to \infty$ and
     - fall steeply without bound or upturn if $f(x) \to -\infty$ as $x \to \infty$.

Comments

- The definition above implies that f is a polynomial function if $f(x)$ is a polynomial.
- The graph of any polynomial function f will:
  - be smooth and unbroken,
  - contain no more than $n - 1$ turning points, where n represents the degree of $f(x)$, and
  - extend infinitely far upward or downward as $x \to \infty$ and as $x \to -\infty$.
- The domain of a polynomial function is all real numbers.
- We can determine the behavior of $f(x)$ as $x \to \infty$ and as $x \to -\infty$ using the signs of values of $f(x)$ that we find in step (2) of procedure B for large positive and negative values of x.
- We may omit tick marks on a y-axis to which a graph bears reference if we intend only to sketch the graph crudely and we plot no points on the graph above or below the x-axis.

EXAMPLE 1
Sketch the graph of the polynomial function with the given definition.

A. $f(x) = 3(x-2)(x-5)^2$

SOLUTION
Intercepts

- $f(x) = 0 \Leftrightarrow 3(x-2)(x-5)^2 = 0$
  $\Leftrightarrow x - 2 = 0$ OR $x - 5 = 0$
  $\Leftrightarrow x = 2$ $\qquad \Leftrightarrow x = 5$

The x-intercepts are 2 and 5.

- $f(0) = 3(0-2)(0-5)^2$
  $= 3(-2)(25)$
  $= -150$

The y-intercept is -150.

Sign Chart

$f(x)$	$-$	$+$	$+$
3	$+$	$+$	$+$
$x-2$	$-$	$+$	$+$
$(x-5)^2$	$+$	$+$	$+$
	2	5	

Behavior of $f(x)$

- $f(x) < 0$ when $x < 2$
  $\Rightarrow f(x) \to -\infty$ as $x \to -\infty$
- $f(x) > 0$ when $x > 5$
  $\Rightarrow f(x) \to \infty$ as $x \to \infty$

Graph

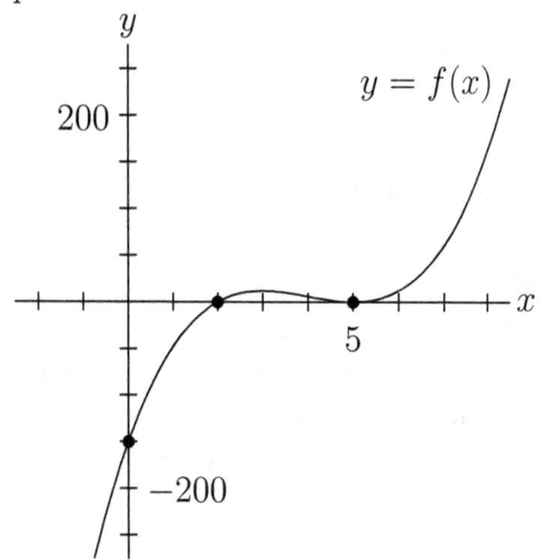

□

B. $f(x) = \frac{1}{6}x(x-3)(x+2)(x+6)$

c. $f(x) = -\frac{1}{2}x^4 + 2x^3 + 6x^2$

Solution

Note

$$\begin{aligned} f(x) &= (-1/2)(x^4 - 4x^3 - 12x^2) \\ &= (-1/2)(x^2)(x^2 - 4x - 12) \\ &= (-1/2)(x^2)(x - 6)(x + 2) \end{aligned}$$

Intercepts

- $f(x) = 0$
  $\Leftrightarrow (-1/2)(x^2)(x - 6)(x + 2) = 0$
  $\Leftrightarrow x = 0$ or $x = 6$ or $x = -2$
  The x-intercepts are -2, 0, and 6.

- $f(0) = (-1/2)(0)^4 + 2(0)^3 + 6(0)^2$
  $= 0$
  The y-intercept is 0.

Sign Chart

$f(x)$	$-$	$+$	$+$	$-$
$-1/2$	$-$	$-$	$-$	$-$
x^2	$+$	$+$	$+$	$+$
$x-6$	$-$	$-$	$-$	$+$
$x+2$	$-$	$+$	$+$	$+$
	-2		0	6

Behavior of $f(x)$

- $f(x) < 0$ when $x < -2$
  $\Rightarrow f(x) \to -\infty$ as $x \to -\infty$
- $f(x) < 0$ when $x > 6$
  $\Rightarrow f(x) \to -\infty$ as $x \to \infty$

Graph

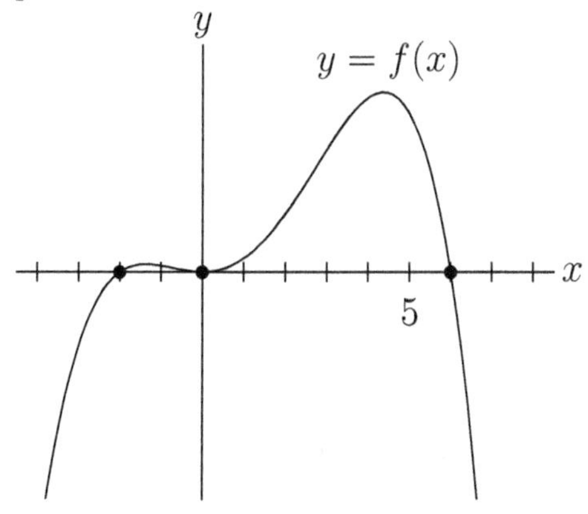

□

d. $f(x) = -x^3 + x^2 + 13x + 14$
[Hint: One zero of $f(x)$ is -2.]

Section 99: Asymptotes

Definitions

- An *asymptote* of a curve is a line that a particle tracing the curve approaches as a coordinate of its location increases or decreases without bound.
- An *oblique* asymptote is an asymptote that is neither horizontal nor vertical.

Notation

- "$x \to a$" means that the value of x approaches a.
- "$x \to a^-$" means that the value of x approaches a from the left.
- "$x \to a^+$" means that the value of x approaches a from the right.

Graphical Relationships

Suppose f and g are functions and A, B, C, and D are real constants.
If $g(x) = A \cdot f(B(x - C)) + D$ for all x in the domain of g, then:

- if the line $x = a$ is a vertical asymptote of the graph of f, then the line $x = a/B + C$ is a vertical asymptote of the graph of g;
- if the line $y = b$ is a horizontal asymptote of the graph of f, then the line $y = Ab + D$ is a horizontal asymptote of the graph of g.

Graphical Relationships

Suppose r is a function defined by $r(x) = f(x)/g(x)$, where $f(x)$ and $g(x)$ are polynomials with degrees m and n and leading coefficients c_m and d_n, respectively. Then:

- the line $x = a$ is a vertical asymptote of the graph of r if $g(a) = 0$ and $f(a) \neq 0$;
- the line $y = 0$ is a horizontal asymptote of the graph of r if $m < n$;
- the line $y = c_m/d_n$ is a horizontal asymptote of the graph of r if $m = n$;
- the line $y = q(x)$ is an oblique asymptote of the graph of r if $m = n + 1$,
  where $q(x)$ represents the quotient yielded by dividing $f(x)$ by $g(x)$.

Proofs (informal)

Suppose r is a function defined by $r(x) = f(x)/g(x)$, where:

- ∘ $f(x) = c_m x^m + c_{m-1} x^{m-1} + \ldots + c_1 x + c_0$ for some constants $c_0, c_1, c_2, \ldots, c_m \in \mathbf{R}$
- ∘ $g(x) = d_n x^n + d_{n-1} x^{n-1} + \ldots + d_1 x + d_0$ for some constants $d_0, d_1, d_2, \ldots, d_n \in \mathbf{R}$

- Suppose $g(a) = 0$ and $f(a) \neq 0$.
  Thus $f(a) = k$ for some $k \neq 0$.
  But since polynomial functions are continuous (i.e., their graphs are unbroken):
  $f(x) \to k$ and $g(x) \to 0$ as $x \to a$
  $\Rightarrow f(x)/g(x) \to \infty$ or $f(x)/g(x) \to -\infty$ as $x \to a$
  $\Rightarrow r(x) \to \infty$ or $r(x) \to -\infty$ as $x \to a$

So the y-coordinate of a particle that traces the graph of r increases or decreases without bound as its x-coordinate approaches a (i.e., as the particle approaches the line $x = a$). Thus the vertical line $x = a$ is an asymptote of the graph of r.

- Suppose $m < n$.
  Observe then that:

$$\begin{aligned}
|r(x) - 0| &= \left|\frac{c_m x^m + c_{m-1}x^{m-1} + \ldots + c_1 x + c_0}{d_n x^n + d_{n-1}x^{n-1} + \ldots + d_1 x + d_0} \cdot \frac{1/x^n}{1/x^n} - 0\right| \\
&= \left|\frac{c_m/x^{n-m} + c_{m-1}/x^{n-m+1} + \ldots + c_1/x^{n-1} + c_0/x^n}{d_n + d_{n-1}/x + \ldots + d_1/x^{n-1} + d_0/x^n}\right| \\
&\to \left|\frac{0 + 0 + \ldots + 0 + 0}{d_n + 0 + \ldots + 0 + 0}\right| \text{ as } x \to \pm\infty \\
&\to 0 \text{ as } x \to \pm\infty
\end{aligned}$$

  So the y-coordinate of a particle that traces the graph of r approaches 0 (i.e., the particle approaches the line $y = 0$) as its x-coordinate increases or decreases without bound. Thus the horizontal line $y = 0$ is an asymptote of the graph of r.

- Suppose $m = n$.
  Observe then that:

$$\begin{aligned}
|r(x) - c_m/d_n| &= \left|\frac{c_m x^m + c_{m-1}x^{m-1} + \ldots + c_1 x + c_0}{d_n x^n + d_{n-1}x^{n-1} + \ldots + d_1 x + d_0} \cdot \frac{1/x^n}{1/x^n} - c_m/d_n\right| \\
&= \left|\frac{c_m + c_{m-1}/x + \ldots + c_1/x^{n-1} + c_0/x^n}{d_n + d_{n-1}/x + \ldots + d_1/x^{n-1} + d_0/x^n} - c_m/d_n\right| \\
&\to \left|\frac{c_m + 0 + \ldots + 0 + 0}{d_n + 0 + \ldots + 0 + 0} - c_m/d_n\right| \text{ as } x \to \pm\infty \\
&\to |c_m/d_n - c_m/d_n| \text{ as } x \to \pm\infty \\
&\to 0 \text{ as } x \to \pm\infty
\end{aligned}$$

  So the y-coordinate of a particle that traces the graph approaches c_m/d_n (i.e., the particle approaches the line $y = c_m/d_n$) as its x-coordinate increases or decreases without bound. Thus the horizontal line $y = c_m/d_n$ is an asymptote of the graph of r.

- Suppose $m = n + 1$,
  So $r(x) = q(x) + v(x)/g(x)$ for some polynomials $q(x)$ and $v(x)$,
  where $\deg(q(x)) = 1$ and $\deg(v(x)) < \deg(g(x))$.
  Observe then that:

$$\begin{aligned}
|r(x) - q(x)| &= |q(x) + v(x)/g(x) - q(x)| \\
&= |v(x)/g(x)| \\
&\to 0 \text{ as } x \to \pm\infty
\end{aligned}$$

  So the y-coordinate of a particle that traces the graph of r approaches $q(x)$ (i.e., the particle approaches the line $y = q(x)$) as its x-coordinate increases or decreases without bound. Thus the line $y = q(x)$ is an asymptote of the graph of r.

⊡

Comments

- Oblique asymptotes are also called *slant* asymptotes.
- We represent asymptotes graphically as dashed lines.
- Before we attempt to identify the vertical asymptotes of the graph of a function, we should ensure that its numerator and denominator have no common zeros by factorizing both these expressions completely and cancelling their common factors.
- We can use a sign chart to determine the direction (up or down) that a particle that traces the graph of a function takes near a vertical asymptote.
- If $r(x) = f(x)/g(x)$ and $\deg(f(x)) > \deg(g(x)) + 1$, then the graph of r will approach neither a horizontal nor an oblique asymptote because the graph of $y = q(x)$, where $q(x)$ represents the quotient yielded by dividing $f(x)$ by $g(x)$, will not be linear.
- The way in which we translate asymptotes of one graph to asymptotes of another graph is similar to the way in which we translate points on one graph to points on another graph.
- To identify an asymptote, we must state the equation of a line.
- A function r defined by $r(x) = f(x)/g(x)$, where $f(x)$ and $g(x)$ are polynomials:
  - may have more than one vertical asymptote and
  - will have at most one horizontal or oblique asymptote, since the graph of r will approach the same line when $x \to \infty$ as when $x \to -\infty$.

Example 1

Find the equations of the asymptotes of the graph of each function.

A. $r(x) = (6x^2 - 7x)/(4x^2 - 13x + 9)$

Solution

Notes

- $4x^2 - 13x + 9 = 0$

  $\Leftrightarrow (4x - 9)(x - 1) = 0$

  $\Leftrightarrow 4x - 9 = 0 \quad \text{OR } x - 1 = 0$

  $\quad \Leftrightarrow 4x = 9 \qquad \Leftrightarrow x = 1$

  $\quad \Leftrightarrow x = 9/4$
- $\deg(6x^2 - 7x) = \deg(4x^2 - 13x + 9)$, since:
  - $\deg(6x^2 - 7x) = 2$
  - $\deg(4x^2 - 13x + 9) = 2$

Asymptotes

The graph of r has:

- vertical asymptotes with equations $x = 1$ and $x = 9/4$;
- a horizontal asymptote with equation $y = 6/4$, or $y = 3/2$.

□

B. $r(x) = (8x + 3)/(4x^2 + 12x - 7)$

C. $r(x) = 5/(3x+2)$

SOLUTION
Notes
- $3x + 2 = 0 \Leftrightarrow 3x = -2$
  $\Leftrightarrow x = -2/3$
- $\deg(5) < \deg(3x+2)$, since:
  - $\deg(5) = 0$
  - $\deg(3x+2) = 1$

Asymptotes
The graph of r has:
- a vertical asymptote with equation $x = -2/3$;
- a horizontal asymptote with equation $y = 0$.

□

D. $r(x) = 1000/(x-5)$

E. $r(x) = (3x^2 - 16x + 1)/(x - 7)$

SOLUTION
Notes
- $x - 7 = 0 \Leftrightarrow x = 7$
- $\deg(3x^2 - 16x + 1) = \deg(x - 7) + 1$, since:
  - $\deg(3x^2 - 16x + 1) = 2$
  - $\deg(x - 7) = 1$
- dividing numerator by denominator synthetically yields:

$$\begin{array}{r|rrr} 7 & 3 & -16 & 1 \\ & & 21 & 35 \\ \hline & 3 & 5 & \boxed{36} \end{array}$$

- $r(x) = 3x + 5 + 36/(x - 7)$

Asymptotes
The graph of r has:
- a vertical asymptote with equation $x = 7$;
- an oblique asymptote with equation $y = 3x + 5$.

□

F. $r(x) = (5x^3 + 3x^2 - 7x - 4)/(x^2 - 4)$
[Hint: Use polynomial division to find the oblique asymptote.]

G. $r(x) = (2x^2 + 11x + 12)/(x + 4)$

Solution

Notes

- $r(x) = (2x + 3)\cancel{(x+4)}/\cancel{(x+4)}$
  $= 2x + 3$
- (domain of r) $= \{\, x \in \mathbf{R} \mid x + 4 \neq 0 \,\}$
  $= \{\, x \in \mathbf{R} \mid x \neq -4 \,\}$

Implication

The graph of r is a broken line with equation $y = 2x + 3$ and hole at $x = -4$. This graph has:

- no vertical asymptotes;
- no horizontal or oblique asymptotes (other than the line $y = 2x + 3$ itself).

□

H. $r(x) = (x - 8)/(8 - x)$

Example 2

Suppose f and g are functions where:

- $g(x) = -f(3x - 3) + 2$
- the graph of f has:
  - the vertical asymptote $x = -6$;
  - the horizontal asymptote $y = 4$.

Translate the asymptotes given above to the graph of g.

Solution

Note

$g(x) = (-1) \cdot f(3(x - 1)) + 2$

Translations

- $x = -6 \mapsto x = -6/3 + 1$
  $\Rightarrow x = -2 + 1$
  $\Rightarrow x = -1$
- $y = 4 \mapsto y = (-1) \cdot 4 + 2$
  $\Rightarrow y = -4 + 2$
  $\Rightarrow y = -2$

Asymptotes

The graph of g has:

- the vertical asymptote $x = -1$;
- the horizontal asymptote $y = -2$.

□

Example 3

Suppose f and g are functions where:

- $g(x) = 5 \cdot f(2x + 6) - 9$
- the graph of f has:
  - the vertical asymptote $x = 5$
  - the horizontal asymptote $y = -3$

Translate the asymptotes given above to the graph of g.

Section 100: Rational Functions

Definition

A function r is *rational* if r is defined by $r(x) = f(x)/g(x)$ for some polynomials $f(x)$ and $g(x)$ with $g(x) \neq 0$.

Procedure

To sketch the graph of a rational function r where $r(x) = f(x)/g(x)$, we:

1. factorize both numerator and denominator of $r(x)$ and cancel their common factors;
2. determine the x-intercepts of the graph of r;
3. determine the equations of the asymptotes of the graph of r;
4. determine the signs of the values of $r(x)$ between the zeros of $f(x)$ and $g(x)$;
5. determine the values of $r(x)$ at various values of x;
6. sketch the graph of r with the characteristics from steps (1) through (5).

Comment

The domain of a rational function will be all real numbers at which the denominator of the expression that defines it does not equal zero.

Example 1

Sketch the graph of the rational function with the given definition.

A. $r(x) = (21 - 7x)/(4x - 12)$

Solution

Notes

- $r(x) = [-7\cancel{(-3+x)}]/[4\cancel{(x-3)}]$
  $= -7/4$
- $4x - 12 = 0 \Leftrightarrow 4x/4 = 12/4$
  $\Leftrightarrow x = 3$
- (domain of r) $= \{\, x \in \mathbf{R} \mid x \neq 3 \,\}$

Graph

The graph of r is a broken line with equation $y = -7/4$ and hole at $x = 3$:

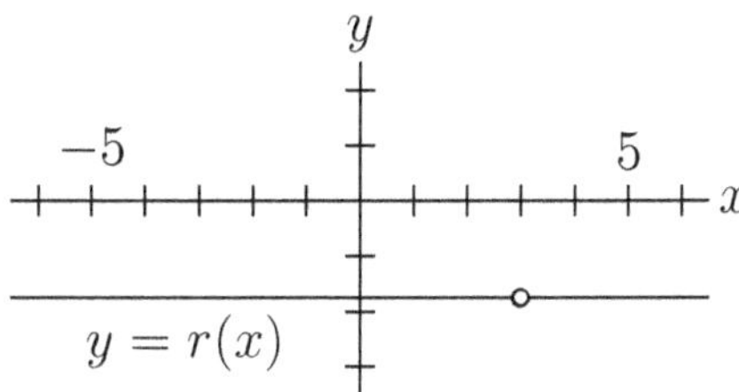

□

B. $r(x) = (2x^2 + 11x + 12)/(x + 4)$

c. $r(x) = 8/(x+4)$

Solution

x-Intercepts

$8/(x+4) \neq 0$ for all $x \in \mathbf{R}$

The graph has no x-intercepts.

Asymptotes

Note that:

- $x + 4 = 0 \Leftrightarrow x = -4$
- $\deg(8) < \deg(x+4)$

The graph has:

- the vertical asymptote $x = -4$;
- the horizontal asymptote $y = 0$.

Sign Chart

$r(x)$	$-$	$+$
8	$+$	$+$
$x+4$	$-$	$+$
	-4	

Table of Values

x	-12	-8	-5	-3	0	4
$r(x)$	-1	-2	-8	8	2	1

Graph

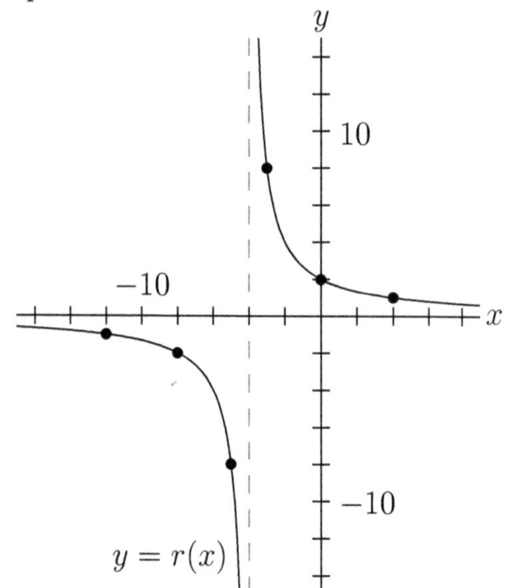

□

d. $r(x) = 12/(x-2)^2$

E. $r(x) = (2x+1)/(x-3)$

SOLUTION

x-Intercepts

$$(2x+1)/(x-3) = 0 \Leftrightarrow 2x+1 = 0$$
$$\Leftrightarrow x = -1/2$$

The x-intercept is $-1/2$.

Asymptotes

Note that:

- $x - 3 = 0 \Leftrightarrow x = 3$
- $\deg(2x+1) = \deg(x-3)$
- $y = 2/1 \Leftrightarrow y = 2$

The graph has:

- the vertical asymptote $x = 3$;
- the horizontal asymptote $y = 2$.

Sign Chart

$r(x)$	$+$	$-$	$+$
$2x+1$	$-$	$+$	$+$
$x-3$	$-$	$-$	$+$
	$-\frac{1}{2}$	3	

Table of Values

x	-4	0	2	4	7	10
$r(x)$	1	-0.3	-5	9	3.8	3

Graph

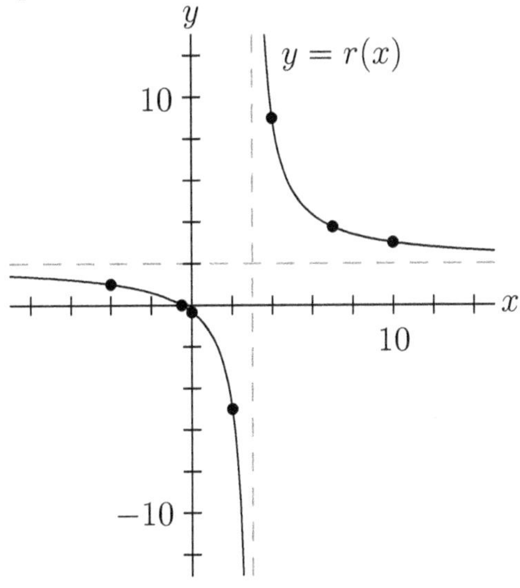

□

F. $r(x) = x/(2x+2)$

G. $r(x) = (3x - 1)/[(x-1)(x+2)]$

Solution

x-Intercepts

$(3x - 1)/[(x-1)(x+2)] = 0$

$\Leftrightarrow 3x - 1 = 0$

$\Leftrightarrow x = 1/3$

The x-intercept is $1/3$.

Asymptotes

Note that:

- $(x-1)(x+2) = 0$
  $\Leftrightarrow x = 1$ OR $x = -2$
- $\deg(3x-1) < \deg[(x-1)(x+2)]$

The graph has:

- vertical asymptotes $x = -2$ and $x = 1$;
- the horizontal asymptote $y = 0$.

Sign Chart

$r(x)$	$-$	$+$	$-$	$+$
$3x-1$	$-$	$-$	$+$	$+$
$x-1$	$-$	$-$	$-$	$+$
$x+2$	$-$	$+$	$+$	$+$

Boundary points: -2, $\frac{1}{3}$, 1

Table of Values

x	-6	-3	-1	0	2	5
$r(x)$	-0.7	-2.5	2	0.5	1.25	0.5

Graph

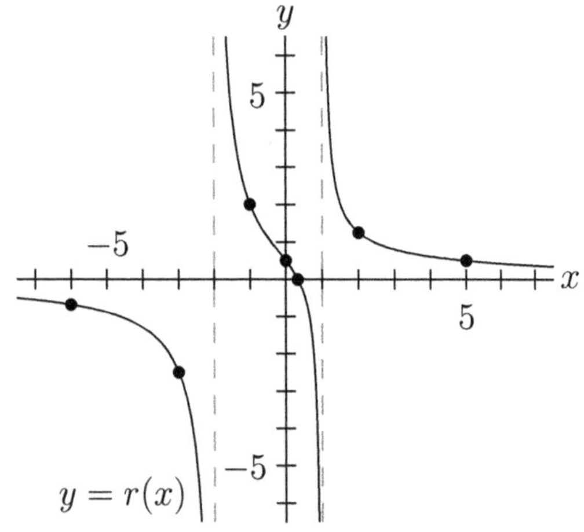

□

H. $r(x) = (-2x^2 - 3x + 12)/(x^2 - 7x + 10)$

I. $r(x) = (x^3 + 2x^2)/(x^2 - 9)$

SOLUTION

Note

$r(x) = [x^2(x+2)]/[(x-3)(x+3)]$

x-Intercepts

$[x^2(x+2)]/[(x-3)(x+3)] = 0$

$\Leftrightarrow x^2(x+2) = 0$

$\Leftrightarrow x^2 = 0$ OR $x + 2 = 0$

$\Leftrightarrow x = 0$ $\Leftrightarrow x = -2$

The x-intercepts are 0 and -2.

Asymptotes

Note that:

- $(x-3)(x+3) = 0$
  $\Leftrightarrow x - 3 = 0$ OR $x + 3 = 0$
  $\Leftrightarrow x = 3$ $\Leftrightarrow x = -3$
- $\deg(x^3 + 2x^2) = \deg(x^2 - 9) + 1$
- dividing numerator by denominator yields:

$$\begin{array}{r|l} & x + 2 \\ x^2 - 9 & x^3 + 2x^2 + 0x + 0 \\ & x^3 \quad\quad - 9x \\ \hline & 2x^2 + 9x \\ & 2x^2 \quad\quad - 18 \\ \hline & 9x + 18 \end{array}$$

- $r(x) = x + 2 + (9x + 18)/(x^2 - 9)$

The graph has:

- vertical asymptotes $x = -3$ and $x = 3$;
- the oblique asymptote $y = x + 2$.

Sign Chart

$r(x)$	$-$	$+$	$-$	$-$	$+$
x^2	$+$	$+$	$+$	$+$	$+$
$x+2$	$-$	$-$	$+$	$+$	$+$
$x-3$	$-$	$-$	$-$	$-$	$+$
$x+3$	$-$	$+$	$+$	$+$	$+$

Break points: -3, -2, 0, 3

Tables of Values

x	-8	-6	-4	-1
$r(x)$	-7.0	-5.3	-4.6	-0.1

x	2	4	6	8
$r(x)$	-3.2	13.7	10.7	11.6

Graph

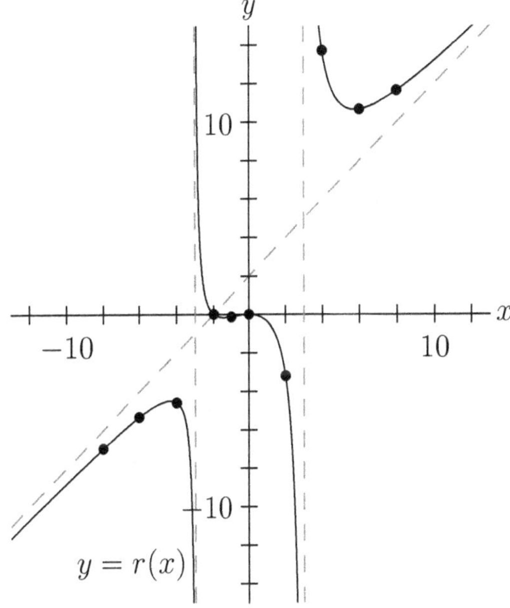

□

Section 101: Irrational Exponents

Definition

Suppose $b \in [0, \infty)$, $m \in \mathbf{R} \setminus \mathbf{Q}$, and $x \in \mathbf{Q}$.
Then b^m is the number that b^x approaches as x approaches m.

Comments

- For $b \in [0, \infty)$, observe that:
  - we defined b^m above for any $m \in \mathbf{R} \setminus \mathbf{Q}$, and
  - we defined b^m previously (see page 182) for any $m \in \mathbf{Q}$.

  Thus we have now defined b^m for any $m \in \mathbf{R}$.
- The properties of exponents we saw on pages 77 and 85 hold true not only for $m, n \in \mathbf{Q}$, but also for $m, n \in \mathbf{R}$ where $a, b \in (0, \infty)$.

Example 1

Approximate each numerical expression to five decimal places using rational exponents and a calculator.

A. $5^{\sqrt{2}}$
[Hint: $\sqrt{2} \approx 1.414213562373$]

Solution

The rational numbers in the sequence

$$1, 1.4, 1.41, 1.414, 1.4142, \ldots$$

suggested by the hint will approach $\sqrt{2}$, so we evaluate 5^x for the numbers x in this sequence (using a calculator) until successive values of 5^x agree to sufficient precision:

x	5^x
1	5.00000
1.4	9.51827
1.41	9.67270
1.414	9.73517
1.4142	9.73831
1.41421	9.73846
1.414213	9.73851
1.4142135	9.73852
1.41421356	9.73852

Thus $5^{\sqrt{2}} \approx 9.73852$.

□

B. $7^{\sqrt{3}}$
[Hint: $\sqrt{3} \approx 1.732050807569$]

Section 102: Exponential Functions

Definition

A function f is *exponential* if defined by $f(x) = b^x$ for some constant $b \in (0,1) \cup (1,\infty)$.

Graphs

The graph of the exponential function with base b will resemble one of the graphs below.

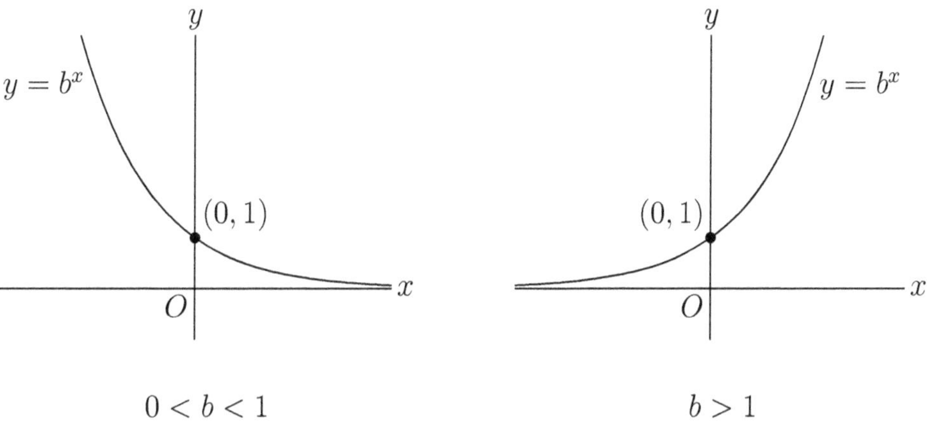

Procedure

To solve some equations that involve exponential functions, we can:

1. convert the expressions on both sides of the equation to the same base;
2. equate the exponents on that base;
3. solve the equation resulting from step (2).

Comments

- The domain of an exponential function with base b is all real numbers.
- The range of an exponential function with base b is $(0,\infty)$.
- The x-axis (that is, the line $y = 0$) is a horizontal asymptote of the graph of $y = b^x$ because $b^x \to 0$ either as $x \to \infty$ (if $0 < b < 1$) or as $x \to -\infty$ (if $b > 1$).

Example 1

Solve each equation.

A. $2^{x+3} = 32$

Solution

$$2^{x+3} = 32 \Leftrightarrow 2^{x+3} = 2^5$$
$$\Leftrightarrow x + 3 = 5$$
$$\Leftrightarrow x + 3 - 3 = 5 - 3$$
$$\Leftrightarrow x = 2$$

Set of solutions: $\{2\}$

□

B. $6^{2x-7} = 216$

C. $25^{3x} = 125^{x-8}$

SOLUTION

$$\begin{aligned} 25^{3x} = 125^{x-8} &\Leftrightarrow ((5)^2)^{3x} = ((5)^3)^{x-8} \\ &\Leftrightarrow (5)^{6x} = (5)^{3x-24} \\ &\Leftrightarrow 6x = 3x - 24 \\ &\Leftrightarrow -3x + 6x = -24 \\ &\Leftrightarrow 3x = -24 \\ &\Leftrightarrow 3x/3 = -24/3 \\ &\Leftrightarrow x = -8 \end{aligned}$$

Set of solutions: $\{-8\}$

□

D. $(1/81)^{x-1} = 3^{x-6}$

E. $8^{5x} = 1/32$

SOLUTION

$$\begin{aligned} 8^{5x} = 1/32 &\Leftrightarrow ((2)^3)^{5x} = 1/(2)^5 \\ &\Leftrightarrow (2)^{15x} = (2)^{-5} \\ &\Leftrightarrow 15x = -5 \\ &\Leftrightarrow 15x/15 = -5/15 \\ &\Leftrightarrow x = -1/3 \end{aligned}$$

Set of solutions: $\{-1/3\}$

□

F. $(2^{x+3})(2^{3x}) = 8$

G. $11^x = -4$

SOLUTION

The set of solutions is $\emptyset$
(since $11^x > 0$ for all $x \in \mathbf{R}$).

□

H. $2^{3x+1} = 0$

I. $2x^2(5^x) - 11x(5^x) + 12(5^x) = 0$

SOLUTION

$$\begin{aligned} &2x^2(5^x) - 11x(5^x) + 12(5^x) = 0 \\ &\Leftrightarrow 5^x(2x^2 - 11x + 12) = 0 \\ &\Leftrightarrow 5^x(2x - 3)(x - 4) = 0 \\ &\Leftrightarrow \cancel{5^x = 0} \quad \text{X} \\ &\text{OR } 2x - 3 = 0 \\ &\qquad \Leftrightarrow 2x = 3 \\ &\qquad \Leftrightarrow x = 3/2 \\ &\text{OR } x - 4 = 0 \\ &\qquad \Leftrightarrow x = 4 \end{aligned}$$

Set of solutions: $\{3/2, 4\}$

□

J. $x^2(7^x) = 16(7^x)$

EXAMPLE 2

Determine the exponential function whose graph passes through the given point.

A. $(2, 16)$

SOLUTION

Let f denote the function in question.
So $f(x) = b^x$ for some $b \in (0,1) \cup (1, \infty)$.

$$f(2) = 16 \Rightarrow b^2 = 16 \\ \Rightarrow b = \sqrt{16} \\ = 4$$

Thus $f(x) = 4^x$.

□

B. $(-3, 1/27)$

EXAMPLE 3

Draw the graph of each function.

A. $f(x) = (1/2)^x$

SOLUTION

Asymptote

The graph has a horizontal asymptote with equation $y = 0$.

Table of Values

x	$f(x)$
-3	$(1/2)^{-3} = 8$
0	$(1/2)^0 = 1$
2	$(1/2)^2 = 1/4 = 0.25$

Graph

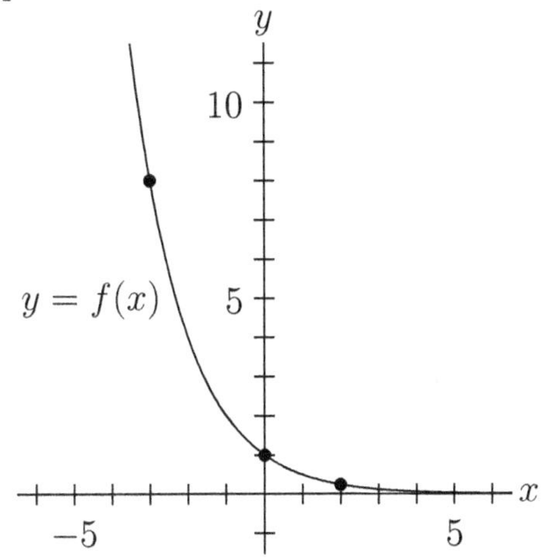

□

B. $f(x) = 2^x$

C. $f(x) = 3^{x+1} - 4$

SOLUTION

Declaration

Let us define g by $g(x) = 3^x$.
Thus $f(x) = g(x+1) - 4$.

Asymptote of Graph of g

The graph of g has a horizontal asymptote with equation $y = 0$.

Table of Values for g

x	$g(x)$
-2	$(3)^{-2} = 1/9 \approx 0.11$
0	$(3)^0 = 1$
2	$(3)^2 = 9$

Translations from g to f

- $y = 0 \mapsto y = 0 - 4$
  $\Rightarrow y = -4$
- $(-2, 0.11) \mapsto (-2-1, 0.11-4)$
  $= (-3, -3.89)$
- $(0, 1) \mapsto (0-1, 1-4)$
  $= (-1, -3)$
- $(2, 9) \mapsto (2-1, 9-4)$
  $= (1, 5)$

Graph of f

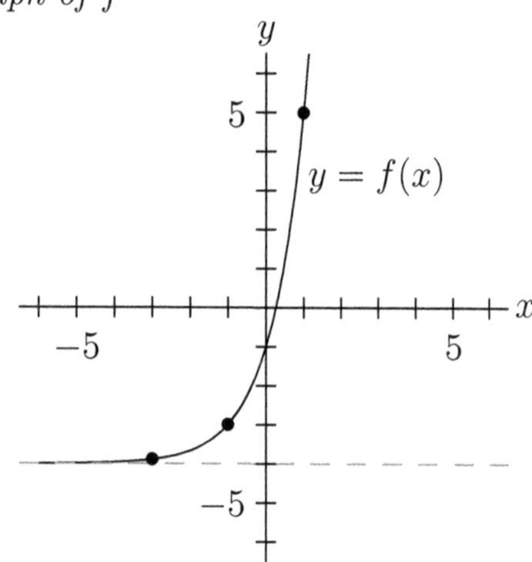

□

D. $f(x) = -(0.1)^{x-2}$

Section 103: Logarithmic Functions

Definitions

- The *logarithmic function with base b* is the inverse of the exponential function with base b.
- The *logarithm with base b of x*, denoted $\log_b x$, is the value of the logarithmic function with base b at x.
- The *natural* logarithm of x, denoted $\ln x$, is the logarithm with base e of x.
- The *common* logarithm of x, denoted $\log x$, is the logarithm with base 10 of x.

Graphs

The graph of the logarithmic function with base b will resemble one of the graphs below.

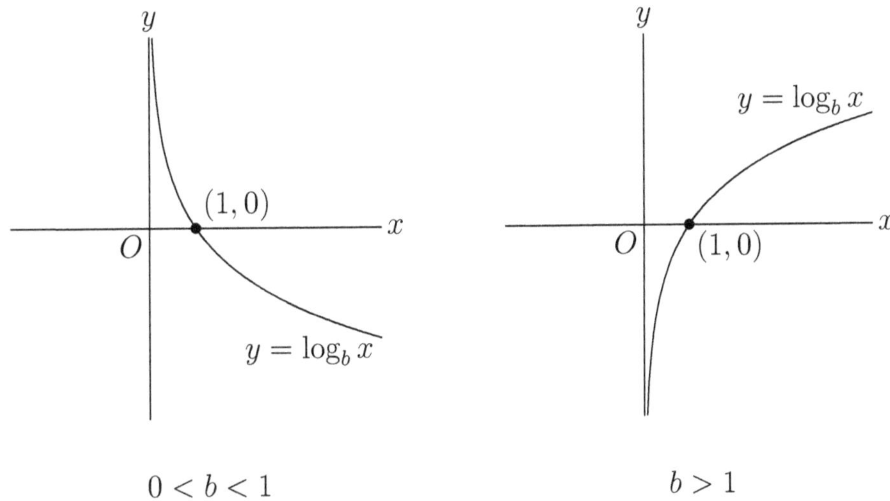

$0 < b < 1$ $\qquad$ $b > 1$

Properties

Suppose $b \in (0,1) \cup (1,\infty)$, $x \in (0,\infty)$, and $y \in \mathbf{R}$.
Then:

- $\log_b x = y \Leftrightarrow b^y = x$
- $\ln x = y \Leftrightarrow e^y = x$
- $\log x = y \Leftrightarrow 10^y = x$

Proofs (first property only)

Suppose $b \in (0,1) \cup (1,\infty)$, $x \in (0,\infty)$, and $y \in \mathbf{R}$.
Let us define functions f and g by $f(x) = \log_b x$ and $g(x) = b^x$.

$$\begin{aligned} \log_b x = y &\Leftrightarrow f(x) = y \\ &\Leftrightarrow g(y) = x \quad \text{(since } f \text{ and } g \text{ are inverses of each other)} \\ &\Leftrightarrow b^y = x \end{aligned}$$

⊡

Terminology

Recall that the equations $\log_b x = y$ and $b^y = x$ are algebraically equivalent.

- The logarithmic form of the equation is $\log_b x = y$.
- The exponential form of the equation is $b^y = x$.

PROPERTIES

Suppose $b \in (0,1) \cup (1,\infty)$, $x \in (0,\infty)$, and $y \in \mathbf{R}$.
Then:

- $\log_b 1 = 0$
- $\log_b b = 1$
- $\log_b b^y = y$
- $b^{\log_b x} = x$

PROOFS

Suppose $b \in (0,1) \cup (1,\infty)$, $x \in (0,\infty)$, and $y \in \mathbf{R}$.

- $b^0 = 1 \Rightarrow \log_b 1 = 0$ (conversion to logarithmic form)
- $b^1 = b \Rightarrow \log_b b = 1$ (conversion to logarithmic form)
- $b^y = b^y \Rightarrow \log_b b^y = y$ (conversion to logarithmic form)
- $\log_b x = \log_b x \Rightarrow b^{\log_b x} = x$ (conversion to exponential form)

⊡

COMMENTS

- The domain of a logarithmic function with base b is $(0,\infty)$.
- The range of a logarithmic function with base b is all real numbers.
- We read the expression "$\log_b x$" aloud as "log base b of x".
- A function f is logarithmic if defined by $f(x) = \log_b x$ for some constant $b \in (0,1) \cup (1,\infty)$.
- Inverses of exponential functions exist because exponential functions are one-to-one.
- The graph of the logarithmic function with base b is the reflection through the line $y = x$ of the graph of the exponential function with base b.
- The y-axis (that is, the line $x = 0$) is a vertical asymptote of the graph of $y = \log_b x$ because either $\log_b x \to \infty$ (if $0 < b < 1$) or $\log_b x \to -\infty$ (if $b > 1$) as $x \to 0^+$.
- We can interpret the logarithm with base b of x as the exponent to which we would raise b to obtain x; in other words, $\log_b x$ is the value of the exponent y at which $b^y = x$.
- We can also write the natural and common logarithms of x as $\log_e x$ and $\log_{10} x$, respectively.

EXAMPLE 1

Evaluate each logarithm.

A. $\log_4 16$
B. $\log_{216} 6$
C. $\log_7 7^{10}$
D. $\log_{12} 1$

SOLUTION

A. 2 (since $4^2 = 16$)
B. 1/3 (since $216^{1/3} = \sqrt[3]{216} = 6$)
C. 10 (since $7^{10} = 7^{10}$)
D. 0 (since $12^0 = 1$)

□

D. $\log_2 32$
E. $\log_5 1/625$
F. $\log_{11} \sqrt{11}$
G. $\log_{15} 15$

EXAMPLE 2
Determine the domain $\mathbb{D}$ of each function.

A. $f(x) = \log_8(5x - 2)$

SOLUTION
Note
$5x - 2 > 0 \Leftrightarrow 5x > 2$
$\Leftrightarrow x > 2/5$

Domain
$\mathbb{D} = \{ x \in \mathbf{R} \mid 5x - 2 > 0 \}$
$= (2/5, \infty)$

□

B. $f(x) = \log_3(7 - x)$

EXAMPLE 3
Draw the graph of each function.

A. $f(x) = \log_3 x$

SOLUTION
Asymptote
The graph has a vertical asymptote with equation $x = 0$.

Table of Values

x	$f(x)$
9	$\log_3 9 = 2$
1	$\log_3 1 = 0$
1/9	$\log_3(1/9) = -2$

Graph

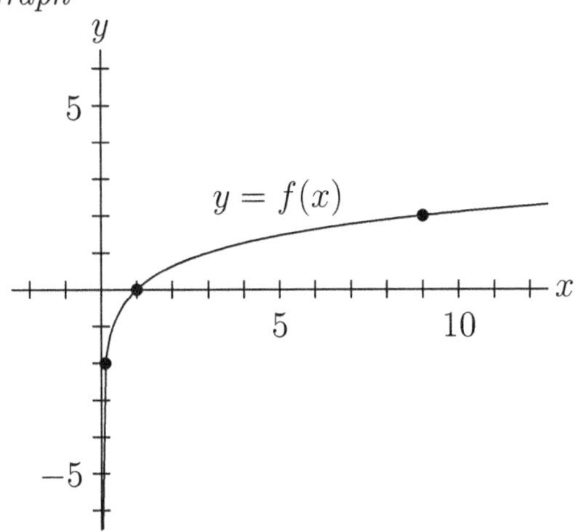

□

B. $f(x) = \log_{1/3} x$

C. $f(x) = -\log_{0.5}(x-1)+4$

SOLUTION

Declaration

Let us define g by $g(x) = \log_{0.5} x$.
Thus $f(x) = -g(x-1)+4$.

Asymptote of Graph of g

The graph of g has a vertical asymptote with equation $x = 0$.

Table of Values for g

x	$g(x)$
0.25	$\log_{0.5} 0.25 = 2$
1	$\log_{0.5} 1 = 0$
4	$\log_{0.5} 4 = -2$

Translations from g to f

- $x = 0 \mapsto x = 0+1$
  $\Rightarrow x = 1$
- $(0.25, 2) \mapsto (0.25+1, -(2)+4)$
  $= (1.25, 2)$
- $(1, 0) \mapsto (1+1, -(0)+4)$
  $= (2, 4)$
- $(4, -2) \mapsto (4+1, -(-2)+4)$
  $= (5, 6)$

Graph of f

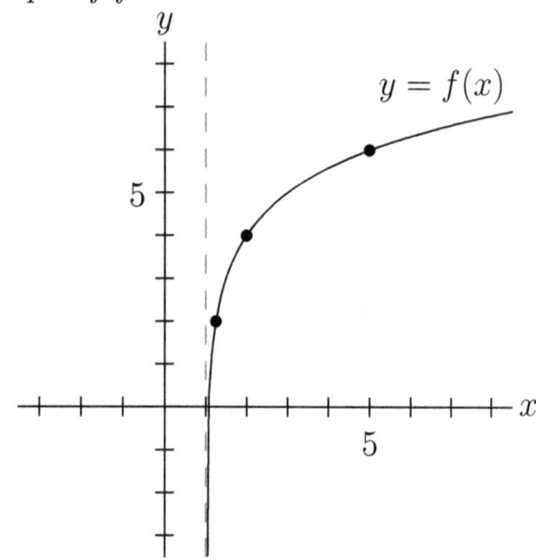

□

D. $f(x) = 3\log_2(x+3) - 5$

EXAMPLE 4
Write each equation in exponential form.

A. $\log_6(1/216) = -3$
B. $\log_{12} x = 20$
C. $\log 100 = 2$
D. $\log_7 343 = 3$
E. $\log_{10000} 0.1 = -1/4$
F. $\ln x = 7$

SOLUTION
A. $6^{-3} = 1/216$
B. $12^{20} = x$
C. $10^2 = 100$

□

EXAMPLE 5
Write each equation in logarithmic form.

A. $3^2 = 9$
B. $125^{(1/3)} = 5$
C. $e^5 = e^5$
D. $4^{-1} = 0.25$
E. $11^x = 70$
F. $10^3 = 1000$

SOLUTION
A. $\log_3 9 = 2$
B. $\log_{125} 5 = 1/3$
C. $\ln e^5 = 5$

□

Section 104: Properties of Logarithms

Properties of Logarithms

Suppose $a, b \in (0,1) \cup (1,\infty)$, $r, s \in (0,\infty)$, and $t \in \mathbf{R}$.
Then:

- $\log_b(r \cdot s) = \log_b r + \log_b s$
- $\log_b(r/s) = \log_b r - \log_b s$
- $\log_b r^t = t \log_b r$
- $\log_b r = (\log_a r)/(\log_a b)$

Proofs

Suppose $a, b \in (0,1) \cup (1,\infty)$, $r, s \in (0,\infty)$, and $t \in \mathbf{R}$.

- $r \cdot s = b^{\log_b r} \cdot b^{\log_b s}$
  $\Rightarrow r \cdot s = b^{\log_b r + \log_b s}$
  $\Rightarrow \log_b(r \cdot s) = \log_b r + \log_b s$

- $r/s = (b^{\log_b r})/(b^{\log_b s})$
  $\Rightarrow r/s = b^{\log_b r - \log_b s}$
  $\Rightarrow \log_b(r/s) = \log_b r - \log_b s$

- $r = b^{\log_b r} \Rightarrow r^t = (b^{\log_b r})^t$
  $\Rightarrow r^t = b^{t \log_b r}$
  $\Rightarrow \log_b r^t = t \log_b r$

- $b^{\log_b r} = r \Rightarrow \log_a b^{\log_b r} = \log_a r$
  $\Rightarrow (\log_b r)(\log_a b) = \log_a r$
  $\Rightarrow \log_b r = (\log_a r)/(\log_a b)$

$\boxdot$

Comment

The formula $\log_b r = (\log_a r)/(\log_a b)$ is called the "Change of Base Formula".

Example 1

Use the Change of Base Formula to rewrite each expression in terms of natural logarithms, common logarithms, and logarithms with base 2.

A. $\log_5 30$

Solution

- $\log_5 30 = (\ln 30)/(\ln 5)$
- $\log_5 30 = (\log 30)/(\log 5)$
- $\log_5 30 = (\log_2 30)/(\log_2 5)$

$\square$

B. $\log_3 100$

C. $\log_{32} 8$

SOLUTION

- $\log_{32} 8 = (\ln 8)/(\ln 32)$
- $\log_{32} 8 = (\log 8)/(\log 32)$
- $\log_{32} 8 = (\log_2 8)/(\log_2 32)$
  $= 3/5$

□

D. $\log_b 17$

EXAMPLE 2

Use the Change of Base Formula to rewrite the expression $\log_7 0.003$ in terms of natural logarithms, then use a calculator to approximate the value of the expression to five decimal places.

SOLUTION

$$\begin{aligned}\log_7 0.003 &= (\ln 0.003)/(\ln 7)\\ &\approx (-5.80914)/(-1.94591)\\ &\approx -2.98531\end{aligned}$$

□

EXAMPLE 3

Use the Change of Base Formula to rewrite the expression $\log_4 5000$ in terms of common logarithms, then use a calculator to approximate the value of the expression to five decimal places.

EXAMPLE 4

Write the given expression as a sum and/or difference of logarithms with no exponents.

A. $\log_3 7x$

SOLUTION

$\log_3 7x = \log_3 7 + \log_3 x$

□

B. $\log_8(y/x)$

C. $\ln x^2$

SOLUTION

$\ln x^2 = 2\ln x$

□

D. $\log 8^m$

E. $\log_2(8x^5/z)$

SOLUTION

$$\begin{aligned}\log_2(8x^5/z) &= \log_2(8x^5) - \log_2 z\\ &= \log_2 8 + \log_2 x^5 - \log_2 z\\ &= 3 + 5\log_2 x - \log_2 z\end{aligned}$$

□

F. $\ln \sqrt{xy^3z^5}$

G. $\log \sqrt[5]{(x-2)/(c^7m^4)}$

SOLUTION

$$\begin{aligned}&\log \sqrt[5]{(x-2)/(c^7m^4)}\\ &\quad= \log[(x-2)/(c^7m^4)]^{(1/5)}\\ &\quad= (1/5)\log[(x-2)/(c^7m^4)]\\ &\quad= (1/5)(\log(x-2) - \log(c^7m^4))\\ &\quad= (1/5)(\log(x-2) - (\log c^7 + \log m^4))\\ &\quad= (1/5)(\log(x-2) - (7\log c + 4\log m))\\ &\quad= (1/5)(\log(x-2) - 7\log c - 4\log m)\\ &\quad= \tfrac{1}{5}\log(x-2) - \tfrac{7}{5}\log c - \tfrac{4}{5}\log m\end{aligned}$$

□

H. $\log_3[(x+2)^2/(3x-10)^7]$

EXAMPLE 5

Write the given expression as a single logarithm.

A. $\log_2 m - \log_2 z$

SOLUTION

$\log_2 m - \log_2 z = \log_2(m/z)$

□

B. $\log_3 5 + \log_3 7$

C. $n \log_b 3$

SOLUTION

$n \log_b 3 = \log_b 3^n$

□

D. $11 \log_2 6$

E. $5\log x + \log y - 10\log z$

SOLUTION

$$\begin{aligned}&5\log x + \log y - 10\log z\\ &\quad= \log x^5 + \log y - \log z^{10}\\ &\quad= \log x^5y - \log z^{10}\\ &\quad= \log(x^5y/z^{10})\end{aligned}$$

□

F. $\ln(x^2-9) - \ln(x+3)$

G. $2\log_7(x+1) + 8\log_7(x-3)$

SOLUTION

$$\begin{aligned}&2\log_7(x+1) + 8\log_7(x-3)\\ &\quad= \log_7(x+1)^2 + \log_7(x-3)^8\\ &\quad= \log_7((x+1)^2(x-3)^8)\end{aligned}$$

□

H. $\log_3 x - \log_4 y$

SOLUTION

The two logarithms in the expression do not have the same base, so we cannot combine them into a single logarithm.

□

EXAMPLE 6

Rewrite each expression in terms of common logarithms of prime numbers.

A. $\log_{0.3} 5$

SOLUTION

$$\begin{aligned}\log_{0.3} 15 &= (\log 15)/(\log 0.3)\\ &= (\log(3 \cdot 5))/(\log(3/10))\\ &= (\log 3 + \log 5)/(\log 3 - \log 10)\\ &= (\log 3 + \log 5)/(\log 3 - 1)\\ &= (\log 3 + \log 5)/(-1 + \log 3)\end{aligned}$$

□

B. $\log_{1/9} 16$

Section 105: Exponential Equations

Procedure A (preferred)

To solve an equation that contains an exponent with a variable, we can:

1. apply algebraic properties to the equation until it takes the form $b^{f(x)} = m$;
2. take the logarithms with base b of both sides of the equation of step (1);
3. solve the equation $f(x) = \log_b m$ for x.

Procedure B (alternative)

To solve an equation that contains an exponent with a variable, we can:

1. apply algebraic properties to the equation until it takes the form $b^{f(x)} = m$;
2. take the natural logarithms of both sides of the equation of step (1);
3. apply the property $\ln b^{f(x)} = f(x) \ln b$ to the left side of the equation of step (2);
4. divide both sides of the equation of step (3) by $\ln b$;
5. solve the equation $f(x) = (\ln m)/(\ln b)$ for x.

Comments

- Both procedures above have comparative advantages:
  - procedure A consists of fewer steps and yields solutions in a simpler form, while
  - procedure B yields solutions in a form that we can enter more easily into a calculator for approximation.
- In step (2) of procedure A we effectively convert the equation $b^{f(x)} = m$ from exponential form to logarithmic form.
- We can use common logarithms instead of natural logarithms in procedure B.

Example 1

Solve each equation using procedure A.

A. $3^x = 10$

Solution

$3^x = 10 \Leftrightarrow \log_3 3^x = \log_3 10$

$\Leftrightarrow x = \log_3 10$

Set of solutions: $\{\log_3 10\}$ □

B. $7^{2x+1} = 54$

C. $4^{x^2} = 71$

Solution

$4^{x^2} = 71 \Leftrightarrow \log_4 4^{x^2} = \log_4 71$

$\Leftrightarrow x^2 = \log_4 71$

$\Leftrightarrow x = \pm\sqrt{\log_4 71}$

Set of solutions: $\{-\sqrt{\log_4 71}, \sqrt{\log_4 71}\}$ □

D. $24^{|x|} = 670$

E. $10^2 \cdot 10^{-3x} = 22$

Solution

$10^2 \cdot 10^{-3x} = 22$

$\Leftrightarrow 10^{2-3x} = 22$

$\Leftrightarrow \log 10^{2-3x} = \log 22$

$\Leftrightarrow 2 - 3x = \log 22$

$\Leftrightarrow -3x = -2 + \log 22$

$\Leftrightarrow x = (-2 + \log 22)/(-3)$

$= -(-2 + \log 22)/3$

$= (2 - \log 22)/3$

Set of solutions: $\{(2 - \log 22)/3\}$

□

F. $8e^x = 44$

G. $8 + 12^{x/4} = 47$

Solution

$8 + 12^{x/4} = 47$

$\Leftrightarrow 12^{x/4} = -8 + 47$

$= 39$

$\Leftrightarrow \log_{12} 12^{x/4} = \log_{12} 39$

$\Leftrightarrow x/4 = \log_{12} 39$

$\Leftrightarrow x = 4 \log_{12} 39$

Set of solutions: $\{4 \log_{12} 39\}$

□

H. $11^{2x}/11^{x+14} = 33$

I. $3(5^{2x}) - 13(5^x) + 14 = 0$

Solution

$3(5^{2x}) - 13(5^x) + 14 = 0$

$\Leftrightarrow 3(5^x)^2 - 13(5^x) + 14 = 0$

$\Leftrightarrow 3u^2 - 13u + 14 = 0 \quad \text{where } u = 5^x$

$\Leftrightarrow (3u - 7)(u - 2) = 0$

$\Leftrightarrow 3u - 7 = 0$

$\Leftrightarrow 3u = 7$

$\Leftrightarrow u = 7/3$

$\Leftrightarrow 5^x = 7/3$

$\Leftrightarrow x = \log_5(7/3)$

OR $u - 2 = 0$

$\Leftrightarrow u = 2$

$\Leftrightarrow 5^x = 2$

$\Leftrightarrow x = \log_5 2$

Set of solutions: $\{\log_5 2, \log_5(7/3)\}$

□

J. $x^2 \cdot 9^x - 6x \cdot 9^x - 7 \cdot 9^x = 0$

K. $6^{2x+4} = 11^{x-5}$

Solution

$6^{2x+4} = 11^{x-5}$

$\Leftrightarrow \log_6 6^{2x+4} = \log_6 11^{x-5}$

$\Leftrightarrow 2x + 4 = (x-5)\log_6 11$

$= x\log_6 11 - 5\log_6 11$

$\Leftrightarrow 2x + 4 + 5\log_6 11 = x\log_6 11$

$\Leftrightarrow 4 + 5\log_6 11 = -2x + x\log_6 11$

$\Leftrightarrow 4 + 5\log_6 11 = x(-2 + \log_6 11)$

$\Leftrightarrow (4 + 5\log_6 11)/(-2 + \log_6 11) = x$

Set of solutions:

$\{(4 + 5\log_6 11)/(-2 + \log_6 11)\}$.

□

L. $2^{10x+9} = 16^{x+3}$

Example 2

Solve each equation using procedure B.

A. $8^x = 2/9$

Solution

$8^x = 2/9 \Leftrightarrow \ln 8^x = \ln(2/9)$

$\Leftrightarrow x\ln 8 = \ln 2 - \ln 9$

$\Leftrightarrow x = (\ln 2 - \ln 9)/(\ln 8)$

Set of solutions: $\{(\ln 2 - \ln 9)/(\ln 8)\}$

□

B. $5^x = 7e$

C. $2^{x+3} = 75$

Solution

$2^{x+3} = 75 \Leftrightarrow \ln 2^{x+3} = \ln 75$

$\Leftrightarrow (x+3)\ln 2 = \ln 75$

$\Leftrightarrow x\ln 2 + 3\ln 2 = \ln 75$

$\Leftrightarrow x\ln 2 = \ln 75 - 3\ln 2$

$\Leftrightarrow x = (\ln 75 - 3\ln 2)/(\ln 2)$

Set of solutions: $\{(\ln 75 - 3\ln 2)/(\ln 2)\}$

□

D. $3^{7x} = 28$

Section 106: Logarithmic Equations

Procedure

To solve an equation that contains a logarithm of an expression with a variable, we:

1. apply algebraic properties to the equation until it takes the form $\log_b f(x) = m$;
2. raise both sides of the equation of step (1) as exponents on the base b;
3. solve the equation $f(x) = b^m$ for x.

Comments

- In step (2) of the procedure above we effectively convert the equation $\log_b f(x) = m$ from logarithmic form to exponential form.
- We must reject all potential solutions for which an argument of a logarithm evaluates to a nonpositive number or a base of a logarithm evaluates to a nonpositive number or 1.

Example 1

Solve each equation.

A. $\log_7 x = 2$

Solution

Assuming $x \in (0, \infty)$:

$\log_7 x = 2 \Leftrightarrow 7^{\log_7 x} = 7^2$

$\Leftrightarrow x = 49$

Set of solutions: $\{49\}$

□

B. $\log_2(x+7) = 3$

C. $\log_{x-3} 625 = 4$

Solution

Conditions for Solution

$x - 3 > 0$ AND $x - 3 \neq 1$

$\Leftrightarrow x > 3$ AND $x \neq 4$

$\Leftrightarrow x \in (3,4) \cup (4, \infty)$

Set of Solutions

Assuming $x \in (3,4) \cup (4, \infty)$:

$\log_{x-3} 625 = 4$

$\Leftrightarrow (x-3)^{\log_{x-3} 625} = (x-3)^4$

$\Leftrightarrow 625 = (x-3)^4$

$\Leftrightarrow x - 3 = \pm\sqrt[4]{625}$

$= \pm 5$

$\Leftrightarrow x = 3 + 5$ OR $x = 3 - 5$

$\Leftrightarrow x = 8$ $\quad \Leftrightarrow \cancel{x = -2}$ X

Set of solutions: $\{8\}$

□

D. $\log_{7x} 512 = 3$

E. $\log_5 3 + \log_5 6x = 2$

SOLUTION

Condition for Solution

$6x > 0 \Leftrightarrow 6x/6 > 0/6$
$\Leftrightarrow x > 0$
$\Leftrightarrow x \in (0, \infty)$

Set of Solutions

Assuming $x \in (0, \infty)$:

$\log_5 3 + \log_5 6x = 2$
$\Leftrightarrow \log_5(3 \cdot 6x) = 2$
$\Leftrightarrow \log_5 18x = 2$
$\Leftrightarrow 5^{\log_5 18x} = 5^2$
$\Leftrightarrow 18x = 25$
$\Leftrightarrow x = 25/18$

Set of solutions: $\{25/18\}$

□

F. $x + 1 = (\log_{100} 23)(\log_{23} 10000)$

G. $2 \ln x + \ln 3 = \ln(8 - 10x)$

SOLUTION

Conditions for Solution

$x > 0$ AND $8 - 10x > 0$
$\Leftrightarrow x > 0$ AND $-10x > -8$
$\Leftrightarrow x > 0$ AND $x < -8/-10$
$\Leftrightarrow x > 0$ AND $x < 4/5$
$\Leftrightarrow 0 < x < 4/5$
$\Leftrightarrow x \in (0, 4/5)$

Set of Solutions

Assuming $x \in (0, 4/5)$:

$2 \ln x + \ln 3 = \ln(8 - 10x)$
$\Leftrightarrow \ln x^2 + \ln 3 = \ln(8 - 10x)$
$\Leftrightarrow \ln(x^2 \cdot 3) = \ln(8 - 10x)$
$\Leftrightarrow \ln 3x^2 = \ln(8 - 10x)$
$\Leftrightarrow e^{\ln 3x^2} = e^{\ln(8-10x)}$
$\Leftrightarrow 3x^2 = 8 - 10x$
$\Leftrightarrow 3x^2 + 10x - 8 = 0$
$\Leftrightarrow (3x - 2)(x + 4) = 0$
$\Leftrightarrow 3x - 2 = 0$ OR $x + 4 = 0$
$\Leftrightarrow 3x = 2$ $\quad\Leftrightarrow \sout{x = -4}$ X
$\Leftrightarrow x = 2/3$

Set of solutions: $\{2/3\}$

□

H. $\log_4(25x - 13) - \log_4(x + 2) = 2$

Section 107: Exponential Inequalities

Properties (proofs omitted)
Suppose $m, n \in (0, \infty)$.
Then:

- $m < n \Leftrightarrow \log_b m > \log_b n \quad$ if $0 < b < 1$
- $m < n \Leftrightarrow \log_b m < \log_b n \quad$ if $b > 1$

Procedure A (preferred)
To solve an inequality of the form $b^{f(x)} < m$, we can:

1. take the logarithms with base b of both sides of the inequality and simplify to:
   a. $f(x) > \log_b m \quad$ if $0 < b < 1$
   b. $f(x) < \log_b m \quad$ if $b > 1$;
2. solve the inequality of step (1) for x.

Procedure B (alternative)
To solve an inequality of the form $b^{f(x)} < m$, we can:

1. take the natural logarithms of both sides of the inequality;
2. rewrite the inequality of step (1) as $f(x) \ln b < \ln m$;
3. divide both sides of the inequality of step (2) by $\ln b$ and simplify to:
   a. $f(x) > (\ln m)/(\ln b) \quad$ if $0 < b < 1$
   b. $f(x) < (\ln m)/(\ln b) \quad$ if $b > 1$;
4. solve the inequality of step (3) for x.

Comments

- We can also follow procedure A or B to solve an inequality of the form $b^{f(x)} > m$, except we reverse the direction of the inequality sign in each step.
- We can use common logarithms instead of natural logarithms in procedure B.

Example 1
Solve each inequality using procedure A.

A. $7 - 0.3^x < 2$

Solution
$7 - 0.3^x < 2$
$\Leftrightarrow -0.3^x < -7 + 2$
$\Leftrightarrow -0.3^x < -5$
$\Leftrightarrow (-1)(-0.3^x) > (-1)(-5)$
$\Leftrightarrow 0.3^x > 5$
$\Leftrightarrow \log_{0.3} 0.3^x < \log_{0.3} 5$
$\Leftrightarrow x < \log_{0.3} 5$
Set of solutions: $(-\infty, \log_{0.3} 5)$

□

B. $(3/2)(4^{5x+1} + 10) > 18$

C. $6^{x+7} \geq 2$

SOLUTION

$$\begin{aligned} 6^{x+7} \geq 2 &\Leftrightarrow \log_6 6^{x+7} \geq \log_6 2 \\ &\Leftrightarrow x + 7 \geq \log_6 2 \\ &\Leftrightarrow x \geq -7 + \log_6 2 \end{aligned}$$

Set of solutions: $[-7 + \log_6 2, \infty)$

□

D. $0.5^{10-x} \leq 200$

EXAMPLE 2

Solve each inequality using procedure B.

A. $11^{x/3} > 500$

SOLUTION

$$\begin{aligned} &11^{x/3} > 500 \\ &\Leftrightarrow \ln 11^{x/3} > \ln 500 \\ &\Leftrightarrow (x/3) \ln 11 > \ln 500 \\ &\Leftrightarrow ((x/3)\cancel{\ln 11})/(\cancel{\ln 11}) \\ &\qquad > (\ln 500)/(\ln 11) \\ &\Leftrightarrow x/3 > (\ln 500)/(\ln 11) \\ &\Leftrightarrow (\cancel{3})(x/\cancel{3}) > (3)(\ln 500)/(\ln 11) \\ &\Leftrightarrow x > (3 \ln 500)/(\ln 11) \end{aligned}$$

Set of solutions: $((3 \ln 500)/(\ln 11), \infty)$

□

B. $(1/4)^{5x} < 16$

C. $(1/3)^{12x} \leq 81$

SOLUTION

$$\begin{aligned} &(1/3)^{12x} \leq 81 \\ &\Leftrightarrow \ln(1/3)^{12x} \leq \ln 81 \\ &\Leftrightarrow 12x \ln(1/3) \leq \ln 81 \\ &\Leftrightarrow (12x\cancel{\ln(1/3)})/(\cancel{\ln(1/3)}) \\ &\qquad \geq (\ln 81)/(\ln(1/3)) \\ &\Leftrightarrow 12x \geq (\ln 81)/(\ln(1/3)) \\ &\Leftrightarrow 12x \geq (\ln(3)^4)/(\ln 1 - \ln 3) \\ &\Leftrightarrow 12x \geq (4 \ln 3)/(0 - \ln 3) \\ &\Leftrightarrow 12x \geq (4\cancel{\ln 3})/(-\cancel{\ln 3}) \\ &\Leftrightarrow 12x \geq -4 \\ &\Leftrightarrow 12x/12 \geq -4/12 \\ &\Leftrightarrow x \geq -1/3 \end{aligned}$$

Set of solutions: $[-1/3, \infty)$

□

D. $3^{x+8} \geq 90$

Section 108: Logarithmic Inequalities

PROPERTIES (proofs omitted)
Suppose $m, n \in \mathbf{R}$.
Then:
- $m < n \Leftrightarrow b^m > b^n$ if $0 < b < 1$
- $m < n \Leftrightarrow b^m < b^n$ if $b > 1$

PROCEDURE
To solve an inequality of the form $\log_b f(x) < m$, we:
1. raise both sides of the inequality as exponents on the base b and simplify to:
   a. $f(x) > b^m$ if $0 < b < 1$
   b. $f(x) < b^m$ if $b > 1$;
2. solve the inequality of step (1) for x.

COMMENTS
- We can also follow the procedure above to solve an inequality of the form $\log_b f(x) > m$, except we reverse the direction of the inequality sign in step (1).
- We must reject all potential solutions for which an argument of a logarithm evaluates to a nonpositive number or a base of a logarithm evaluates to a nonpositive number or 1.

EXAMPLE 1
Solve each inequality.

A. $\log_{0.7}(x - 5) > 2$

SOLUTION
Condition for Solutions
$x - 5 > 0 \Leftrightarrow x > 5$
$\qquad \Leftrightarrow x \in (5, \infty)$

Note
Assuming $x \in (5, \infty)$:
$\log_{0.7}(x - 5) > 2$
$\Leftrightarrow (0.7)^{\log_{0.7}(x-5)} < (0.7)^2$
$\Leftrightarrow x - 5 < 0.49$
$\Leftrightarrow x < 0.49 + 5$
$\Leftrightarrow x < 5.49$

Set of Solutions
$(\text{set of solutions}) = (-\infty, 5.49) \cap (5, \infty)$
$\qquad = (5, 5.49)$

□

B. $\log_2(13 - x) \leq 3$

C. $\log[(x+2)(x+3)] \le \log 2$

SOLUTION

Zeros of Factors of Argument

- $x+2=0 \Leftrightarrow x=-2$
- $x+3=0 \Leftrightarrow x=-3$

Sign Chart for Argument

$(x+2)(x+3)$	$+$	$-$	$+$
$x+2$	$-$	$-$	$+$
$x+3$	$-$	$+$	$+$
	-3	-2	

Condition for Solutions

$(x+2)(x+3) > 0$
$\Leftrightarrow x \in (-\infty,-3) \cup (-2,\infty)$

Note

Assuming $x \in (-\infty,-3) \cup (-2,\infty)$:
$\log[(x+2)(x+3)] \le \log 2$
$\Leftrightarrow \log(x^2+5x+6) \le \log 2$
$\Leftrightarrow 10^{\log(x^2+5x+6)} \le 10^{\log 2}$
$\Leftrightarrow x^2+5x+6 \le 2$
$\Leftrightarrow x^2+5x+6-2 \le 0$
$\Leftrightarrow x^2+5x+4 \le 0$
$\Leftrightarrow (x+4)(x+1) \le 0$

Zeros of Factors in Inequality

- $x+4 \Leftrightarrow x=-4$
- $x+1 \Leftrightarrow x=-1$

Sign Chart for Inequality

$(x+4)(x+1)$	$+$	$-$	$+$
$x+4$	$-$	$+$	$+$
$x+1$	$-$	$-$	$+$
	-4	-1	

Set of Solutions

(set of solutions)
$= [-4,-1] \cap [(-\infty,-3) \cup (-2,\infty)]$
$= [-4,-3) \cup (-2,-1]$

□

D. $\ln(x^2-3) < 0$

E. $\log_2(x-1)+\log_2 x > 1$

SOLUTION

Conditions for Solutions

$x-1>0$ AND $x>0$

$\Leftrightarrow x>1$ AND $x>0$

$\Leftrightarrow x>1$

$\Leftrightarrow x\in(1,\infty)$

Note

Assuming $x\in(1,\infty)$:

$\log_2(x-1)+\log_2 x>1$

$\Leftrightarrow \log_2(x-1)(x)>1$

$\Leftrightarrow \log_2(x^2-x)>1$

$\Leftrightarrow 2^{\log_2(x^2-x)}>2^1$

$\Leftrightarrow x^2-x>2$

$\Leftrightarrow x^2-x-2>0$

$\Leftrightarrow (x-2)(x+1)>0$

Zeros of the Factors

- $x-2=0\Leftrightarrow x=2$
- $x+1=0\Leftrightarrow x=-1$

Sign Chart

$(x-2)(x+1)$	$+$	$-$	$+$
$x-2$	$-$	$-$	$+$
$x+1$	$-$	$+$	$+$

-1 2

Set of Solutions

(set of solutions)

$=[(-\infty,-1)\cup(2,\infty)]\cap(1,\infty)$

$=(2,\infty)$

□

F. $\log_3(10x+14)-\log_3(x+1)\geq 2$

Section 109: Compound Interest

Formula A

Suppose a person earns interest periodically on a single deposit he has made in an account.
Let P denote the amount of money he deposited initially in the account,
r denote the periodic rate of interest he earns on the money in the account,
t denote the number of periods of time he has held the money in the account, and
A_t denote the amount of money he has accumulated in the account after t periods of time.
Then $A_t = P(1+r)^t$.

Proof (by induction)

Suppose P, r, t, and A_t denote the quantities described above.

I: Note that:

- $A_0 = P$
- $P(1+r)^0 = P \cdot 1$
  $= P$

Thus $A_t = P(1+r)^t$ for $t = 0$.

II: Suppose $A_t = P(1+r)^t$ for some $t \in \mathbf{N} \cup \{0\}$.

$$\begin{aligned} A_{t+1} &= A_t + A_t \cdot r \\ &= A_t \cdot (1+r) \\ &= P(1+r)^t \cdot (1+r) \\ &= P(1+r)^{t+1} \end{aligned}$$

Thus $A_{t+1} = P(1+r)^{t+1}$ if $A_t = P(1+r)^t$.

By cases I and II, $A_t = P(1+r)^t$ for all $t \in \mathbf{N} \cup \{0\}$.

⊡

Formula B

Suppose a person earns interest periodically on a single deposit he has made in an account.
Let P denote the amount of money he deposited initially in the account,
R denote the annual rate of interest he earns on the money in the account,
n denote the number of times the interest is compounded per year,
y denote the number of years he has held the money in the account, and
A_y denote the amount of money he has accumulated in the account after y years.
Then $A_y = P(1+R/n)^{ny}$.

Proof

Suppose P, R, n, y, and A_y denote the quantities described above.
Note that:

- the periodic rate of interest is R/n;
- the number of periods of time for which the money has been invested is ny.

Thus $A_y = P(1+R/n)^{ny}$ by formula A.

⊡

FORMULA C

Suppose a person earns interest that is compounded continuously on a single deposit he has made in an account.

Let P denote the amount of money he deposited initially in the account,

R denote the annual rate of interest he earns on the money in the account,

y denote the number of years he has held the money in the account, and

A_y denote the amount of money he has accumulated in the account after y years.

Then $A_y = Pe^{Ry}$.

PROOF

Suppose P, R, y, and A_y denote the quantities described above,

n denotes the number of times the interest is compounded per year,

and $m = n/R$.

Note that:

- $(1 + R/n)^n = (1 + 1/(n/R))^{(n/R)R}$
  $= (1 + 1/m)^{mR}$
  $= ((1 + 1/m)^m)^R$
  $\to e^R$ as $m \to \infty$
- $m \to \infty$ as $n \to \infty$
- $A_y = P(1 + R/n)^{ny}$ by formula B

Hence:

$(1 + R/n)^n \to e^R$ as $n \to \infty \Rightarrow P((1 + R/n)^n)^y \to P(e^R)^y$ as $n \to \infty$

$\Rightarrow P(1 + R/n)^{ny} \to Pe^{Ry}$ as $n \to \infty$

$\Rightarrow A_y \to Pe^{Ry}$ as $n \to \infty$

Thus $A_y = Pe^{Ry}$ if the interest is compounded continuously.

⊡

RULE OF 72

The amount of money in an account that bears interest at an annual rate of $x\%$ will double in approximately $72/x$ years.

DEFINITIONS

- The *principal* of an account is the amount of the initial deposit into it.
- The *annual percentage rate* (APR) on an account is the annual rate of interest that it bears.
- The *annual percentage yield* (APY) on an account is the annual percentage change in the amount of money in the account.

COMMENT

- In the formulas above, assume the person makes one deposit initially and none subsequently.
- In the examples to follow, assume the variables P, R, n, y, and A_y represent the quantities described in formulas B and C.
- We should use a calculator for exercises of the kinds in this section.

EXAMPLE 1

A man places \$1000 in an investment and earns an annual rate of interest on the investment of 7%. Find the value of the investment after 20 years for each frequency of compounding. [Hint: For this investment, $P = 1000$, $R = 0.07$, and $y = 20$.]

A. monthly

SOLUTION

For $n = 12$:

$$\begin{aligned} A_{20} &= 1000(1 + 0.07/12)^{(12)(20)} \\ &\approx 4038.74 \end{aligned}$$

Its value will be \$4,038.74.

□

B. weekly

C. continuously

SOLUTION

$$\begin{aligned} A_{20} &= 1000e^{(0.07)(20)} \\ &= 1000e^{1.4} \\ &\approx 4055.20 \end{aligned}$$

Its value will be \$4,055.20.

□

D. annually

EXAMPLE 2

How long will it take for an investment to triple in value if it bears interest at an annual rate of 8% compounded quarterly?

SOLUTION

For $R = 0.08$ and $n = 4$:

$$\begin{aligned} &A_y = 3P \\ &\Rightarrow P(1 + R/n)^{ny} = 3P \\ &\Rightarrow P(1 + 0.08/4)^{4y} = 3P \\ &\Rightarrow P(1 + 0.02)^{4y} = 3P \\ &\Rightarrow \cancel{P}(1.02)^{4y}/\cancel{P} = 3\cancel{P}/\cancel{P} \\ &\Rightarrow (1.02)^{4y} = 3 \\ &\Rightarrow \ln(1.02)^{4y} = \ln 3 \\ &\Rightarrow 4y \ln 1.02 = \ln 3 \\ &\Rightarrow (y \cdot \cancel{4 \ln 1.02})/(\cancel{4 \ln 1.02}) \\ &\qquad = (\ln 3)/(4 \ln 1.02) \\ &\Rightarrow y = (\ln 3)/(4 \ln 1.02) \\ &\qquad \approx 13.8695 \end{aligned}$$

It will take about 13.9 years.

□

EXAMPLE 3

A miser placed \$7000 in a savings account on the first day of 1982. He earned interest on the funds at an annual rate of 6% compounded monthly. In what year had the amount of money in the account grown to \$8375?

Example 4
If a new car depreciates in value from \$15000 to \$300 in 25 years, then at what annual rate does the car depreciate?

Solution
For $A_y = 300$, $P = 15000$, $y = 25$, $n = 1$:

$$\begin{aligned}
A_y &= P(1 + R/n)^{ny} \\
\Rightarrow 300 &= 15000(1 + R/1)^{(1)(25)} \\
&= 15000(1 + R)^{25} \\
\Rightarrow 300/15000 &= (1 + R)^{25} \\
\Rightarrow 0.02 &= (1 + R)^{25} \\
\Rightarrow \sqrt[25]{0.02} &= 1 + R \\
\Rightarrow -1 + \sqrt[25]{0.02} &= R \\
\Rightarrow -0.144852 &\approx R \\
\Rightarrow R &\approx -14.49\%
\end{aligned}$$

The rate of depreciation is about 14.49%. □

Example 5
A woman placed \$500 in a certificate of deposit on which interest was compounded daily. After three years, the value of the CD had grown to \$621.47. What was the annual rate of interest on the CD?

Example 6
A tycoon purchased a painting in 1950 for \$17,500. He sold the painting in 1990 for \$105,900. What annual rate of return did he earn on the investment, assuming continuous compounding?

Solution
Substitutions

- $y = 1990 - 1950$
  $= 40$
- $P = 17500$
- $A_y = 105900$

Value

$$\begin{aligned}
A_y &= Pe^{Ry} \\
\Rightarrow 105900 &= 17500e^{R(40)} \\
\Rightarrow 105900/17500 &= \cancel{17500}e^{R(40)}/\cancel{17500} \\
\Rightarrow 105900/17500 &= e^{40R} \\
\Rightarrow \ln(105900/17500) &= 40R \\
\Rightarrow (\ln(105900/17500))/40 &= \cancel{40}R/\cancel{40} \\
\Rightarrow 0.045007 &\approx R \\
\Rightarrow R &\approx 4.50\%
\end{aligned}$$

His rate of return was about 4.50%. □

Example 7
If values of real estate in a certain town appreciate annually at a rate of 10% compounded continuously, then how long will it take a piece of real estate in the town to quadruple in value?

EXAMPLE 8

How much money must someone invest now at an annual rate of interest of 5% compounded weekly if he would like to accumulate $5000 in ten years?

SOLUTION

For $A_y = 5000$, $R = 0.05$, $n = 52$, $y = 10$:

$A_y = P(1 + R/n)^{ny}$

$\Rightarrow 5000 = P(1 + 0.05/52)^{(52)(10)}$

$\Rightarrow 5000 \approx P(1.648325)$

$\Rightarrow 5000/1.648325 \approx P(\cancel{1.648325})/\cancel{1.648325}$

$\Rightarrow 3033.38 \approx P$

He must invest $3,033.38 now.

□

EXAMPLE 9

A banker pays interest at a rate of 13% compounded semiannually on a certain account. How much money must you place in the account now if you wish to accumulate $400 in two years?

EXAMPLE 10

Use the Rule of 72 to estimate the number of years required for an investment to double in value at the given annual rate of interest.

A. 8%

SOLUTION

$y \approx 72/8$

$= 9$

□

B. 3%

Section 110: Exponential Growth and Decay

Definitions

- A substance grows (or decays) *exponentially* if the percentage increase (or decrease) in the amount of the substance is the same over any interval of time of a given length.
- The *doubling time* of a substance that grows exponentially is the length of time required for the substance to double in amount.
- The *half-life* of a substance that decays exponentially is the length of time required for the substance to disintegrate by half.

Formula

Suppose Q_t is a function that returns the amount in existence at time t of a substance that grows or decays exponentially.
Then $Q_t = Q_0 e^{kt}$ for some constants $Q_0, k \in \mathbf{R}$.

Comments

- In the formula above:
  - Q_0 represents the initial amount of the substance.
  - k is called the *constant of exponential growth* (or *decay*) of the substance, where:
    - if $k > 0$, then the substance grows (that is, its amount increases) over time;
    - if $k < 0$, then the substance decays (that is, its amount decreases) over time.
- Populations can also grow or decay exponentially.

Example 1

If the formula $Q_t = 15e^{0.027t}$ returns the number of bacteria (in millions) alive in a dish at time t, then determine each of the following:

- the constant of growth for the bacteria;
- the initial number of bacteria;
- the number of bacteria at $t = 5$.

Solution

- $k = 0.027$

- $Q_0 = 15$
  The initial number of bacteria in the dish is 15 million (or 15,000,000).

- $Q_5 = 15e^{0.027(5)}$
  ≈ 17.168
  The number of bacteria at $t = 5$ is about 17.168 million (or 17,168,000).

□

Example 2

The number of rabbits in a hutch grows exponentially from an initial population of 4 with a constant of growth of 0.85, where time is measured in months.
Determine each of the following:

- a formula for the number of rabbits in the hutch after t months;
- the number of rabbits in the hutch after ten months.

EXAMPLE 3

Suppose the constant of exponential decay for cesium-137 is −0.023, where time is measured in years.

A. Determine the half-life of cesium-137.
B. If only 19% of an original amount of cesium-137 is now present, then how long has the sample been decaying?
C. What percent of an amount of cesium-137 will remain after 100 years?

SOLUTION

Let Q_t denote the function that represents the amount of cesium-137 present in a sample at time t.

Note that $Q_t = Q_o e^{-0.023t}$ for some Q_0.

A. $Q_t = (1/2)Q_0$

$$\Rightarrow Q_0 e^{-0.023t} = (1/2)Q_0$$
$$\Rightarrow (\cancel{Q_0} e^{-0.023t})/\cancel{Q_0} = ((1/2)\cancel{Q_0})/\cancel{Q_0}$$
$$\Rightarrow e^{-0.023t} = 1/2$$
$$\Rightarrow \ln e^{-0.023t} = \ln(1/2)$$
$$\Rightarrow -0.023t = \ln(1/2)$$
$$\Rightarrow t = (\ln(1/2))/(-0.023)$$
$$\approx 30.1368$$

The half-life is about 30 years.

B. $Q_t = 0.19 \cdot Q_0$

$$\Rightarrow Q_0 e^{-0.023t} = 0.19 \cdot Q_0$$
$$\Rightarrow (Q_0 e^{-0.023t})/Q_0 = (0.19 \cdot Q_0)/Q_0$$
$$\Rightarrow e^{-0.023t} = 0.19$$
$$\Rightarrow \ln e^{-0.023t} = \ln 0.19$$
$$\Rightarrow -0.023t = \ln 0.19$$
$$\Rightarrow t = (\ln 0.19)/(-0.023)$$
$$\approx 72.2057$$

The length of decay is about 72 years.

C. $Q_{100}/Q_0 = \cancel{Q_0} e^{-0.023(100)}/\cancel{Q_0}$

$$= e^{-2.3}$$
$$\approx 0.100259$$
$$\approx 10.03\%$$

Roughly 10.03% of the substance will remain after 100 years.

□

EXAMPLE 4

Suppose the initial mass of cobalt-60 in a sample is 80 grams and the half-life for cobalt-60 is 5.3 years.

A. Determine a formula for the amount of cobalt-60 in the sample after t years.
B. Determine the length of time it will take for the mass to decay to 10 grams.
C. What percent of the initial mass will remain after 10.6 years?

EXAMPLE 5

In 1980, the population of Los Angeles was 2.97 million; in 2000, its population was 3.69 million. If the rate of growth remains the same, in what year will the population of the city reach 5 million?

SOLUTION

Declarations

Let P_t denote the function that represents the population (in millions) of Los Angeles t years after 1980.

Notes

- in the year 1980, $t = 0$
- in the year 2000, $t = 2000 - 1980 = 20$
- $P_t = 2.97e^{kt}$ for some constant k

Constant of Growth

$P_{20} = 3.69$
$\Rightarrow 2.97e^{k(20)} = 3.69$
$\Rightarrow 2.97e^{20k}/2.97 = 3.69/2.97$
$\Rightarrow e^{20k} = 3.69/2.97$
$\Rightarrow \ln e^{20k} = \ln(3.69/2.97)$
$\Rightarrow 20k = \ln(3.69/2.97)$
$\Rightarrow k = (\ln(3.69/2.97))/20$
$\Rightarrow k \approx 0.010853$

Time Elapsed

$P_t = 5$
$\Rightarrow 2.97e^{kt} = 5$
$\Rightarrow e^{kt} = 5/2.97$
$\Rightarrow \ln e^{kt} = \ln(5/2.97)$
$\Rightarrow kt = \ln(5/2.97)$
$\Rightarrow t = (\ln(5/2.97))/k$
$\approx (\ln(5/2.97))/0.010853$
≈ 47.994

Conclusion

The population of Los Angeles would reach 5 million about 48 years after 1980, or in the year 2028.

□

EXAMPLE 6

Suppose the population of a small town increases from 502 to 793 in twelve years. Find the doubling time of the town.

Section 111: Periodic Functions

Definitions

- A nonconstant function f with domain $\mathbb{D}$ is *periodic* if there exists some $p \in (0, \infty)$ such that $f(x+p) = f(x)$ for all $x \in \mathbb{D}$.
- The *period* of a periodic function f with domain $\mathbb{D}$ is the smallest number $p \in (0, \infty)$ such that $f(x+p) = f(x)$ for all $x \in \mathbb{D}$.
- The *amplitude* of a periodic function f is half the difference of the maximum and minimum of f, assuming those values exist.

Definition

Suppose f and g are periodic functions
    and the graph of g is a horizontal translation of the graph of f.
The *phase shift* of g with reference to f is the number of units of this translation.

Properties (proofs omitted)

Suppose f and g are periodic functions and A, B, C, and D are real constants.
Then:

- if $g(x) = f(x - C)$, then the phase shift of g with reference to f is C;
- if $g(x) = A \cdot f(B(x - C)) + D$, the period of f is p, and the amplitude of f is a, then the period of g is $p/|B|$ and the amplitude of g is $|A|a$.

Comments

- If f is a periodic function with period p and amplitude a, then:
    - the graph of f will consist of a cyclical pattern that repeats every p units;
    - p will equal the horizontal distance between successive peaks or troughs (or more generally, between corresponding points in successive cycles) in the graph of f;
    - a will equal half the vertical distance between a peak and a trough in the graph of f.
- If the mean of the maximum and minimum of a periodic function is zero, then the amplitude of the function will equal both:
    - the maximum of the function and
    - the additive inverse of the minimum of the function.
- The term *period* for a periodic function f with domain $\mathbb{D}$ can also mean an interval with width equal to the smallest number $p \in (0, \infty)$ such that $f(x+p) = f(x)$ for all $x \in \mathbb{D}$.

Example 1

Determine the phase shift of the function g with reference to f, assuming that f is a periodic function.

A. $g(x) = f(x - 7)$

Solution

The phase shift is 7.

□

B. $g(x) = f(x + 100)$

EXAMPLE 2

Determine the period p_g and amplitude a_g of the function g, assuming that f is a periodic function with the given period p_f and amplitude a_f.

A. $p_f = 8$, $a_f = 12$, $g(x) = (5/3) \cdot f(-2x + 7)$

SOLUTION

Note

$g(x) = (5/3) \cdot f(-2(x - 7/2))$

Period

$$\begin{aligned} p_g &= 8/|-2| \\ &= 8/2 \\ &= 4 \end{aligned}$$

Amplitude

$$\begin{aligned} a_g &= |5/3| \cdot 12 \\ &= (5/3)(12) \\ &= 20 \end{aligned}$$

□

B. $p_f = 2\pi$, $a_f = 1$, $g(x) = -9 \cdot f(\pi x/4)$

EXAMPLE 3

Determine the period p and amplitude a of the function with the given graph.

A.

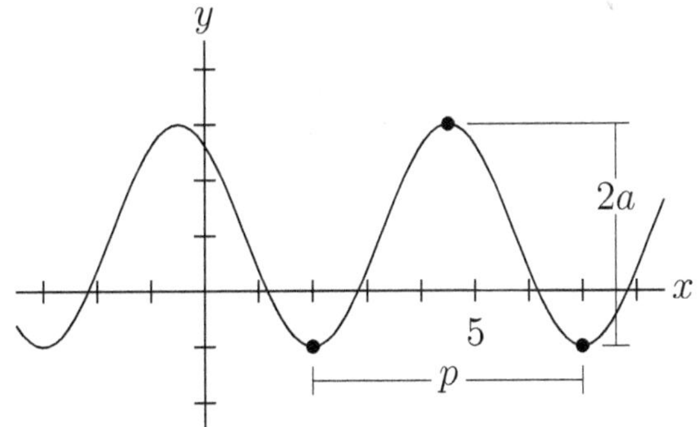

SOLUTION

Period

$$\begin{aligned} p &= 7 - 2 \\ &= 5 \end{aligned}$$

Amplitude

$$\begin{aligned} a &= (1/2)(3 - (-1)) \\ &= (1/2)(4) \\ &= 2 \end{aligned}$$

□

B.

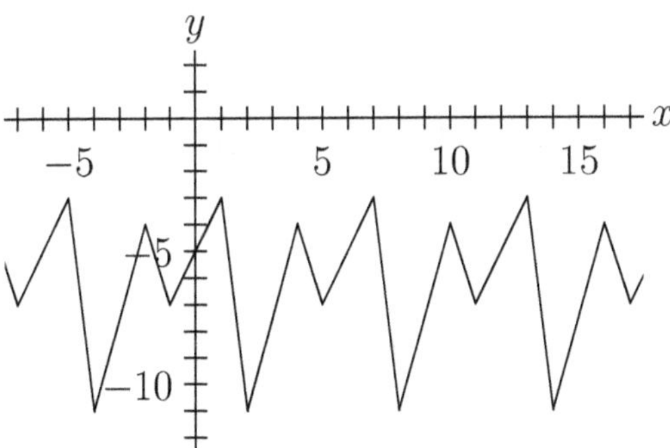

Section 112: Circles

Definitions

- A *circle* is a locus of points in a plane that lie a given distance from a fixed point.
- The *center* of a circle is the fixed point from which all points on the circle are equidistant.
- The *radius* of a circle is the distance between each point on the circle and its center.
- The *unit circle* is the circle with radius 1 and center $(0, 0)$.

Equation

The *standard form* of the equation of the circle that has radius r and center with coordinates (h, k) is $(x - h)^2 + (y - k)^2 = r^2$.

Derivation

Suppose a circle has radius r and center (h, k).
Let (x, y) be the coordinates of a point on the circle.
By definition of circle:

(distance from (x, y) to (h, k)) $= r$

$\Rightarrow \sqrt{(x-h)^2 + (y-k)^2} = r$

$\Rightarrow (\sqrt{(x-h)^2 + (y-k)^2})^2 = (r)^2$

$\Rightarrow (x-h)^2 + (y-k)^2 = r^2$

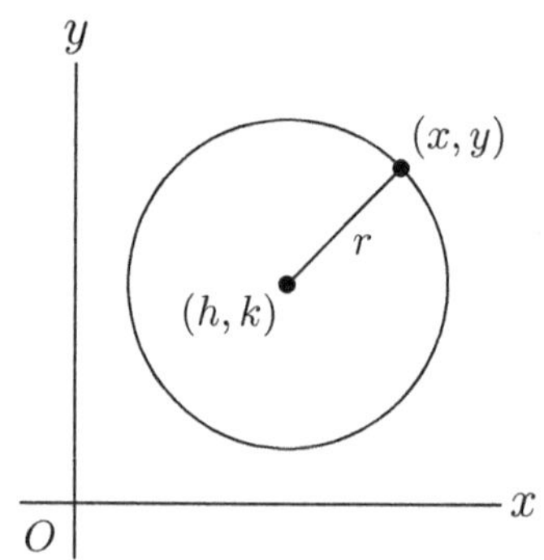

⊡

Example 1

Determine the equation of the circle with the given characteristics. Leave the equation in standard form.

A. center has coordinates $(2, 3)$;
point $(-4, 10)$ lies on the circle

Solution

Note

The equation of the circle will take the form $(x - 2)^2 + (y - 3)^2 = r^2$ for some $r \in \mathbf{R}$.

Value of r^2

$(-4, 10)$ satisfies the equation above

$\Rightarrow (-4 - 2)^2 + (10 - 3)^2 = r^2$

$\Rightarrow (-6)^2 + (7)^2 = r^2$

$\Rightarrow 85 = r^2$

Equation

$(x - 2)^2 + (y - 3)^2 = 85$

□

B. center has coordinates $(-5, -1)$;
y-axis lies tangent to the circle

EXAMPLE 2
Find the center and radius of the circle with the given equation and then draw the circle.

A. $x^2 + (y+6)^2 = 25$

SOLUTION
Note
$x^2 + (y+6)^2 = 25$
$\Leftrightarrow (x-0)^2 + (y-(-6))^2 = (5)^2$

Center and Radius
- The center is $(0, -6)$.
- The radius is 5.

Graph

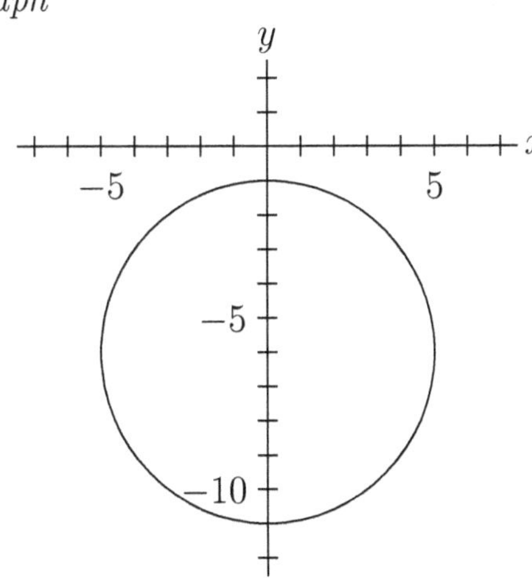

□

B. $(x+3)^2 + (y-1)^2 = 16$

EXAMPLE 3
Draw the graph of the inequality $(x+1)^2 + (y-4)^2 > 4$.

SOLUTION
Notes
- The graph of the equation $(x+1)^2 + (y-4)^2 = 4$ is a circle with center $(-1, 4)$ and radius 2.
- $(0, 0)$ satisfies $(x+1)^2 + (y-4)^2 > 4$.

Procedure
- We draw the circle described above with dashes, thereby dividing the plane into two regions.
- We shade the region of the plane that contains the point $(0, 0)$.

Graph

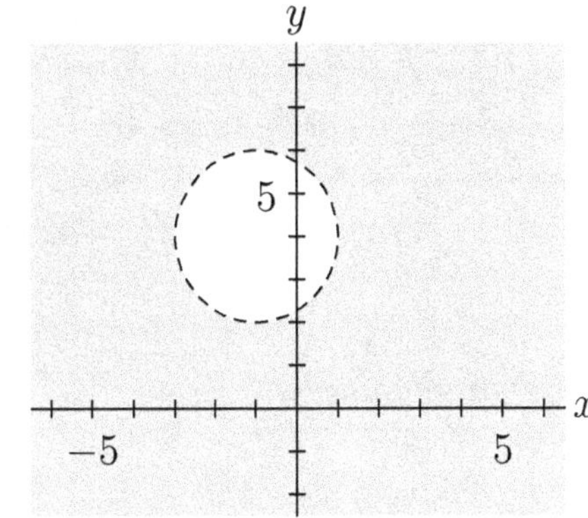

□

EXAMPLE 4

Determine the center and radius of the circle with the given equation.

A. $x^2 + y^2 - 6x + 10y = 15$

SOLUTION

Note

We must first convert the equation into standard form by completing the square:

$$\begin{aligned}
&x^2 + y^2 - 6x + 10y = 15 \\
&\quad \Leftrightarrow x^2 - 6x + y^2 + 10y = 15 \\
&\quad \Leftrightarrow x^2 + (-6)x + (-3)^2 - (-3)^2 \\
&\qquad\qquad + y^2 + 10y + (5)^2 - (5)^2 = 15 \\
&\quad \Leftrightarrow (x-3)^2 - 9 + (y+5)^2 - 25 = 15 \\
&\quad \Leftrightarrow (x-3)^2 + (y+5)^2 = 15 + 9 + 25 \\
&\qquad\qquad = 49 \\
&\quad \Leftrightarrow (x-3)^2 + (y-(-5))^2 = (7)^2
\end{aligned}$$

Center and Radius

- The center is $(3, -5)$.
- The radius is 7.

□

B. $x^2 + y^2 + 2x - 8 = 0$

C. $2x^2 + 2y^2 + 7x + 16y - 3 = 0$

SOLUTION

Note

$$\begin{aligned}
&2x^2 + 2y^2 + 7x + 16y - 3 = 0 \\
&\quad \Leftrightarrow 2x^2 + 7x + 2y^2 + 16y - 3 = 0 \\
&\quad \Leftrightarrow \tfrac{1}{2}(2x^2 + 7x + 2y^2 + 16y - 3) = \tfrac{1}{2}(0) \\
&\quad \Leftrightarrow x^2 + (7/2)x + y^2 + 8y - 3/2 = 0 \\
&\quad \Leftrightarrow x^2 + (7/2)x + (7/4)^2 - (7/4)^2 \\
&\qquad\qquad + y^2 + 8y + (4)^2 - (4)^2 - 3/2 = 0 \\
&\quad \Leftrightarrow (x + 7/4)^2 - 49/16 \\
&\qquad\qquad + (y+4)^2 - 16 - 3/2 = 0 \\
&\quad \Leftrightarrow (x + 7/4)^2 - 49/16 \\
&\qquad\qquad + (y+4)^2 - 256/16 - 24/16 = 0 \\
&\quad \Leftrightarrow (x + 7/4)^2 + (y+4)^2 - 329/16 = 0 \\
&\quad \Leftrightarrow (x + 7/4)^2 + (y+4)^2 = 329/16 \\
&\qquad\qquad = (\sqrt{329}/4)^2
\end{aligned}$$

Center and Radius

- The center is $(-7/4, -4)$.
- The radius is $\sqrt{329}/4$.

□

D. $3x^2 + 3y^2 - 12x - 15y = 10$

Section 113: Angles

DEFINITIONS

- An *angle* is a union of two rays with a common endpoint.
- The *vertex* of an angle is the endpoint of the rays that form it.
- The rays that form an angle are called the *initial side* and the *terminal side* of the angle.
- An angle is in *standard position* if drawn in the Cartesian plane with its vertex at the origin and its initial side along the positive side of the x-axis.

NOTATION

We can denote an angle by placing the angle symbol ($\angle$) to the left of one of the following:

- a sequence of three letters, which denote a point on a ray, the vertex, and a point on the other ray, respectively;
- a single letter, which denotes the vertex.

COMMENTS

- In three-letter notation for an angle, the middle letter always denotes the vertex.
- We can use single-letter notation for an angle only in cases where such uniquely identifies it.
- We often draw between the rays of an angle either:
  - an arrow, directed from its initial side to its terminal side, or
  - an arc.

EXAMPLE 1

Consider the angle shown at right.

Observe that:

- we can denote the angle by $\angle ABC$, $\angle CBA$, or $\angle B$;
- the ray $\overrightarrow{BC}$ is the initial side of the angle;
- the ray $\overrightarrow{BA}$ is the terminal side of the angle.

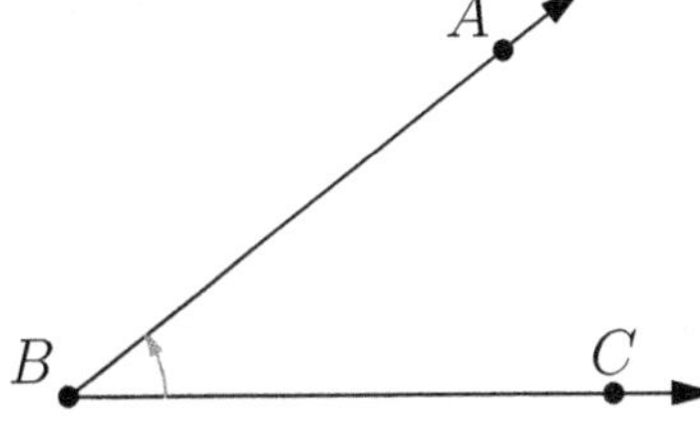

$\square$

Section 114: Measures of Angles in Degrees

Definitions

- A *degree* is 1/360 of one complete rotation of a ray about its endpoint.
- A *minute* is 1/60 of a degree.
- A *second* is 1/60 of a minute.
- The *measure in degrees* of an angle is the amount of rotation of its initial side about its vertex that will bring that side into coincidence with its terminal side. The measure equals
  - $+n$ degrees if this amount is $n/360$ of one complete counterclockwise rotation, or
  - $-n$ degrees if this amount is $n/360$ of one complete clockwise rotation.

Notation

- The symbols °, ′, and ″ mean "degrees", "minutes", and "seconds", respectively.
- We usually denote the measure of an angle with either:
  - the concatenation of the letter m and the notation for the angle, or
  - a Greek letter.

Comments

- We draw an arrow from the initial side to the terminal side of an angle to indicate both the direction and the extent of the rotation that determines the measure of the angle.
- The measure of an angle by definition is:
  - positive if the rotation of the initial side is counterclockwise;
  - negative if the rotation of the initial side is clockwise.
- The figure at right depicts major angles with measures in degrees between 0° and 360°, drawn in standard position. Observe that:
  - the measure of a right angle is 90°;
  - the measure of the angle for which exactly one complete counterclockwise rotation brings the initial side into coincidence with the terminal side is 360°;
  - the terminal sides of two angles drawn in standard position will coincide when the measures of the angles differ by a multiple of 360°.

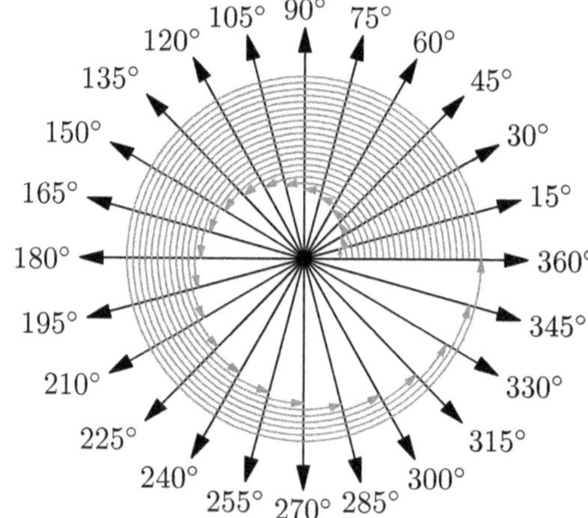

Example 1

The initial side of $\angle ABC$ shown at right would coincide with the terminal side of $\angle ABC$ after one-eighth of one complete counterclockwise rotation about the vertex B. Thus:

$$\begin{aligned} \text{m}\angle ABC &= (1/8)\cdot 360^\circ \\ &= 45^\circ \end{aligned}$$

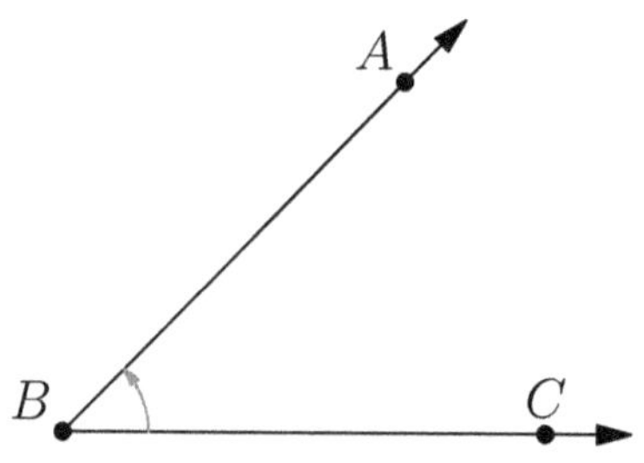

□

EXAMPLE 2
We provide sketches and measures in degrees of three angles below.

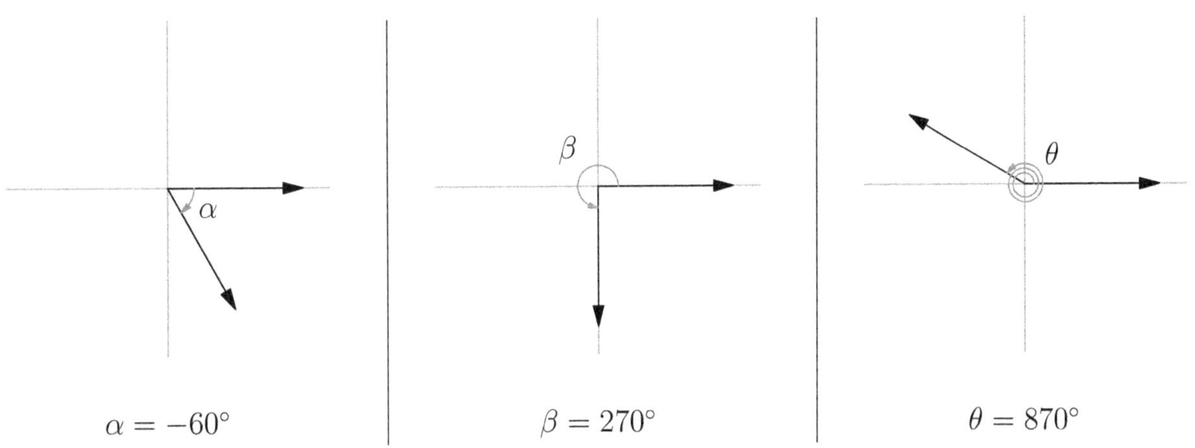

□

EXAMPLE 3
Draw in standard position the angle with the given measure.

A. $210°$

SOLUTION
Procedure

- We draw the initial side of the angle as a ray that lies along the positive side of the x-axis with endpoint at the origin.
- We draw the terminal side of the angle as a ray that coincides with $210°/360°$, or 58.3%, of a full counterclockwise rotation of the initial side.

Figure

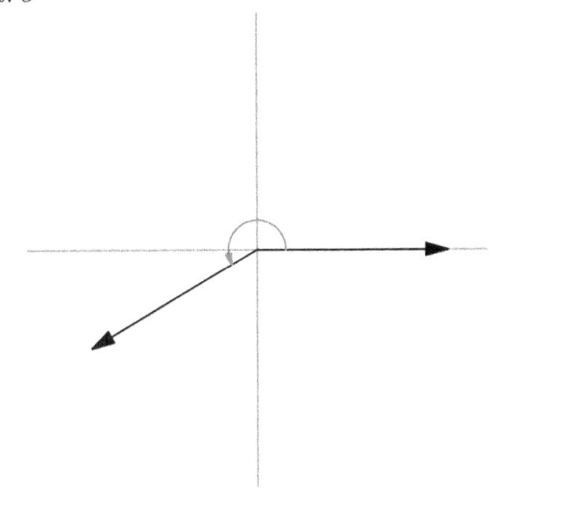

□

B. $135°$

EXAMPLE 4

Find measures, two positive and two negative, of four angles whose terminal sides coincide with that of the angle of given measure. Assume all angles lie in standard position.

A. $190°$

SOLUTION

- $190° + 360° \cdot 1 = 550°$
- $190° + 360° \cdot 2 = 910°$
- $190° - 360° \cdot 1 = -170°$
- $190° - 360° \cdot 2 = -530°$

□

B. $-465°$

EXAMPLE 5

The terminal side of the angle with measure θ coincides with the terminal side of the angle with the given measure when drawn in standard position. Find θ if $0° \leq \theta < 360°$.

A. $-792°$

SOLUTION

Notes

- $(0° - (-792°))/(360°) = 2.2$
- $(360° - (-792°))/(360°) = 3.2$

Implications

- $-792° + 360° \cdot 2.2 = 0°$
- $-792° + 360° \cdot 3.2 = 360°$
- $-792° + 360° \cdot 3 \in (0°, 360°)$

Conclusion

$$\begin{aligned} \theta &= -792° + 360° \cdot 3 \\ &= 288° \end{aligned}$$

□

B. $5640°$

EXAMPLE 6

Convert each measure to the form $d°\, m'\, s''$. Use a calculator if necessary.

A. $31.4158°$

SOLUTION

$$\begin{aligned} 31.4158° &= 31°\,(0.4158°)(60'/1°) \\ &= 31°\,24.948' \\ &= 31°\,24'\,(0.948')(60''/1') \\ &\approx 31°\,24'\,57'' \end{aligned}$$

□

B. $74.755°$

EXAMPLE 7
Convert each measure into degrees, into minutes, and into seconds. Use a calculator.

A. $8^\circ\, 52'\, 11''$

SOLUTION

- $8^\circ\, 52'\, 11'' = 8^\circ + (52')(1^\circ/60') + (11'')(1'/60'')(1^\circ/60') \approx 8.8697^\circ$
- $8^\circ\, 52'\, 11'' = (8^\circ)(60'/1^\circ) + 52' + (11'')(1'/60'') \approx 532.1833'$
- $8^\circ\, 52'\, 11'' = (8^\circ)(60'/1^\circ)(60''/1') + (52')(60''/1') + 11'' = 31{,}931''$

□

B. $-43^\circ\, 20'\, 36''$

Section 115: Measures of Angles in Radians

DEFINITION

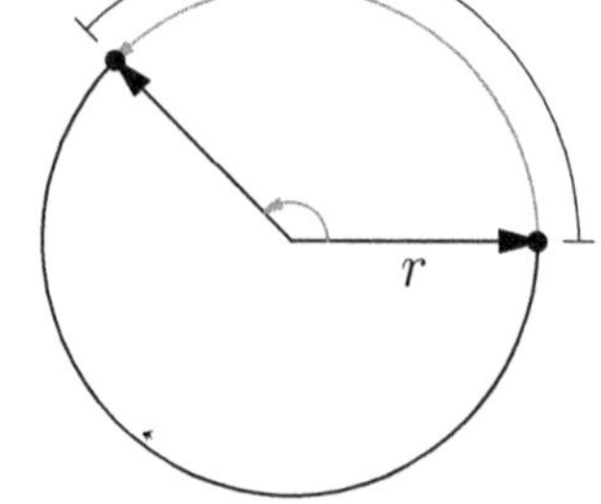

The *measure in radians* of an angle is the ratio or the additive inverse of the ratio of the length of an arc of a circle that subtends the angle to the radius of the circle. The measure equals

- ∘ $+s/r$ radians if the arc is measured counterclockwise, or
- ∘ $-s/r$ radians if the arc is measured clockwise,

where s represents the length of the arc
and r represents the radius of the circle.

COMMENTS

- We assume in the definition above that:
  - ∘ the vertex of the angle lies at the center of the circle;
  - ∘ we measure the arc from the point of intersection of the circle and the initial side of the angle to the point of intersection of the circle and the terminal side of the angle;
  - ∘ we measure the arc and the radius in the same unit.

- The ratio s/r has no units because the units of s and r will cancel upon division. Thus we can omit the unit when we provide the measure of an angle in radians. The label "radian" merely connotes the interpretation of a given measure of an angle as a ratio.

- The size of the circle we use to determine the measure of an angle in radians will not affect the value of the ratio s/r. If we use the unit circle (that is, the circle with radius 1), then s/r will equal s and the measure will equal s or $-s$.

- The length of an arc of a circle will be greater than the circumference of the circle if the arc overlaps the entire circle.

- The values of r and s are always positive; hence for an angle of measure θ:
  - ∘ $\theta > 0 \Leftrightarrow \theta = s/r$
  - ∘ $\theta < 0 \Leftrightarrow \theta = -s/r$

- The figure at right depicts major angles with measures in radians between 0 and 2π, drawn in standard position. Observe that
  - ∘ the measure of a right angle is $\pi/2$ radians;
  - ∘ the measure of the angle subtended by the arc whose length equals the circumference of the circle is 2π radians (assuming counterclockwise measurement of the arc), since $s = 2\pi r$ in such case and so $s/r = 2\pi r/r = 2\pi$;
  - ∘ the terminal sides of two angles drawn in standard position will coincide when the measures of the angles in radians differ by a multiple of 2π.

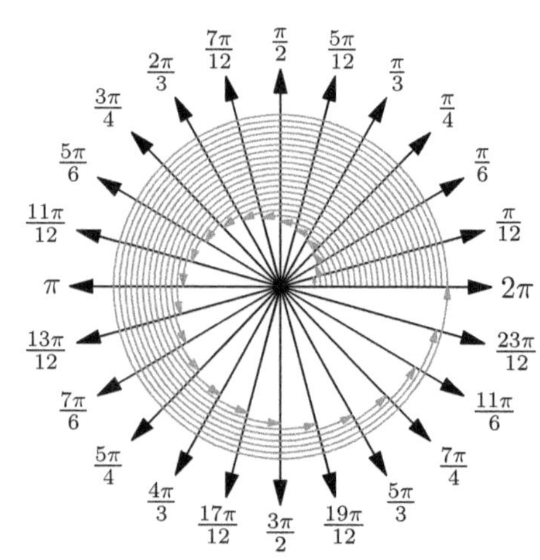

- Make special note of the angles with measure between 0 and $\pi/2$:
  - 0 radians is equivalent to $0°$
  - $\pi/6$ radians is equivalent to $30°$
  - $\pi/4$ radians is equivalent to $45°$
  - $\pi/3$ radians is equivalent to $60°$
  - $\pi/2$ radians is equivalent to $90°$

- The definition of the radian may seem somewhat more natural and less arbitrary than the definition of the degree. We may well ask why we do not define a degree to be 1/100 instead of 1/360 of a complete rotation.

EXAMPLE 1

Suppose that r denotes the radius of a circle centered at the vertex of an angle of measure θ and s denotes the length of the arc that subtends the angle. Find the value of the indicated variable.

A. Suppose $r = 5$ m, $s = 10$ m, and $\theta > 0$.
Find θ.

SOLUTION

$$\begin{aligned}\theta &= s/r \\ &= (10 \not{\text{m}})/(5 \not{\text{m}}) \\ &= 2\end{aligned}$$

□

B. Suppose $r = 6$ ft, $s = 4\pi$ ft, and $\theta < 0$.
Find θ.

C. Suppose $\theta = -\pi/3$ and $r = 18$ yd.
Find s.

SOLUTION

$$\begin{aligned}\theta = -s/r &\Rightarrow -r\theta = s \\ &\Rightarrow -(18 \text{ yd})(-\pi/3) = s \\ &\Rightarrow 18\pi/3 \text{ yd} = s \\ &\Rightarrow s = 6\pi \text{ yd}\end{aligned}$$

□

D. Suppose $\theta = 5$ and $r = 4$ mm.
Find s.

E. Suppose $\theta = -1/7$ and $s = 2$ cm.
Find r.

SOLUTION

$$\begin{aligned}\theta = -s/r &\Rightarrow -r\theta = s \\ &\Rightarrow r = s/(-\theta) \\ &\quad\;\; = (2 \text{ cm})/(-(-1/7)) \\ &\quad\;\; = (2 \text{ cm}) \cdot (7/1) \\ &\quad\;\; = 14 \text{ cm}\end{aligned}$$

□

F. Suppose $\theta = 3\pi$ and $s = 15\pi$.
Find r.

EXAMPLE 2

We provide sketches and measures in radians of three angles below.

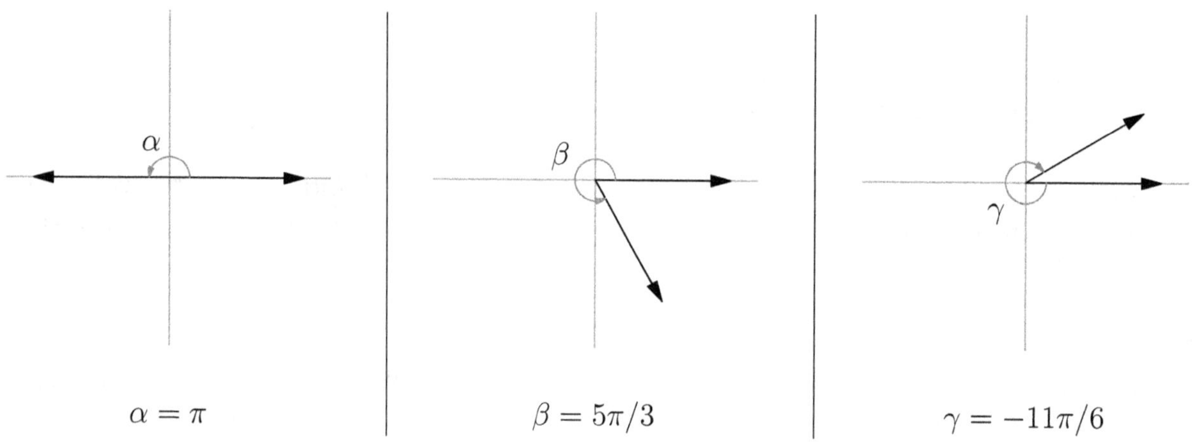

$\square$

EXAMPLE 3

Draw in standard position the angle with the given measure.

A. $5\pi/6$

SOLUTION

Procedure

- We draw the initial side of the angle as a ray that lies along the positive side of the x-axis with endpoint at the origin.
- We draw the terminal side of the angle as a ray that coincides with $(5\pi/6)/(2\pi)$, or 41.7% of a full counterclockwise rotation of the initial side.

Figure

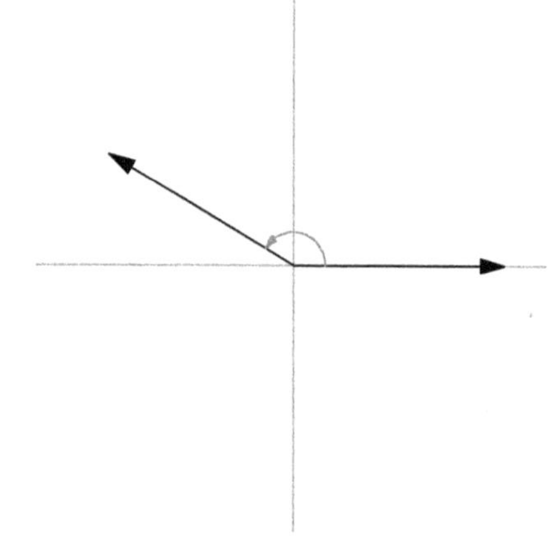

$\square$

B. $-3\pi/4$

EXAMPLE 4
Find measures, two positive and two negative, of four angles whose terminal sides coincide with that of the angle of given measure. Assume all angles lie in standard position.

A. $-5\pi/2$

SOLUTION
- $-5\pi/2 + 2\pi \cdot 1 = -5\pi/2 + 4\pi/2$
  $= -\pi/2$
- $-5\pi/2 + 2\pi \cdot 2 = -5\pi/2 + 8\pi/2$
  $= 3\pi/2$
- $-5\pi/2 + 2\pi \cdot 3 = -5\pi/2 + 12\pi/2$
  $= 7\pi/2$
- $-5\pi/2 - 2\pi \cdot 1 = -5\pi/2 - 4\pi/2$
  $= -9\pi/2$

□

B. $\pi/3$

EXAMPLE 5
The terminal side of the angle with measure θ coincides with the terminal side of the angle with the given measure when drawn in standard position. Find θ if $0 \leq \theta < 2\pi$.

A. $14\pi/3$

SOLUTION
Notes
- $(0 - 14\pi/3)/(2\pi) \approx -2.3$
- $(2\pi - 14\pi/3)/(2\pi) \approx -1.3$

Implications
- $14\pi/3 + (2\pi)(-2.3) \approx 0$
- $14\pi/3 + (2\pi)(-1.3) \approx 2\pi$
- $14\pi/3 + (2\pi)(-2) \in (0, 2\pi)$

Conclusion
$\theta = 14\pi/3 + (2\pi)(-2)$
$= 14\pi/3 - 4\pi$
$= 14\pi/3 - 12\pi/3$
$= 2\pi/3$

□

B. $-9\pi/2$

Section 116: Conversion between Degrees and Radians

PROPERTIES

Suppose θ_d and θ_r represent the measures of an angle in degrees and radians, respectively. Then:

- $\theta_r = \theta_d \cdot (\pi/180°)$
- $\theta_d = \theta_r \cdot (180°/\pi)$

PROOFS

Suppose θ_d and θ_r represent the measures of an angle in degrees and radians, respectively. Note that $\pi = 180°$, so $\pi/180° = 180°/\pi = 1$.
Therefore:

- $\theta_r = \theta_d \Rightarrow \theta_r = \theta_d \cdot 1$
  $\Rightarrow \theta_r = \theta_d \cdot (\pi/180°)$
- $\theta_d = \theta_r \Rightarrow \theta_d = \theta_r \cdot 1$
  $\Rightarrow \theta_d = \theta_r \cdot (180°/\pi)$

⊡

EXAMPLE 1

Convert each measure from degrees to radians.

A. 225°

SOLUTION

$$
\begin{aligned}
225° &= (225°)(\pi/180°) \\
&= 225\pi/180 \\
&= [5(\cancel{45})(\pi)]/[4(\cancel{45})] \\
&= 5\pi/4
\end{aligned}
$$

□

B. 420°

EXAMPLE 2

Convert each measure from radians to degrees.

A. $-31\pi/6$

SOLUTION

$$
\begin{aligned}
-31\pi/6 &= (-31\pi/6)(180°/\pi) \\
&= ((-31 \cdot 180)/6)° \\
&= ((-31 \cdot 30 \cdot \cancel{6})/\cancel{6})° \\
&= -930°
\end{aligned}
$$

□

B. $3\pi/4$

Section 117: Reference Angles

Definition

A *reference angle* is an acute angle formed by the positive or negative side of the x-axis and the terminal side of a given angle drawn in standard position.

Procedure

To determine the measure β of the reference angle for an angle of measure θ, we:

1. draw the angle of measure θ in standard position;
2. use the figure of step (1) to determine an equation that relates θ and β;
3. solve the equation of step (2) for β.

Comments

- With practice, we will be able to identify measures of reference angles immediately without recourse to figures or equations.
- An angle whose terminal side lies on an axis will not have a reference angle.

Example 1

Determine the measure β of the reference angle for the angle of measure θ.

A. $\theta = 2\pi/3$

Solution

Figure

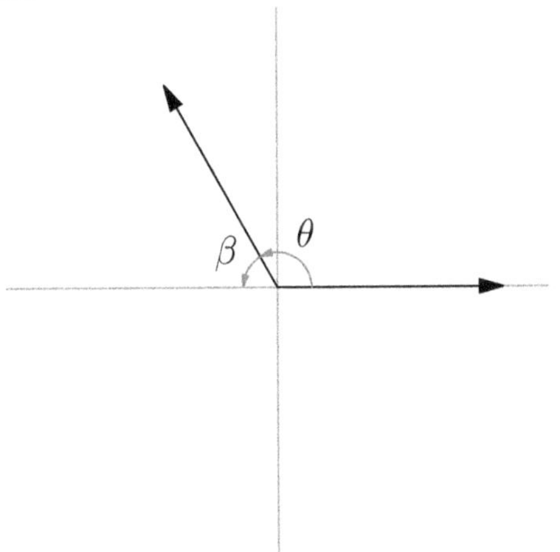

Angle

$$\begin{aligned}\theta + \beta = \pi \Rightarrow \beta &= \pi - \theta \\ &= \pi - 2\pi/3 \\ &= 3\pi/3 - 2\pi/3 \\ &= \pi/3\end{aligned}$$

□

B. $\theta = 7\pi/6$

Solution

Figure

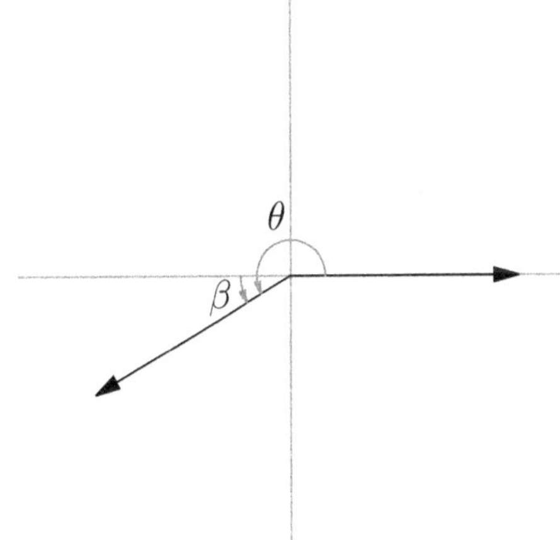

Angle

$$\begin{aligned}\pi + \beta = \theta \Rightarrow \beta &= \theta - \pi \\ &= 7\pi/6 - \pi \\ &= 7\pi/6 - 6\pi/6 \\ &= \pi/6\end{aligned}$$

□

C. $\theta = -3\pi/5$

Solution

Figure

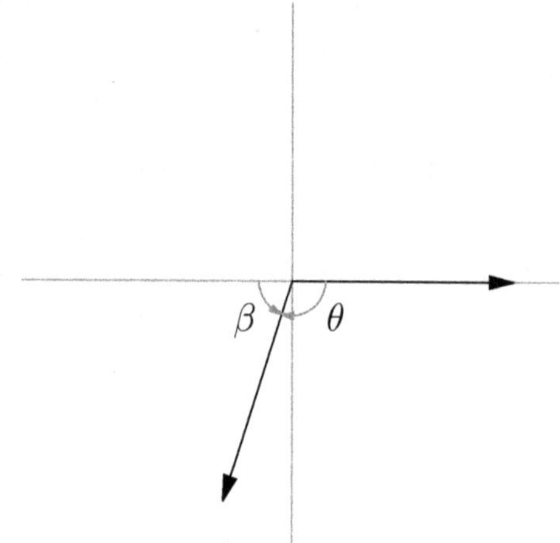

Angle

$$\begin{aligned} \beta + (-\theta) = \pi \Rightarrow \beta &= \pi + \theta \\ &= 5\pi/5 + (-3\pi/5) \\ &= 2\pi/5 \end{aligned}$$

□

D. $\theta = 3\pi/4$

E. $\theta = 13\pi/6$

Solution

Figure

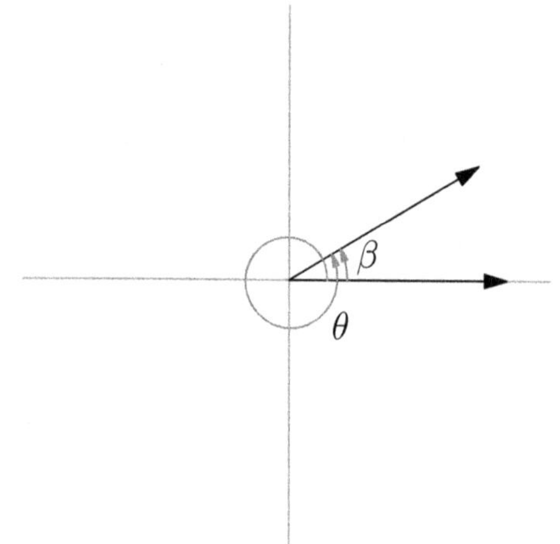

Angle

$$\begin{aligned} 2\pi + \beta = \theta \Rightarrow \beta &= \theta - 2\pi \\ &= 13\pi/6 - 12\pi/6 \\ &= \pi/6 \end{aligned}$$

□

F. $\theta = -\pi/7$

G. $\theta = 15\pi/4$

SOLUTION

Figure

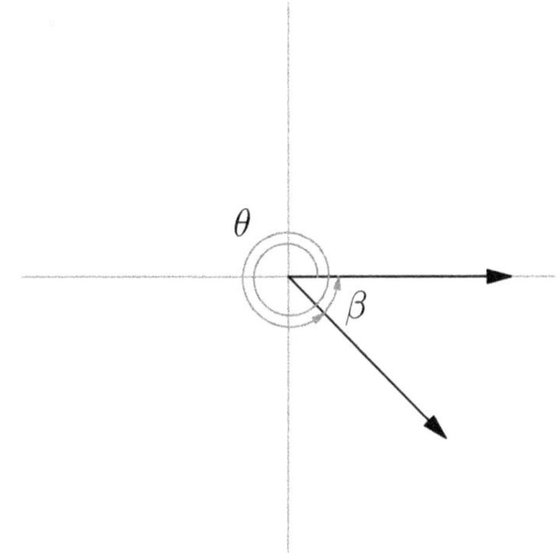

Angle

$$\begin{aligned}\theta + \beta = 2 \cdot 2\pi &\Rightarrow \theta + \beta = 4\pi \\ &\Rightarrow \beta = 4\pi - \theta \\ &\qquad = 4\pi - 15\pi/4 \\ &\qquad = 16\pi/4 - 15\pi/4 \\ &\qquad = \pi/4\end{aligned}$$

□

H. $\theta = -20\pi/3$

Section 118: Trigonometric Functions

Definition

The basic *trigonometric* functions are *cosine*, *sine*, *tangent*, *secant*, *cosecant*, and *cotangent*.

Notation

- We abbreviate the functions listed above as cos, sin, tan, sec, csc, and cot, respectively.
- Within trigonometric functions we usually omit parentheses around arguments of a single term.
- To denote powers of images of arguments of trigonometric functions, we usually place the exponents to the left of those arguments. For example, we write "$(\sin x)^3$" as "$\sin^3 x$".

Definition

Suppose $t \in \mathbf{R}$.
Let (x, y) denote the coordinates of the terminal point of the arc of length $|t|$ traced along the unit circle from the point $(1, 0)$, where the arc is traced in the counterclockwise direction if $t \geq 0$ and in the clockwise direction if $t < 0$.
The values of the trigonometric functions are given by:

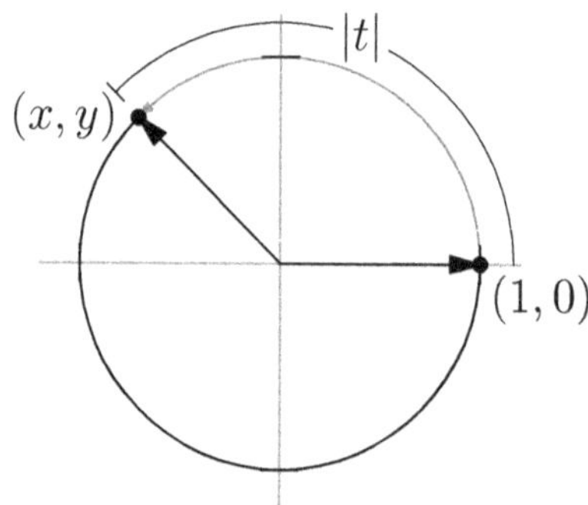

- $\cos t = x$
- $\sin t = y$
- $\tan t = y/x$
- $\sec t = 1/x$
- $\csc t = 1/y$
- $\cot t = x/y$

Definition

Suppose θ denotes the measure of an angle in radians.
Let (x, y) denote the coordinates of the point at which the terminal side of the angle, when drawn in standard position, intersects the unit circle.
The values of the trigonometric functions are given by:

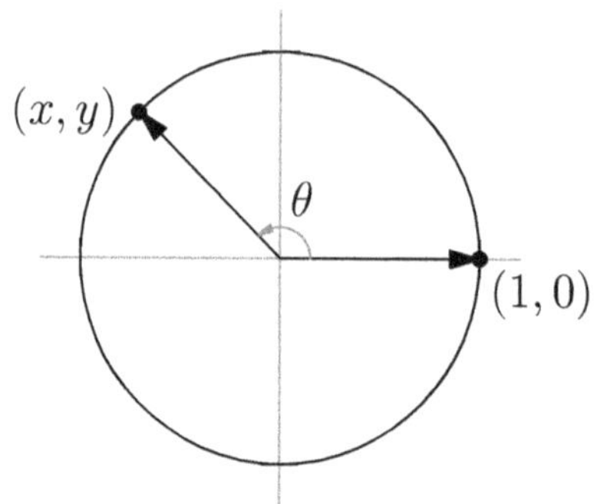

- $\cos \theta = x$
- $\sin \theta = y$
- $\tan \theta = y/x$
- $\sec \theta = 1/x$
- $\csc \theta = 1/y$
- $\cot \theta = x/y$

Domains of Trigonometric Functions

- (domain of cosine) $= \mathbf{R}$
- (domain of sine) $= \mathbf{R}$
- (domain of secant) $= \{\, t \in \mathbf{R} \mid \cos t \neq 0 \,\}$
  $= \{\, t \in \mathbf{R} \mid t \neq \pi/2 + \pi k \ \forall\ k \in \mathbf{Z} \,\}$
- (domain of cosecant) $= \{\, t \in \mathbf{R} \mid \sin t \neq 0 \,\}$
  $= \{\, t \in \mathbf{R} \mid t \neq \pi k \ \forall\ k \in \mathbf{Z} \,\}$
- (domain of tangent) $= \{\, t \in \mathbf{R} \mid \cos t \neq 0 \,\}$
  $= \{\, t \in \mathbf{R} \mid t \neq \pi/2 + \pi k \ \forall\ k \in \mathbf{Z} \,\}$
- (domain of cotangent) $= \{\, t \in \mathbf{R} \mid \sin t \neq 0 \,\}$
  $= \{\, t \in \mathbf{R} \mid t \neq \pi k \ \forall\ k \in \mathbf{Z} \,\}$

RANGES OF TRIGONOMETRIC FUNCTIONS

- (range of cosine) $= \{ x \in \mathbf{R} \mid x$ is the x-coordinate of a point on the unit circle $\}$
  $= [-1, 1]$
- (range of sine) $= \{ y \in \mathbf{R} \mid y$ is the y-coordinate of a point on the unit circle $\}$
  $= [-1, 1]$
- (range of secant) $= \{ 1/x \mid x \in [-1, 0) \cup (0, 1] \}$
  $= (-\infty, -1] \cup [1, \infty)$
- (range of cosecant) $= \{ 1/y \mid y \in [-1, 0) \cup (0, 1] \}$
  $= (-\infty, -1] \cup [1, \infty)$
- (range of tangent) $= \{ y/x \mid y \in [-1, 1]$ AND $x \in [-1, 0) \cup (0, 1] \}$
  $= \mathbf{R}$
- (range of cotangent) $= \{ x/y \mid x \in [-1, 1]$ AND $y \in [-1, 0) \cup (0, 1] \}$
  $= \mathbf{R}$

PROPERTIES (proofs omitted)
Suppose $\theta \in \mathbf{R}$.
Then:

- $\tan\theta = \sin\theta / \cos\theta$
- $\sec\theta = 1/\cos\theta$
- $\csc\theta = 1/\sin\theta$
- $\cot\theta = \cos\theta / \sin\theta$

COMMENT
We have defined the trigonometric functions in two ways because we can regard the arguments of trigonometric functions either as real numbers or as measures of angles. Our definitions are consistent and effectively equivalent because the arc on the unit circle that subtends the angle with measure θ (in radians) drawn in standard position will originate at the point $(1, 0)$ and have length $|\theta|$.

Section 119: Cosine and Sine of Measures of Major Acute Angles

PROPERTY A

$\cos(\pi/6) = \sqrt{3}/2$ and $\sin(\pi/6) = 1/2$

PROOF

To find the coordinates (x, y) of the point at which the terminal side of the angle of measure $\pi/6$ intersects the unit circle, consider the right triangle formed by the rays of the angle and the line segment from (x, y) to $(x, 0)$. In this triangle:

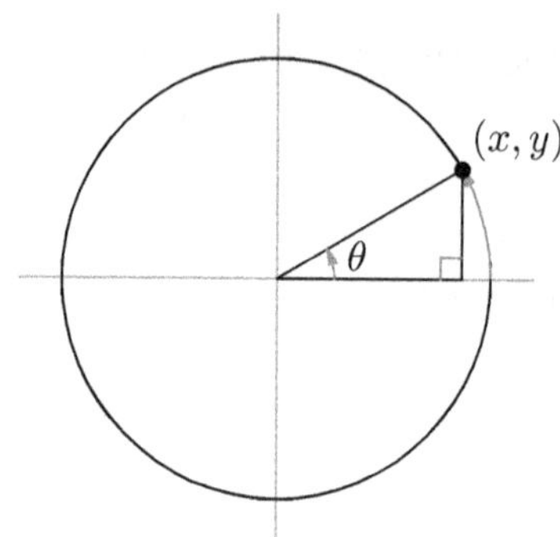

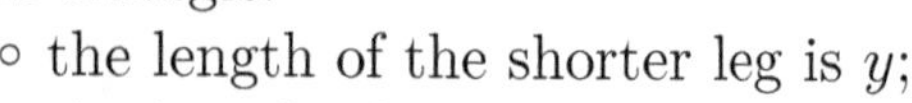

- the length of the shorter leg is y;
- the length of the longer leg is x;
- the length of the hypotenuse is 1.

Furthermore:

- the measure of the given angle is $\pi/6$;
- the measure of the right angle is $\pi/2$;
- the sum of the angles in any triangle is π.

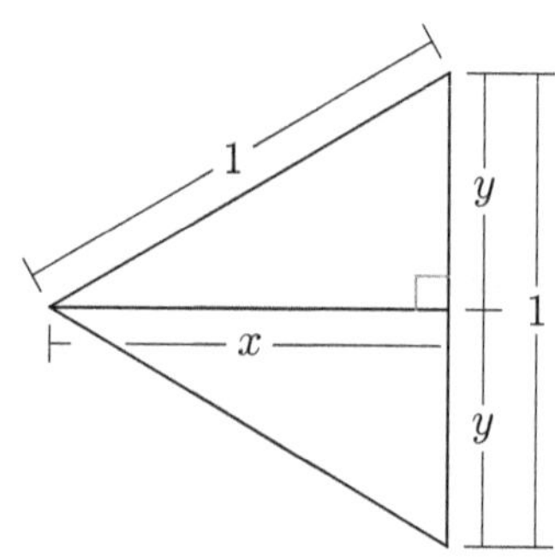

So the measure α of the third angle of the triangle is $\pi/3$, since:

$$\begin{aligned} &\pi/6 + \pi/2 + \alpha = \pi \\ &\Rightarrow \alpha = \pi - \pi/6 - \pi/2 \\ &\quad = 6\pi/6 - \pi/6 - 3\pi/6 \\ &\quad = 2\pi/6 \\ &\quad = \pi/3 \end{aligned}$$

Let us now place two copies of this triangle together to form a larger triangle as shown in the second figure above. All angles of this larger triangle have measure $\pi/3$. Therefore, all the sides of the larger triangle have the same length, namely 1.

But observe that $2y = 1$, so $y = 1/2$. Since we now know the lengths of two sides of the original right triangle, we can apply the Pythagorean Theorem to find the third:

$$\begin{aligned} x^2 + y^2 = (1)^2 &\Rightarrow x^2 + (1/2)^2 = 1 \\ &\Rightarrow x^2 + 1/4 = 1 \\ &\Rightarrow x^2 = 3/4 \\ &\Rightarrow x = \sqrt{3/4} \\ &\qquad = \sqrt{3}/2 \end{aligned}$$

So the terminal side of the angle of measure $\pi/6$ intersects the unit circle at $(\sqrt{3}/2, 1/2)$. Therefore, $\cos(\pi/6) = \sqrt{3}/2$ and $\sin(\pi/6) = 1/2$.

⊡

Property B
$\cos(\pi/4) = \sqrt{2}/2$ and $\sin(\pi/4) = \sqrt{2}/2$

Proof
To find the coordinates (x, y) of the point at which the terminal side of the angle of measure $\pi/4$ intersects the unit circle, consider the right triangle formed by the rays of the angle and the line segment from (x, y) to $(x, 0)$. In this triangle:

- the length of the legs are x and y;
- the length of the hypotenuse is 1.

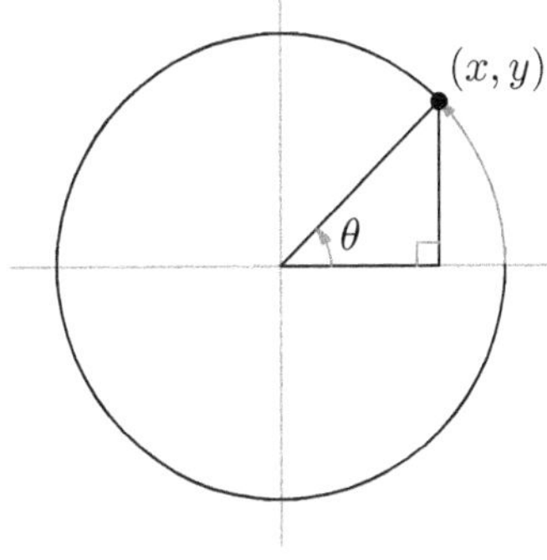

Furthermore:

- the measure of the given angle is $\pi/4$;
- the measure of the right angle is $\pi/2$;
- the sum of the angles in any triangle is π.

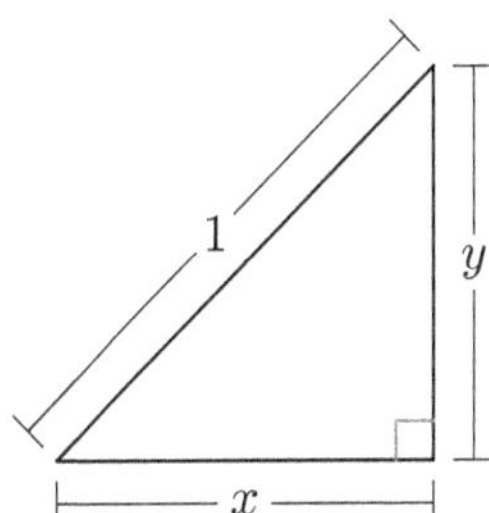

So the measure α of the third angle of the triangle is $\pi/4$, since:

$$\begin{aligned} &\pi/4 + \pi/2 + \alpha = \pi \\ &\Rightarrow \alpha = \pi - \pi/4 - \pi/2 \\ &\quad = 4\pi/4 - \pi/4 - 2\pi/4 \\ &\quad = \pi/4 \end{aligned}$$

But since the two acute angles in the triangle have equal measure, the sides opposite those angles must have equal length. Hence $y = x$. We now apply the Pythagorean Theorem to find this length:

$$\begin{aligned} x^2 + y^2 = (1)^2 &\Rightarrow x^2 + x^2 = 1 \\ &\Rightarrow 2x^2 = 1 \\ &\Rightarrow x^2 = 1/2 \\ &\Rightarrow x = \sqrt{1/2} = \sqrt{2}/2 \\ &\Rightarrow y = \sqrt{2}/2 \end{aligned}$$

So the terminal side of the angle of measure $\pi/4$ intersects the unit circle at $(\sqrt{2}/2, \sqrt{2}/2)$. Therefore, $\cos(\pi/4) = \sqrt{2}/2$ and $\sin(\pi/4) = \sqrt{2}/2$.

⊡

Property C
$\cos(\pi/3) = 1/2$ and $\sin(\pi/3) = \sqrt{3}/2$

Proof
We follow the proof of property A, but align the copies of the inscribed right triangle along the sides of length y rather than the sides of length x. We see then that $2x = 1$, and hence $x = 1/2$ and $y = \sqrt{3}/2$. Therefore, $\cos(\pi/3) = 1/2$ and $\sin(\pi/3) = \sqrt{3}/2$.

⊡

PROPERTY D
$\cos 0 = 1$ and $\sin 0 = 0$

PROOF
The terminal side of the angle of measure 0 intersects the unit circle at $(1, 0)$. Therefore, $\cos 0 = 1$ and $\sin 0 = 0$.

$\boxdot$

PROPERTY E
$\cos(\pi/2) = 0$ and $\sin(\pi/2) = 1$

PROOF
The terminal side of the angle of measure $\pi/2$ intersects the unit circle at $(0, 1)$. Therefore, $\cos(\pi/2) = 0$ and $\sin(\pi/2) = 1$.

$\boxdot$

TABLE
The table below provides the values of cosine, sine, tangent, secant, cosecant, and cotangent at 0, $\pi/6$, $\pi/4$, $\pi/3$, and $\pi/2$:

θ	0	$\pi/6$	$\pi/4$	$\pi/3$	$\pi/2$
$\cos\theta$	1	$\sqrt{3}/2$	$\sqrt{2}/2$	$1/2$	0
$\sin\theta$	0	$1/2$	$\sqrt{2}/2$	$\sqrt{3}/2$	1
$\tan\theta$	0	$\sqrt{3}/3$	1	$\sqrt{3}$	—
$\sec\theta$	1	$2\sqrt{3}/3$	$\sqrt{2}$	2	—
$\csc\theta$	—	2	$\sqrt{2}$	$2\sqrt{3}/3$	1
$\cot\theta$	—	$\sqrt{3}$	1	$\sqrt{3}/3$	0

COMMENT
We can remember easily the values of cosine and sine for the five measures of major angles between 0 and $\pi/2$ in the unsimplified forms given in the table below:

θ	0	$\pi/6$	$\pi/4$	$\pi/3$	$\pi/2$
$\cos\theta$	$\sqrt{4}/2$	$\sqrt{3}/2$	$\sqrt{2}/2$	$\sqrt{1}/2$	$\sqrt{0}/2$
$\sin\theta$	$\sqrt{0}/0$	$\sqrt{1}/2$	$\sqrt{2}/2$	$\sqrt{3}/2$	$\sqrt{4}/2$

Observe that as the values for θ increase through the given measures of the table from 0 to $\pi/2$, the denominators remain constant at 2 but the radicands in the numerators either decrease from 4 to 0 (for cosine) or increase from 0 to 4 (for sine) in increments of 1. While we may keep the unsimplified forms in mind, we should of course use the simplified forms on paper.

Section 120: Evaluating Trigonometric Functions

Properties (proofs omitted)
Suppose θ represents the measure of an angle
and β represents the measure of the reference angle of the angle of measure θ.
If the terminal side of the angle of measure θ lies in:

- quadrant I, then $\cos\theta = \cos\beta$ and $\sin\theta = \sin\beta$
- quadrant II, then $\cos\theta = -\cos\beta$ and $\sin\theta = \sin\beta$
- quadrant III, then $\cos\theta = -\cos\beta$ and $\sin\theta = -\sin\beta$
- quadrant IV, then $\cos\theta = \cos\beta$ and $\sin\theta = -\sin\beta$

Procedure
To evaluate a trigonometric function of θ where the measure of the reference angle of the angle of measure θ equals $\pi/6$, $\pi/4$, or $\pi/3$, we:

1. draw the angle of measure θ and its reference angle in standard position;
2. use symmetric properties of the unit circle and the values of the cosine and the sine of the measure of the reference angle to determine the value of the function at θ.

Comment
We can evaluate $\cos\theta$ and $\sin\theta$ without recourse to reference angles when the terminal side of the angle with measure θ lies on an axis.

Example 1
Evaluate each expression.

A. $\cos(2\pi/3)$

Solution
Figure

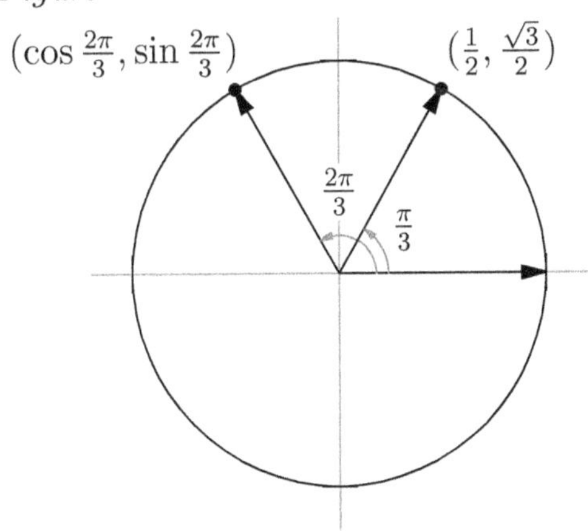

Evaluation

$$\begin{aligned}\cos(2\pi/3) &= -\cos(\pi/3)\\ &= -1/2\end{aligned}$$

□

B. $\sin(5\pi/4)$

C. $\sin(-3\pi)$

SOLUTION
Figure

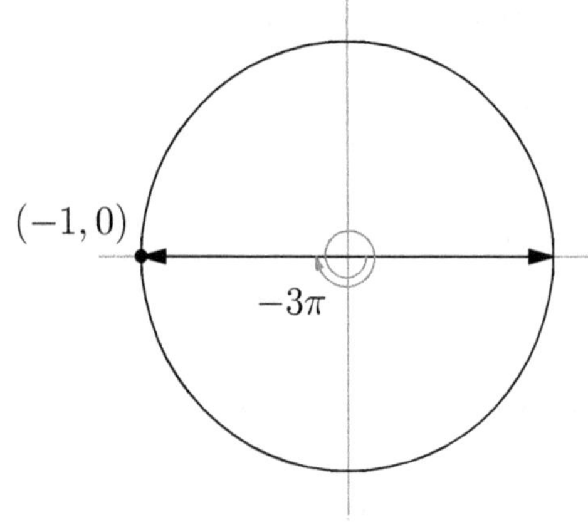

Evaluation
$\sin(-3\pi) = 0$

□

D. $\cos(7\pi/2)$

E. $\cot(-5\pi/6)$

SOLUTION
Figure

$(\frac{\sqrt{3}}{2}, \frac{1}{2})$
$\pi/6$
$-\frac{5\pi}{6}$
$(\cos(-\frac{5\pi}{6}), \sin(-\frac{5\pi}{6}))$

Evaluation

$$\begin{aligned}\cot(-5\pi/6) &= \cos(-5\pi/6)/\sin(-5\pi/6)\\ &= (-\cos(\pi/6))/(-\sin(\pi/6))\\ &= (-\sqrt{3}/2)/(-1/2)\\ &= (-\sqrt{3}/2)\cdot(-2/1)\\ &= \sqrt{3}\end{aligned}$$

□

F. $\sec(17\pi/3)$

G. $\sin(31\pi/2)$

SOLUTION
Figure

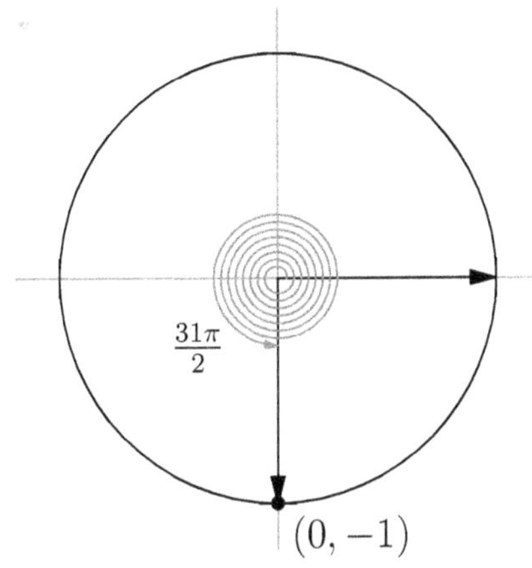

Evaluation
$\sin(31\pi/2) = -1$

□

H. $\cos(-7\pi/4)$

I. $\csc(15\pi/4)$

SOLUTION
Figure

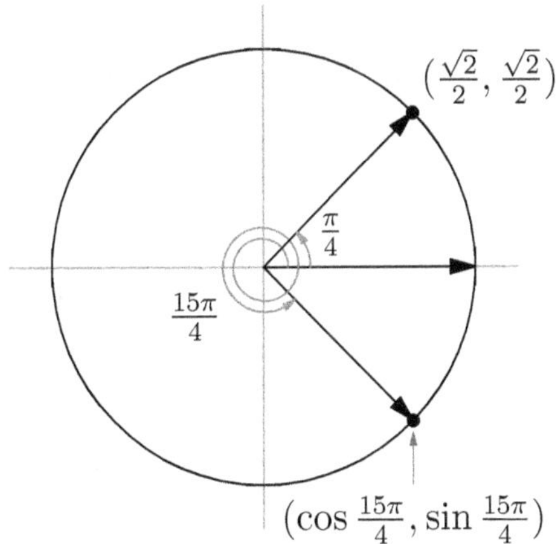

Evaluation

$$\begin{aligned}\csc(15\pi/4) &= 1/\sin(15\pi/4)\\ &= 1/(-\sin(\pi/4))\\ &= 1/(-\sqrt{2}/2)\\ &= -2/\sqrt{2}\\ &= (-2/\sqrt{2})\cdot(\sqrt{2}/\sqrt{2})\\ &= -2\sqrt{2}/2\\ &= -\sqrt{2}\end{aligned}$$

□

J. $\tan(-7\pi/6)$

Section 121: Graphs of the Cosine and Sine Functions

PROPERTY

The functions cosine and sine are periodic with period 2π and amplitude 1

PROOF (for cosine only)

- The cosine function is periodic since $\cos(x + 2\pi) = \cos(x)$ for all $x \in \mathbf{R}$.
- there does not exist any $p \in (0, 2\pi)$ such that $\cos p = 1$
  - $\Rightarrow$ there does not exist any $p \in (0, 2\pi)$ such that $\cos(0 + p) = \cos 0$
  - $\Rightarrow$ 2π is the smallest value for $p \in (0, \infty)$ such that $\cos(x + p) = \cos x$ for all $x \in \mathbf{R}$
  - $\Rightarrow$ (period of cosine) $= 2\pi$
- (range of cosine) $= [-1, 1]$
  - $\Rightarrow$ (amplitude of cosine) $= (1/2)(1 - (-1))$ $= 1$

$\boxdot$

GRAPHS

We can draw the graphs of the cosine and sine functions on the interval $[0, 2\pi]$ using a table of values (see right):

x	$\cos x$	$\sin x$
0	1	0
$\pi/6$	0.866	0.5
$\pi/4$	0.707	0.707
$\pi/3$	0.5	0.866
$\pi/2$	0	1
$2\pi/3$	−0.5	0.866
$3\pi/4$	−0.707	0.707
$5\pi/6$	−0.866	0.5
π	−1	0
$7\pi/6$	−0.866	−0.5
$5\pi/4$	−0.707	−0.707
$4\pi/3$	−0.5	−0.866
$3\pi/2$	0	−1
$5\pi/3$	0.5	−0.866
$7\pi/4$	0.707	−0.707
$11\pi/6$	0.866	−0.5
2π	1	0

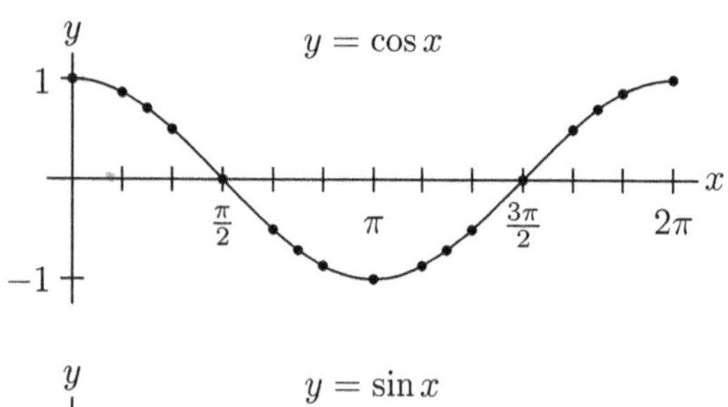

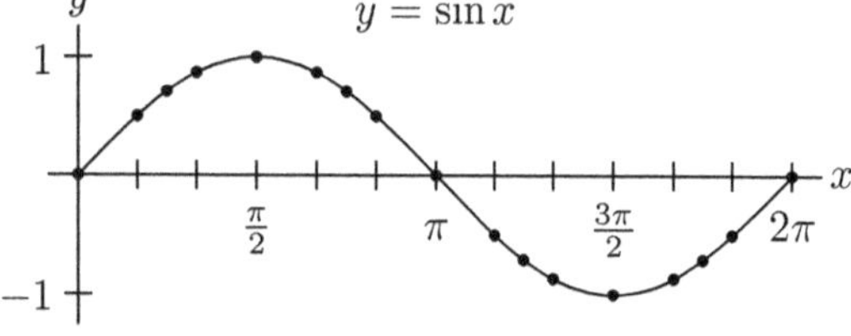

Since the functions have period 2π, we can draw the graphs on a wider interval by repeating the cycle every 2π units:

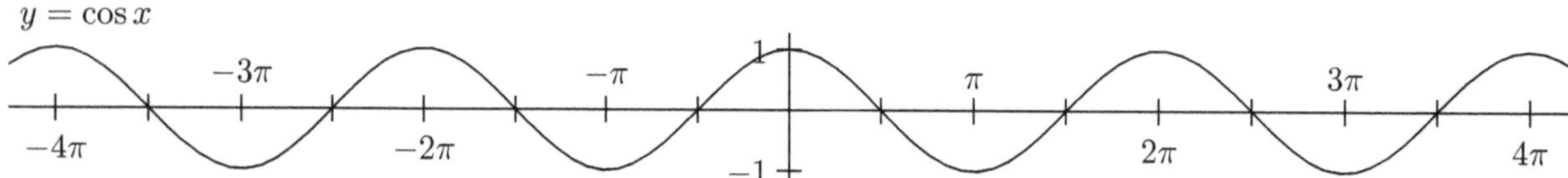

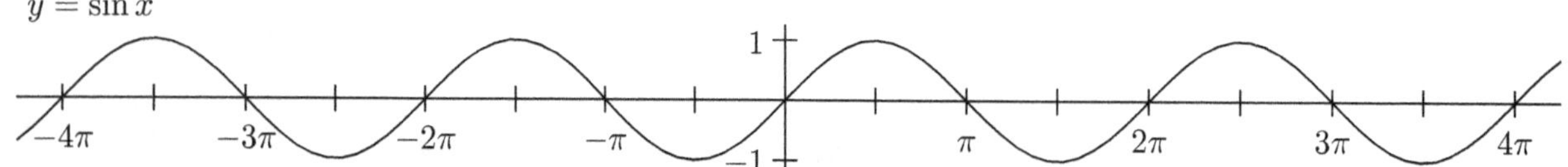

Example 1

State the period p and amplitude a of each function and draw one cycle of its graph.

A. $f(x) = 3\cos x$

Solution

Period

$p = 2\pi$

Amplitude

$a = |3| \cdot 1$
$\quad = 3$

Translations from $y = \cos x$ *to* $y = f(x)$

- $(0\,,1) \mapsto (0\,,3 \cdot 1)$
  $\quad = (0\,,3)$
- $(\pi/2\,,0) \mapsto (\pi/2\,,3 \cdot 0)$
  $\quad = (\pi/2\,,0)$
- $(\pi\,,-1) \mapsto (\pi\,,3 \cdot (-1))$
  $\quad = (\pi\,,-3)$
- $(3\pi/2\,,0) \mapsto (3\pi/2\,,3 \cdot 0)$
  $\quad = (3\pi/2\,,0)$
- $(2\pi\,,1) \mapsto (2\pi\,,3 \cdot 1)$
  $\quad = (2\pi\,,3)$

Graph

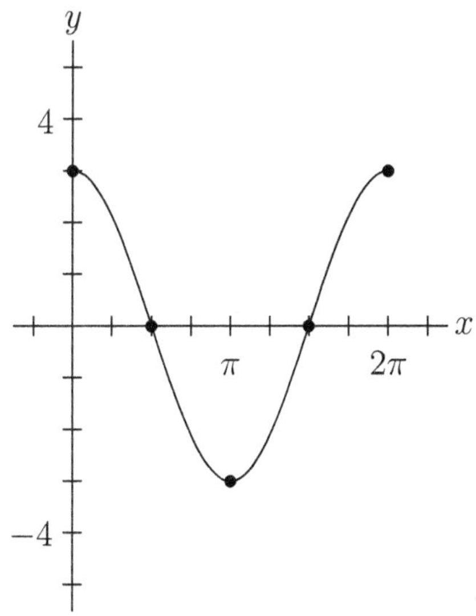

□

B. $f(x) = \sin 2x$

c. $f(x) = -5\sin(\pi x/4)$

Solution

Note

$f(x) = -5\sin((\pi/4)\cdot x)$

Period

$p = (2\pi)/|\pi/4|$
$= 2\pi\cdot(4/\pi)$
$= 8$

Amplitude

$a = |-5|\cdot 1$
$= 5$

Translations from $y = \sin x$ to $y = f(x)$

- $(0, 0) \mapsto (0/(\pi/4), -5\cdot 0)$
  $= (0, 0)$
- $(\pi/2, 1) \mapsto ((\pi/2)/(\pi/4), -5\cdot 1)$
  $= ((\pi/2)(4/\pi), -5)$
  $= (2, -5)$
- $(\pi, 0) \mapsto (\pi/(\pi/4), -5\cdot 0)$
  $= (\pi(4/\pi), 0)$
  $= (4, 0)$
- $(3\pi/2, -1) \mapsto ((3\pi/2)/(\pi/4), -5\cdot(-1))$
  $= ((3\pi/2)(4/\pi), 5)$
  $= (6, 5)$
- $(2\pi, 0) \mapsto (2\pi/(\pi/4), -5\cdot 0)$
  $= ((2\pi)(4/\pi), 0)$
  $= (8, 0)$

Graph

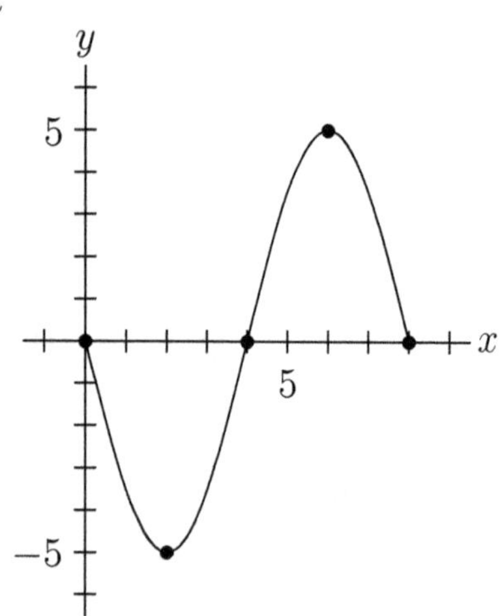

□

d. $f(x) = -4 + 4\cos x$

E. $f(x) = \frac{1}{2}\sin(3x + 3) - 7$

SOLUTION

Note

$$\begin{aligned} f(x) &= \tfrac{1}{2}\sin(3(x+1)) - 7 \\ &= \tfrac{1}{2}\sin(3(x-(-1))) - 7 \end{aligned}$$

Period

$$\begin{aligned} p &= (2\pi)/|3| \\ &= 2\pi/3 \end{aligned}$$

Amplitude

$$\begin{aligned} a &= |(1/2)|(1) \\ &= 1/2 \end{aligned}$$

Translations from $y = \sin x$ to $y = f(x)$

- $(0\,,0) \mapsto (0/3 - 1\,,(1/2)(0) - 7)$
  $= (-1\,,-7)$
- $(\pi/2\,,1) \mapsto ((\pi/2)/3 - 1\,,(1/2)(1) - 7)$
  $= (\pi/6 - 1\,,1/2 - 7)$
  $\approx (-0.48\,,-6.5)$
- $(\pi\,,0) \mapsto (\pi/3 - 1\,,(1/2)(0) - 7)$
  $\approx (0.05\,,-7)$
- $(3\pi/2\,,-1)$
  $\mapsto ((3\pi/2)/3 - 1\,,(1/2)(-1) - 7)$
  $= (\pi/2 - 1\,,-1/2 - 7)$
  $\approx (0.57\,,-7.5)$
- $(2\pi\,,0) \mapsto (2\pi/3 - 1\,,(1/2)(0) - 7)$
  $\approx (1.09\,,-7)$

Graph

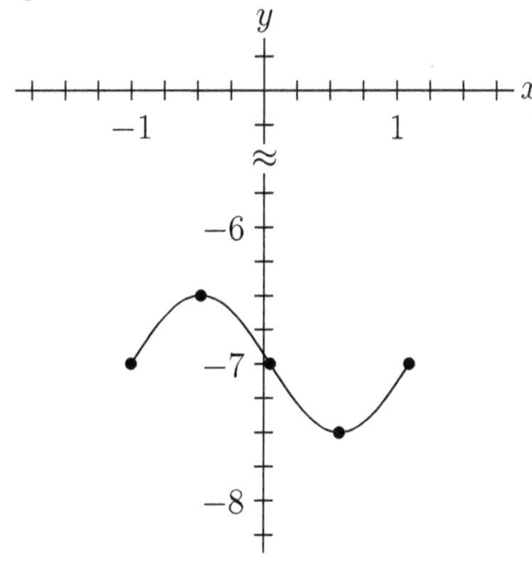

□

F. $f(x) = -6\cos((x + \pi)/2) + 1$

EXAMPLE 2

Suppose the curve at right is the graph of a function f that is defined by the equation $f(x) = A \cdot \cos(B(x - C)) + D$ for some constants $A, B, C, D \in \mathbf{R}$.

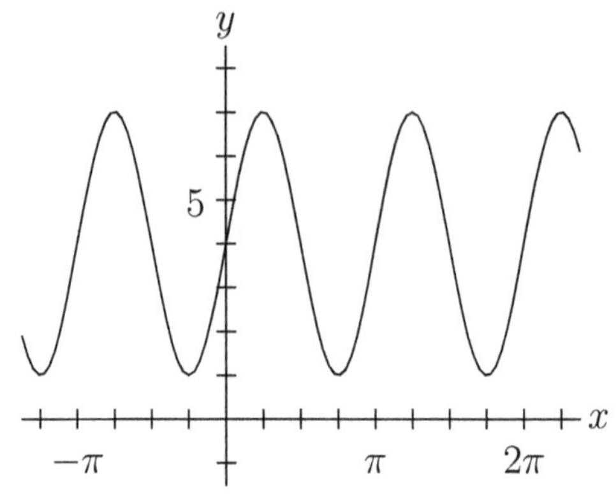

A. Find the period p and amplitude a of f.
B. Find the values of A and B, assuming that both are positive.
C. Find the values of C and D.
D. Write the equation that defines f.

SOLUTION

A. *Notes*

- The graph makes one complete cycle on the interval $[-\pi/4, 3\pi/4]$.
- The maximum of f is 7.
- The minimum of f is 1.

Period

$$p = 3\pi/4 - (-\pi/4) = 4\pi/4 = \pi$$

Amplitude

$$a = (1/2)(7 - 1) = (1/2)(6) = 3$$

B. *Value of A*

$$a = |A| \cdot 1 \Rightarrow 3 = A \cdot 1 \Rightarrow A = 3$$

Value of B

$$p = 2\pi/|B| \Rightarrow \pi = 2\pi/B \Rightarrow \pi B = 2\pi \Rightarrow \pi B/\pi = 2\pi/\pi \Rightarrow B = 2$$

C. *Comment*

Since $A = 3$ and $B = 2$, we now know that $f(x) = 3 \cdot \cos(2(x - C)) + D$. So the graph of f must be the translation right C units and up D units of the graph of $y = 3\cos 2x$.

But peaks translate to peaks here, and thus the point $(0, 1)$ on $y = \cos x$ maps to the point $(0, 3)$ on $y = 3\cos 2x$, which maps to the point $(\pi/4, 7)$ on $y = f(x)$.

(We could also assume $(0, 3)$ maps to the points $(5\pi/4, 7)$ or $(9\pi/4, 7)$ on $y = f(x)$, with different results for C below.)

Value of C

$$C = \pi/4 - 0 = \pi/4$$

Value of D

$$D = 7 - 3 = 4$$

D. *Equation*

$$f(x) = 3 \cdot \cos(2(x - \pi/4)) + 4$$

□

Section 122: Graphs of Other Trigonometric Functions

Lemma A
Suppose $\theta \in \mathbf{R}$.
Then:

- $\cos(\theta + \pi) = -\cos\theta$
- $\sin(\theta + \pi) = -\sin\theta$

Proof
Suppose $\theta \in \mathbf{R}$.
Observe that the points with coordinates $(\cos\theta, \sin\theta)$ and $(\cos(\theta + \pi), \sin(\theta + \pi))$ are symmetric with respect to the origin (see figure at right).
Therefore:

- $\cos(\theta + \pi) = -\cos\theta$
- $\sin(\theta + \pi) = -\sin\theta$

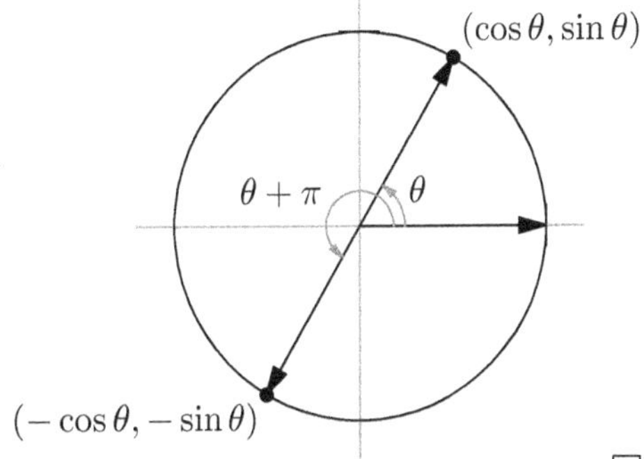

$\boxdot$

Lemma B
Suppose $\theta \in \mathbf{R}$.
Then:

- $\tan(\theta + \pi) = \tan\theta$ (provided $\theta \neq \pi/2 + \pi k$ for any $k \in \mathbf{Z}$)
- $\cot(\theta + \pi) = \cot\theta$ (provided $\theta \neq \pi k$ for any $k \in \mathbf{Z}$)

Proof (for tangent only)
Suppose $\theta \in \{ t \in \mathbf{R} \mid t \neq \pi/2 + \pi k \ \forall\ k \in \mathbf{Z} \}$.
Then:

$$\begin{aligned}\tan(\theta + \pi) &= (\sin(\theta + \pi))/\cos(\theta + \pi) \\ &= (-\sin\theta)/(-\cos\theta) \quad \text{(by Lemma A)} \\ &= \sin\theta/\cos\theta \\ &= \tan\theta\end{aligned}$$

$\boxdot$

Property
The functions tangent and cotangent are periodic with period π.

Proof (for tangent only)

- The tangent function is periodic since $\tan(x + \pi) = \tan x$ for all x in its domain.
- there does not exist any $p \in (0, \pi)$ such that $\tan p = 0$
  $\Rightarrow$ there does not exist any $p \in (0, \pi)$ such that $\tan(0 + p) = \tan 0$
  $\Rightarrow$ π is the smallest value for $p \in (0, \infty)$ such that $\tan(x + p) = \tan x$ for all x
  $\Rightarrow$ (period of tangent) $= \pi$

$\boxdot$

Property (proof omitted)
The functions secant and cosecant are periodic with period 2π.

Comments

- The ranges of secant, cosecant, tangent, and cotangent have no upper or lower bound, so these functions have no amplitudes.
- The graph of a trigonometric function will have a vertical asymptote with equation $x = c$ for every real number c at which that function is not defined.

Graphs
We can draw the graphs of the tangent and cotangent functions on the interval $[0, \pi]$ (see lower right) using a table of values (see lower left), then repeat the cycle every π units to draw the graphs on a wider interval (see bottom):

x	$\tan x$	$\cot x$
0	0	undefined
$\pi/6$	0.577	1.732
$\pi/4$	1	1
$\pi/3$	1.732	0.577
$\pi/2$	undefined	0
$2\pi/3$	−1.732	−0.577
$3\pi/4$	−1	−1
$5\pi/6$	−0.577	−1.732
π	0	undefined

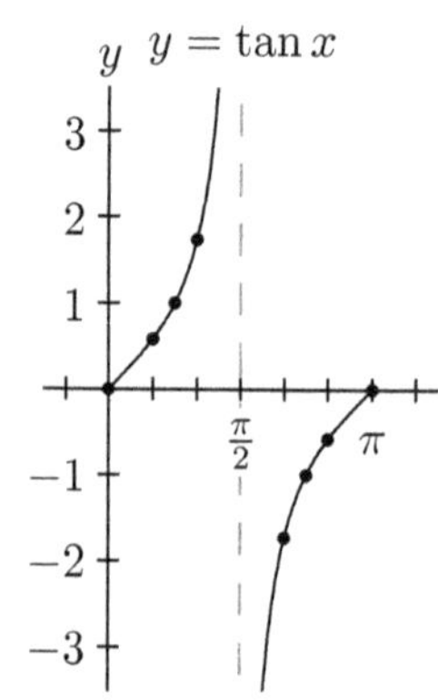

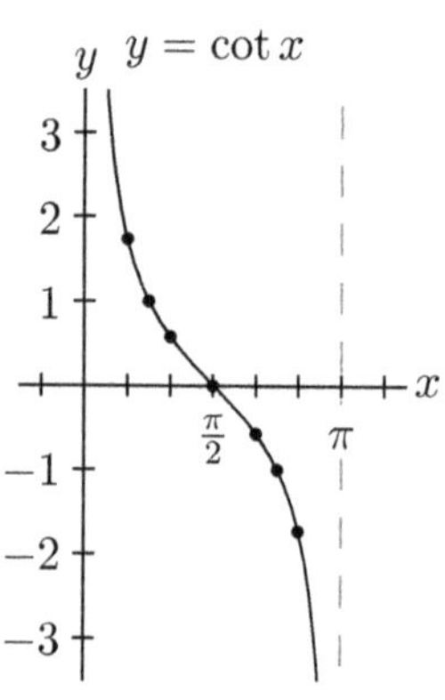

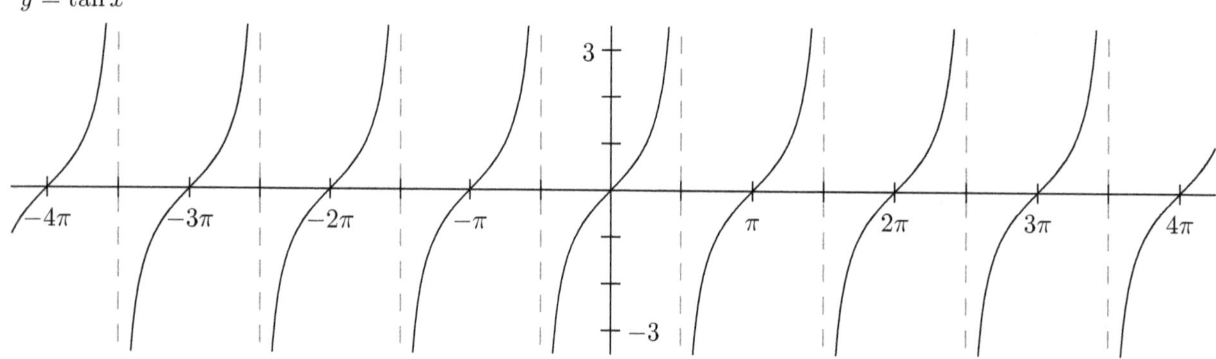

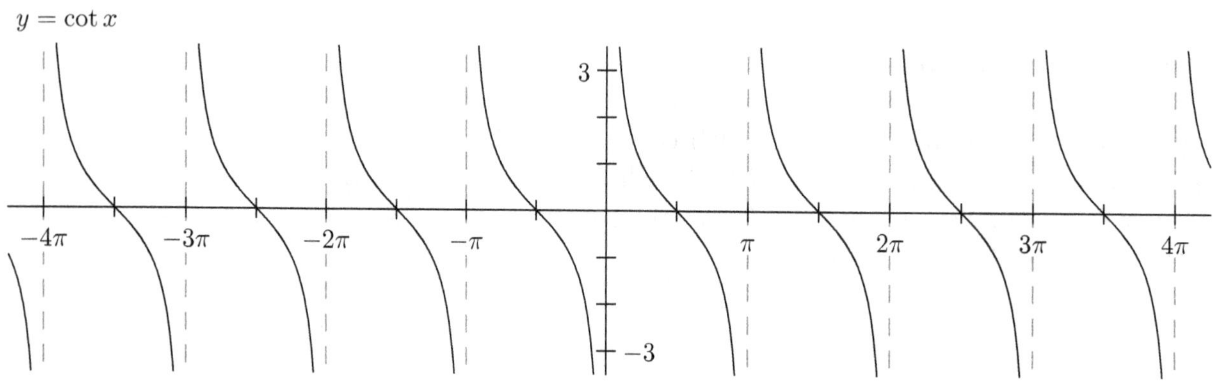

Graphs

We can draw the graphs of the secant and cosecant functions on the interval $[0, 2\pi]$ (see lower right) using a table of values (see lower left), then repeat the cycle every 2π units to draw the graphs on a wider interval (see bottom):

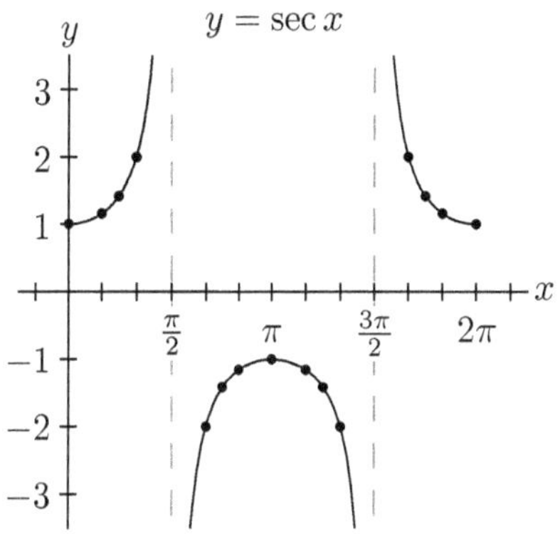

x	$\sec x$	$\csc x$
0	1	undefined
$\pi/6$	1.155	2
$\pi/4$	1.414	1.414
$\pi/3$	2	1.155
$\pi/2$	undefined	1
$2\pi/3$	−2	1.155
$3\pi/4$	−1.414	1.414
$5\pi/6$	−1.155	2
π	−1	undefined
$7\pi/6$	−1.155	−2
$5\pi/4$	−1.414	−1.414
$4\pi/3$	−2	−1.155
$3\pi/2$	undefined	−1
$5\pi/3$	2	−1.155
$7\pi/4$	1.414	−1.414
$11\pi/6$	1.155	−2
2π	1	undefined

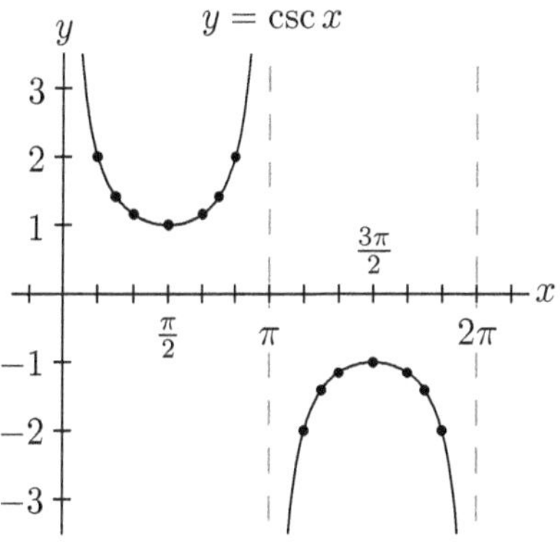

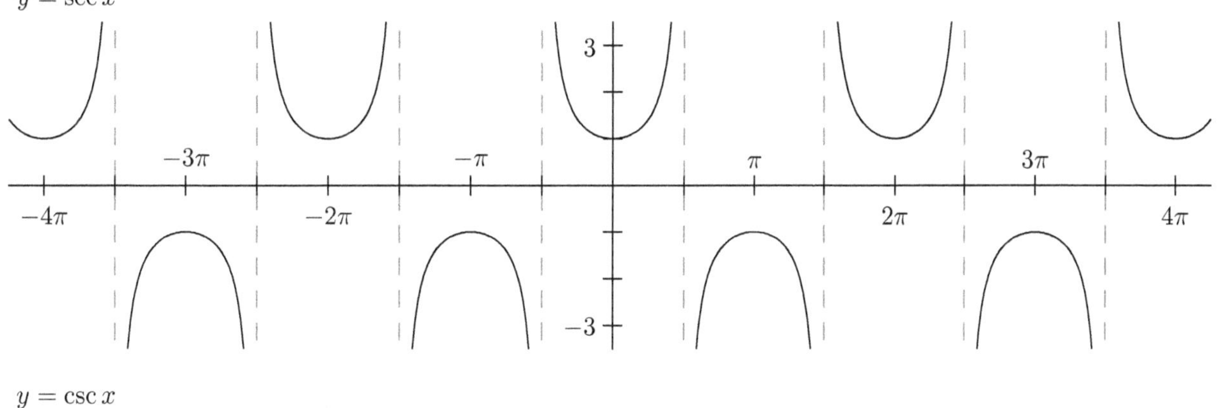

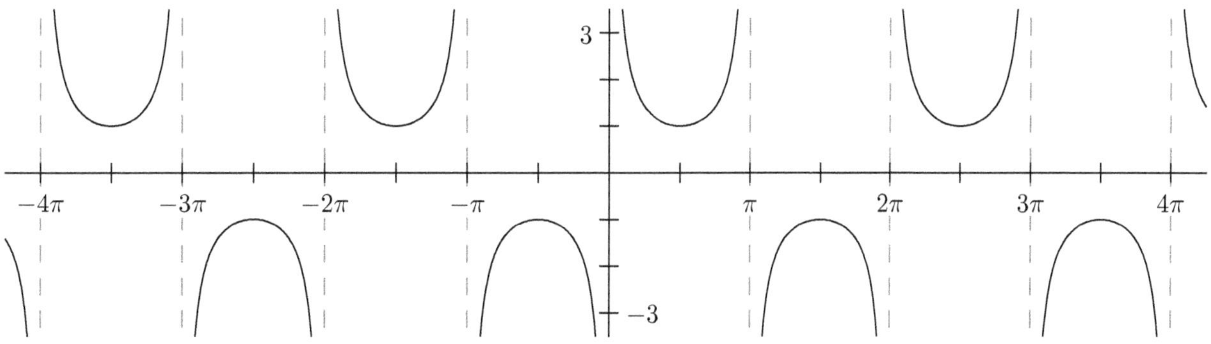

EXAMPLE 1
State the period p of each function and draw one cycle of its graph.

A. $f(x) = -2\csc(x/3) + 2$

SOLUTION
Note
$f(x) = -2\csc((1/3)\cdot x) + 2$

Period
$$\begin{aligned} p &= (2\pi)/|1/3| \\ &= (2\pi)\cdot(3/1) \\ &= 6\pi \end{aligned}$$

Translations from $y = \csc x$ *to* $y = f(x)$
- $x = 0 \mapsto x = 0/(1/3)$ $\Rightarrow x = 0$
- $(\pi/2, 1) \mapsto ((\pi/2)/(1/3), -2\cdot 1 + 2)$ $= ((\pi/2)\cdot(3/1), -2+2)$ $= (3\pi/2, 0)$
- $x = \pi \mapsto x = \pi/(1/3)$ $\Rightarrow x = \pi\cdot(3/1)$ $\Rightarrow x = 3\pi$
- $(3\pi/2, -1)$ $\mapsto ((3\pi/2)/(1/3), -2\cdot(-1)+2)$ $= ((3\pi/2)\cdot(3/1), 2+2)$ $= (9\pi/2, 4)$
- $x = 2\pi \mapsto x = 2\pi/(1/3)$ $\Rightarrow x = 2\pi\cdot(3/1)$ $\Rightarrow x = 6\pi$

Graph

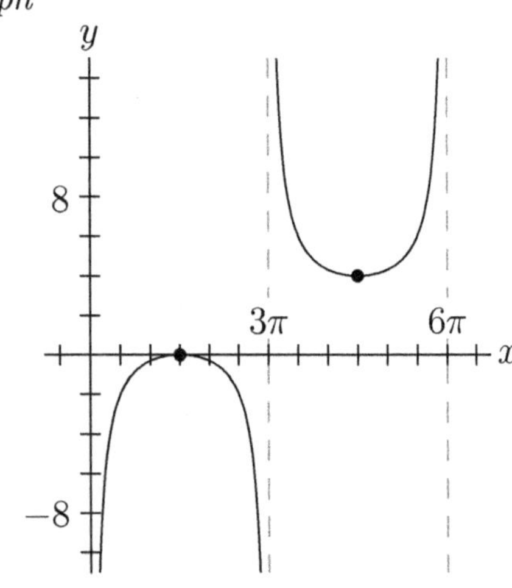

□

B. $f(x) = \frac{1}{2}\sec(x+\pi)$

C. $f(x) = \tan(2x - \pi/2) - 3$

SOLUTION

Note

$f(x) = \tan(2(x - \pi/4)) - 3$

Period

$p = (\pi)/|2|$
$= \pi/2$

Translations from $y = \tan x$ to $y = f(x)$

- $x = \pi/2 \mapsto x = (\pi/2)/2 + \pi/4$
  $\Rightarrow x = \pi/4 + \pi/4$
  $\Rightarrow x = 2\pi/4$
  $\Rightarrow x = \pi/2$
- $(3\pi/4, -1) \mapsto ((3\pi/4)/2 + \pi/4, -1 - 3)$
  $= (3\pi/8 + 2\pi/8, -4)$
  $= (5\pi/8, -4)$
- $(\pi, 0) \mapsto ((\pi)/2 + \pi/4, 0 - 3)$
  $= (2\pi/4 + \pi/4, -3)$
  $= (3\pi/4, -3)$
- $(5\pi/4, 1) \mapsto ((5\pi/4)/2 + \pi/4, 1 - 3)$
  $= (5\pi/8 + 2\pi/8, -2)$
  $= (7\pi/8, -2)$
- $x = 3\pi/2 \mapsto x = (3\pi/2)/2 + \pi/4$
  $\Rightarrow x = 3\pi/4 + \pi/4$
  $\Rightarrow x = 4\pi/4$
  $\Rightarrow x = \pi$

Graph

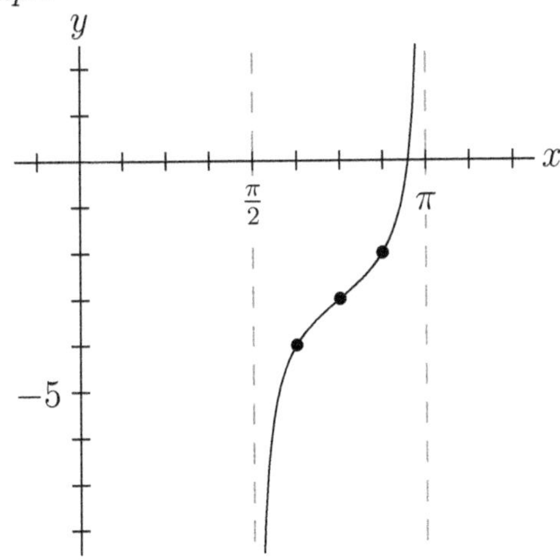

□

D. $f(x) = -\cot(x/2) + 5$

Section 123: Trigonometry of Right Triangles

PROPERTIES

Suppose θ denotes the measure of an acute angle in a right triangle,
x denotes the length of the leg of the triangle adjacent to the angle of measure θ,
y denotes the length of the leg of the triangle opposite the angle of measure θ,
and r denotes the length of the hypotenuse of the triangle.

Then:

- $\cos\theta = x/r$
- $\sin\theta = y/r$
- $\tan\theta = y/x$
- $\sec\theta = r/x$
- $\csc\theta = r/y$
- $\cot\theta = x/y$

PROOF (for cosine, sine, and tangent only)

Suppose θ, x, y, and r denote the quantities described above. Let $\tilde{x}$ and $\tilde{y}$ denote the lengths of the legs of a right triangle with angle of measure θ and hypotenuse of length 1 that lie adjacent to and opposite, respectively, the angle of measure θ.

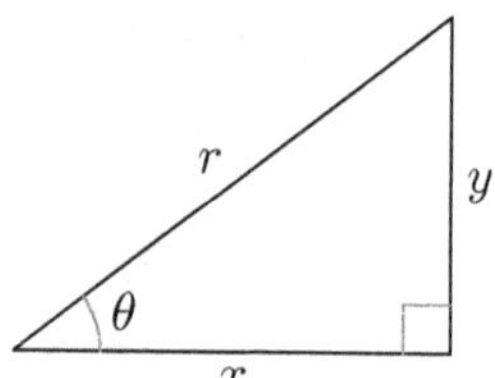

The triangle with side lengths x, y, and r is similar to the triangle with side lengths $\tilde{x}$, $\tilde{y}$, and 1. Therefore:

- $\tilde{x}/1 = x/r \Rightarrow \tilde{x} = x/r$
- $\tilde{y}/1 = y/r \Rightarrow \tilde{y} = y/r$
- $\tilde{y}/\tilde{x} = y/x$

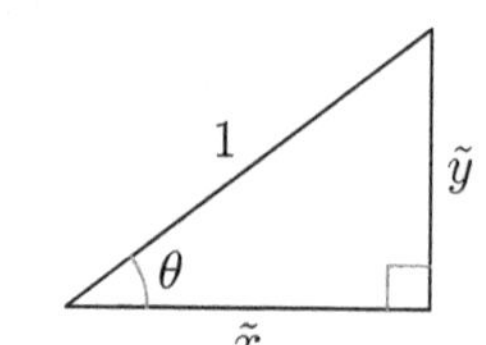

By placing the triangle of lengths x, y, and 1 inside the unit circle, we may also note that:

- $\cos\theta = \tilde{x}$
- $\sin\theta = \tilde{y}$
- $\tan\theta = \tilde{y}/\tilde{x}$

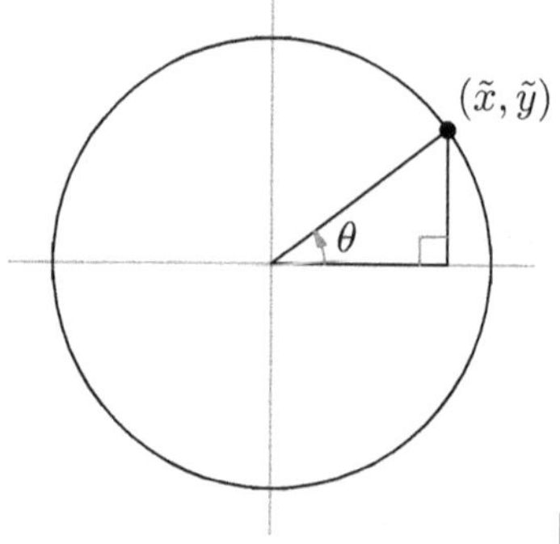

Thus:

- $\cos\theta = x/r$
- $\sin\theta = y/r$
- $\tan\theta = y/x$

⊡

COMMENTS

- We can use the phrase "soh, cah, toa" to help remember these formulas:
  - **s**in θ = **o**pposite/**h**ypotenuse
  - **c**os θ = **a**djacent/**h**ypotenuse
  - **t**an θ = **o**pposite/**a**djacent

- Internal angles of polygons have positive measure by convention, so we will usually draw a nondirectional arc instead of a directional arrow between the sides of such an angle to indicate the extent of its measure.

EXAMPLE 1
Use the given figure to evaluate the six basic trigonometric functions at θ.

A.

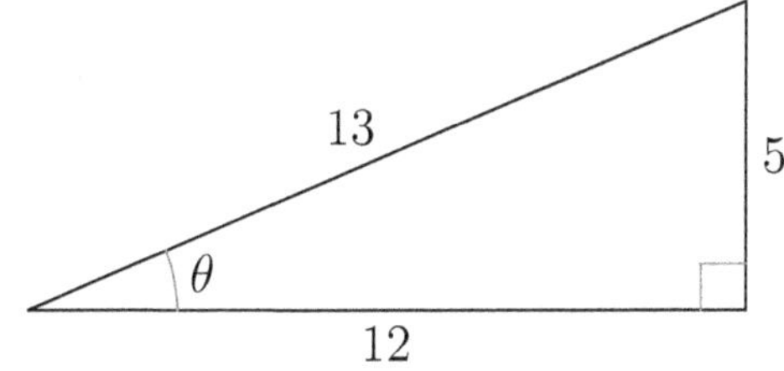

B.

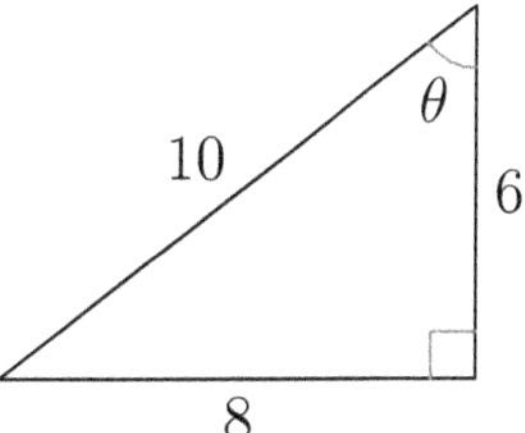

SOLUTION
- $\cos\theta = 12/13$
- $\sin\theta = 5/13$
- $\tan\theta = 5/12$
- $\sec\theta = 13/12$
- $\csc\theta = 13/5$
- $\cot\theta = 12/5$

□

EXAMPLE 2
Determine the unknown lengths of the sides of each triangle.

A.

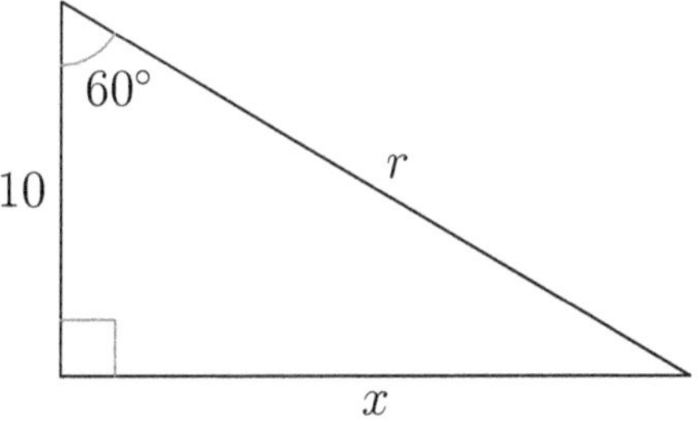

B.

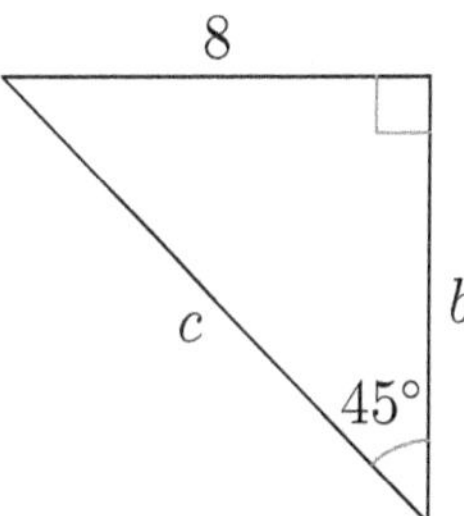

SOLUTION
Value of x
$\tan 60° = x/10$
$\Leftrightarrow 10(\tan 60°) = 10(x/10)$
$\Leftrightarrow 10(\sqrt{3}) = x$
$\Leftrightarrow x = 10\sqrt{3}$

Value of r
$\cos 60° = 10/r$
$\Leftrightarrow r(\cos 60°) = r(10/r)$
$\Leftrightarrow r(1/2) = 10$
$\Leftrightarrow r(1/2)(2) = 10(2)$
$\Leftrightarrow r = 20$

□

EXAMPLE 3
Use the given information to evaluate the six basic trigonometric functions at θ. Assume that $0° < \theta < 90°$.

A. $\cos\theta = 3/7$

SOLUTION
Figure
For aid, we draw a right triangle with:
- acute angle of measure θ;
- hypotenuse of length 7;
- leg of length 3;
- leg of unknown length s.

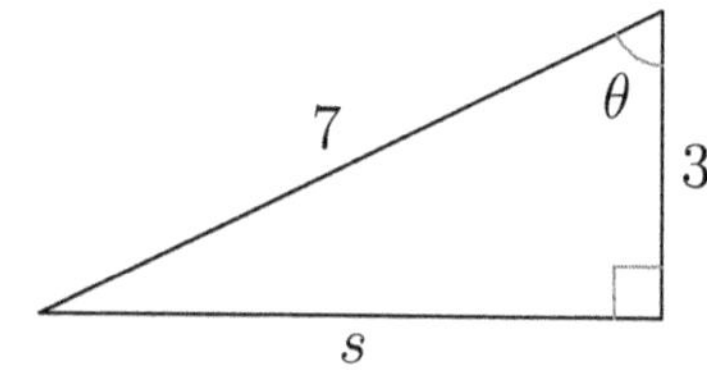

Value of s

$$\begin{aligned}(3)^2 + s^2 = (7)^2 &\Leftrightarrow 9 + s^2 = 49 \\ &\Leftrightarrow s^2 = 49 - 9 \\ &\Leftrightarrow s = \sqrt{40} \\ &\qquad = 2\sqrt{10}\end{aligned}$$

Values of Functions
- $\cos\theta = 3/7$ (as given)
- $\sin\theta = s/7$
  $= 2\sqrt{10}/7$
- $\tan\theta = s/3$
  $= 2\sqrt{10}/3$
- $\sec\theta = 7/3$
- $\csc\theta = 7/s$
  $= 7/(2\sqrt{10})$
  $= [7/(2\sqrt{10})] \cdot [\sqrt{10}/\sqrt{10}]$
  $= 7\sqrt{10}/(2 \cdot 10)$
  $= 7\sqrt{10}/20$
- $\cot\theta = 3/s$
  $= 3/(2\sqrt{10})$
  $= [3/(2\sqrt{10})] \cdot [\sqrt{10}/\sqrt{10}]$
  $= 3\sqrt{10}/(2 \cdot 10)$
  $= 3\sqrt{10}/20$

□

B. $\tan\theta = 2/3$

EXAMPLE 4
A woman hauls garbage four kilometers over a straight road from the base of a mountain to the top of the mountain. If the angle of elevation of the road is $5°$, then how high is the mountain? Use a calculator to approximate the height to the nearest thousandth of a kilometer.

SOLUTION
Declaration
Let h denote the height of the mountain (in kilometers).

Figure
We draw a picture (not necessarily to scale) for aid:

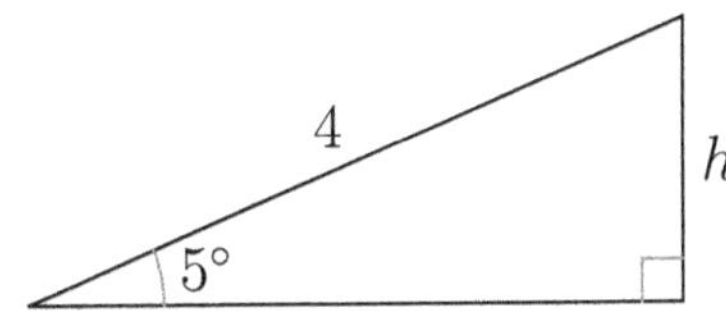

Value
$\sin 5° = h/4$
$\Rightarrow 4(\sin 5°) = 4(h/4)$
$\Rightarrow 4\sin 5° = h$
$\Rightarrow h = 4\sin 5°$
$\approx 4(0.0871557)$
≈ 0.349
The height of the mountain is roughly 0.349 km (or 349 m).

□

EXAMPLE 5
Suppose a man at the top of a lighthouse sees a ship in the ocean. If the lighthouse is 150 feet tall (with its base at sea level) and the angle of depression from the top of the lighthouse to the ship is $0.6°$, then what is the distance from the base of the lighthouse to the ship? Use a calculator to approximate the distance to the nearest foot.

Section 124: Trigonometric Identities

Property A
Suppose $\theta \in \mathbf{R}$.
Then:

- $\sin^2\theta + \cos^2\theta = 1$
- $\tan^2\theta + 1 = \sec^2\theta$ (provided $\theta \neq \pi/2 + \pi k$ for any $k \in \mathbf{Z}$)
- $1 + \cot^2\theta = \csc^2\theta$ (provided $\theta \neq \pi k$ for any $k \in \mathbf{Z}$)

Proofs
Suppose $\theta \in \mathbf{R}$.

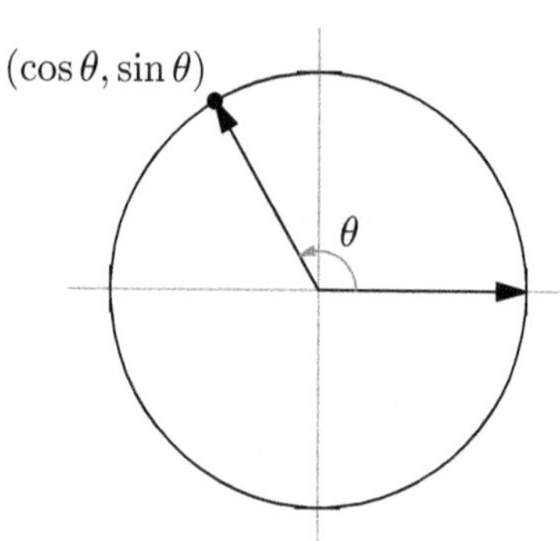

- the point $(\cos\theta, \sin\theta)$ lies on the unit circle
  $\Rightarrow (\cos\theta, \sin\theta)$ satisfies $x^2 + y^2 = 1$
  $\Rightarrow (\cos\theta)^2 + (\sin\theta)^2 = 1$
  $\Rightarrow \sin^2\theta + \cos^2\theta = 1$

- $\sin^2\theta + \cos^2\theta = 1 \Rightarrow (\sin^2\theta + \cos^2\theta)/(\cos^2\theta) = 1/\cos^2\theta$
  $\Rightarrow \sin^2\theta/\cos^2\theta + \cos^2\theta/\cos^2\theta = 1/\cos^2\theta$
  $\Rightarrow (\sin\theta/\cos\theta)^2 + 1 = (1/\cos\theta)^2$
  $\Rightarrow \tan^2\theta + 1 = \sec^2\theta$

- $\sin^2\theta + \cos^2\theta = 1 \Rightarrow (\sin^2\theta + \cos^2\theta)/(\sin^2\theta) = 1/\sin^2\theta$
  $\Rightarrow \sin^2\theta/\sin^2\theta + \cos^2\theta/\sin^2\theta = 1/\sin^2\theta$
  $\Rightarrow 1 + (\cos\theta/\sin\theta)^2 = (1/\sin\theta)^2$
  $\Rightarrow 1 + \cot^2\theta = \csc^2\theta$

⊡

Property B
Suppose $\theta \in \mathbf{R}$.
Then:

- $\cos(-\theta) = \cos\theta$
- $\sin(-\theta) = -\sin\theta$
- $\tan(-\theta) = -\tan\theta$ (provided $\theta \neq \pi/2 + \pi k$ for any $k \in \mathbf{Z}$)

Proofs
Suppose $\theta \in \mathbf{R}$.
The points $(\cos\theta, \sin\theta)$ and $(\cos(-\theta), \sin(-\theta))$ are symmetric with respect to the x-axis (see figure at right).
Therefore:

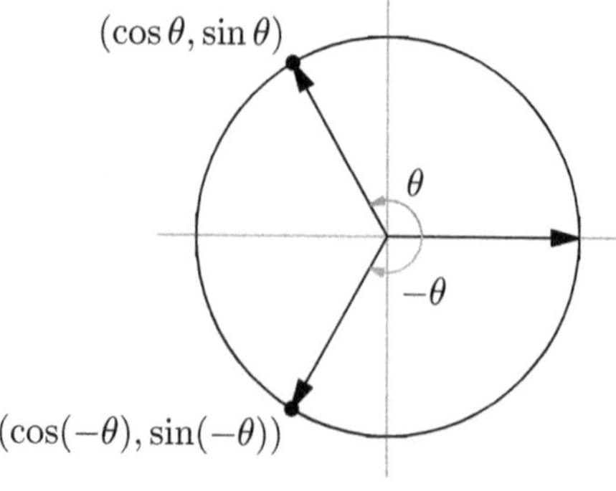

- $\cos(-\theta) = \cos\theta$
- $\sin(-\theta) = -\sin\theta$
- $\tan(-\theta) = \sin(-\theta)/\cos(-\theta)$
  $= -\sin\theta/\cos\theta$
  $= -\tan\theta$

⊡

COMMENTS

- The trigonometric identities in property A can take multiple forms. Equivalent forms of the identity $\sin^2\theta + \cos^2\theta = 1$ include the following, where the symbol "$\pm$" means either "+" or "−" (depending on θ), but not both "+" and "−":
  - $\sin^2\theta = 1 - \cos^2\theta$
  - $\sin\theta = \pm\sqrt{1 - \cos^2\theta}$
  - $\cos^2\theta = 1 - \sin^2\theta$
  - $\cos\theta = \pm\sqrt{1 - \sin^2\theta}$
- We can often simplify trigonometric expressions after rewriting them in terms of cosine and sine. Such simplifications yield other trigonometric identities.
- Unless stated otherwise, assume that variables in arguments of trigonometric functions within identities represent real numbers at which those functions are defined.

EXAMPLE 1

Prove each identity.

A. $\csc^2 x + \sec^2 x = \csc^2 x \sec^2 x$

PROOF

$$\begin{aligned}
&\csc^2 x + \sec^2 x \\
&\quad = (1/\sin x)^2 + (1/\cos x)^2 \\
&\quad = 1/\sin^2 x + 1/\cos^2 x \\
&\quad = (1/\sin^2 x)\cdot(\cos^2 x/\cos^2 x) \\
&\qquad\qquad + (1/\cos^2 x)\cdot(\sin^2 x/\sin^2 x) \\
&\quad = \cos^2 x/(\sin^2 x\cos^2 x) \\
&\qquad\qquad + \sin^2 x/(\sin^2 x\cos^2 x) \\
&\quad = (\cos^2 x + \sin^2 x)/(\sin^2 x\cos^2 x) \\
&\quad = 1/(\sin^2 x\cos^2 x) \\
&\quad = (1/\sin^2 x)\cdot(1/\cos^2 x) \\
&\quad = \csc^2 x\sec^2 x
\end{aligned}$$

⊡

B. $\tan\theta\sin\theta = \sec\theta - \cos\theta$

C. $(\csc\theta)(\csc\theta + \sin(-\theta)) = \cot^2\theta$

PROOF

$$\begin{aligned}
&(\csc\theta)(\csc\theta + \sin(-\theta)) \\
&\quad = (\csc\theta)(\csc\theta + (-\sin\theta)) \\
&\quad = (1/\sin\theta)(1/\sin\theta - \sin\theta) \\
&\quad = (1/\sin\theta)(1/\sin\theta - \sin^2\theta/\sin\theta) \\
&\quad = (1/\sin\theta)((1 - \sin^2\theta)/\sin\theta) \\
&\quad = (1 - \sin^2\theta)/\sin^2\theta \\
&\quad = \cos^2\theta/\sin^2\theta \\
&\quad = (\cos\theta/\sin\theta)^2 \\
&\quad = \cot^2\theta
\end{aligned}$$

⊡

D. $\sin t/(1 - \cos t) = (1 + \cos t)/\sin t$

E. $(\cos\beta - \sin\beta)^2 + 2\sin\beta\cos\beta = 1$

PROOF

$$\begin{aligned}&(\cos\beta - \sin\beta)^2 + 2\sin\beta\cos\beta\\ &\quad = (\cos\beta)^2 - \cancel{2\cos\beta\sin\beta} + (\sin\beta)^2\\ &\qquad\qquad + \cancel{2\sin\beta\cos\beta}\\ &\quad = \cos^2\beta + \sin^2\beta\\ &\quad = 1\end{aligned}$$

⊡

F. $\cos^2 x - \sin^2 x = 1 - 2\sin^2 x$

EXAMPLE 2

Simplify the given expressions.

A. $\sin^3\theta + \cos^2\theta\sin\theta$

SOLUTION

$$\begin{aligned}\sin^3\theta + \cos^2\theta\sin\theta &= (\sin\theta)(\sin^2\theta + \cos^2\theta)\\ &= (\sin\theta)(1)\\ &= \sin\theta\end{aligned}$$

□

B. $\cos x \sin x \sec x$

C. $\sin t/(\csc t + \cot t)$

SOLUTION

$$\begin{aligned}&\sin t/(\csc t + \cot t)\\ &\quad = \sin t/(1/\sin t + \cos t/\sin t)\\ &\quad = \sin t/[(1 + \cos t)/\sin t]\\ &\quad = \sin t \cdot [\sin t/(1 + \cos t)]\\ &\quad = \sin^2 t/(1 + \cos t)\\ &\quad = (1 - \cos^2 t)/(1 + \cos t)\\ &\quad = [(1 - \cos t)\cancel{(1 + \cos t)}]/\cancel{(1 + \cos t)}\\ &\quad = 1 - \cos t\end{aligned}$$

□

D. $\cos\theta/\sec\theta + \sin\theta/\csc\theta$

E. $\sin t + \cos t\cot t$

SOLUTION

$$\begin{aligned}&\sin t + \cos t\cot t\\ &\quad = (\sin t)(\sin t/\sin t) + (\cos t)(\cos t/\sin t)\\ &\quad = \sin^2 t/\sin t + \cos^2 t/\sin t\\ &\quad = (\sin^2 t + \cos^2 t)/\sin t\\ &\quad = 1/\sin t\\ &\quad = \csc t\end{aligned}$$

□

F. $(1 - \cos^2\theta)(1 + \cot^2\theta)$

EXAMPLE 3
Suppose $\tan\theta = -5/12$ and $\pi/2 < \theta < \pi$.

A. Find $\cos\theta$ using a trigonometric identity.
B. Find $\cos\theta$ using a right triangle and a reference angle.

SOLUTIONS
A. *Note*
$\pi/2 < \theta < \pi \Rightarrow \sec\theta < 0$

Value
$\tan^2\theta + 1 = \sec^2\theta$
$\Rightarrow (-5/12)^2 + 1 = \sec^2\theta$
$\Rightarrow 25/144 + 144/144 = \sec^2\theta$
$\Rightarrow 169/144 = \sec^2\theta$
$\Rightarrow -\sqrt{169/144} = \sec\theta$
$\Rightarrow -13/12 = 1/\cos\theta$
$\Rightarrow (-13/12)\cos\theta = 1$
$\Rightarrow \cos\theta = 1/(-13/12)$
$= -12/13$

B. *Declaration*
Let α denote the reference angle of the angle of measure θ.

Notes
- $\pi/2 < \theta < \pi \Rightarrow \cos\theta = -\cos\alpha$
- $\pi/2 < \theta < \pi \Rightarrow \tan\theta = -\tan\alpha$
  $\Rightarrow \tan\alpha = -\tan\theta$
  $= 5/12$

Figure

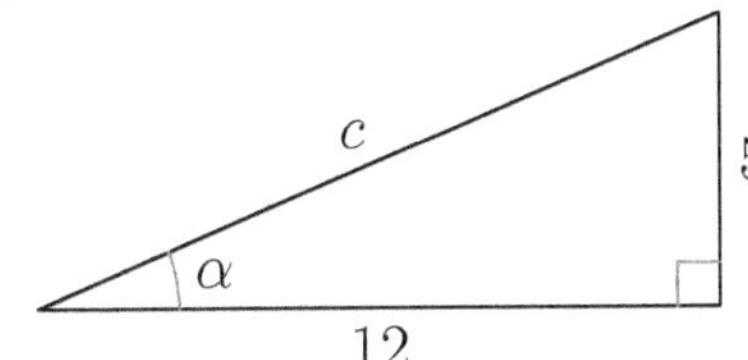

Value
$(5)^2 + (12)^2 = c^2 \Rightarrow 169 = c^2$
$\Rightarrow c = \sqrt{169} = 13$
$\Rightarrow \cos\alpha = 12/13$
$\Rightarrow \cos\theta = -12/13$

□

EXAMPLE 4
Suppose $\sin\theta = -6/10$ and $\pi < \theta < 3\pi/2$.

A. Find $\cot\theta$ using a trigonometric identity.
B. Find $\cot\theta$ using a right triangle and a reference angle.

Section 125: Addition and Subtraction Formulas

Formula A
Suppose $s, t \in \mathbf{R}$.
Then $\cos(s-t) = \cos s \cos t + \sin s \sin t$.

Proof
Suppose $s, t \in \mathbf{R}$.
Let O denote the origin,
P denote the point with coordinates $(\cos s, \sin s)$,
Q denote the point with coordinates $(\cos t, \sin t)$,
R denote the point with coordinates $(\cos(s-t), \sin(s-t))$, and
S denote the point with coordinates $(1, 0)$.

Observe that $\triangle ORS$ is simply a rotation of $\triangle OPQ$, as the figures below indicate.

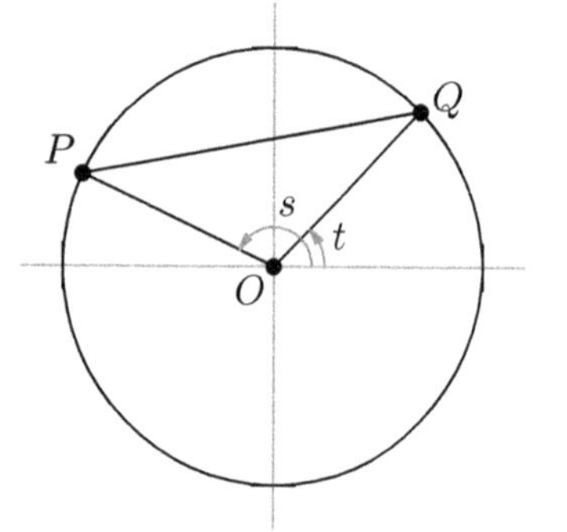

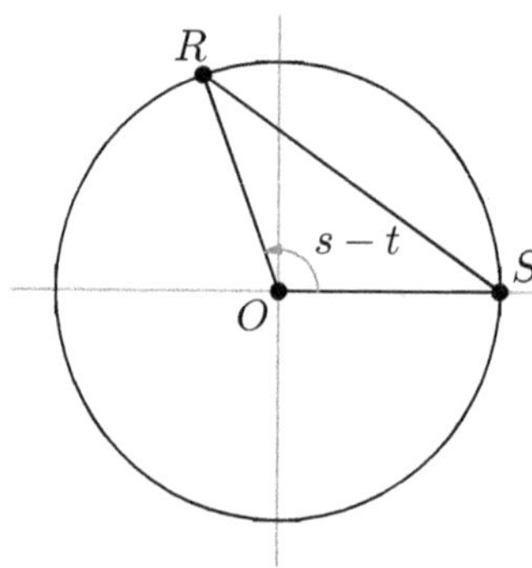

Thus the distance between R and S is the same as the distance between P and Q.
But:

$d(R, S) = d(P, Q)$
$\Rightarrow \sqrt{(\cos(s-t)-1)^2 + (\sin(s-t)-0)^2} = \sqrt{(\cos s - \cos t)^2 + (\sin s - \sin t)^2}$
$\Rightarrow \left(\sqrt{(\cos(s-t)-1)^2 + (\sin(s-t)-0)^2}\right)^2 = \left(\sqrt{(\cos s - \cos t)^2 + (\sin s - \sin t)^2}\right)^2$
$\Rightarrow (\cos(s-t)-1)^2 + (\sin(s-t)-0)^2 = (\cos s - \cos t)^2 + (\sin s - \sin t)^2$
$\Rightarrow \cos^2(s-t) - 2\cos(s-t) + 1 + \sin^2(s-t)$
$= \cos^2 s - 2\cos s \cos t + \cos^2 t + \sin^2 s - 2\sin s \sin t + \sin^2 t$
$\Rightarrow (\cos^2(s-t) + \sin^2(s-t)) - 2\cos(s-t) + 1$
$= (\cos^2 s + \sin^2 s) - 2\cos s \cos t + (\cos^2 t + \sin^2 t) - 2\sin s \sin t$
$\Rightarrow 1 - 2\cos(s-t) + 1 = 1 - 2\cos s \cos t + 1 - 2\sin s \sin t$
$\Rightarrow 2 - 2\cos(s-t) = 2 - 2\cos s \cos t - 2\sin s \sin t$
$\Rightarrow -2 + 2 - 2\cos(s-t) = -2 + 2 - 2\cos s \cos t - 2\sin s \sin t$
$\Rightarrow -2\cos(s-t) = -2\cos s \cos t - 2\sin s \sin t$
$\Rightarrow -2(\cos(s-t)) = -2(\cos s \cos t + \sin s \sin t)$
$\Rightarrow -2(\cos(s-t))/(-2) = -2(\cos s \cos t + \sin s \sin t)/(-2)$
$\Rightarrow \cos(s-t) = \cos s \cos t + \sin s \sin t$

⊡

Lemma
Suppose $x \in \mathbf{R}$.
Then:
- $\cos(\pi/2 - x) = \sin x$
- $\sin(\pi/2 - x) = \cos x$

Proofs
Suppose $x \in \mathbf{R}$.

- $$\begin{aligned}\cos(\pi/2 - x) &= \cos(\pi/2)\cos x + \sin(\pi/2)\sin x \quad \text{(by formula A)}\\ &= (0)(\cos x) + (1)(\sin x)\\ &= \sin x\end{aligned}$$

- $$\begin{aligned}\sin(\pi/2 - x) &= \cos(\pi/2 - (\pi/2 - x)) \quad \text{(by the first identity of this lemma)}\\ &= \cos(\cancel{\pi/2} - \cancel{\pi/2} + x)\\ &= \cos x\end{aligned}$$

⊡

Addition and Subtraction Formulas
Suppose $s, t \in \mathbf{R}$.
Then:
- $\cos(s - t) = \cos s \cos t + \sin s \sin t$ (restated from formula A)
- $\cos(s + t) = \cos s \cos t - \sin s \sin t$
- $\sin(s - t) = \sin s \cos t - \cos s \sin t$
- $\sin(s + t) = \sin s \cos t + \cos s \sin t$
- $\tan(s - t) = (\tan s - \tan t)/(1 + \tan s \tan t)$
  (provided $s, t, s - t \in \{x \in \mathbf{R} \mid x \neq \pi/2 + \pi k \ \forall\, k \in \mathbf{Z}\}$)
- $\tan(s + t) = (\tan s + \tan t)/(1 - \tan s \tan t)$
  (provided $s, t, s + t \in \{x \in \mathbf{R} \mid x \neq \pi/2 + \pi k \ \forall\, k \in \mathbf{Z}\}$)

Proofs (final five formulas only)
Suppose $s, t \in \mathbf{R}$.

- $$\begin{aligned}\cos(s + t) &= \cos(s - (-t))\\ &= \cos s \cos(-t) + \sin s \sin(-t)\\ &= \cos s \cos t + (\sin s)(-\sin t)\\ &= \cos s \cos t - \sin s \sin t\end{aligned}$$

- $$\begin{aligned}\sin(s - t) &= \cos(\pi/2 - (s - t))\\ &= \cos((\pi/2 - s) + t)\\ &= \cos(\pi/2 - s)\cos t - \sin(\pi/2 - s)\sin t\\ &= \sin s \cos t - \cos s \sin t\end{aligned}$$

- $$\begin{aligned}\sin(s + t) &= \sin(s - (-t))\\ &= \sin s \cos(-t) - \cos s \sin(-t)\\ &= \sin s \cos t - (\cos s)(-\sin t)\\ &= \sin s \cos t + \cos s \sin t\end{aligned}$$

- $$\begin{aligned}\tan(s-t) &= \sin(s-t)/\cos(s-t)\\ &= (\sin s\cos t - \cos s\sin t)/(\cos s\cos t + \sin s\sin t)\\ &= [(\sin s\cos t - \cos s\sin t)/(\cos s\cos t + \sin s\sin t)]\cdot[1/(\cos s\cos t)]/[1/(\cos s\cos t)]\\ &= [(\sin s\,\cancel{\cos t})/(\cos s\,\cancel{\cos t}) - (\cancel{\cos s}\sin t)/(\cancel{\cos s}\cos t)]\\ &\qquad /[(\cancel{\cos s\cos t})/(\cancel{\cos s\cos t}) + (\sin s\sin t)/(\cos s\cos t)]\\ &= [\sin s/\cos s - \sin t/\cos t]/[1 + (\sin s/\cos s)(\sin t/\cos t)]\\ &= (\tan s - \tan t)/(1+\tan s\tan t)\end{aligned}$$

- $$\begin{aligned}\tan(s+t) &= \tan(s-(-t))\\ &= [\tan s - \tan(-t)]/[1 + (\tan s)(\tan(-t))]\\ &= [\tan s - (-\tan t)]/[1 + (\tan s)(-\tan t)]\\ &= (\tan s + \tan t)/(1 - \tan s\tan t)\end{aligned}$$

$\boxdot$

FORMULA B

Suppose $a, b, x, \theta \in \mathbf{R}$ where $\cos\theta = a/\sqrt{a^2+b^2}$ and $\sin\theta = b/\sqrt{a^2+b^2}$.
Then $a\sin x + b\cos x = \sqrt{a^2+b^2}\cdot\sin(x+\theta)$.

PROOF

Suppose $a, b, x, \theta \in \mathbf{R}$ where $\cos\theta = a/\sqrt{a^2+b^2}$ and $\sin\theta = b/\sqrt{a^2+b^2}$.

$$\begin{aligned}\sqrt{a^2+b^2}\cdot\sin(x+\theta) &= \sqrt{a^2+b^2}\cdot(\sin x\cos\theta + \cos x\sin\theta)\\ &= \sqrt{a^2+b^2}\cdot[(\sin x)(a/\sqrt{a^2+b^2}) + (\cos x)(b/\sqrt{a^2+b^2})]\\ &= (\cancel{\sqrt{a^2+b^2}})(\sin x)(a/\cancel{\sqrt{a^2+b^2}}) + (\cancel{\sqrt{a^2+b^2}})(\cos x)(b/\cancel{\sqrt{a^2+b^2}})\\ &= a\sin x + b\cos x\end{aligned}$$

$\boxdot$

EXAMPLE 1

Prove each identity.

A. $\cos(x+\pi) = -\cos x$

PROOF

$$\begin{aligned}\cos(x+\pi) &= \cos x\cos\pi - \sin x\sin\pi\\ &= (\cos x)(-1) - (\sin x)(0)\\ &= -\cos x\end{aligned}$$

$\boxdot$

B. $\tan(x-\pi) = \tan x$

C. $\sin 2t = 2\sin t\cos t$

PROOF

$$\begin{aligned}\sin 2t &= \sin(t+t)\\ &= \sin t\cos t + \cos t\sin t\\ &= \sin t\cos t + \sin t\cos t\\ &= 2\sin t\cos t\end{aligned}$$

$\boxdot$

D. $\cos 2t = \cos^2 t - \sin^2 t$

E. $(\cos(x-y))/(\cos x \sin y) = \cot y + \tan x$

Proof

$$\begin{aligned}&(\cos(x-y))/(\cos x \sin y)\\&\quad = (\cos x \cos y + \sin x \sin y)/(\cos x \sin y)\\&\quad = (\cancel{\cos x} \cos y)/(\cancel{\cos x} \sin y)\\&\qquad\qquad + (\sin x \cancel{\sin y})/(\cos x \cancel{\sin y})\\&\quad = \cos y/\sin y + \sin x/\cos x\\&\quad = \cot y + \tan x\end{aligned}$$

⊡

F. $\tan 2t = (2\tan t)/(1-\tan^2 t)$

G. $\cot(\pi/2-\theta) = \tan\theta$

Proof

$$\begin{aligned}\cot(\pi/2-\theta) &= \cos(\pi/2-\theta)/\sin(\pi/2-\theta)\\&= \sin\theta/\cos\theta\\&= \tan\theta\end{aligned}$$

⊡

H. $\sec(m-\pi/2) = \csc m$

Example 2

Use an addition formula to determine the exact value of the given expression.

A. $\sin(\pi/12)$

Solution

$$\begin{aligned}&\sin(\pi/12)\\&\quad = \sin(3\pi/12 - 2\pi/12)\\&\quad = \sin(\pi/4-\pi/6)\\&\quad = \sin(\pi/4)\cos(\pi/6) - \cos(\pi/4)\sin(\pi/6)\\&\quad = (\sqrt{2}/2)(\sqrt{3}/2) - (\sqrt{2}/2)(1/2)\\&\quad = \sqrt{6}/4 - \sqrt{2}/4\\&\quad = (\sqrt{6}-\sqrt{2})/4\end{aligned}$$

□

B. $\cos 105°$

Example 3

Simplify each expression.

A. $(\tan 2t + \tan 3t)/(1 - \tan 2t \tan 3t)$

Solution

$$\begin{aligned}&(\tan 2t + \tan 3t)/(1 - \tan 2t \tan 3t)\\&\quad = \tan(2t+3t)\\&\quad = \tan 5t\end{aligned}$$

□

B. $\cos\theta\cos 5\theta + \sin 5\theta \sin\theta$

C. $\cos(3\pi/10)\cos(\pi/5) - \sin(3\pi/10)\sin(\pi/5)$

Solution

$\cos(3\pi/10)\cos(\pi/5) - \sin(3\pi/10)\sin(\pi/5)$
$= \cos(3\pi/10 + \pi/5)$
$= \cos(3\pi/10 + 2\pi/10)$
$= \cos(5\pi/10)$
$= \cos(\pi/2)$
$= 0$

□

D. $\sin(\pi/6)\cos(\pi/3) + \cos(\pi/6)\sin(\pi/3)$

Example 4

Rewrite each expression in terms of sine using formula B.

A. $5\sin x - 5\cos x$

Solution

Notes

- $5\sin x - 5\cos x = 5\sin x + (-5)\cos x$
- $\sqrt{(5)^2 + (-5)^2} = \sqrt{50}$
  $= 5\sqrt{2}$
- $5/\sqrt{(5)^2 + (-5)^2} = \not{5}/(\not{5}\sqrt{2})$
  $= (1/\sqrt{2})(\sqrt{2}/\sqrt{2})$
  $= \sqrt{2}/2$
- $-5/\sqrt{(5)^2 + (-5)^2} = -\sqrt{2}/2$
- $\cos\theta = \sqrt{2}/2$ AND $\sin\theta = -\sqrt{2}/2$ AND $\theta \in [0, 2\pi)$
  $\Rightarrow \theta = 7\pi/4$

Expression

$5\sin x - 5\cos x = (5\sqrt{2})\sin(x + 7\pi/4)$

□

B. $\sin x + \sqrt{3}\cos x$

C. $(\sqrt{3}/2)(\sin x) + (1/2)(\cos x)$

Solution

Notes

- $(\sqrt{3}/2)^2 + (1/2)^2 = 3/4 + 1/4$
  $= 1$
- $\cos\theta = \frac{\sqrt{3}}{2}$ AND $\sin\theta = \frac{1}{2}$ AND $\theta \in [0, 2\pi)$
  $\Rightarrow \theta = \pi/6$

Expression

$(\sqrt{3}/2)(\sin x) + (1/2)(\cos x) = \sin(x + \pi/6)$

□

D. $\cos x - \sin x$

EXAMPLE 5

Find $\sin(\alpha - \beta)$ if:

- $\cos\alpha = 3/5$
- $\sin\beta = 24/25$
- $0 < \alpha < \pi/2$
- $\pi/2 < \beta < \pi$

SOLUTION

Notes

- $0 < \alpha < \pi/2 \Rightarrow \sin\alpha > 0$
- $\pi/2 < \beta < \pi \Rightarrow \cos\beta < 0$

Value of $\sin\alpha$

$$\begin{aligned}
&\cos^2\alpha + \sin^2\alpha = 1\\
&\quad\Rightarrow \sin^2\alpha = 1 - \cos^2\alpha\\
&\quad\Rightarrow \sin\alpha = +\sqrt{1-\cos^2\alpha}\\
&\qquad\qquad = \sqrt{1-(3/5)^2}\\
&\qquad\qquad = \sqrt{25/25 - 9/25}\\
&\qquad\qquad = \sqrt{16/25}\\
&\qquad\qquad = 4/5
\end{aligned}$$

Value of $\cos\beta$

$$\begin{aligned}
&\cos^2\beta + \sin^2\beta = 1\\
&\quad\Rightarrow \cos^2\beta = 1 - \sin^2\beta\\
&\quad\Rightarrow \cos\beta = -\sqrt{1-\sin^2\beta}\\
&\qquad\qquad = -\sqrt{1-(24/25)^2}\\
&\qquad\qquad = -\sqrt{625/625 - 576/625}\\
&\qquad\qquad = -\sqrt{49/25}\\
&\qquad\qquad = -7/25
\end{aligned}$$

Value of $\sin(\alpha - \beta)$

$$\begin{aligned}
&\sin(\alpha-\beta)\\
&\quad = \sin\alpha\cos\beta - \cos\alpha\sin\beta\\
&\quad = (4/5)(-7/25) - (3/5)(24/25)\\
&\quad = -28/125 - 72/125\\
&\quad = -100/125\\
&\quad = -4/5
\end{aligned}$$

□

EXAMPLE 6

Find $\cos(\alpha + \beta)$ if:

- $\sin\alpha = 3/5$
- $\cos\beta = -5/13$
- $\pi/2 < \alpha < \pi$
- $\pi < \beta < 3\pi/2$

Section 126: Double-Angle and Half-Angle Formulas

Double-Angle Formulas
Suppose $t \in \mathbf{R}$.
Then:

- $\cos 2t = \cos^2 t - \sin^2 t$
- $\sin 2t = 2 \sin t \cos t$
- $\tan 2t = (2 \tan t)/(1 - \tan^2 t)$ (provided $t, 2t \in \{\, x \in \mathbf{R} \mid x \neq \pi/2 + \pi k \ \forall\ k \in \mathbf{Z}\,\}$)

Proofs
See examples 1c, 1d, and 1f on pages 380 and 381.

$\boxdot$

Formulas
Suppose $t \in \mathbf{R}$.
Then:

- $\cos 2t = 2\cos^2 t - 1$
- $\cos 2t = 1 - 2\sin^2 t$
- $\cos^2 t = (1 + \cos 2t)/2$
- $\sin^2 t = (1 - \cos 2t)/2$
- $\tan^2 t = (1 - \cos 2t)/(1 + \cos 2t)$ (provided $t \in \{\, x \in \mathbf{R} \mid x \neq \pi/2 + \pi k \ \forall\ k \in \mathbf{Z}\,\}$)

Proofs
Suppose $t \in \mathbf{R}$.

- $$\begin{aligned}\cos 2t &= \cos^2 t - \sin^2 t \\ &= \cos^2 t - (1 - \cos^2 t) \\ &= 2\cos^2 t - 1\end{aligned}$$

- $$\begin{aligned}\cos 2t &= \cos^2 t - \sin^2 t \\ &= (1 - \sin^2 t) - \sin^2 t \\ &= 1 - 2\sin^2 t\end{aligned}$$

- $$\begin{aligned}\cos 2t = 2\cos^2 t - 1 &\Rightarrow 1 + \cos 2t = 2\cos^2 t \\ &\Rightarrow (1 + \cos 2t)/2 = (\not{2}\cos^2 t)/\not{2} \\ &\Rightarrow \cos^2 t = (1 + \cos 2t)/2\end{aligned}$$

- $$\begin{aligned}\cos 2t = 1 - 2\sin^2 t &\Rightarrow -1 + \cos 2t = -2\sin^2 t \\ &\Rightarrow -(1 - \cos 2t)/(-2) = (-2\sin^2 t)/(-2) \\ &\Rightarrow \sin^2 t = (1 - \cos 2t)/2\end{aligned}$$

- $$\begin{aligned}\tan^2 t &= (\sin^2 t)/(\cos^2 t) \\ &= [(1 - \cos 2t)/2]/[(1 + \cos 2t)/2] \\ &= [(1 - \cos 2t)/2] \cdot [2/(1 + \cos 2t)] \\ &= (1 - \cos 2t)/(1 + \cos 2t)\end{aligned}$$

$\boxdot$

Half-Angle Formulas

Suppose $t \in \mathbf{R}$.

Then:

- $\cos(t/2) = \pm\sqrt{(1+\cos t)/2}$
- $\sin(t/2) = \pm\sqrt{(1-\cos t)/2}$
- $\tan(t/2) = (1-\cos t)/(\sin t)$ (provided $t \neq \pi k$ for any $k \in \mathbf{Z}$)
- $\tan(t/2) = (\sin t)/(1+\cos t)$ (provided $t \neq \pi + 2\pi k$ for any $k \in \mathbf{Z}$)

Proofs

Suppose $t \in \mathbf{R}$.

- $$\begin{aligned} \cos^2(t/2) &= (1+\cos(2\cdot t/2))/2 \\ &= (1+\cos t)/2 \\ \Rightarrow \cos(t/2) &= \pm\sqrt{(1+\cos t)/2} \end{aligned}$$

- $$\begin{aligned} \sin^2(t/2) &= (1-\cos(2\cdot t/2))/2 \\ &= (1-\cos t)/2 \\ \Rightarrow \sin(t/2) &= \pm\sqrt{(1-\cos t)/2} \end{aligned}$$

- $$\begin{aligned} (1-\cos 2t)/(\sin 2t) &= (1-(1-2\sin^2 t))/(2\sin t\cos t) \\ &= (\not{2}\sin^2 t)/(\not{2}\sin t\cos t) \\ &= (\cancel{\sin t}\sin t)/(\cancel{\sin t}\cos t) \\ &= \tan t \\ \Rightarrow \tan(t/2) &= (1-\cos(2\cdot t/2))/(\sin(2\cdot t/2)) \\ &= (1-\cos t)/(\sin t) \end{aligned}$$

- $$\begin{aligned} (\sin 2t)/(1+\cos 2t) &= (2\sin t\cos t)/(1+(2\cos^2 t-1)) \\ &= (\not{2}\sin t\cos t)/(\not{2}\cos^2 t) \\ &= (\sin t\,\cancel{\cos t})/(\cos t\,\cancel{\cos t}) \\ &= \tan t \\ \Rightarrow \tan(t/2) &= (\sin(2\cdot t/2))/(1+\cos(2\cdot t/2)) \\ &= (\sin t)/(1+\cos t) \end{aligned}$$

⊡

Example 1

Prove each identity.

A. $1+\cos 2\theta = \cot\theta\sin 2\theta$

Proof

$$\begin{aligned} \cot\theta\sin 2\theta &= (\cos\theta/\cancel{\sin\theta})(2\cancel{\sin\theta}\cos\theta) \\ &= 2\cos^2\theta \\ &= 2[(1+\cos 2\theta)/2] \\ &= 1+\cos 2\theta \end{aligned}$$

⊡

B. $\cot t = (1+\cos 2t)/\sin 2t$

EXAMPLE 2
Use the figure below to find $\cos 2\theta$.

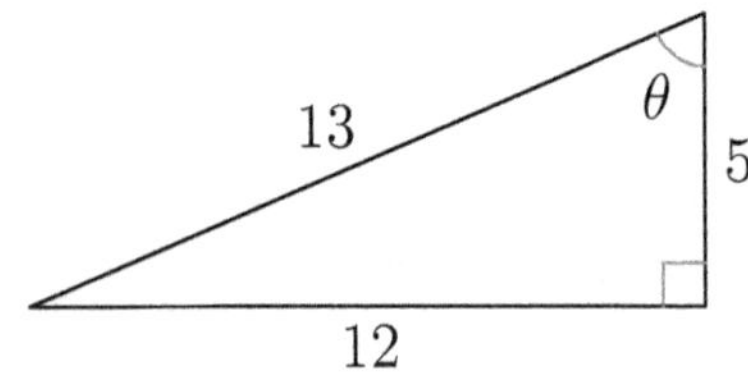

SOLUTION

$$\begin{aligned}\cos 2\theta &= \cos^2\theta - \sin^2\theta \\ &= (5/13)^2 - (12/13)^2 \\ &= 25/169 - 144/169 \\ &= -119/169\end{aligned}$$

□

EXAMPLE 3
Use the figure below to find $\tan(\beta/2)$.

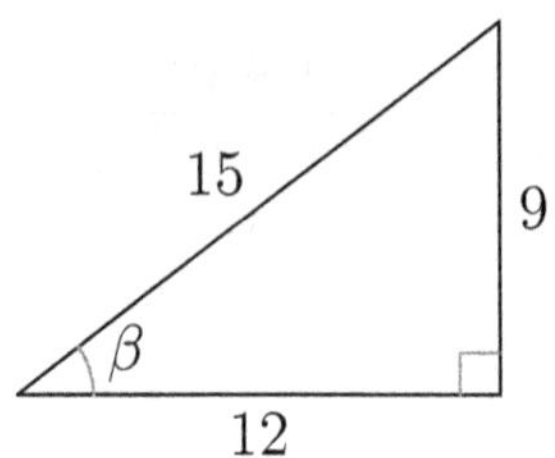

EXAMPLE 4
Suppose $\sin t = -3/5$ and $\pi < t < 3\pi/2$.
Find $\sin(t/2)$.

SOLUTION

Notes

- $\pi < t < 3\pi/2 \Rightarrow \cos t < 0$
- $\pi < t < 3\pi/2$
  $\Rightarrow \pi/2 < t/2 < (3\pi/2)/2$
  $\Rightarrow \pi/2 < t/2 < 3\pi/4$
  $\Rightarrow \sin(t/2) > 0$

Value of $\cos t$

$$\begin{aligned}\cos t &= -\sqrt{1 - \sin^2 t} \\ &= -\sqrt{1 - (-3/5)^2} \\ &= -\sqrt{25/25 - 9/25} \\ &= -\sqrt{16/25} \\ &= -4/5\end{aligned}$$

Value of $\sin(t/2)$

$$\begin{aligned}\sin(t/2) &= +\sqrt{(1 - \cos t)/2} \\ &= \sqrt{(5/5 - (-4/5))/2} \\ &= \sqrt{(9/5) \cdot (1/2)} \\ &= \sqrt{9/10} \\ &= 3/\sqrt{10} \\ &= (3/\sqrt{10}) \cdot (\sqrt{10}/\sqrt{10}) \\ &= 3\sqrt{10}/10\end{aligned}$$

□

EXAMPLE 5
Suppose $\cos\alpha = 7/25$ and $3\pi/2 < \alpha < 2\pi$.
Find $\tan 2\alpha$.

EXAMPLE 6

Use a half-angle formula to determine the exact value of the given expression.

A. $\tan \pi/8$

SOLUTION

$$\begin{aligned}\tan \pi/8 &= \tan((\pi/4)/2) \\ &= (1-\cos \pi/4)/(\sin \pi/4) \\ &= (1-\sqrt{2}/2)/(\sqrt{2}/2) \\ &= 1/(\sqrt{2}/2) - (\cancel{\sqrt{2}/2})/(\cancel{\sqrt{2}/2}) \\ &= 2/\sqrt{2} - 1 \\ &= (2/\sqrt{2})(\sqrt{2}/\sqrt{2}) - 1 \\ &= \cancel{2}\sqrt{2}/\cancel{2} - 1 \\ &= -1 + \sqrt{2}\end{aligned}$$

□

B. $\sin 15°$

EXAMPLE 7

Write each expression in a form that contains no exponents greater than 1.

A. $\cos^4 t$

SOLUTION

$$\begin{aligned}\cos^4 t &= (\cos^2 t)^2 \\ &= [(1+\cos 2t)/2]^2 \\ &= (1+\cos 2t)^2/(2)^2 \\ &= [(1)^2 + 2(1)(\cos 2t) + (\cos 2t)^2]/4 \\ &= [1 + 2\cos 2t + \cos^2 2t]/4 \\ &= [1 + 2\cos 2t + (1+\cos 2(2t))/2]/4 \\ &= [1 + 2\cos 2t + (1+\cos 4t)/2]/4 \\ &= (1/4)[1 + 2\cos 2t + (1+\cos 4t)/2] \\ &= (1/4)[2/2 + (4\cos 2t)/2 \\ &\qquad + (1+\cos 4t)/2] \\ &= (1/4)[(2 + 4\cos 2t + 1 + \cos 4t)/2] \\ &= (1/4)[(3 + 4\cos 2t + \cos 4t)/2] \\ &= (3 + 4\cos 2t + \cos 4t)/8\end{aligned}$$

□

B. $\cos^2\alpha \sin^2\alpha$

Section 127: Product-to-Sum and Sum-to-Product Formulas

Product-to-Sum Formulas
Suppose $s, t \in \mathbf{R}$.
Then:

- $\cos s \cos t = (1/2)[\cos(s-t) + \cos(s+t)]$
- $\cos s \sin t = (1/2)[\sin(s+t) - \sin(s-t)]$
- $\sin s \sin t = (1/2)[\cos(s-t) - \cos(s+t)]$

Proofs
Suppose $s, t \in \mathbf{R}$.

- $\cos(s-t) + \cos(s+t) = (\cos s \cos t + \cancel{\sin s \sin t}) + (\cos s \cos t - \cancel{\sin s \sin t})$
  $= 2 \cos s \cos t$
  $\Rightarrow (1/2)[\cos(s-t) + \cos(s+t)] = (1/2)(2 \cos s \cos t)$
  $\Rightarrow (1/2)[\cos(s-t) + \cos(s+t)] = \cos s \cos t$

- $\sin(s+t) - \sin(s-t) = (\cancel{\sin s \cos t} + \cos s \sin t) - (\cancel{\sin s \cos t} - \cos s \sin t)$
  $= 2 \cos s \sin t$
  $\Rightarrow (1/2)[\sin(s+t) - \sin(s-t)] = (1/2)(2 \cos s \sin t)$
  $\Rightarrow (1/2)[\sin(s+t) - \sin(s-t)] = \cos s \sin t$

- $\cos(s-t) - \cos(s+t) = (\cancel{\cos s \cos t} + \sin s \sin t) - (\cancel{\cos s \cos t} - \sin s \sin t)$
  $= 2 \sin s \sin t$
  $\Rightarrow (1/2)[\cos(s-t) - \cos(s+t)] = (1/2)(2 \sin s \sin t)$
  $\Rightarrow (1/2)[\cos(s-t) - \cos(s+t)] = \sin s \sin t$

$\boxdot$

Sum-to-Product Formulas
Suppose $s, t \in \mathbf{R}$.
Then:

- $\cos s + \cos t = 2 \cos[(s+t)/2] \cos[(s-t)/2]$
- $\cos s - \cos t = -2 \sin[(s+t)/2] \sin[(s-t)/2]$
- $\sin s + \sin t = 2 \sin[(s+t)/2] \cos[(s-t)/2]$
- $\sin s - \sin t = 2 \cos[(s+t)/2] \sin[(s-t)/2]$

Proofs (first formula only)
Suppose $s, t \in \mathbf{R}$.
$2 \cos[(s+t)/2] \cos[(s-t)/2]$
$= 2 \cdot (1/2)[\cos((s+t)/2 - (s-t)/2) + \cos((s+t)/2 + (s-t)/2)]$
$= \cos(((s+t) - (s-t))/2) + \cos(((s+t) + (s-t))/2)$
$= \cos((\cancel{s} + t - \cancel{s} + t)/2) + \cos((s + \cancel{t} + s - \cancel{t})/2)$
$= \cos(2t/2) + \cos(2s/2)$
$= \cos t + \cos s$
$= \cos s + \cos t$

$\boxdot$

EXAMPLE 1
Prove each identity.

A. $\dfrac{\cos 7t - \cos 3t}{\sin 7t - \sin 3t} = -\tan 5t$

PROOF

$$\begin{aligned}
&\frac{\cos 7t - \cos 3t}{\sin 7t - \sin 3t} \\
&\quad = \frac{-2\sin[(7t+3t)/2]\sin[(7t-3t)/2]}{2\cos[(7t+3t)/2]\sin[(7t-3t)/2]} \\
&\quad = -\frac{2\sin(10t/2)\sin(4t/2)}{2\cos(10t/2)\sin(4t/2)} \\
&\quad = -\frac{\cancel{2}\sin 5t\,\cancel{\sin 2t}}{\cancel{2}\cos 5t\,\cancel{\sin 2t}} \\
&\quad = -\frac{\sin 5t}{\cos 5t} \\
&\quad = -\tan 5t
\end{aligned}$$

⊡

B. $\sin(x+y)\sin(x-y) = \sin^2 x - \sin^2 y$

EXAMPLE 2
Convert each product to a sum or difference and simplify.

A. $\cos 3t \cos 2t$

SOLUTION

$$\begin{aligned}
&\cos 3t \cos 2t \\
&\quad = (1/2)[\cos(3t-2t) + \cos(3t+2t)] \\
&\quad = (1/2)[\cos t + \cos 5t] \\
&\quad = (1/2)\cos t + (1/2)\cos 5t
\end{aligned}$$

□

B. $\sin\theta \cos 7\theta$

C. $\cos 75° \sin 15°$

SOLUTION

$$\begin{aligned}
&\cos 75° \sin 15° \\
&\quad = (1/2)[\sin(75° + 15°) - \sin(75° - 15°)] \\
&\quad = (1/2)[\sin 90° - \sin 60°] \\
&\quad = (1/2)[1 - \sqrt{3}/2] \\
&\quad = (1/2)[2/2 - \sqrt{3}/2] \\
&\quad = (1/2)[(2-\sqrt{3})/2] \\
&\quad = (2-\sqrt{3})/4
\end{aligned}$$

□

D. $\sin(\pi/8)\sin(3\pi/8)$

EXAMPLE 3
Convert each sum or difference to a product and simplify.

A. $\sin 50° - \sin 10°$

SOLUTION
$\sin 50° - \sin 10°$
$= 2\cos[(50° + 10°)/2]\sin[(50° - 10°)/2]$
$= 2\cos(60°/2)\sin(40°/2)$
$= 2\cos 30° \sin 20°$
$= 2(\sqrt{3}/2)(\sin 20°)$
$= \sqrt{3}\sin 20°$

□

B. $\cos(x + \pi/6) - \cos(x - \pi/6)$

C. $\cos 2\pi + \cos 5\pi$

SOLUTION
$\cos 2\pi + \cos 5\pi$
$= 2\cos[(2\pi + 5\pi)/2]\cos[(2\pi - 5\pi)/2]$
$= 2\cos(7\pi/2)\cos(-3\pi/2)$
$= 2(0)(0)$
$= 0$

□

D. $\sin 37.5° + \sin 7.5°$

Section 128: Inverse Trigonometric Functions

Definition

We define the *restricted* cosine, sine, tangent, secant, cosecant, and cotangent functions, denoted $\cos_r$, $\sin_r$, $\tan_r$, $\sec_r$, $\csc_r$, and $\cot_r$, respectively, by:

- $\cos_r x = \cos x$ for all $x \in [0, \pi]$
- $\sin_r x = \sin x$ for all $x \in [-\pi/2, \pi/2]$
- $\tan_r x = \tan x$ for all $x \in (-\pi/2, \pi/2)$
- $\sec_r x = \sec x$ for all $x \in [0, \pi/2) \cup (\pi/2, \pi]$
- $\csc_r x = \csc x$ for all $x \in [-\pi/2, 0) \cup (0, \pi/2]$
- $\cot_r x = \cot x$ for all $x \in (0, \pi)$

Definition

The *inverse* cosine, sine, tangent, secant, cosecant, and cotangent functions, denoted $\cos^{-1}$, $\sin^{-1}$, $\tan^{-1}$, $\sec^{-1}$, $\csc^{-1}$, and $\cot^{-1}$, respectively, are the inverses of the restricted cosine, sine, tangent, secant, cosecant, and cotangent functions, respectively.

Domains of Inverse Trigonometric Functions

The domains of the inverse trigonometric functions are the ranges of the restricted trigonometric functions:

- (domain of inverse cosine) $= [-1, 1]$
- (domain of inverse sine) $= [-1, 1]$
- (domain of inverse tangent) $= \mathbf{R}$
- (domain of inverse secant) $= (-\infty, -1] \cup [1, \infty)$
- (domain of inverse cosecant) $= (-\infty, -1] \cup [1, \infty)$
- (domain of inverse cotangent) $= \mathbf{R}$

Ranges of Inverse Trigonometric Functions

The ranges of the inverse trigonometric functions are the domains of the restricted trigonometric functions:

- (range of inverse cosine) $= [0, \pi]$
- (range of inverse sine) $= [-\pi/2, \pi/2]$
- (range of inverse tangent) $= (-\pi/2, \pi/2)$
- (range of inverse secant) $= [0, \pi/2) \cup (\pi/2, \pi]$
- (range of inverse cosecant) $= [-\pi/2, 0) \cup (0, \pi/2]$
- (range of inverse cotangent) $= (0, \pi)$

Notation

The inverse trigonometric functions have alternative names and notations:

- inverse cosine is also called *arccosine* and denoted arccos;
- inverse sine is also called *arcsine* and denoted arcsin;
- inverse tangent is also called *arctangent* and denoted arctan;
- inverse secant is also called *arcsecant* and denoted arcsec;
- inverse cosecant is also called *arccosecant* and denoted arccsc;
- inverse cotangent is also called *arccotangent* and denoted arccot.

GRAPHS

The graphs of the inverse trigonometric functions are the reflections of the graphs of the restricted trigonometric functions through the line $y = x$:

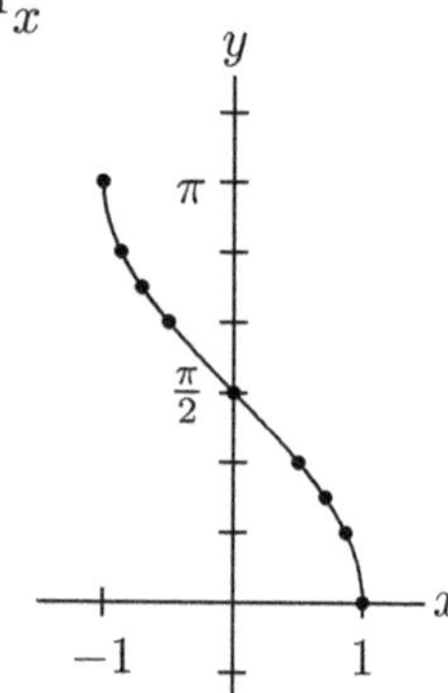

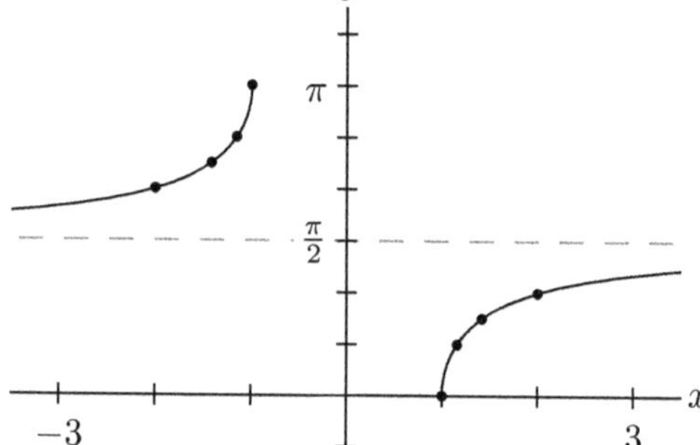

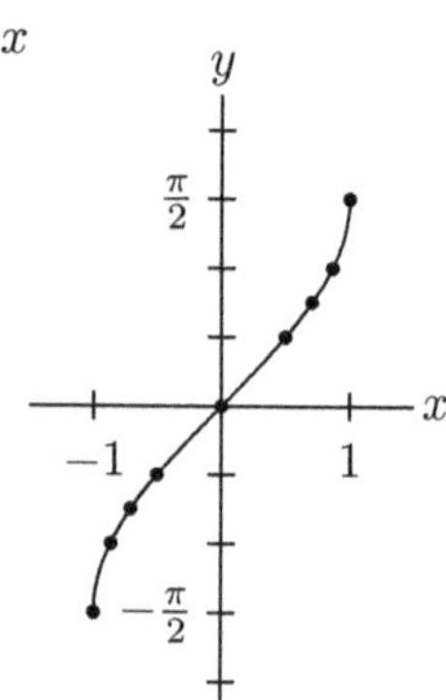

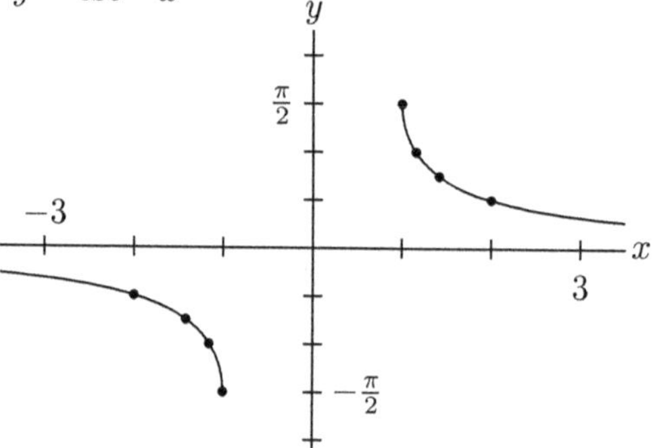

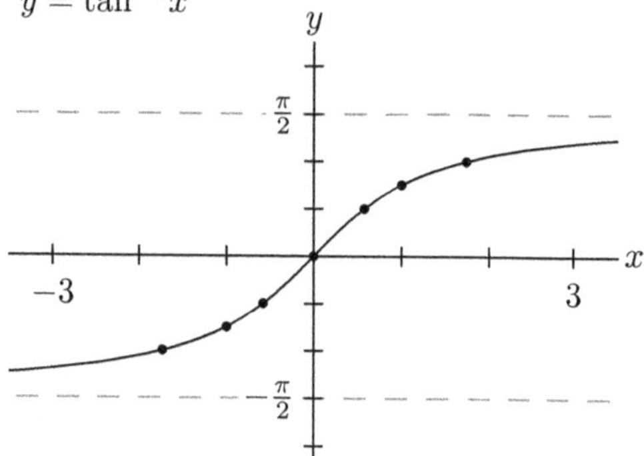

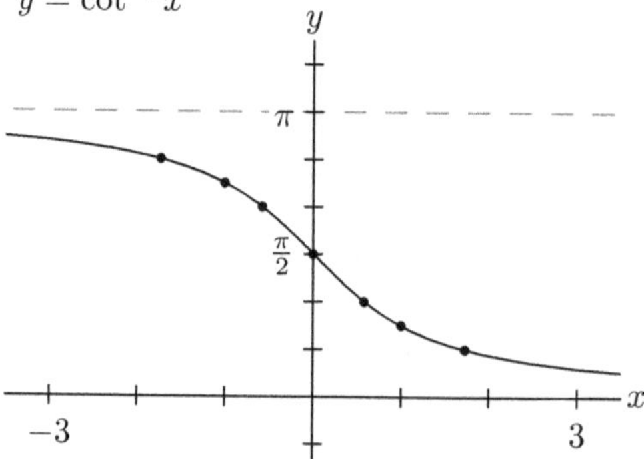

PROPERTIES

- $\cos^{-1} y = x \Leftrightarrow \cos x = y$ for all $x \in [0, \pi]$
- $\sin^{-1} y = x \Leftrightarrow \sin x = y$ for all $x \in [-\pi/2, \pi/2]$
- $\tan^{-1} y = x \Leftrightarrow \tan x = y$ for all $x \in (-\pi/2, \pi/2)$
- $\sec^{-1} y = x \Leftrightarrow \sec x = y$ for all $x \in [0, \pi/2) \cup (\pi/2, \pi]$
- $\csc^{-1} y = x \Leftrightarrow \csc x = y$ for all $x \in [-\pi/2, 0) \cup (0, \pi/2]$
- $\cot^{-1} y = x \Leftrightarrow \cot x = y$ for all $x \in (0, \pi)$

Properties

- $\cos^{-1}x$ is the unique number in the interval $[0, \pi]$ whose cosine is x
- $\sin^{-1}x$ is the unique number in the interval $[-\pi/2, \pi/2]$ whose sine is x
- $\tan^{-1}x$ is the unique number in the interval $(-\pi/2, \pi/2)$ whose tangent is x
- $\sec^{-1}x$ is the unique number in the interval $[0, \pi/2) \cup (\pi/2, \pi]$ whose secant is x
- $\csc^{-1}x$ is the unique number in the interval $[-\pi/2, 0) \cup (0, \pi/2]$ whose cosecant is x
- $\cot^{-1}x$ is the unique number in the interval $(0, \pi)$ whose cotangent is x

Properties

- $\cos(\cos^{-1}x) = x$ for all $x \in [-1, 1]$
- $\sin(\sin^{-1}x) = x$ for all $x \in [-1, 1]$
- $\tan(\tan^{-1}x) = x$ for all $x \in \mathbf{R}$
- $\sec(\sec^{-1}x) = x$ for all $x \in (-\infty, -1] \cup [1, \infty)$
- $\csc(\csc^{-1}x) = x$ for all $x \in (-\infty, -1] \cup [1, \infty)$
- $\cot(\cot^{-1}x) = x$ for all $x \in \mathbf{R}$

Properties

- $\cos^{-1}(\cos x) = x$ for all $x \in [0, \pi]$
- $\sin^{-1}(\sin x) = x$ for all $x \in [-\pi/2, \pi/2]$
- $\tan^{-1}(\tan x) = x$ for all $x \in (-\pi/2, \pi/2)$
- $\sec^{-1}(\sec x) = x$ for all $x \in [0, \pi/2) \cup (\pi/2, \pi]$
- $\csc^{-1}(\csc x) = x$ for all $x \in [-\pi/2, 0) \cup (0, \pi/2]$
- $\cot^{-1}(\cot x) = x$ for all $x \in (0, \pi)$

Properties

- $\sec^{-1}x = \cos^{-1}(1/x)$ for all $x \in (-\infty, -1] \cup [1, \infty)$
- $\csc^{-1}x = \sin^{-1}(1/x)$ for all $x \in (-\infty, -1] \cup [1, \infty)$

Property

$\sin^{-1}x + \cos^{-1}x = \pi/2$ for all $x \in [-1, 1]$

Comments

- Inverses of periodic functions do not exist because periodic functions are not one-to-one. The notations $\cos^{-1}$, et al., can mislead because the functions they represent are not inverses of basic trigonometric functions, but inverses of restricted trigonometric functions.

- The domains to which we have restricted the trigonometric functions in order to define corresponding inverse functions are not the only legitimate ones. We have provided here the standard definitions for the restricted cosine, sine, tangent, and cotangent functions, but alternative definitions for the restricted secant and cosecant functions include:
  - $\sec_r x = \sec x$ for all $x \in [0, \pi/2) \cup [\pi, 3\pi/2)$
  - $\csc_r x = \csc x$ for all $x \in (0, \pi/2] \cup (\pi, 3\pi/2]$

EXAMPLE 1
Evaluate each expression.

A. $\sin^{-1}(1/2)$

SOLUTION
$\sin^{-1}(1/2) = \pi/6$
(since $\sin \pi/6 = 1/2$)

☐

B. $\cos^{-1}(-1)$

C. $\arccos 2$

SOLUTION
$\arccos 2$ is undefined
(since the domain of arccos is $[-1, 1]$)

☐

D. $\arcsin 0$

E. $\tan^{-1}(1)$

SOLUTION
$\tan^{-1}(1) = \pi/4$
(since $\tan \pi/4 = 1$)

☐

F. $\arctan \sqrt{3}$

G. $\tan(\arctan(-5))$

SOLUTION
$\tan(\arctan(-5)) = -5$

☐

H. $\arccos(\cos(\pi/8))$

I. $\cos^{-1}(\cos 2\pi)$

SOLUTION
$$\begin{aligned}\cos^{-1}(\cos 2\pi) &= \cos^{-1} 1\\ &= 0\end{aligned}$$

☐

J. $\sin(\sin^{-1}(3/2))$

K. $\sin(\cos^{-1}(-\sqrt{3}/2))$

SOLUTION
$$\begin{aligned}\sin(\cos^{-1}(-\sqrt{3}/2)) &= \sin(5\pi/6)\\ &= 1/2\end{aligned}$$

☐

L. $\sin^{-1}(\tan 225°)$

M. $\csc^{-1}(\sqrt{2})$

Solution
$\csc^{-1}(\sqrt{2}) = \pi/4$

□

N. $\sec^{-1}(2)$

O. $\sec(\cot^{-1}(1) + \sin^{-1}(1))$

Solution

$$\begin{aligned}\sec(\cot^{-1}(1) + \sin^{-1}(1)) &= \sec(\pi/4 + \pi/2)\\ &= \sec(\pi/4 + 2\pi/4)\\ &= \sec(3\pi/4)\\ &= -\sqrt{2}\end{aligned}$$

□

P. $\cos(\csc^{-1}(2/\sqrt{3})) - \sin(\cot^{-1}(\sqrt{3}))$

Example 2
Use right triangles to evaluate each expression.

A. $\tan(\arcsin(3/5))$

Solution
Declaration
Let $\theta = \arcsin(3/5)$.

Notes
- $0 < \theta < \pi/2$
- $\sin\theta = 3/5$

Figure

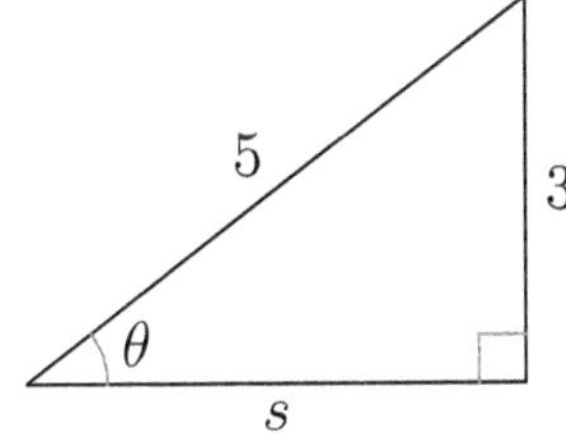

Value

$$\begin{aligned}&(3)^2 + s^2 = (5)^2\\ &\Rightarrow s^2 = (5)^2 - (3)^2\\ &\qquad = 16\\ &\Rightarrow s = \sqrt{16}\\ &\qquad = 4\\ &\Rightarrow \tan\theta = 3/4\\ &\Rightarrow \tan(\arcsin(3/5)) = 3/4\end{aligned}$$

□

B. $\cos(\tan^{-1}(4))$

EXAMPLE 3
Write the expression $\sin(\cos^{-1}x)$ in a form that contains no trigonometric functions. Assume $0 < x < 1$.

SOLUTION
Declaration
Let $\theta = \cos^{-1}x$.

Notes
- $0 < \theta < \pi/2 \quad (\text{since } 0 < x < 1)$
- $\cos\theta = x$
  $= x/1$

Figure

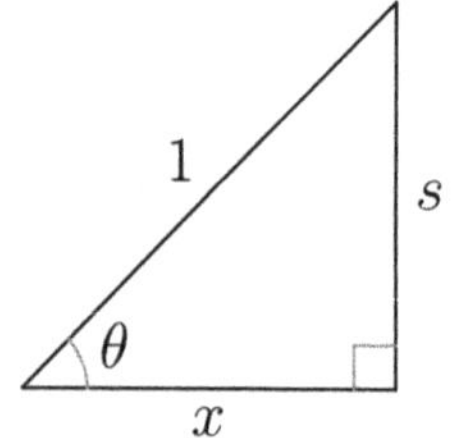

Value

$$
\begin{aligned}
&x^2 + s^2 = (1)^2 \\
&\quad\Rightarrow s^2 = (1)^2 - x^2 \\
&\qquad\quad = 1 - x^2 \\
&\quad\Rightarrow s = \sqrt{1 - x^2} \\
&\quad\Rightarrow \sin\theta = \sqrt{1 - x^2}/1 \\
&\qquad\quad = \sqrt{1 - x^2} \\
&\quad\Rightarrow \sin(\cos^{-1}x) = \sqrt{1 - x^2}
\end{aligned}
$$

□

EXAMPLE 4
Write the expression $\tan(\sec^{-1}(y/5))$ in a form that contains no trigonometric functions. Assume $y > 5$.

Section 129: Trigonometric Equations

Procedure

To solve an equation of the form $f(x) = c$ where f is a periodic function with period p, we:

1. identify all solutions s_i to the equation within an interval of length p;
2. state the set of solutions as the union of sets $\{\, s_i + pk \mid k \in \mathbf{Z} \,\}$ (in recognition that $f(x) = c$ if and only if $f(x + pk) = c$ for any integer k).

Example 1

Solve each equation. Leave your solutions in units of radians.

A. $\cos\theta = \sqrt{3}/2$

Solution

Notes

- The period of cosine is 2π.
- An interval of width 2π is $[0, 2\pi)$.

Solutions for One Period

$\cos\theta = \sqrt{3}/2$ AND $\theta \in [0, 2\pi)$

$\Leftrightarrow \theta = \pi/6$ OR $\theta = 11\pi/6$

Set of Solutions

Set of solutions:

$\{\, \pi/6 + 2\pi k \mid k \in \mathbf{Z} \,\} \cup \{\, 11\pi/6 + 2\pi k \mid k \in \mathbf{Z} \,\}$ □

B. $\sin x = -\sqrt{2}/2$

C. $\tan t = -\sqrt{3}$

Solution

Notes

- The period of tangent is π.
- An interval of width π is $[0, \pi)$.

Solution for One Period

$\tan t = -\sqrt{3}$ AND $t \in [0, \pi)$

$\Leftrightarrow t = 2\pi/3$

Set of Solutions

Set of solutions: $\{\, 2\pi/3 + \pi k \mid k \in \mathbf{Z} \,\}$ □

D. $\cot\theta = 1$

E. $\sin\alpha\tan\alpha - \sin\alpha = 0$

SOLUTION

Values of Functions

$\sin\alpha\ \tan\alpha - \sin\alpha = 0$

$\iff (\sin\alpha)(\tan\alpha - 1) = 0$

$\iff \sin\alpha = 0$ OR $\tan\alpha = 1$

Solutions for One Period

- $\sin\alpha = 0$ AND $\alpha \in [0, 2\pi)$
  $\iff \alpha = 0$ OR $\alpha = \pi$
- $\tan\alpha = 1$ AND $\alpha \in [0, \pi)$
  $\iff \alpha = \pi/4$

Set of Solutions

The set of solutions is

$$\{\, 0 + 2\pi k \mid k \in \mathbf{Z} \,\} \cup \{\, \pi + 2\pi k \mid k \in \mathbf{Z} \,\} \cup \{\, \pi/4 + \pi k \mid k \in \mathbf{Z} \,\},$$

which simplifies to

$$\{\, \pi k \mid k \in \mathbf{Z} \,\} \cup \{\, \pi/4 + \pi k \mid k \in \mathbf{Z} \,\}.$$

□

F. $2\cos^2 x - 3\cos x + 1 = 0$

G. $\csc^2 x - 2 = 0$

SOLUTION

Values of Functions

$\csc^2 x - 2 = 0$

$\iff \csc^2 x = 2$

$\iff \csc x = \sqrt{2}$ OR $\csc x = -\sqrt{2}$

Solutions for One Period

- $\csc x = \sqrt{2}$ AND $x \in [0, 2\pi)$
  $\iff x = \pi/4$ OR $x = 3\pi/4$
- $\csc x = -\sqrt{2}$ AND $x \in [0, 2\pi)$
  $\iff x = 5\pi/4$ OR $x = 7\pi/4$

Set of Solutions

The set of solutions is

$$\{\, \pi/4 + 2\pi k \mid k \in \mathbf{Z} \,\} \cup \{\, 3\pi/4 + 2\pi k \mid k \in \mathbf{Z} \,\} \cup \{\, 5\pi/4 + 2\pi k \mid k \in \mathbf{Z} \,\} \cup \{\, 7\pi/4 + 2\pi k \mid k \in \mathbf{Z} \,\},$$

which simplifies to

$$\{\, \pi/4 + (\pi/2)k \mid k \in \mathbf{Z} \,\}.$$

□

H. $2\cos^2\beta + 11\sin\beta = 7$

EXAMPLE 2
Solve each equation in units of degrees and then determine the solutions of the equation that lie in the interval $[0°, 360°)$.

A. $\sin 3\theta = -\sqrt{3}/2$

SOLUTION
Notes
- The period of sine is $360°$.
- An interval of width $360°$ is $[0°, 360°)$.

Values of Expression for One Period
$\sin 3\theta = -\sqrt{3}/2$ AND $3\theta \in [0°, 360°)$
$\Leftrightarrow 3\theta = 240°$ OR $3\theta = 300°$

General Solutions
Assume $k \in \mathbf{Z}$.
- $3\theta = 240° + 360°k$
  $\Leftrightarrow (1/3)(3\theta) = (1/3)(240° + 360°k)$
  $\Leftrightarrow \theta = (1/3)(240°) + (1/3)(360°k)$
  $\Leftrightarrow \theta = 80° + 120°k$
- $3\theta = 300° + 360°k$
  $\Leftrightarrow (1/3)(3\theta) = (1/3)(300° + 360°k)$
  $\Leftrightarrow \theta = (1/3)(300°) + (1/3)(360°k)$
  $\Leftrightarrow \theta = 100° + 120°k$

Set of Solutions
The set of solutions is
$$\{\, 80° + 120°k \mid k \in \mathbf{Z} \,\} \cup \{\, 100° + 120°k \mid k \in \mathbf{Z} \,\}.$$

Substitutions
- $80° + 120° \cdot 0 = 80°$
- $80° + 120° \cdot 1 = 200°$
- $80° + 120° \cdot 2 = 320°$
- $80° + 120° \cdot 3 = 440°$ X (too big)
- $100° + 120° \cdot 0 = 100°$
- $100° + 120° \cdot 1 = 220°$
- $100° + 120° \cdot 2 = 340°$
- $100° + 120° \cdot 3 = 460°$ X (too big)

Solutions in Given Interval
The set of solutions in $[0°, 360°)$ is
$\{80°, 100°, 200°, 220°, 320°, 340°\}$. □

B. $\cos(\theta/2) = 0$

Example 3
Find approximations for the solutions of each equation that lie in the interval $[0°, 360°)$. Use a calculator and round your approximations to two decimal places.

A. $\sin x = -0.78$

Solution
Solutions for One Period

- $\sin x = -0.78$ AND $x \in [-90°, 90°]$
  $\Rightarrow x = \sin^{-1}(-0.78)$
  $\Rightarrow x \approx -51.26°$ (by calculator)
- $\sin x = -0.78$ AND $x \in (90°, 270°)$
  $\Rightarrow x \approx 180° + 51.26°$ (refer to figure)
  $\Rightarrow x \approx 231.26°$

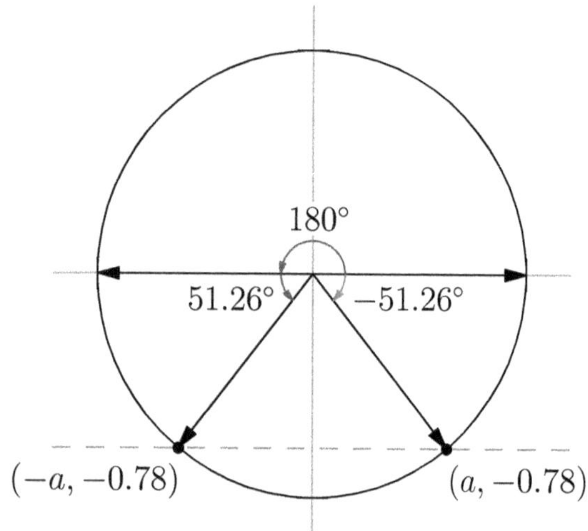

Set of Solutions
The set of solutions is

$$\{\,-51.26° + 360°k \mid k \in \mathbf{Z}\,\} \cup \{\,231.26° + 360°k \mid k \in \mathbf{Z}\,\}.$$

Substitutions

- $-51.26° + 360° \cdot 0 = -51.26°$ X
- $-51.26° + 360° \cdot 1 = 308.74°$
- $231.26° + 360° \cdot 0 = 231.26°$

Solutions in Given Interval
The set of solutions in $[0°, 360°)$ is
$\{231.26°,\ 308.74°\}$.

□

B. $\cos x = -0.81$

EXAMPLE 4

Find approximations for the solutions of each statement that lie in the interval $[0°, 360°)$. Use a calculator and round your approximations to two decimal places.

A. $\cos x = 0.8$ AND $\sin x = -0.6$

SOLUTION

$\cos x = 0.8$ AND $\sin x = -0.6$

$$\Leftrightarrow x = 360° + \sin^{-1}(-0.6)$$
$$\approx 360° + 36.87°$$
$$= 323.13°$$

The set of sols. in $[0°, 360°)$ is $\{323.13°\}$. □

B. $\cos x = -\sqrt{21}/5$ AND $\sin x = -2/5$

C. $\cos x = -5/6$ AND $\sin x = \sqrt{11}/6$

SOLUTION

$\cos x = -5/6$ AND $\sin x = \sqrt{11}/6$

$$\Leftrightarrow x = \cos^{-1}(-5/6)$$
$$\approx 146.44°$$

The set of sols. in $[0°, 360°)$ is $\{146.44°\}$. □

D. $\cos x = \sqrt{30}/7$ AND $\sin x = \sqrt{19}/7$

EXAMPLE 5

Rewrite each expression in terms of sine using formula B on page 380. Use a calculator where necessary.

A. $-2\sin x - 3\cos x$

SOLUTION

Notes

- $-2\sin x - 3\cos x = (-2)\sin x + (-3)\cos x$
- $\sqrt{(-2)^2 + (-3)^2} = \sqrt{4+9}$
  $= \sqrt{13}$
- $\cos\theta = -2/\sqrt{13}$ AND $\sin\theta = -3/\sqrt{13}$ AND $\theta \in [0, 2\pi)$

$$\Rightarrow \theta = \pi + \cos^{-1}(2/\sqrt{13})$$
$$\approx 4.1244$$

Expression

$-2\sin x - 3\cos x = (\sqrt{13})\sin(x + 4.1244)$ □

B. $5\sin x - 4\cos x$

EXAMPLE 6

Find approximations for the solutions of each equation that lie in the interval $[0, 2\pi)$. Use a calculator and round your approximations to four decimal places.

A. $\sec q = 1.37$

SOLUTION

Notes

- $\sec q = 1.37 \Leftrightarrow 1/\cos q = 1.37$
  $\Leftrightarrow 1 = 1.37 \cos q$
  $\Leftrightarrow 1/1.37 = \cos q$
- The period of cosine is 2π.
- An interval of width 2π is $[0, 2\pi)$.

Solutions for One Period

- $\cos q = 1/1.37$ AND $q \in [0, \pi]$
  $\Rightarrow q = \cos^{-1}(1/1.37)$
  $\Rightarrow q \approx 0.7525$ (by calculator)
- $\cos q = 1/1.37$ AND $q \in (\pi, 2\pi)$
  $\Rightarrow q \approx -0.7525 + 2\pi$ (refer to figure)
  $\Rightarrow q \approx 5.5307$

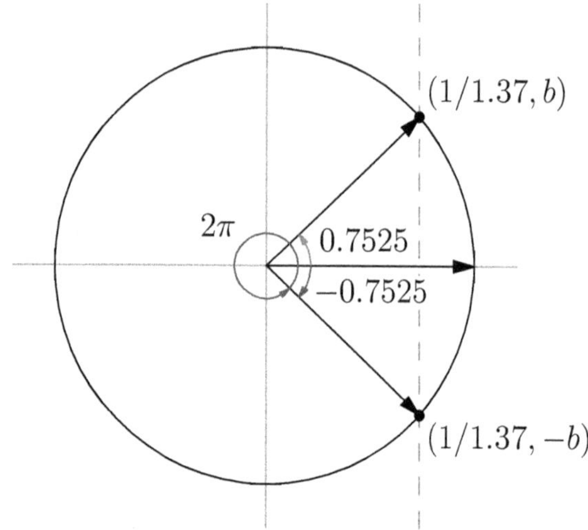

Solutions in Given Interval

The set of solutions in $[0, 2\pi)$ is $\{0.7525,\ 5.5307\}$.

Comment

We found the solutions in $[0, 2\pi)$ immediately, so we did not need to substitute values for k in more general expressions. □

B. $\csc \beta = 4.35$

C. $\cot 2\theta = 0.44$

SOLUTION

Notes

- $\cot 2\theta = 0.44 \Leftrightarrow 1/\tan 2\theta = 0.44$
  $\Leftrightarrow 1 = 0.44 \tan 2\theta$
  $\Leftrightarrow 1/0.44 = \tan 2\theta$
- The period of tangent is π.
- An interval of width π is $(-\pi/2, \pi/2)$.

Value of Expression for One Period

$\tan 2\theta = 1/0.44$ AND $\theta \in (-\pi/2, \pi/2)$
$\Rightarrow 2\theta = \tan^{-1}(1/0.44)$
$\Rightarrow 2\theta \approx 1.1563$ (by calculator)

General Solution

$2\theta = 1.1563 + \pi k$
$\Leftrightarrow (1/2)(2\theta) = (1/2)(1.1563 + \pi k)$
$\Leftrightarrow \theta = (1/2)(1.1563) + (1/2)(\pi k)$
$\Leftrightarrow \theta = 0.5781 + (\pi/2)k$

Set of Solutions

The set of solutions is
$\{0.5781 + (\pi/2)k \mid k \in \mathbf{Z}\}$.

Substitutions

- $0.5781 + (\pi/2) \cdot 0 = 0.5781$
- $0.5781 + (\pi/2) \cdot 1 = 2.1489$
- $0.5781 + (\pi/2) \cdot 2 = 3.7197$
- $0.5781 + (\pi/2) \cdot 3 = 5.2905$
- $0.5781 + (\pi/2) \cdot 4 = 6.8613$ X

Solutions in Given Interval

The set of solutions in $[0, 2\pi)$ is
$\{0.5781, 2.1489, 3.7197, 5.2905\}$.

□

D. $\tan 4t = -1.63$

Section 130: Area of a Triangle

Lemma (proofs omitted)
Suppose $x \in \mathbf{R}$.
Then:

- $\cos(\pi - x) = -\cos x$
- $\sin(\pi - x) = \sin x$

Formula
Suppose θ represents the measure of an angle of a given triangle and
a and b represent the lengths of the sides of the triangle that form that angle.
Then the area A of the triangle is given by the formula $A = (1/2)ab\sin\theta$.

Proof
Suppose a, b, θ, and A represent the quantities described above for a given triangle.
Let h represent the height of the triangle when the side with length b serves as its base.
We prove the formula for all possible values of θ.

- Suppose $0 < \theta < \pi/2$.
  Note that $\sin\theta = h/a$, so $h = a\sin\theta$.
  Thus:
  $$\begin{aligned} A &= (1/2)(\text{base})(\text{height}) \\ &= (1/2)(b)(h) \\ &= (1/2)(b)(a\sin\theta) \\ &= (1/2)ab\sin\theta \end{aligned}$$

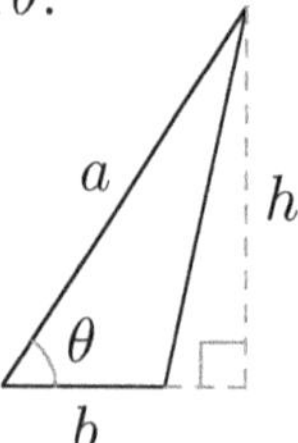

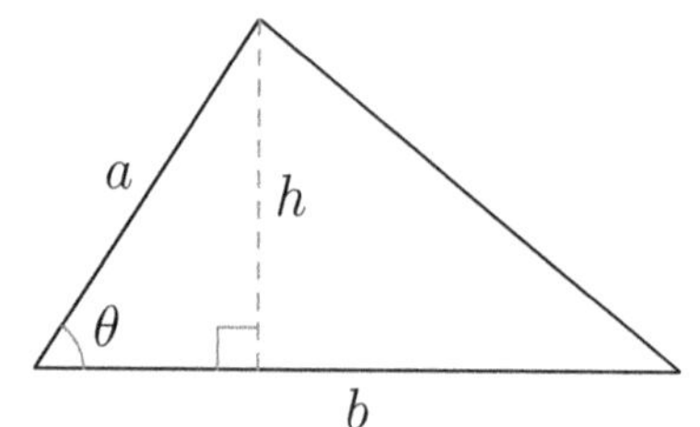

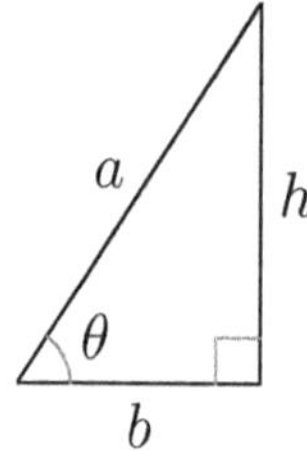

- Suppose $\theta = \pi/2$.
  Then:
  $$\begin{aligned} A &= (1/2)(\text{base})(\text{height}) \\ &= (1/2)(b)(a) \\ &= (1/2)ab \cdot 1 \\ &= (1/2)ab \cdot \sin(\pi/2) \\ &= (1/2)ab\sin\theta \end{aligned}$$

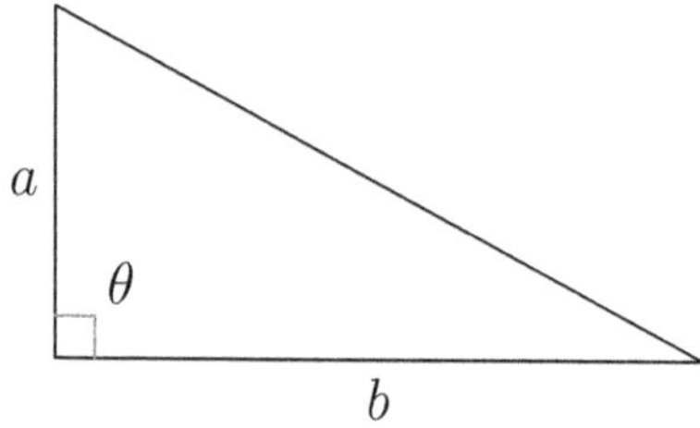

- Suppose $\pi/2 < \theta < \pi$.
  Let ϕ represent the measure of the angle supplementary to the angle of measure θ.
  Note that $\sin\phi = h/a$, so $h = a\sin\phi$.
  Thus:
  $$\begin{aligned} A &= (1/2)(\text{base})(\text{height}) \\ &= (1/2)(b)(h) \\ &= (1/2)(b)(a\sin\phi) \\ &= (1/2)(b)(a\sin(\pi - \theta)) \\ &= (1/2)(b)(a\sin\theta) \\ &= (1/2)ab\sin\theta \end{aligned}$$

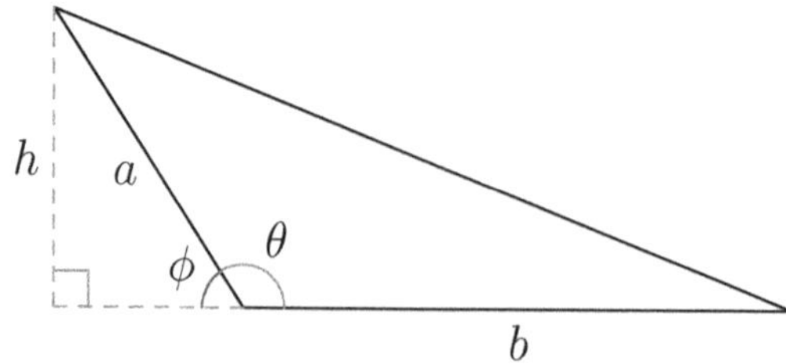

⊡

COMMENT
In the formula on the previous page, we can substitute measures for θ in either radians or degrees.

EXAMPLE 1
Determine the area of the region shaded in the figure below. Assume that the radius of the circle is 1 unit.

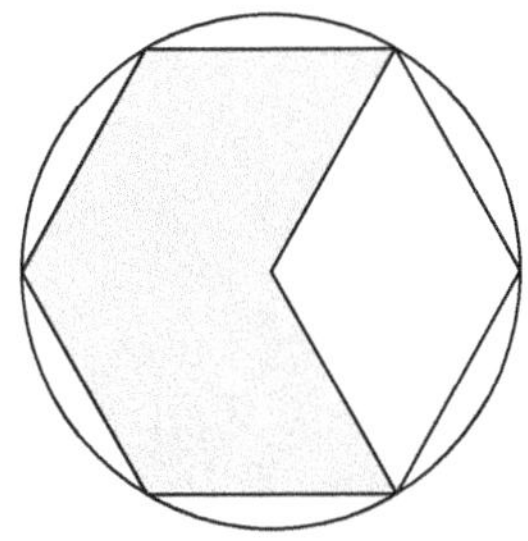

SOLUTION
Approach
We divide the hexagon into six congruent triangles, each of which has an angle with vertex at the center of the circle.

Notes
For each of the angles of the congruent triangles with vertex at the center of the circle:

- its measure equals $360°/6$, or $60°$;
- both of the sides that form it have length 1.

Area of Each Triangle

$$\begin{aligned}(\text{area of triangle}) &= (1/2)(1)(1)(\sin 60°)\\ &= (1/2)(\sqrt{3}/2)\\ &= \sqrt{3}/4\end{aligned}$$

Area of Shaded Region

$$\begin{aligned}(\text{area of shaded region}) &= 4 \cdot \sqrt{3}/4\\ &= \sqrt{3}\end{aligned}$$

□

EXAMPLE 2
Determine the area of the triangle shown below. Use a calculator and round your solution to two decimal places.

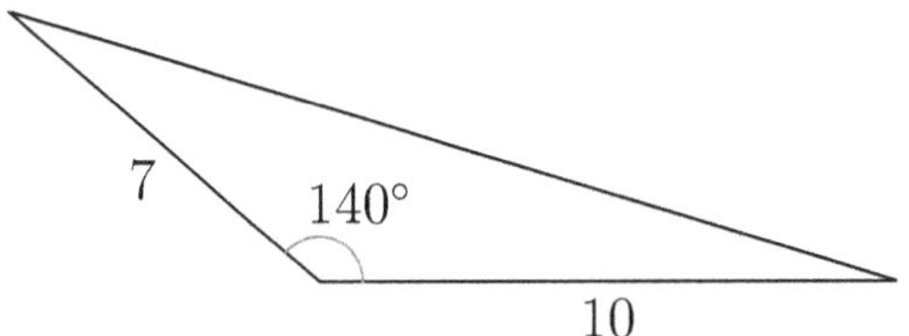

Section 131: Area of a Circular Sector

FORMULA

If the vertex of an angle of measure θ (where $0 < \theta < 2\pi$) lies at the center of a circle of radius r, then the area A of the sector bounded by the circle and the sides of the angle is given by the formula $A = (1/2)r^2\theta$.

PROOF

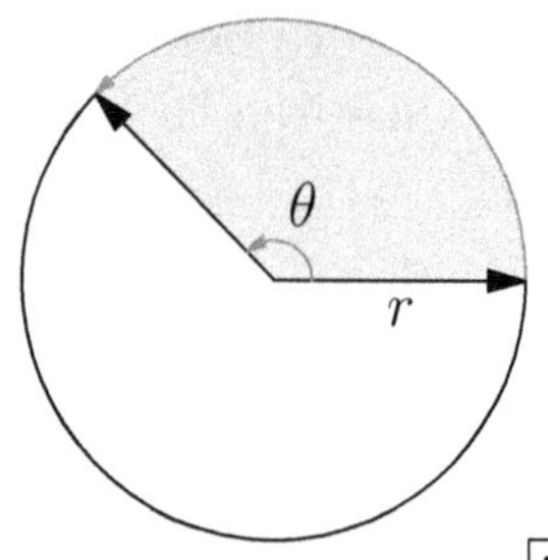

Suppose θ, r, and A represent the quantities described above.
Then:

$A/(\pi r^2) = \theta/(2\pi)$ (by properties of geometry)

$\Rightarrow [A/(\cancel{\pi r^2})] \cdot \cancel{\pi r^2} = [\theta/(2\cancel{\pi})] \cdot \cancel{\pi} r^2$

$\Rightarrow A = (1/2)r^2\theta$

□

COMMENTS

- In the formula above, we can substitute measures for θ in radians only, not degrees.
- In the examples to follow, assume θ, r, and A represent the quantities described above.

EXAMPLE 1

Find the volume of a piece of pie cut from a circular pan if:

- the pan is 1 inch deep;
- the pan is 10 inches in diameter;
- the central angle of the piece is $30°$.

SOLUTION

Notes

- $r = (1/2)(10)$
  $= 5$
- $\theta = (30°)(\pi/180°)$
  $= \pi/6$

Volume of Piece of Pie

For $r = 5$ and $\theta = \pi/6$:

$(\text{volume}) = (\text{area of top})(\text{depth})$
$= ((1/2)r^2\theta)(\text{depth})$
$\Rightarrow (\text{volume}) = (1/2)(5)^2(\pi/6)(1)$
$= (1/2)(25)(\pi/6)$
$= 25\pi/12$
≈ 6.545 (by calculator)

The volume of the piece of pie is about 6.545 cubic inches.

□

EXAMPLE 2

Determine the area of the region shaded in the figure below. Assume the triangle is equilateral and the radius of the circle is 1 unit.

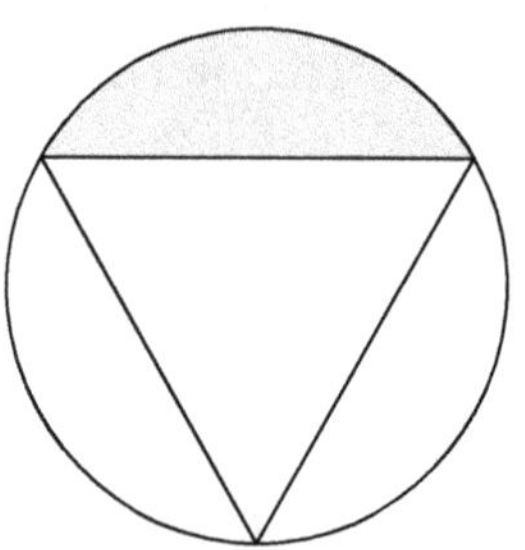

Example 3

Determine the area of the sector bounded by the sides of the angle of measure θ and the circle of radius r centered at the vertex of the angle.

A. $\theta = 3\pi/4$, $r = 8$ cm

Solution

For $\theta = 3\pi/4$ and $r = 8$:

$A = (1/2)r^2\theta$

$$\begin{aligned}\Rightarrow A &= (1/2)(8)^2(3\pi/4)\\ &= (1/2)(64)(3\pi/4)\\ &= (32)(3\pi/4)\\ &= (8)(3\pi)\\ &= 24\pi\end{aligned}$$

The area is 24π square centimeters.

□

B. $\theta = 11\pi/6$, $r = 3$ m

Section 132: The Law of Sines

The Law of Sines
Suppose α, β, and γ represent the measures of the angles of a given triangle and a, b, and c represent the lengths of the sides opposite those angles, respectively. Then $(\sin\alpha)/a = (\sin\beta)/b = (\sin\gamma)/c$.

Proof
Suppose a, b, c, α, β, and γ represent the lengths and measures described above and depicted at right.
We have seen that:

- (area of triangle) $= (1/2)bc\sin\alpha$
- (area of triangle) $= (1/2)ac\sin\beta$
- (area of triangle) $= (1/2)ab\sin\gamma$

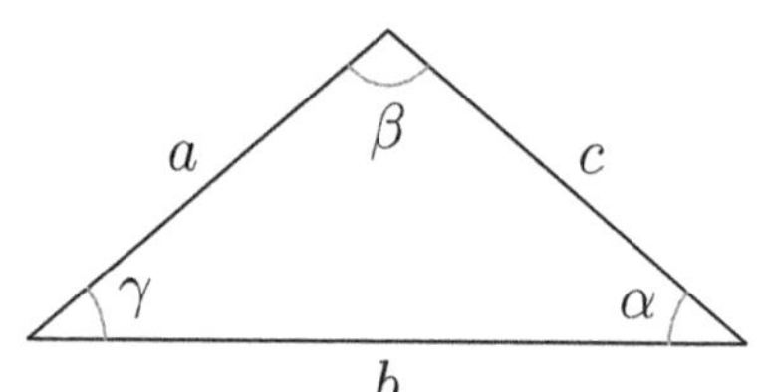

Therefore:

$$\frac{bc\sin\alpha}{2} = \frac{ac\sin\beta}{2} = \frac{ab\sin\gamma}{2}$$

$$\Rightarrow \frac{\cancel{b}\cancel{c}\sin\alpha}{\cancel{2}}\cdot\frac{\cancel{2}}{a\cancel{b}\cancel{c}} = \frac{\cancel{a}\cancel{c}\sin\beta}{\cancel{2}}\cdot\frac{\cancel{2}}{\cancel{a}b\cancel{c}} = \frac{\cancel{a}\cancel{b}\sin\gamma}{\cancel{2}}\cdot\frac{\cancel{2}}{\cancel{a}\cancel{b}c}$$

$$\Rightarrow \frac{\sin\alpha}{a} = \frac{\sin\beta}{b} = \frac{\sin\gamma}{c}$$

⊡

Comment
In both the Law of Sines and the Law of Cosines (see next section), we can substitute for α, β, and γ measures in either radians or degrees.

Example 1
In the following exercises, assume that a, b, and c represent the lengths of the sides of a triangle that lie opposite angles of the same triangle of measures α, β, and γ, respectively. Use a calculator and round your solutions to two decimal places.

A. Suppose $\alpha = 37°$, $\beta = 111°$, and $a = 5$. Find b.

Solution
$(\sin 37°)/5 = (\sin 111°)/b$
$\Rightarrow 5b\cdot(\sin 37°)/5 = 5b\cdot(\sin 111°)/b$
$\Rightarrow b\sin 37° = 5\sin 111°$
$\Rightarrow b = (5\sin 111°)/\sin 37°$
≈ 7.76

□

B. Suppose $\alpha = 73°$, $\gamma = 9°$, and $c = 8$. Find a.

C. Suppose $\beta = 50°$, $b = 7.0$, and $c = 8.6$. Find all possible values of γ and draw for each value of γ a triangle that depicts the relationships between the given variables.

SOLUTION

Value of $\sin\gamma$

$(\sin\gamma)/8.6 = (\sin 50°)/7.0$

$\Rightarrow \sin\gamma = 8.6 \cdot (\sin 50°)/7.0$

≈ 0.9411

Values of γ

We find all values for γ in $[0, 180°]$:

- $\sin\gamma \approx 0.9411$ AND $\gamma \in [0°, 90°]$
  $\Rightarrow \gamma \approx \sin^{-1}(0.9411)$
  $\Rightarrow \gamma \approx 70.24°$
- $\sin\gamma \approx 0.9411$ AND $\gamma \in (90°, 180°]$
  $\Rightarrow \gamma \approx 180° - 70.24°$
  $\Rightarrow \gamma \approx 109.76°$

So the possible values of γ are roughly $70.24°$ and $109.76°$.

Figures

For $\gamma \approx 70.24°$:

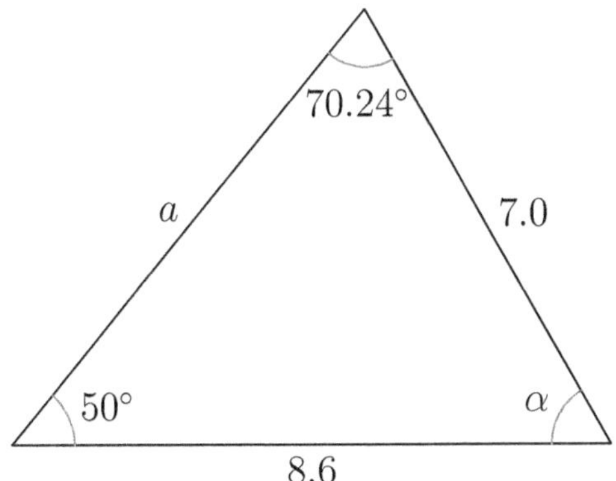

For $\gamma \approx 109.76°$:

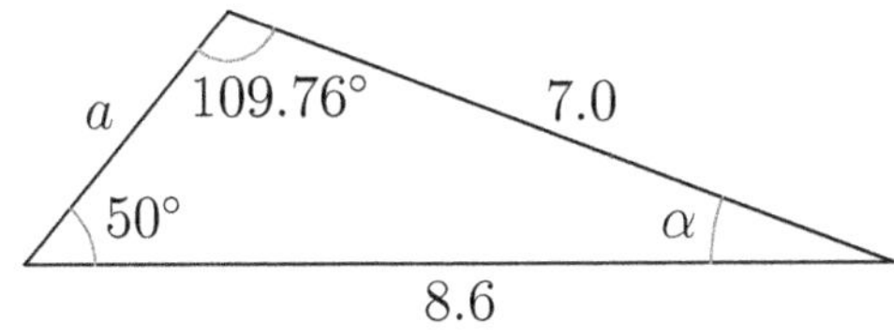

□

D. Suppose $\alpha = 98°$, $a = 7.2$, and $b = 4.8$. Find all possible values of β and draw for each value of β a triangle that depicts the relationships between the given variables.

Section 133: The Law of Cosines

THE LAW OF COSINES (proof omitted)
Suppose α, β, and γ represent the measures of the angles of a given triangle and a, b, and c represent the lengths of the sides opposite those angles, respectively.
Then:

- $a^2 = b^2 + c^2 - 2bc\cos\alpha$
- $b^2 = c^2 + a^2 - 2ca\cos\beta$
- $c^2 = a^2 + b^2 - 2ab\cos\gamma$

COMMENT
When we wish to find the measure α of some angle of a triangle, we seek a value in $(0°, 180°)$. But observe that:

- $\cos\alpha = m$ AND $\alpha \in (0°, 180°) \Leftrightarrow \alpha = \cos^{-1} m$
- $\sin\alpha = m$ AND $\alpha \in (0°, 180°) \Leftrightarrow \alpha = \sin^{-1} m$ OR $\alpha = 180° - \sin^{-1} m$

Thus the Law of Cosines will yield a single solution for α, while the Law of Sines will yield two possible solutions for α.

In fact, given information sufficient to apply the Law of Sines but insufficient to apply the Law of Cosines, both of the two possible measures for α that the Law of Sines determines may truly correspond to angles in triangles; however, given information sufficient to apply both the Law of Sines and the Law of Cosines, only one of the two possible measures for α that the Law of Sines determines will be valid, namely the single measure yielded by the Law of Cosines. We can avoid uncertainty in such a situation by using the latter law rather than the former.

EXAMPLE 1
In the following exercises, assume that a, b, and c represent the lengths of the sides of a triangle that lie opposite angles of the same triangle of measures α, β, and γ, respectively. Use a calculator and round your solutions to two decimal places.

A. Suppose $a = 9.0$, $b = 30.4$, and $c = 27.5$. Find β.

SOLUTION

$$b^2 = c^2 + a^2 - 2ca\cos\beta$$
$$\Rightarrow b^2 + 2ca\cos\beta = c^2 + a^2$$
$$\Rightarrow 2ca\cos\beta = c^2 + a^2 - b^2$$
$$\Rightarrow \cos\beta = (c^2 + a^2 - b^2)/(2ca)$$
$$= \frac{(27.5)^2 + (9.0)^2 - (30.4)^2}{2(27.5)(9.0)}$$
$$\approx -0.1756$$
$$\Rightarrow \beta \approx \cos^{-1}(-0.1756)$$
$$\approx 100.11°$$

□

B. Suppose $a = 9.0$, $b = 30.4$, and $c = 27.5$. Find γ.

C. Suppose $\gamma = 41°$, $a = 6.5$, and $b = 8.1$. Find c and draw a triangle that depicts the relationships between the given variables.

SOLUTION

Value of c

$$\begin{aligned} c^2 &= a^2 + b^2 - 2ab\cos\gamma \\ &= (6.5)^2 + (8.1)^2 - 2(6.5)(8.1)\cos 41° \\ &\approx 28.3891 \\ \Rightarrow c &\approx \sqrt{28.3891} \\ &\approx 5.33 \end{aligned}$$

Figure

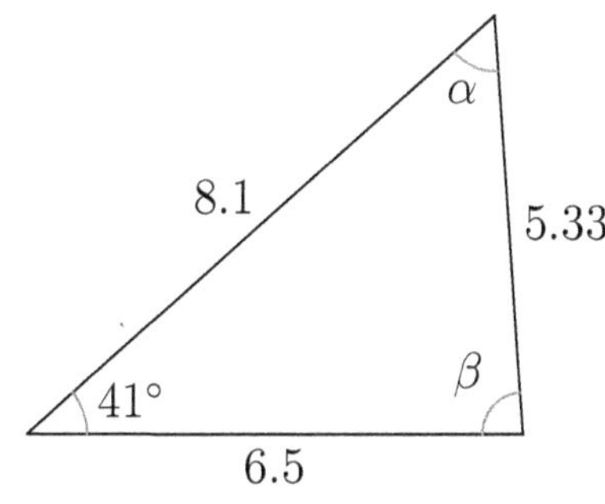

□

D. Suppose $\alpha = 41°$, $b = 7.3$, and $c = 3.7$. Find a and draw a triangle that depicts the relationships between the given variables.

Section 134: Heron's Formula

Heron's Formula

Suppose a, b, and c represent the lengths of the sides of a triangle and $s = (1/2)(a+b+c)$. Then the area A of the triangle is given by the formula $A = \sqrt{s(s-a)(s-b)(s-c)}$.

Proof

Suppose a, b, c, s, and A represent the quantities described above.
Let γ denote the measure of the angle of the triangle opposite the side of length c.
Note that:

- $A = (1/2)ab\sin\gamma$
$$\begin{aligned} \Rightarrow A^2 &= [(1/2)ab\sin\gamma]^2 \\ &= (1/2)^2a^2b^2\sin^2\gamma \\ &= (1/4)a^2b^2(1-\cos^2\gamma) \\ &= (1/4)a^2b^2(1-\cos\gamma)(1+\cos\gamma) \end{aligned}$$
- $c^2 = a^2+b^2-2ab\cos\gamma$
$$\Rightarrow 2ab\cos\gamma = a^2+b^2-c^2$$
$$\Rightarrow \cos\gamma = (a^2+b^2-c^2)/(2ab)$$

Therefore:

- $$\begin{aligned} 1-\cos\gamma &= 1-(a^2+b^2-c^2)/(2ab) \\ &= (2ab)/(2ab)-(a^2+b^2-c^2)/(2ab) \\ &= [2ab-(a^2+b^2-c^2)]/(2ab) \\ &= -[-2ab+(a^2+b^2-c^2)]/(2ab) \\ &= -[a^2-2ab+b^2-c^2]/(2ab) \\ &= -[(a-b)^2-c^2]/(2ab) \\ &= -[(a-b-c)(a-b+c)]/(2ab) \\ &= [(-a+b+c)(a-b+c)]/(2ab) \end{aligned}$$
- $$\begin{aligned} 1+\cos\gamma &= 1+(a^2+b^2-c^2)/(2ab) \\ &= (2ab)/(2ab)+(a^2+b^2-c^2)/(2ab) \\ &= [2ab+(a^2+b^2-c^2)]/(2ab) \\ &= [a^2+2ab+b^2-c^2]/(2ab) \\ &= [(a+b)^2-c^2]/(2ab) \\ &= [(a+b-c)(a+b+c)]/(2ab) \end{aligned}$$

Hence:

$$\begin{aligned} A^2 &= (1/4)a^2b^2\cdot[[(-a+b+c)(a-b+c)]/(2ab)]\cdot[[(a+b-c)(a+b+c)]/(2ab)] \\ &= [(1/2)^2\cancel{a^2b^2}]/(4\cancel{a^2b^2})\cdot(-a+b+c)(a-b+c)(a+b-c)(a+b+c) \\ &= (1/2)^2\cdot(1/4)\cdot(-a+b+c+a-a)(a-b+c+b-b)(a+b-c+c-c)(a+b+c) \\ &= (1/2)^4\cdot(a+b+c-2a)(a+b+c-2b)(a+b+c-2c)(a+b+c) \\ &= (1/2)(a+b+c-2a)\cdot(1/2)(a+b+c-2b)\cdot(1/2)(a+b+c-2c)\cdot(1/2)(a+b+c) \\ &= [(1/2)(a+b+c)-a]\cdot[(1/2)(a+b+c)-b]\cdot[(1/2)(a+b+c)-c]\cdot[(1/2)(a+b+c)] \\ &= (s-a)(s-b)(s-c)(s) \\ \Rightarrow A &= \sqrt{s(s-a)(s-b)(s-c)} \end{aligned}$$

⊡

EXAMPLE 1

Given the lengths of the sides of a triangle:

- find the area of the triangle using Heron's Formula,
- find the measure θ (in degrees) of the angle opposite the side of the third length, and
- find the area of the triangle using the formula $A = (1/2)ab\sin\theta$.

Use a calculator and round your answers to two decimal places. Assume the variables A, a, b, c, and s represent the quantities described in the statement of Heron's Formula.

A. lengths of sides: 5, 9, 10

SOLUTION

- *Value of s*

For $a = 5$, $b = 9$, and $c = 10$:

$$s = (1/2)(a+b+c)$$
$$\Rightarrow s = (1/2)(5+9+10) = (1/2)(24) = 12$$

Area of Triangle by Heron's Formula

$$A = \sqrt{s(s-a)(s-b)(s-c)}$$
$$\Rightarrow A = \sqrt{12(12-5)(12-9)(12-10)} = \sqrt{12(7)(3)(2)} = \sqrt{504} \approx 22.45$$

- *Value of θ*

$$c^2 = a^2 + b^2 - 2ab\cos\theta$$
$$\Rightarrow (10)^2 = (5)^2 + (9)^2 - 2(5)(9)\cos\theta$$
$$\Rightarrow 100 = 25 + 81 - 90\cos\theta$$
$$\Rightarrow -25 + (-81) + 100 = -90\cos\theta$$
$$\Rightarrow -6 = -90\cos\theta$$
$$\Rightarrow (-6)/(-90) = (-90\cos\theta)/(-90)$$
$$\Rightarrow 6/90 = \cos\theta$$
$$\Rightarrow \theta = \cos^{-1}(6/90) = \cos^{-1}(1/15) \approx 86.18^\circ$$

- *Area of Triangle by Formula with Sine*

$$A = (1/2)ab\sin\theta$$
$$\Rightarrow A \approx (1/2)(5)(9)\sin 86.18^\circ \approx 22.45$$

□

B. lengths of sides: 7, 14, 20

Section 135: Conic Sections

Definition

A *conic section* is a cross-section of the surface of a right circular double cone.

Comment

The following curves are conic sections:

- circles
- parabolas
- ellipses
- hyperbolas

We may consider the following geometric objects to be degenerate conic sections:

- lines
- points

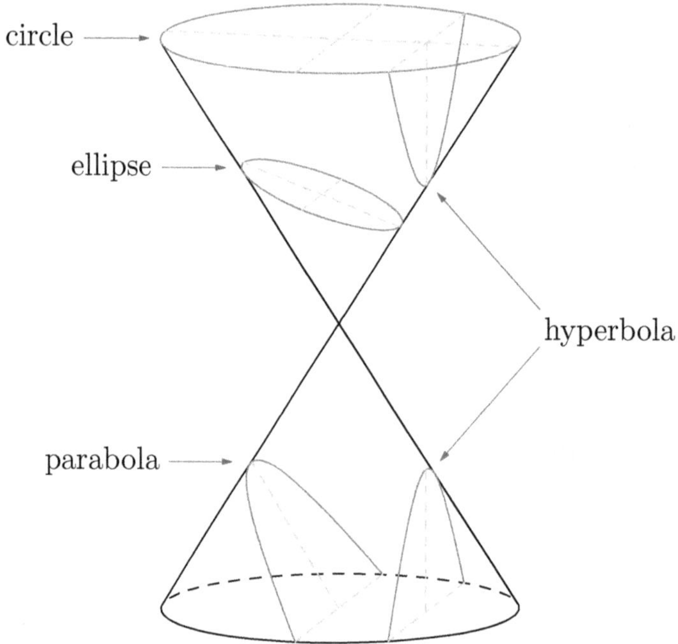

Section 136: Parabolas

Definitions

- A *parabola* is a locus of points in a plane that lie the same distance from a fixed line (called the *directrix*) as from a fixed point (called the *focus*) not on the line.
- The *axis of symmetry* of a parabola is the line that runs perpendicular to its directrix and passes through its focus.
- The *vertex* of a parabola is the point at which it intersects its axis of symmetry.
- The *latus rectum* of a parabola is the line segment with endpoints on the parabola that runs parallel to its directrix and passes through its focus.
- The *focal diameter* of a parabola is the length of its latus rectum.

Property

The focal diameter of a parabola equals twice the distance from its focus to its directrix.

Proof

Given a parabola, note that:
- its focus lies at the midpoint of its latus rectum;
- both endpoints of its latus rectum (as points on the parabola) lie the same distance from its focus as from its directrix;
- all points on its latus rectum (including its focus and the endpoints of the latus rectum) lie the same distance from its directrix.

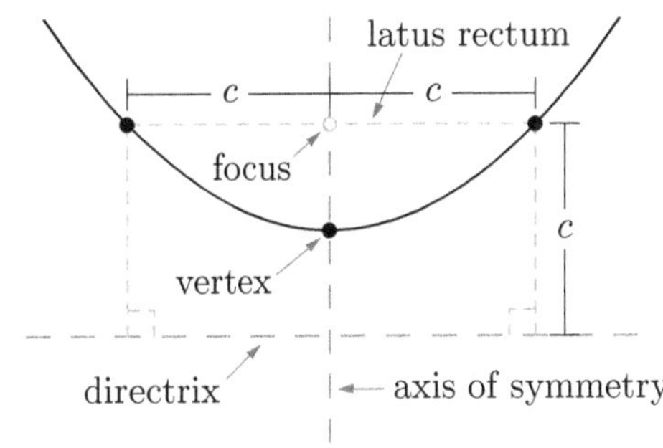

Therefore:

$$\begin{aligned}\text{focal diameter of a parabola} &= \text{distance between the endpoints of its latus rectum}\\ &= 2\cdot(\text{distance from one such endpoint to its focus})\\ &= 2\cdot(\text{distance from one such endpoint to its directrix})\\ &= 2\cdot(\text{distance from its focus to its directrix})\end{aligned}$$

⊡

Properties

- The focal diameter of the parabola with focus $(0, p)$ and directrix $y = -p$ is $4|p|$.
- The focal diameter of the parabola with focus $(p, 0)$ and directrix $x = -p$ is $4|p|$.

Proof (first property only)

Consider the parabola with focus $(0, p)$ and directrix $y = -p$.

For this parabola:

$$\begin{aligned}(\text{focal diameter}) &= 2\cdot(\text{distance between focus and directrix})\\ &= 2\cdot|p-(-p)|\\ &= 2\cdot|2p|\\ &= 2\cdot 2|p|\\ &= 4|p|\end{aligned}$$

⊡

EQUATION
The *standard form* of the equation of the parabola with vertex at coordinates $(0,0)$ and focus at coordinates $(0,p)$ is $x^2 = 4py$.

DERIVATION
Consider the parabola with vertex $(0,0)$ and focus $(0,p)$.
Note that its directrix must then have the equation $y = -p$.
Let (x,y) denote the coordinates of a point on the parabola,
d_1 denote the distance from (x,y) to its focus,
and d_2 denote the distance from (x,y) to its directrix.

By the definition of parabola:

$$
\begin{aligned}
d_1 &= d_2 \\
&\Rightarrow \sqrt{(x-0)^2 + (y-p)^2} = |y-(-p)| \\
&\Rightarrow \left(\sqrt{x^2+(y-p)^2}\,\right)^2 = |y+p|^2 \\
&\Rightarrow x^2 + (y-p)^2 = (y+p)^2 \\
&\Rightarrow x^2 + y^2 - 2py + p^2 = y^2 + 2py + p^2 \\
&\Rightarrow x^2 + \cancel{y^2} - 2py + \cancel{p^2} - \cancel{y^2} - \cancel{p^2} = 2py \\
&\Rightarrow x^2 = 2py + 2py \\
&\Rightarrow x^2 = 4py
\end{aligned}
$$

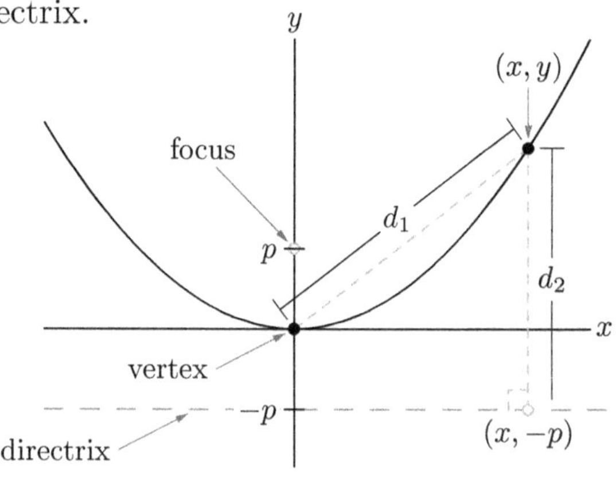

□

EQUATION
The *standard form* of the equation of the parabola with vertex at coordinates $(0,0)$ and focus at coordinates $(p,0)$ is $y^2 = 4px$.

DERIVATION
Consider the parabola with vertex $(0,0)$ and focus $(p,0)$.
Note that its directrix must then have the equation $x = -p$.
Let (x,y) denote the coordinates of a point on the parabola,
d_1 denote the distance from (x,y) to its focus,
and d_2 denote the distance from (x,y) to its directrix.

By definition of parabola:

$$
\begin{aligned}
d_1 &= d_2 \\
&\Rightarrow \sqrt{(x-p)^2 + (y-0)^2} = |x-(-p)| \\
&\Rightarrow \left(\sqrt{(x-p)^2+y^2}\right)^2 = |x+p|^2 \\
&\Rightarrow (x-p)^2 + y^2 = (x+p)^2 \\
&\Rightarrow x^2 - 2px + p^2 + y^2 = x^2 + 2px + p^2 \\
&\Rightarrow \cancel{x^2} - 2px + \cancel{p^2} + y^2 - \cancel{x^2} - \cancel{p^2} = 2px \\
&\Rightarrow y^2 = 2px + 2px \\
&\Rightarrow y^2 = 4px
\end{aligned}
$$

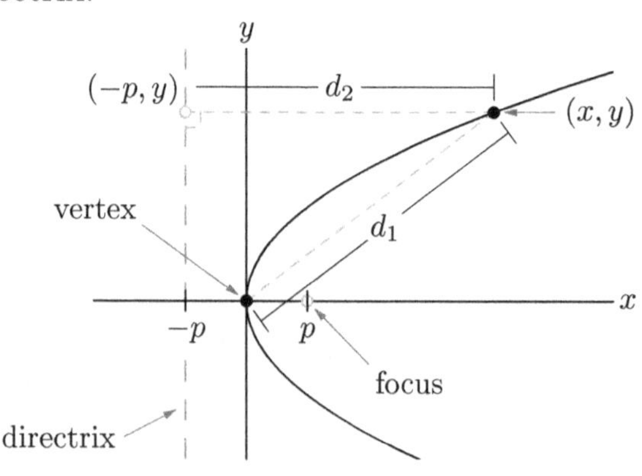

□

COMMENTS

Recall from page 275 that:

- a parabola is a rotation of the graph of a quadratic function;
- a parabola is symmetric with respect to its axis of symmetry.

EXAMPLE 1

Determine the equation of the parabola with vertex $(0, 0)$ with the given characteristics.

A. focus has coordinates $(0, 6)$

SOLUTION

$x^2 = 4(6)y \Leftrightarrow x^2 = 24y$

☐

B. focus has coordinates $(-2, 0)$

C. directrix has equation $x = 3$

SOLUTION

Note

The focus has coordinates $(-3, 0)$.

Equation

$y^2 = 4(-3)x \Leftrightarrow y^2 = -12x$

☐

D. directrix has equation $y = -5$

E. focus lies on the x-axis;
point $(12, 4)$ lies on the parabola

SOLUTION

Note

the focus lies on the x-axis
$\Rightarrow$ the focus lies at $(p, 0)$ for some p
$\Rightarrow$ the equation is $y^2 = 4px$

Value of p

$(12, 4)$ satisfies $y^2 = 4px$
$\Rightarrow (4)^2 = 4p(12)$
$\Rightarrow 16 = 48p$
$\Rightarrow 16/48 = 48p/48$
$\Rightarrow 1/3 = p$

Equation

$y^2 = 4(1/3)x \Leftrightarrow y^2 = 4x/3$

☐

F. focus lies on the y-axis;
focus lies below the x-axis;
distance from focus to directrix is 3

G. focus lies on the y-axis;
focus lies below the x-axis;
focal diameter $= 7$

SOLUTION

Notes

- the focus lies on the y-axis
  $\Rightarrow$ the focus is $(0, p)$ for some p
- the focus lies below the x-axis
  $\Rightarrow p < 0$

Value of p

focal diameter $= 7$

$\Rightarrow 4|p| = 7$
$\Rightarrow 4(-p) = 7 \quad \text{(since } p < 0)$
$\Rightarrow -4p/(-4) = 7/(-4)$
$\Rightarrow p = -7/4$

Equation

$x^2 = 4(-7/4)y \Leftrightarrow x^2 = -7y$

□

H. directrix has a negative x-intercept;
focal diameter $= 6$

EXAMPLE 2

Identify the focus, directrix, vertex, and focal diameter of the parabola with the given equation.

A. $y^2 = 2x$

SOLUTION

Value of p

$y^2 = 2x$ AND $y^2 = 4px$

$\Rightarrow 4px = 2x$
$\Rightarrow 4p = 2$
$\Rightarrow \cancel{4}p/\cancel{4} = 2/4$
$\Rightarrow p = 1/2$

Characteristics

- the focus has coordinates $(1/2, 0)$
- the directrix has equation $x = -1/2$
- the vertex has coordinates $(0, 0)$
- focal diameter $= 4 \cdot |1/2|$
  $= 2$

□

B. $x^2 = 4y$

C. $x^2 - 6x - 8y - 11 = 0$

SOLUTION

Note

$x^2 - 6x - 8y - 11 = 0$
$\Leftrightarrow x^2 - 6x + 9 - 9 - 8y - 11 = 0$
$\Leftrightarrow (x + (-3))^2 - 9 - 8y - 11 = 0$
$\Leftrightarrow (x - 3)^2 - 8y - 20 = 0$
$\Leftrightarrow (x - 3)^2 = 8y + 20$
$\Leftrightarrow (x - 3)^2 = 8(y + 20/8)$
$\Leftrightarrow (x - 3)^2 = 8(y - (-5/2))$

Implication

The parabola with given equation is a shift right 3 units and down 5/2 units of the parabola with the equation $x^2 = 8y$.

Value of p

$x^2 = 8y$ AND $x^2 = 4py$
$\Rightarrow 4py = 8y$
$\Rightarrow 4p = 8$
$\Rightarrow 4p/4 = 8/4$
$\Rightarrow p = 2$

Characteristics of Parabola to Translate

- the focus has coordinates $(0, 2)$
- the directrix has equation $y = -2$
- the vertex has coordinates $(0, 0)$

Translations to Parabola in Question

- $(0\,,2) \to (0+3\,,2-5/2)$
  $= (3\,,-1/2)$
- $y = -2 \to y = -2 - 5/2$
  $\Rightarrow y = -9/2$
- $(0\,,0) \to (0+3\,,0-5/2)$
  $= (3\,,-5/2)$

Characteristics of Parabola in Question

- the focus has coordinates $(3, -1/2)$
- the directrix has equation $y = -9/2$
- the vertex has coordinates $(3, -5/2)$
- focal diameter $= 4(2)$
  $= 8$

□

D. $y^2 + 2y = 4x + 11$

EXAMPLE 3
Draw the parabola with the given characteristics.

A. focus has coordinates $(1, 0)$;
directrix has equation $x = -1$;
vertex has coordinates $(0, 0)$;
focal diameter $= 4$

SOLUTION

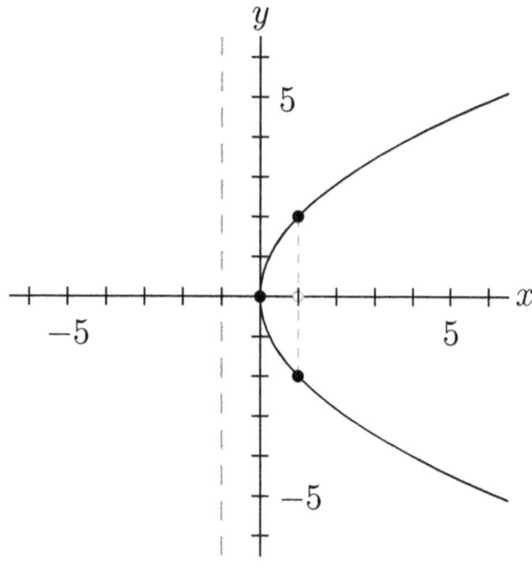

□

B. focus has coordinates $(0, -2)$;
directrix has equation $y = 2$;
vertex has coordinates $(0, 0)$;
focal diameter $= 8$

C. focus has coordinates $(3, -1/2)$;
directrix has equation $y = -9/2$;
vertex has coordinates $(3, -5/2)$;
focal diameter $= 8$

SOLUTION

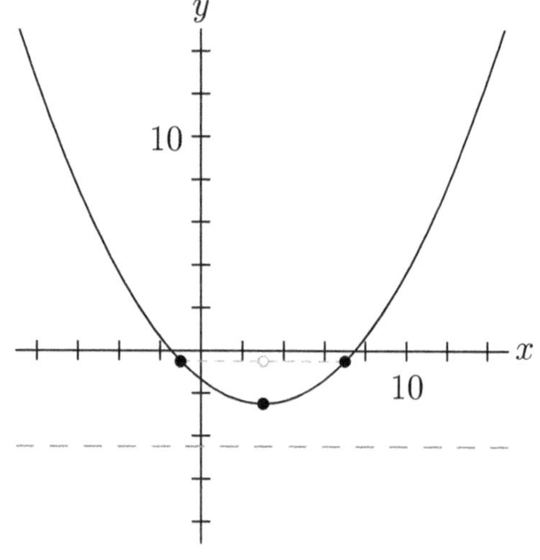

□

D. focus has coordinates $(-5/2, 2)$;
directrix has equation $x = 1/2$;
vertex has coordinates $(-1, 2)$;
focal diameter $= 6$

Section 137: Ellipses

Definitions

- An *ellipse* is a locus of points in a plane, each with the same sum of its distances from two fixed points (called the *foci*).
- The *center* of an ellipse is the midpoint of the line segment that joins its foci.
- The *focal axis* of an ellipse is the line that passes through its foci.
- The *vertices* of an ellipse are the points at which it intersects its focal axis.
- The *major axis* of an ellipse is the line segment that joins its vertices.
- The *minor axis* of an ellipse is the line segment with endpoints on the ellipse that runs perpendicular to its major axis and passes through its center.
- The *eccentricity* of an ellipse is the ratio of the distance between its foci to the length of its major axis.

Properties

The ellipse with foci $(-c, 0)$ and $(c, 0)$ has

- center $(0, 0)$
- vertices $(-a, 0)$ and $(a, 0)$
- major axis of length $2a$
- minor axis of length $2b$
- eccentricity of magnitude c/a

where a equals half the sum of the distances from the foci to a point on the ellipse. and b equals $\sqrt{a^2 - c^2}$.

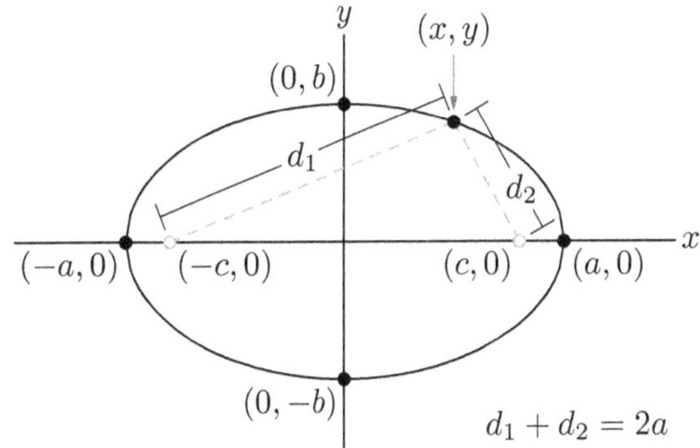

Proofs

- The foci of the ellipse have coordinates $(-c, 0)$ and $(c, 0)$.
  Therefore, the center of the ellipse has coordinates $((-c + c)/2, (0 + 0)/2)$, or $(0, 0)$.

- The ellipse has equation $x^2/a^2 + y^2/b^2 = 1$ (see next page), so its x-intercepts are $-a$ and a.
  But the x-axis is the focal axis of the ellipse, so its vertices must be $(-a, 0)$ and $(a, 0)$.

- The vertices of the ellipse have coordinates $(-a, 0)$ and $(a, 0)$.
  Therefore, the length of its major axis equals $a - (-a)$, or $2a$.

- The y-axis runs perpendicular to the major axis of the ellipse and passes through its center.
  Thus the endpoints of the minor axis of the ellipse lie at its intersection with the y-axis.
  But the ellipse has equation $x^2/a^2 + y^2/b^2 = 1$, so its y-intercepts are $-b$ and b.
  Therefore, the length of minor axis equals $b - (-b)$, or $2b$.

- The distance between the foci of the ellipse is $c - (-c) = 2c$.
  Therefore, the eccentricity is $2c/2a$, or c/a.

⊡

EQUATION

The *standard form* of the equation of the ellipse with foci at coordinates $(-c, 0)$ and $(c, 0)$ is $x^2/a^2 + y^2/b^2 = 1$,

where a equals half the sum of the distances from the foci to a point on the ellipse and b equals $\sqrt{a^2 - c^2}$.

DERIVATION

Consider the ellipse with foci $(-c, 0)$ and $(c, 0)$.

Let (x, y) denote the coordinates of a point on the ellipse,
a denote half the sum of the distances from the foci to (x, y),
and b equal $\sqrt{a^2 - c^2}$.

Then:

$$\begin{aligned}
&(1/2) \cdot [(\text{distance from } (x,y) \text{ to } (-c,0)) + (\text{distance from } (x,y) \text{ to } (c,0))] = a \\
&\Rightarrow (\text{distance from } (x,y) \text{ to } (-c,0)) + (\text{distance from } (x,y) \text{ to } (c,0)) = 2a \\
&\Rightarrow \sqrt{(x-(-c))^2 + (y-0)^2} + \sqrt{(x-c)^2 + (y-0)^2} = 2a \\
&\Rightarrow \sqrt{(x+c)^2 + y^2} + \sqrt{(x-c)^2 + y^2} = 2a \\
&\Rightarrow \sqrt{(x+c)^2 + y^2} = 2a - \sqrt{(x-c)^2 + y^2} \\
&\Rightarrow (\sqrt{(x+c)^2 + y^2}\,)^2 = (2a - \sqrt{(x-c)^2 + y^2}\,)^2 \\
&\Rightarrow (x+c)^2 + y^2 = (2a)^2 - 2(2a)(\sqrt{(x-c)^2 + y^2}\,) + (\sqrt{(x-c)^2 + y^2}\,)^2 \\
&\qquad\qquad\qquad\;\; = 4a^2 - 4a\sqrt{(x-c)^2 + y^2} + (x-c)^2 + y^2 \\
&\Rightarrow x^2 + 2cx + c^2 + y^2 = 4a^2 - 4a\sqrt{(x-c)^2 + y^2} + x^2 - 2cx + c^2 + y^2 \\
&\Rightarrow 2cx = 4a^2 - 4a\sqrt{(x-c)^2 + y^2} + \cancel{x^2} - 2cx + \cancel{c^2} + \cancel{y^2} - \cancel{x^2} - \cancel{c^2} - \cancel{y^2} \\
&\qquad\quad = 4a^2 - 4a\sqrt{(x-c)^2 + y^2} - 2cx \\
&\Rightarrow 2cx + 2cx - 4a^2 = -4a\sqrt{(x-c)^2 + y^2} \\
&\Rightarrow 4cx - 4a^2 = -4a\sqrt{(x-c)^2 + y^2} \\
&\Rightarrow 4(cx - a^2) = 4(-a\sqrt{(x-c)^2 + y^2}\,) \\
&\Rightarrow 4(cx - a^2)/4 = 4(-a\sqrt{(x-c)^2 + y^2}\,)/4 \\
&\Rightarrow cx - a^2 = -a\sqrt{(x-c)^2 + y^2} \\
&\Rightarrow (cx - a^2)^2 = (-a\sqrt{(x-c)^2 + y^2}\,)^2 \\
&\Rightarrow (cx)^2 - 2(cx)a^2 + (a^2)^2 = (-a)^2(\sqrt{(x-c)^2 + y^2}\,)^2 \\
&\Rightarrow c^2x^2 - 2a^2cx + a^4 = a^2((x-c)^2 + y^2) \\
&\qquad\qquad\qquad\qquad\quad\;\; = a^2(x^2 - 2cx + c^2 + y^2) \\
&\qquad\qquad\qquad\qquad\quad\;\; = a^2x^2 - 2a^2cx + a^2c^2 + a^2y^2 \\
&\Rightarrow c^2x^2 + a^4 = a^2x^2 - \cancel{2a^2cx} + a^2c^2 + a^2y^2 + \cancel{2a^2cx} \\
&\Rightarrow c^2x^2 + a^4 = a^2x^2 + a^2c^2 + a^2y^2 \\
&\Rightarrow a^4 - a^2c^2 = a^2x^2 - c^2x^2 + a^2y^2 \\
&\Rightarrow a^2(a^2 - c^2) = (a^2 - c^2)x^2 + a^2y^2 \\
&\Rightarrow a^2(\sqrt{a^2 - c^2}\,)^2 = (\sqrt{a^2 - c^2}\,)^2x^2 + a^2y^2 \\
&\Rightarrow a^2b^2 = b^2x^2 + a^2y^2 \\
&\Rightarrow a^2b^2/(a^2b^2) = (b^2x^2 + a^2y^2)/(a^2b^2) \\
&\qquad\qquad\qquad\quad\; = (b^2x^2)/(a^2b^2) + (a^2y^2)/(a^2b^2) \\
&\Rightarrow 1 = x^2/a^2 + y^2/b^2
\end{aligned}$$

⊡

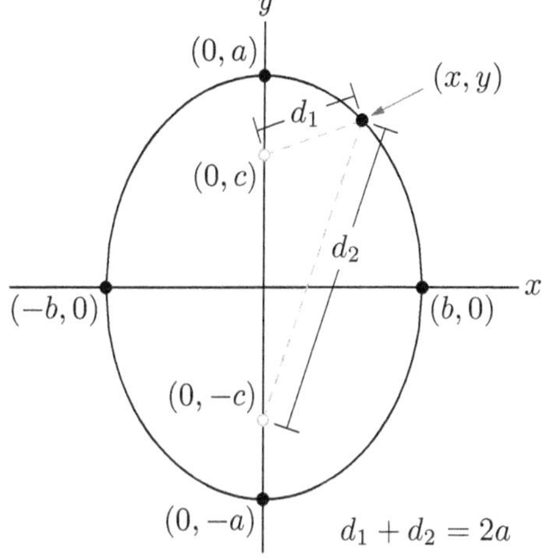

PROPERTIES (proofs omitted)
The ellipse with foci $(0, -c)$ and $(0, c)$ has
- center $(0, 0)$
- vertices $(0, -a)$ and $(0, a)$
- major axis of length $2a$
- minor axis of length $2b$
- eccentricity of magnitude c/a

where a equals half the sum of the distances from the foci to a point on the ellipse
and b equals $\sqrt{a^2 - c^2}$.

EQUATION (derivation omitted)
The *standard form* of the equation of the ellipse with foci at coordinates $(0, -c)$ and $(0, c)$ is $x^2/b^2 + y^2/a^2 = 1$,
where a equals half the sum of the distances from the foci to a point on the ellipse
and b equals $\sqrt{a^2 - c^2}$.

COMMENTS
- For the ellipse with equation $x^2/m^2 + y^2/n^2 = 1$:
  - the major axis will be horizontal if $m^2 > n^2$;
  - the major axis will be vertical if $n^2 > m^2$.
- The graph of the equation $(x - h)^2/m^2 + (y - k)^2/n^2 = 1$ (as a translation of the graph of the equation $x^2/m^2 + y^2/n^2 = 1$) is also an ellipse.
- Assume that $c > 0$ wherever we suppose that $-c$ and c represent coordinates of the foci of ellipses or hyperbolas (see next section).
- $b = \sqrt{a^2 - c^2} \Leftrightarrow c = \sqrt{a^2 - b^2} \Leftrightarrow a = \sqrt{b^2 + c^2}$ (since a, b, and c are positive as defined)

EXAMPLE 1
Consider the equation $9x^2 - 18x + (1/2)y^2 - 8y + 40 = 0$.
Rewrite this equation of an ellipse in the form $(x - h)^2/m^2 + (y - k)^2/n^2 = 1$.

SOLUTION
We complete the square for both the x- and y-terms:
$9x^2 - 18x + (1/2)y^2 - 8y + 40 = 0$
$\Leftrightarrow 9(x^2 - 2x) + (1/2)(y^2 - 16y) + 40 = 0$
$\Leftrightarrow 9(x^2 - 2x + (-1)^2 - (-1)^2) + (1/2)(y^2 - 16y + (-8)^2 - (-8)^2) + 40 = 0$
$\Leftrightarrow 9((x - 1)^2 - 1) + (1/2)((y - 8)^2 - 64) + 40 = 0$
$\Leftrightarrow 9(x - 1)^2 - 9 + (1/2)(y - 8)^2 - 32 + 40 = 0$
$\Leftrightarrow 9(x - 1)^2 + (1/2)(y - 8)^2 - 1 = 0$
$\Leftrightarrow (x - 1)^2/(1/9) + (y - 8)^2/(2) = 1$
$\Leftrightarrow (x - 1)^2/(1/3)^2 + (y - 8)^2/(\sqrt{2})^2 = 1$

□

EXAMPLE 2
Prove that the x-intercepts of the graph of the equation $x^2/a^2 + y^2/b^2 = 1$ are $-a$ and a.

PROOF
Suppose $x^2/a^2 + y^2/b^2 = 1$.
Then:

$$\begin{aligned} y = 0 &\Rightarrow x^2/a^2 + (0)^2/b^2 = 1 \\ &\Rightarrow x^2/a^2 = 1 \\ &\Rightarrow x^2 = a^2 \\ &\Rightarrow x = \pm\sqrt{a^2} \\ &\quad\;\; = \pm a \end{aligned}$$

So the x-intercepts are $-a$ and a.

⊡

EXAMPLE 3
Find the y-coordinates of all points on the graph of the equation $x^2/10 + y^2/50 = 1$ with an x-coordinate of 3.

EXAMPLE 4
Identify the foci, vertices, center, eccentricity, and lengths of the major and minor axes of the ellipse with the given equation.

A. $25x^2 + 4y^2 = 100$

SOLUTION
Note

$$\begin{aligned} &25x^2 + 4y^2 = 100 \\ &\Leftrightarrow (25x^2 + 4y^2)/100 = 100/100 \\ &\Leftrightarrow 25x^2/100 + 4y^2/100 = 1 \\ &\Leftrightarrow x^2/4 + y^2/25 = 1 \\ &\Leftrightarrow x^2/(2^2) + y^2/(5^2) = 1 \\ &\Leftrightarrow x^2/b^2 + y^2/a^2 = 1 \end{aligned}$$

where $a = 5$ and $b = 2$

Value of c
For $a = 5$ and $b = 2$:

$$c = \sqrt{a^2 - b^2} \Rightarrow c = \sqrt{(5)^2 - (2)^2} = \sqrt{21}$$

Characteristics
- the foci have coordinates $(0, \pm\sqrt{21})$
- the vertices have coordinates $(0, \pm 5)$
- the center has coordinates $(0, 0)$
- length of major axis $= 2(5) = 10$
- length of minor axis $= 2(2) = 4$
- eccentricity $= \sqrt{21}/5$

□

B. $9x^2 + 16y^2 = 9$

C. $(x+4)^2/9+(y-2)^2=1$

Solution

Notes

- The ellipse with given equation is a shift left 4 units and up 2 units of the ellipse with the equation $x^2/9+y^2=1$.
- $x^2/9+y^2=1$
  $\Leftrightarrow x^2/(3^2)+y^2/(1^2)=1$
  $\Leftrightarrow x^2/a^2+y^2/b^2=1$
  where $a=3$ and $b=1$

Value of c

For $a=3$ and $b=1$:

$$c=\sqrt{a^2-b^2} \Rightarrow c=\sqrt{9-1} = \sqrt{8} = 2\sqrt{2}$$

Characteristics of Ellipse to Translate

- the foci have coordinates $(\pm 2\sqrt{2},0)$
- the vertices have coordinates $(\pm 3,0)$
- the center has coordinates $(0,0)$

Translations to Ellipse in Question

- $(2\sqrt{2},0)\mapsto(2\sqrt{2}-4,0+2)$
- $(-2\sqrt{2},0)\mapsto(-2\sqrt{2}-4,0+2)$
- $(3,0)\mapsto(3-4,0+2) = (-1,2)$
- $(-3,0)\mapsto(-3-4,0+2) = (-7,2)$
- $(0,0)\mapsto(0-4,0+2) = (-4,2)$

Characteristics of Ellipse in Question

- the foci have coordinates
  $(-4+2\sqrt{2},2)\approx(-1.17,2)$ and
  $(-4-2\sqrt{2},2)\approx(-6.83,2)$
- the vertices have coordinates $(-1,2)$ and $(-7,2)$
- the center has coordinates $(-4,2)$
- length of major axis $=2(3)=6$
- length of minor axis $=2(1)=2$
- eccentricity $=2\sqrt{2}/3$

□

D. $(x-2)^2/25+(y-3)^2/36=1$

Example 5

Determine the equation of the ellipse with the given characteristics.

A. foci: $(-3,-8)$ and $(7,-8)$;
vertices: $(-11,-8)$ and $(15,-8)$

Solution

Note

Since the foci of the ellipse lie on the same horizontal line, its major axis will be horizontal and its equation will take the form $(x-h)^2/a^2+(y-k)^2/b^2=1$ for some $a,b,h,k \in \mathbf{R}$.

Value of a

$$\begin{aligned} 2a &= \text{length of major axis} \\ \Rightarrow a &= (\text{length of major axis})/2 \\ &= (15-(-11))/2 \\ &= 13 \end{aligned}$$

Value of c

$$\begin{aligned} 2c &= \text{distance between the foci} \\ \Rightarrow c &= (\text{distance between the foci})/2 \\ &= (7-(-3))/2 \\ &= 5 \end{aligned}$$

Value of b

For $a=13$ and $c=5$:

$$\begin{aligned} b=\sqrt{a^2-c^2} \Rightarrow b &= \sqrt{(13)^2-(5)^2} \\ &= \sqrt{169-25} \\ &= \sqrt{144} \\ &= 12 \end{aligned}$$

Values of h and k

$$\begin{aligned} (h,k) &= \text{coordinates of center of ellipse} \\ &= \text{midpoint of the line segment that joins the foci} \\ &= ((-3+7)/2,(-8+(-8))/2) \\ &= (4/2,-16/2) \\ &= (2,-8) \end{aligned}$$

Equation

$$(x-2)^2/(13)^2+(y-(-8))^2/(12)^2=1$$
$$\Leftrightarrow (x-2)^2/169+(y+8)^2/144=1$$

□

B. endpoints of major axis: $(1,-12)$, $(1,18)$;
endpoints of minor axis: $(-4,3)$, $(6,3)$

C. vertices lie on the y-axis;
center lies at the origin;
point $(1, 3)$ lies on the ellipse;
length of the major axis is three times the length of the minor axis

SOLUTION

Notes

- Since the vertices of the ellipse lie on the y-axis, the major axis of the ellipse will be vertical and its equation will take the form $(x-h)^2/b^2 + (y-k)^2/a^2 = 1$ for some $a, b, h, k \in \mathbf{R}$.
- $2a =$ length of major axis
- $2b =$ length of minor axis

Values of h and k

$$\begin{aligned}(h, k) &= \text{coordinates of center of ellipse}\\ &= (0, 0)\end{aligned}$$

Substitution

$$\begin{aligned}2a = 3(2b) &\Rightarrow \cancel{2}a/\cancel{2} = 3(\cancel{2}b)/\cancel{2}\\ &\Rightarrow a = 3b\end{aligned}$$

Value of b^2

the point $(1, 3)$ lies on the ellipse

$$\begin{aligned}&\Rightarrow (1-h)^2/b^2 + (3-k)^2/a^2 = 1\\ &\Rightarrow (1-0)^2/b^2 + (3-0)^2/(3b)^2 = 1\\ &\Rightarrow 1/b^2 + 9/(9b^2) = 1\\ &\Rightarrow 1/b^2 + 1/b^2 = 1\\ &\Rightarrow 2/b^2 = 1\\ &\Rightarrow (2/b^2)(b^2) = (1)(b^2)\\ &\Rightarrow 2 = b^2\end{aligned}$$

Value of a^2

$$\begin{aligned}a = 3b \Rightarrow a^2 &= (3b)^2\\ &= 9b^2\\ &= 9(2)\\ &= 18\end{aligned}$$

Equation

$$\begin{aligned}&(x-0)^2/2 + (y-0)^2/18 = 1\\ &\Leftrightarrow x^2/2 + y^2/18 = 1\end{aligned}$$

□

D. center lies at the origin;
eccentricity is $2/3$;
one vertex has coordinates $(-9, 0)$

EXAMPLE 6
Draw the ellipse with the given characteristics.

A. center has coordinates $(0, 0)$;
foci lie on the y-axis;
length of major axis $= 10$;
length of minor axis $= 4$

SOLUTION

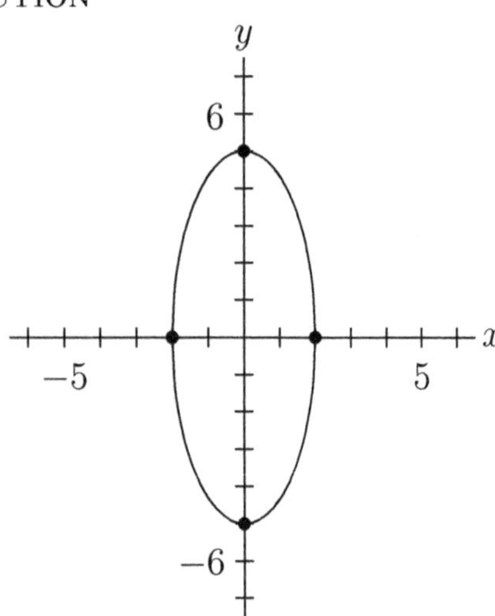

☐

B. center has coordinates $(0, 0)$;
foci lie on the x-axis;
length of major axis $= 9$;
length of minor axis $= 6$

C. center has coordinates $(-4, 2)$;
foci lie on the same horizontal line;
length of major axis $= 6$;
length of minor axis $= 3$

SOLUTION

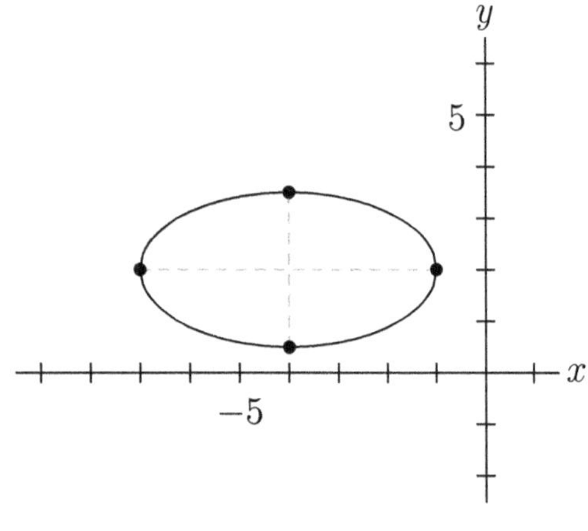

☐

D. center has coordinates $(2, -5)$;
foci lie on the same vertical line;
length of major axis $= 14$;
length of minor axis $= 8$

Section 138: Hyperbolas

Definitions

- A *hyperbola* is a locus of points in a plane, each with the same difference of its distances from two fixed points (called the *foci*).
- The *center* of a hyperbola is the midpoint of the line segment that joins its foci.
- The *focal axis* of a hyperbola is the line that passes through its foci.
- The *vertices* of a hyperbola are the points at which it intersects its focal axis.
- The *transverse axis* of a hyperbola is the line segment that joins its vertices.
- The *conjugate axis* of a hyperbola is the line segment that runs perpendicular to its transverse axis, has midpoint at its center, and has length equal to the square root of the difference of the square of the distance between its foci and the square of the length of its transverse axis.
- The *eccentricity* of a hyperbola is the ratio of the distance between its foci to the length of its transverse axis.

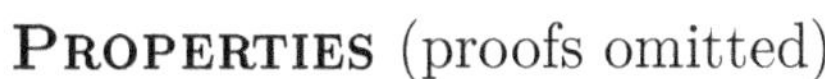

Properties (proofs omitted)

The hyperbola with foci $(-c, 0)$ and $(c, 0)$ has
- center $(0, 0)$
- vertices $(-a, 0)$ and $(a, 0)$
- transverse axis of length $2a$
- conjugate axis of length $2b$
- eccentricity of magnitude c/a

where a equals half the difference of the distances from the foci to a point on the hyperbola and b equals $\sqrt{c^2 - a^2}$.

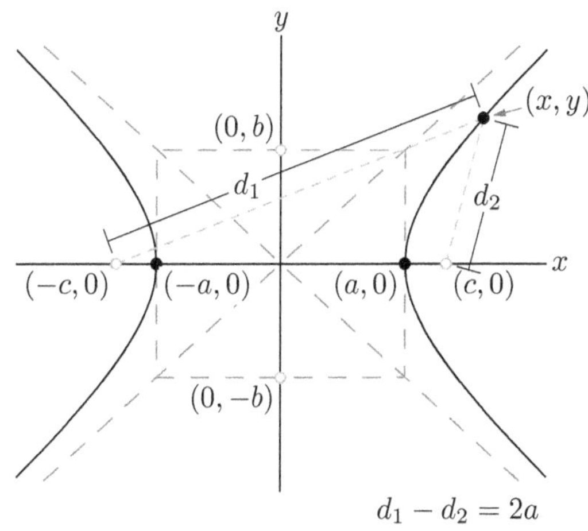

$d_1 - d_2 = 2a$

Equation (derivation omitted)

The *standard form* of the equation of the hyperbola with foci at coordinates $(-c, 0)$ and $(c, 0)$ is $x^2/a^2 - y^2/b^2 = 1$,

where a equals half the difference of the distances from the foci to a point on the hyperbola and b equals $\sqrt{c^2 - a^2}$.

Properties (proofs omitted)

The hyperbola with foci $(0, -c)$ and $(0, c)$ has
- center $(0, 0)$
- vertices $(0, -a)$ and $(0, a)$
- transverse axis of length $2a$
- conjugate axis of length $2b$
- eccentricity of magnitude c/a

where a equals half the difference of the distances from the foci to a point on the hyperbola and b equals $\sqrt{c^2 - a^2}$.

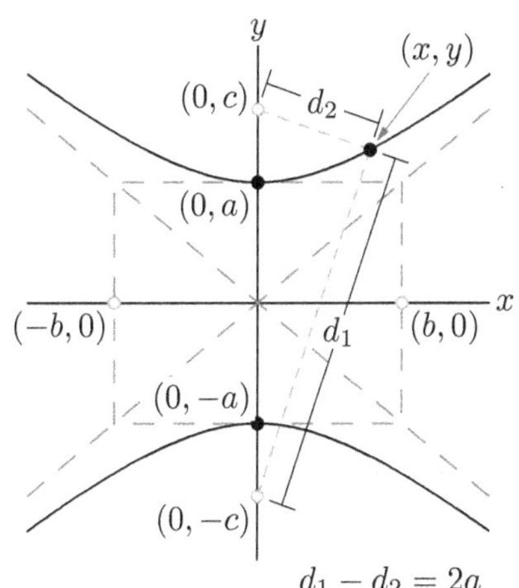

$d_1 - d_2 = 2a$

EQUATION (derivation omitted)
The *standard form* of the equation of the hyperbola with foci at coordinates $(0, -c)$ and $(0, c)$ is $y^2/a^2 - x^2/b^2 = 1$,
where a equals half the difference of the distances from the foci to a point on the hyperbola and b equals $\sqrt{c^2 - a^2}$.

PROCEDURE
To draw the graph of $x^2/a^2 - y^2/b^2 = 1$ or $y^2/a^2 - x^2/b^2 = 1$, we:

1. plot the vertices of the hyperbola;
2. draw a rectangle of dimensions $2a$ by $2b$ whose edges of length $2b$ run parallel to the conjugate axis of the hyperbola and have midpoints at the vertices of the hyperbola;
3. draw dashed lines through opposite corners of the rectangle of step (2);
4. draw a hyperbola through its vertices using the dashed lines of step (3) as asymptotes.

COMMENTS

- Proofs of the properties and derivations of the equations for the hyperbola follow closely the corresponding proofs and derivations for the ellipse.
- The transverse axis of the hyperbola corresponds loosely to the major axis of the ellipse and the conjugate axis of the hyperbola corresponds loosely to the minor axis of the ellipse.
- The graphs of the equations $(x-h)^2/a^2 - (y-k)^2/b^2 = 1$ and $(y-k)^2/a^2 - (x-h)^2/b^2 = 1$ (as translations of the graphs of the equations $x^2/a^2 - y^2/b^2 = 1$ and $y^2/a^2 - x^2/b^2 = 1$, respectively) are also hyperbolas.
- The dashed lines of step (3) of the procedure above will have the equations:
  - $y = (b/a)x$ and $y = -(b/a)x$ if the foci lie on the x-axis;
  - $y = (a/b)x$ and $y = -(a/b)x$ if the foci lie on the y-axis.
- $b = \sqrt{c^2 - a^2} \Leftrightarrow a = \sqrt{c^2 - b^2} \Leftrightarrow c = \sqrt{a^2 + b^2}$ (since a, b, and c are positive as defined)

EXAMPLE 1
Rewrite the equation of each hyperbola in either the form $(x-h)^2/a^2 - (y-k)^2/b^2 = 1$ or the form $(y-k)^2/a^2 - (x-h)^2/b^2 = 1$.

A. $4x^2 + 16x - 5y^2 - 4 = 0$

SOLUTION
$4x^2 + 16x - 5y^2 - 4 = 0$
$\Leftrightarrow 4(x^2 + 4x) - 5y^2 = 0 + 4$
$\Leftrightarrow 4(x^2 + 4x + (2)^2 - (2)^2) - 5y^2 = 4$
$\Leftrightarrow 4((x+2)^2 - 4) - 5y^2 = 4$
$\Leftrightarrow 4(x+2)^2 - 16 - 5y^2 = 4$
$\Leftrightarrow 4(x+2)^2 - 5y^2 = 4 + 16$
$\Leftrightarrow [4(x+2)^2 - 5y^2]/20 = 20/20$
$\Leftrightarrow 4(x+2)^2/20 - 5y^2/20 = 1$
$\Leftrightarrow (x+2)^2/5 - y^2/4 = 1$
$\Leftrightarrow (x+2)^2/(\sqrt{5})^2 - y^2/(2)^2 = 1$
□

B. $12y^2 - 24y - x^2 - 8x - 7 = 0$

EXAMPLE 2
Find the x-coordinates of all points on the graph of the equation $x^2/27 - y^2/8 = 1$ with a y-coordinate of 4.

SOLUTION

$$\begin{aligned} y = 4 &\Rightarrow x^2/27 - (4)^2/8 = 1 \\ &\Rightarrow x^2/27 = 1 + (4)^2/8 \\ &\qquad\qquad = 3 \\ &\Rightarrow x^2 = 3(27) \\ &\Rightarrow x = \pm\sqrt{3(27)} \\ &\quad x = \pm\sqrt{81} \\ &\quad\; = \pm 9 \end{aligned}$$

The x-coordinates are -9 and 9.

□

EXAMPLE 3
Find the y-intercepts of the graph of the equation $(y-3)^2/13 - (x+6)^2/3 = 1$.

EXAMPLE 4
Identify the foci, vertices, center, eccentricity, and lengths of the transverse and conjugate axes of the hyperbola with the given equation.

A. $-49x^2 + 9y^2 = 441$

SOLUTION

Note

$$\begin{aligned} &-49x^2 + 9y^2 = 441 \\ &\Leftrightarrow (-49x^2 + 9y^2)/441 = 441/441 \\ &\Leftrightarrow -49x^2/441 + 9y^2/441 = 1 \\ &\Leftrightarrow -x^2/9 + y^2/49 = 1 \\ &\Leftrightarrow y^2/(7^2) - x^2/(3^2) = 1 \\ &\Leftrightarrow y^2/a^2 - x^2/b^2 = 1 \\ &\qquad \text{where } a = 7 \text{ and } b = 3 \end{aligned}$$

Value of c

For $a = 7$ and $b = 3$:

$$\begin{aligned} c = \sqrt{a^2 + b^2} \Rightarrow c &= \sqrt{(7)^2 + (3)^2} \\ &= \sqrt{58} \end{aligned}$$

Characteristics

- the foci have coordinates $(0, \pm\sqrt{58})$
- the vertices have coordinates $(0, \pm 7)$
- the center has coordinates $(0, 0)$
- length of transverse axis $= 2(7) = 14$
- length of conjugate axis $= 2(3) = 6$
- eccentricity $= \sqrt{58}/7$

□

B. $4x^2 - 25y^2 = 100$

C. $(x+4)^2/36 - (y+1)^2/25 = 1$

SOLUTION

Notes

- The hyperbola with given equation is a shift left 4 units and down 1 unit of the hyperbola $x^2/36 - y^2/25 = 1$.
- $x^2/36 - y^2/25 = 1$
  $\Leftrightarrow x^2/(6^2) - y^2/(5^2) = 1$
  $\Leftrightarrow x^2/a^2 - y^2/b^2 = 1$
  where $a = 6$ and $b = 5$

Value of c

For $a = 6$ and $b = 5$:

$$c = \sqrt{a^2+b^2} \Rightarrow c = \sqrt{(6)^2+(5)^2} = \sqrt{61}$$

Characteristics of Hyperbola to Translate

- the foci have coordinates $(\pm\sqrt{61}, 0)$
- the vertices have coordinates $(\pm 6, 0)$
- the center has coordinates $(0, 0)$

Translations to Hyperbola in Question

- $(\sqrt{61}, 0) \mapsto (\sqrt{61} - 4, 0 - 1) = (-4 + \sqrt{61}, -1)$
- $(-\sqrt{61}, 0) \mapsto (-\sqrt{61} - 4, 0 - 1) = (-4 - \sqrt{61}, -1)$
- $(6, 0) \mapsto (6 - 4, 0 - 1) = (2, -1)$
- $(-6, 0) \mapsto (-6 - 4, 0 - 1) = (-10, -1)$
- $(0, 0) \mapsto (0 - 4, 0 - 1) = (-4, -1)$

Characteristics of Hyperbola in Question

- the foci have coordinates
  $(-4+\sqrt{61}, -1) \approx (3.81, -1)$ and
  $(-4-\sqrt{61}, -1) \approx (-11.81, -1)$
- the vertices have coordinates $(2, -1)$ and $(-10, -1)$
- the center has coordinates $(-4, -1)$
- length of transverse axis $= 2(6) = 12$
- length of conjugate axis $= 2(5) = 10$
- eccentricity $= \sqrt{61}/6$

□

D. $(y+3)^2/4 - (x-5)^2 = 1$

Example 5

Determine the equation of the hyperbola with the given characteristics.

A. vertices lie on the same vertical line;
y-coordinates of vertices: 1 and 13;
asymptotes: $y = 2x + 11$, $y = -2x + 3$

Solution

Note

Since the vertices of the hyperbola lie on the same vertical line, its transverse axis will be vertical and its equation will take the form $(y-k)^2/a^2 - (x-h)^2/b^2 = 1$ for some $a, b, h, k \in \mathbf{R}$.

Value of a

$$\begin{aligned} 2a &= \text{length of transverse axis} \\ \Rightarrow a &= (\text{length of transverse axis})/2 \\ &= (13-1)/2 \\ &= 6 \end{aligned}$$

Value of b

the slopes of the asymptotes are ± 2

$$\begin{aligned} &\Rightarrow a/b = 2 \\ &\Rightarrow 6/b = 2 \\ &\Rightarrow 6 = 2b \\ &\Rightarrow 6/2 = 2b/2 \\ &\Rightarrow 3 = b \end{aligned}$$

Values of h and k

the asymptotes intersect at center (h, k)

$$\begin{aligned} &\Rightarrow 2h + 11 = -2h + 3 \\ &\Rightarrow 2h + 2h = 3 - 11 \\ &\Rightarrow 4h = -8 \\ &\Rightarrow 4h/4 = -8/4 \\ &\Rightarrow h = -2 \\ &\Rightarrow k = 2(-2) + 11 \\ & = 7 \end{aligned}$$

Equation

$$\begin{aligned} &(y-7)^2/(6)^2 - (x-(-2))^2/(3)^2 = 1 \\ &\Leftrightarrow (y-7)^2/36 - (x+2)^2/9 = 1 \end{aligned}$$

□

B. center: $(3, 1)$;
length of transverse axis: 4;
length of conjugate axis: 18;
foci lie on the same horizontal line

C. foci: $(-14, -1)$ and $(6, -1)$;
eccentricity: $5/2$

SOLUTION

Note

The foci of the hyperbola lie on the same horizontal line, so its transverse axis will be horizontal and its equation will take the form $(x-h)^2/a^2 - (y-k)^2/b^2 = 1$ for some $a, b, h, k \in \mathbf{R}$.

Value of c

$2c =$ distance between the foci

$$\begin{aligned} \Rightarrow c &= (\text{distance between the foci})/2 \\ &= (6 - (-14))/2 \\ &= 10 \end{aligned}$$

Value of a

eccentricity $= c/a$

$$\begin{aligned} &\Rightarrow 5/2 = 10/a \\ &\Rightarrow (5/2)(a) = 10 \\ &\Rightarrow (2/5)(5/2)(a) = (2/5)(10) \\ &\Rightarrow a = 4 \end{aligned}$$

Value of b

For $a = 4$ and $c = 10$:

$$\begin{aligned} b = \sqrt{c^2 - a^2} \Rightarrow b &= \sqrt{(10)^2 - (4)^2} \\ &= \sqrt{100 - 16} \\ &= \sqrt{84} \end{aligned}$$

Values of h and k

$$\begin{aligned} (h, k) &= \text{coordinates of center of hyperbola} \\ &= \text{midpoint of the line segment that joins the foci} \\ &= ((-14+6)/2, (-1+(-1))/2) \\ &= (-8/2, -2/2) \\ &= (-4, -1) \end{aligned}$$

Equation

$$(x - (-4))^2/4^2 - (y - (-1))^2/(\sqrt{84})^2 = 1$$
$$\Leftrightarrow (x+4)^2/16 - (y+1)^2/84 = 1$$

□

D. vertices: $(-3, 1)$ and $(-3, 5)$;
hyperbola passes through $(5, 9)$

EXAMPLE 6
Draw the hyperbola with the given characteristics.

A. center has coordinates $(0,0)$;
foci lie on the y-axis;
length of transverse axis $= 10$;
length of conjugate axis $= 6$

SOLUTION

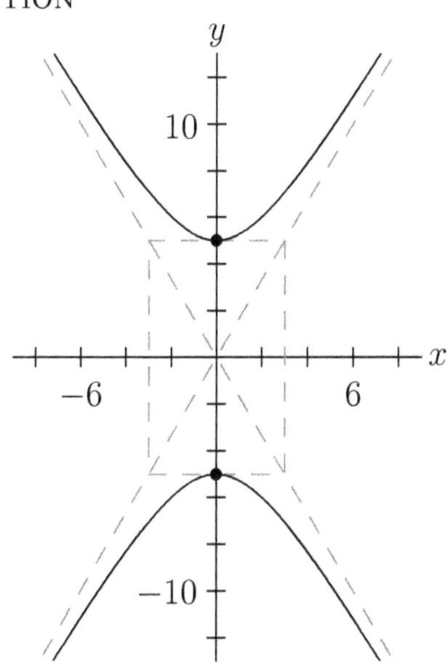

□

B. center has coordinates $(0,0)$;
foci lie on the x-axis;
length of transverse axis $= 12$;
length of conjugate axis $= 8$

C. center has coordinates $(1,3)$;
foci lie on the same horizontal line;
length of transverse axis $= 8$;
length of conjugate axis $= 5$

SOLUTION

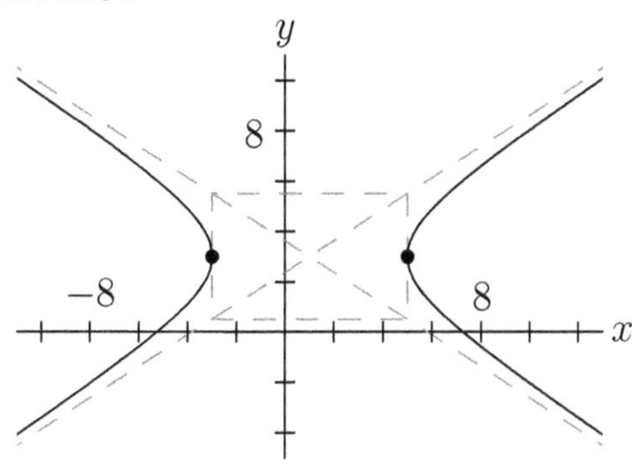

□

D. center has coordinates $(-3,-3)$;
foci lie on the same vertical line;
length of transverse axis $= 9$;
length of conjugate axis $= 4$

Section 139: Parametric Equations

DEFINITIONS

- A *parametric equation* is an equation that equates a variable to an expression with a parameter.
- To *eliminate a parameter* from a set of parametric equations means to determine a single equation that does not contain the parameter but relates all variables defined within the set in the same way that the parametric equations relate them.

COMMENTS

- A set of parametric equations with parameter t is a set of equations of the form $q_i = f_i(t)$ for some variables q_i and functions f_i.
- The graph of a set of parametric equations $\{x = f(t),\ y = g(t)\}$ is the locus of all points in the set $\{\,(x, y) \in \mathbf{R}^2 \mid x = f(t) \text{ AND } y = g(t) \text{ for some } t \in \mathbf{R}\,\}$. Each value of the parameter t will generate a point (x, y) on the graph.
- The graph of a set of parametric equations is identical to the graph of the equation yielded by eliminating the parameter from the set.
- Though we can use any variable as a parameter, we conventionally use t.

EXAMPLE 1

The set of equations $\{x = 5t^4,\ y = -2t,\ z = \sqrt{t}\}$ constitutes a set of parametric equations with parameter t.

□

EXAMPLE 2

For each set of parametric equations, determine the point (x, y) on its graph that corresponds to the given value of the parameter.

A. $\{x = r^3,\ y = r^2 + 5\}$; $r = 2$

SOLUTION

For $r = 2$:

$$\begin{aligned}(x, y) &= ((2)^3, (2)^2 + 5)\\ &= (8, 9)\end{aligned}$$

□

B. $\{x = \cos 2m,\ y = \sin 3m\}$; $m = \pi/4$

EXAMPLE 3

Eliminate the parameter from each set of parametric equations.

A. $\{x = t + 2,\ y = t^2 - 1\}$

SOLUTION

$$\begin{aligned}x = t + 2 &\Rightarrow x - 2 = t\\ &\Rightarrow y = (x - 2)^2 - 1\end{aligned}$$

□

B. $\{x = 2t,\ y = t + 3\}$

C. $\{x = 3\sin t,\ y = 2\cos t\}$

SOLUTION

Notes

- $x = 3\sin t \Leftrightarrow x/3 = \sin t$
- $y = 2\cos t \Leftrightarrow y/2 = \cos t$

Equation

$\sin^2 t + \cos^2 t = 1$

$\Rightarrow (x/3)^2 + (y/2)^2 = 1$

$\Rightarrow x^2/9 + y^2/4 = 1$

□

D. $\{x = 5\cos t - 4,\ y = 5\sin t - 1\}$

EXAMPLE 4

Draw the graph of each set of parametric equations.

A. $\{x = t + 2,\ y = t^2 - 1\}$

SOLUTION

Table of Values

t	x	y
-3	$-3+2=-1$	$(-3)^2-1=\ \ 8$
-2	$-2+2=\ \ 0$	$(-2)^2-1=\ \ 3$
-1	$-1+2=\ \ 1$	$(-1)^2-1=\ \ 0$
0	$0+2=\ \ 2$	$(0)^2-1=-1$
1	$1+2=\ \ 3$	$(1)^2-1=\ \ 0$
2	$2+2=\ \ 4$	$(2)^2-1=\ \ 3$
3	$3+2=\ \ 5$	$(3)^2-1=\ \ 8$

Graph

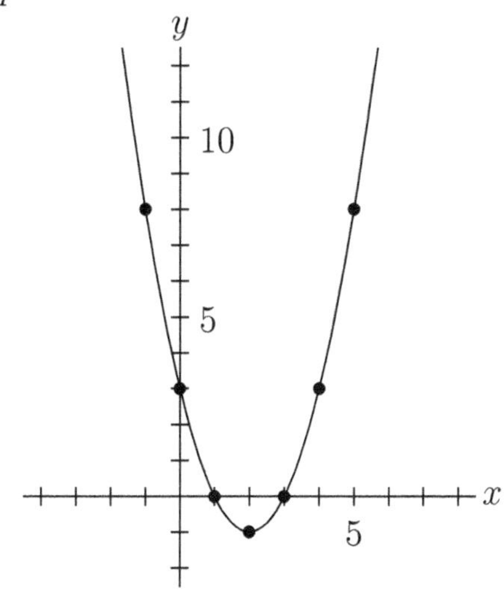

□

B. $\{x = 2t,\ y = t + 3\}$

C. $\{x = 3\sin t,\ y = 2\cos t\}$

Solution

Table of Values

t	x	y
0	$3\sin 0 = 0$	$2\cos 0 = 2$
$\frac{\pi}{12}$	$3\sin\frac{\pi}{12} \approx 0.78$	$2\cos\frac{\pi}{12} \approx 1.93$
$\frac{\pi}{6}$	$3\sin\frac{\pi}{6} = 1.5$	$2\cos\frac{\pi}{6} \approx 1.73$
$\frac{\pi}{4}$	$3\sin\frac{\pi}{4} \approx 2.12$	$2\cos\frac{\pi}{4} \approx 1.41$
$\frac{\pi}{3}$	$3\sin\frac{\pi}{3} \approx 2.60$	$2\cos\frac{\pi}{3} = 1$
$\frac{5\pi}{12}$	$3\sin\frac{5\pi}{12} \approx 2.90$	$2\cos\frac{5\pi}{12} \approx 0.52$
$\frac{\pi}{2}$	$3\sin\frac{\pi}{2} = 3$	$2\cos\frac{\pi}{2} = 0$
$\frac{7\pi}{12}$	$3\sin\frac{7\pi}{12} \approx 2.90$	$2\cos\frac{7\pi}{12} \approx -0.52$
$\frac{2\pi}{3}$	$3\sin\frac{2\pi}{3} \approx 2.60$	$2\cos\frac{2\pi}{3} = -1$
$\frac{3\pi}{4}$	$3\sin\frac{3\pi}{4} \approx 2.12$	$2\cos\frac{3\pi}{4} \approx -1.41$
$\frac{5\pi}{6}$	$3\sin\frac{5\pi}{6} = 1.5$	$2\cos\frac{5\pi}{6} \approx -1.73$
$\frac{11\pi}{12}$	$3\sin\frac{11\pi}{12} \approx 0.78$	$2\cos\frac{11\pi}{12} \approx -1.93$
π	$3\sin\pi = 0$	$2\cos\pi = -2$
$\frac{13\pi}{12}$	$3\sin\frac{13\pi}{12} \approx -0.78$	$2\cos\frac{13\pi}{12} \approx -1.93$
$\frac{7\pi}{6}$	$3\sin\frac{7\pi}{6} = -1.5$	$2\cos\frac{7\pi}{6} \approx -1.73$
$\frac{5\pi}{4}$	$3\sin\frac{5\pi}{4} \approx -2.12$	$2\cos\frac{5\pi}{4} \approx -1.41$
$\frac{4\pi}{3}$	$3\sin\frac{4\pi}{3} \approx -2.60$	$2\cos\frac{4\pi}{3} = -1$
$\frac{17\pi}{12}$	$3\sin\frac{17\pi}{12} \approx -2.90$	$2\cos\frac{17\pi}{12} \approx -0.52$
$\frac{3\pi}{2}$	$3\sin\frac{3\pi}{2} = -3$	$2\cos\frac{3\pi}{2} = 0$
$\frac{19\pi}{12}$	$3\sin\frac{19\pi}{12} \approx -2.90$	$2\cos\frac{19\pi}{12} \approx 0.52$
$\frac{5\pi}{3}$	$3\sin\frac{5\pi}{3} \approx -2.60$	$2\cos\frac{5\pi}{3} = 1$
$\frac{7\pi}{4}$	$3\sin\frac{7\pi}{4} \approx -2.12$	$2\cos\frac{7\pi}{4} \approx 1.41$
$\frac{11\pi}{6}$	$3\sin\frac{11\pi}{6} = -1.5$	$2\cos\frac{11\pi}{6} \approx 1.73$
$\frac{23\pi}{12}$	$3\sin\frac{23\pi}{12} \approx -0.78$	$2\cos\frac{23\pi}{12} \approx 1.93$
2π	$3\sin 2\pi = 0$	$2\cos 2\pi = 2$

Graph

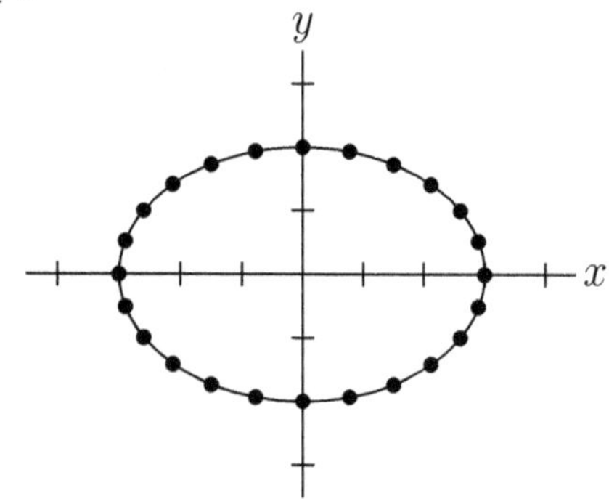

□

D. $\{x = 5\cos t - 4,\ y = 5\sin t - 1\}$

Section 140: Polar Coordinates

Definitions

- The *rectangular coordinate system* is a system in which the location of a point in the plane is given as a pair of coordinates that bear reference to a set of axes.
- The *polar coordinate system* is a system in which the location of a point in the plane is given as a pair of coordinates that bear reference to a ray called the *polar axis*.
- The *pole* is the endpoint of the polar axis.

Procedure

To depict points with given polar coordinates in a plane with given polar axis, we:

1. draw circles with centers at the pole and radii at even increments;
2. draw rays with endpoints at the pole that form familiar angles with the polar axis;
3. identify for each point with polar coordinates (r, θ) its proper location in the plane, where the point should lie $|r|$ units from the pole on the ray that forms an angle:
   - of measure θ with the polar axis if $r \geq 0$, or
   - of measure $\theta + \pi$ with the polar axis if $r < 0$;
4. plot each point in the plane in reference to the circles and rays of steps (1) and (2).

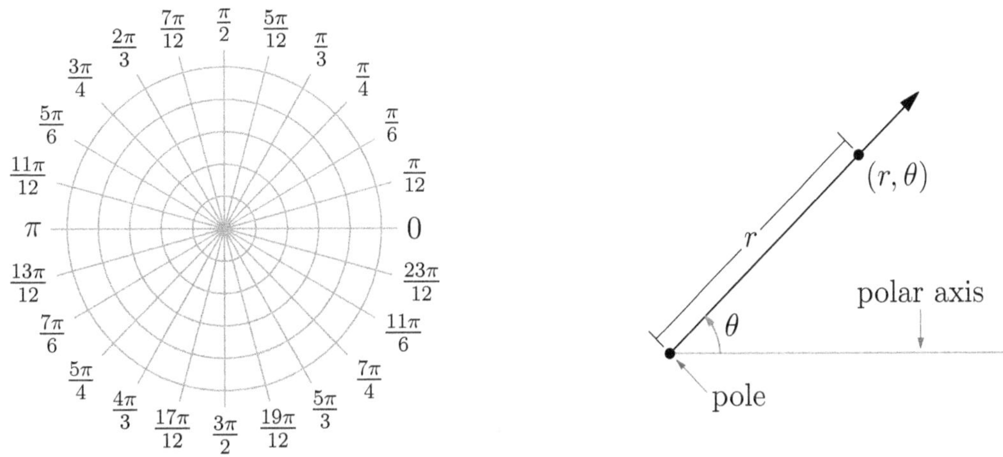

Comments

- The point with polar coordinates $(0, \theta)$ lies at the pole for all measures θ.
- We can indicate locations of points in the polar coordinate system in an infinite number of ways since the ray that forms an angle of measure θ with the polar axis will coincide with any ray that forms an angle of measure $\theta + 2\pi k$ with the polar axis for some integer k.
- Unless stated otherwise, we will assume that the polar axis is the ray with endpoint at the origin that passes through the point with rectangular coordinates $(1, 0)$.
- Equations can relate polar coordinates r and θ just as they can relate rectangular coordinates x and y. The graph of such a polar equation will be the locus of all points whose polar coordinates correspond to solutions (r, θ) of that equation.

PROPERTIES

Suppose a point has rectangular coordinates (x, y) and polar coordinates (r, θ).
Then:

- $x = r\cos\theta$
- $y = r\sin\theta$
- $\tan\theta = y/x$ (provided $x \neq 0$)
- $r^2 = x^2 + y^2$

PROOFS

Suppose a point has rectangular coordinates (x, y) and polar coordinates (r, θ).
Note that the point must then also have the polar coordinates $(-r, \theta + \pi)$ (see case II).

- We prove $x = r\cos\theta$ for the cases $r > 0$, $r < 0$, and $r = 0$ separately:

I: $r > 0 \Rightarrow |x|/r = |\cos\theta|/1$ (see figure below)
$\Rightarrow |x|/r = |\cos\theta|$
$\Rightarrow |x| = r|\cos\theta|$
$\Rightarrow x = r\cos\theta$ OR $-x = r(-\cos\theta)$ (since x and $\cos\theta$ have the same sign)
$\Rightarrow x = r\cos\theta$

II: $r < 0 \Rightarrow -r > 0$
$\Rightarrow x = -r\cos(\theta + \pi)$ (by case I and the note above)
$\Rightarrow x = (-r)(-\cos\theta)$
$\Rightarrow x = r\cos\theta$

III: $r = 0 \Rightarrow x = 0$
$\Rightarrow x = 0 \cdot \cos\theta$
$\Rightarrow x = r\cos\theta$

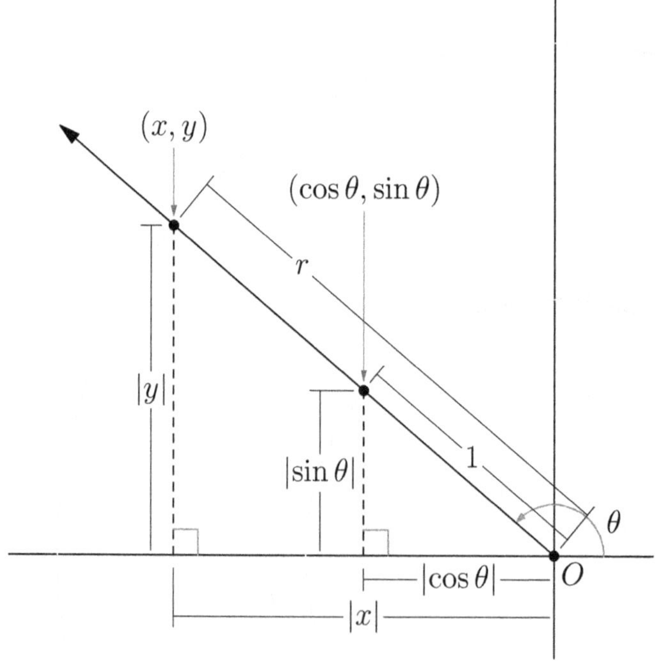

- We omit the proof that $y = r\sin\theta$ (it resembles that for $x = r\cos\theta$).

- $x = r\cos\theta$ AND $y = r\sin\theta$
$\Rightarrow y/x = (\not{r}\sin\theta)/(\not{r}\cos\theta)$
$= (\sin\theta)/(\cos\theta)$
$= \tan\theta$

- $x^2 + y^2 = (r\cos\theta)^2 + (r\sin\theta)^2$
$= r^2\cos^2\theta + r^2\sin^2\theta$
$= r^2(\cos^2\theta + \sin^2\theta)$
$= r^2(1)$
$= r^2$

⊡

EXAMPLE 1

Plot the points with the given polar coordinates in the planes below. Assume the circles drawn have radii 1, 2, 3, 4, and 5.

A. $A(2, \pi/4)$, $B(4, 3\pi/2)$, $C(3, -4\pi/3)$

SOLUTION

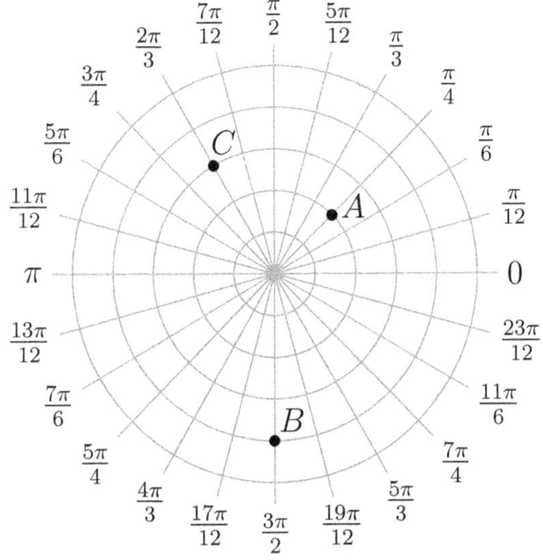

□

B. $D(1, 2\pi/3)$, $E(0, 11\pi/6)$, $F(-5, 5\pi/4)$

SOLUTION

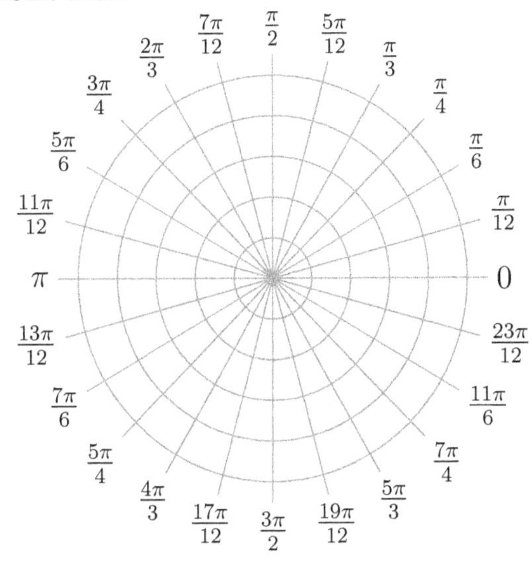

EXAMPLE 2

Find the polar coordinates (r, θ) of the point with rectangular coordinates (x, y), where r and θ satisfy the conditions $0 \le \theta < 2\pi$ and $r > 0$.

A. $(x, y) = (21, 7\sqrt{3})$

SOLUTION

Values of θ and r

- $x = 21$ AND $y = 7\sqrt{3}$ AND $0 \le \theta < 2\pi$
  $\Rightarrow \tan\theta = 7\sqrt{3}/21$ AND $0 < \theta < \pi/2$
  $\Rightarrow \tan\theta = \sqrt{3}/3$ AND $0 < \theta < \pi/2$
  $\Rightarrow \theta = \tan^{-1}(\sqrt{3}/3)$
  $= \pi/6$
- $x = 21$ AND $y = 7\sqrt{3}$ AND $r > 0$
  $\Rightarrow r^2 = (21)^2 + (7\sqrt{3})^2$ AND $r > 0$
  $\Rightarrow r^2 = 588$ AND $r > 0$
  $\Rightarrow r = \sqrt{588}$
  $= 14\sqrt{3}$

Pair of Coordinates

$(r, \theta) = (14\sqrt{3}, \pi/6)$

□

B. $(x, y) = (-2, -2)$

EXAMPLE 3

Find the rectangular coordinates (x, y) of the point with polar coordinates (r, θ).

A. $(r, \theta) = (3, 7\pi/6)$

SOLUTION

$r = 3$ AND $\theta = 7\pi/6$

$$\begin{aligned} \Rightarrow (x, y) &= (3\cos(7\pi/6), 3\sin(7\pi/6)) \\ &= (3(-\sqrt{3}/2), 3(-1/2)) \\ &= (-3\sqrt{3}/2, -3/2) \end{aligned}$$

□

B. $(r, \theta) = (4, 3\pi/4)$

EXAMPLE 4

Identify two other pairs (r, θ) of polar coordinates, one with $r > 0$ and one with $r < 0$, that designate the location of the point with the given polar coordinates.

A. $(3, 7\pi/6)$

SOLUTION

Notes

- $(3, 7\pi/6 + 2\pi) = (3, 7\pi/6 + 12\pi/6)$
  $= (3, 19\pi/6)$
- $(-3, 7\pi/6 - \pi) = (-3, 7\pi/6 - 6\pi/6)$
  $= (-3, \pi/6)$

Conclusion

The pairs $(3, 19\pi/6)$ and $(-3, \pi/6)$ designate the same location as $(3, 7\pi/6)$.

□

B. $B(4, 3\pi/4)$

EXAMPLE 5

Convert the given equation from rectangular form to polar form. Leave your answers in the form $r = f(\theta)$ or $r^2 = f(\theta)$ for some function f.

A. $x^2 + y^2 = 25$

SOLUTION

$$\begin{aligned} x^2 + y^2 = 25 &\Rightarrow r^2 = 25 \\ &\Rightarrow r = \sqrt{25} \\ &\Rightarrow r = 5 \end{aligned}$$

(The answer $r = -5$ is acceptable also, but we need not write both equations.)

□

B. $y = 3$

C. $6xy = 11$

SOLUTION

$$\begin{aligned} 6xy = 11 &\Rightarrow 6(r\cos\theta)(r\sin\theta) = 11 \\ &\Rightarrow 6r^2\cos\theta\sin\theta = 11 \\ &\Rightarrow 3r^2(2\cos\theta\sin\theta) = 11 \\ &\Rightarrow 3r^2\sin 2\theta = 11 \\ &\Rightarrow r^2 = 11/(3\sin 2\theta) \end{aligned}$$

□

D. $2x + y = 7$

EXAMPLE 6

Convert the given equation from polar form to rectangular form and describe its graph.

A. $\theta = 3\pi/4$

SOLUTION

Rectangular Form

$\theta = 3\pi/4$

$\Rightarrow \tan\theta = \tan(3\pi/4)$

$\Rightarrow y/x = -1$

$\Rightarrow (y/x)(x) = (-1)(x)$

$\Rightarrow y = -x$

Description of Graph

The graph is the line with slope -1 that passes through the origin.

□

B. $r = 5$

C. $r = 6\sin\theta$

SOLUTION

Rectangular Form

$r = 6\sin\theta$

$\Rightarrow r \cdot r = r \cdot 6\sin\theta$

$\Rightarrow r^2 = 6r\sin\theta$

$\Rightarrow x^2 + y^2 = 6y$

$\Rightarrow x^2 + y^2 - 6y = 0$

$\Rightarrow x^2 + y^2 - 6y + (-3)^2 - (-3)^2 = 0$

$\Rightarrow x^2 + (y-3)^2 - 9 = 0$

$\Rightarrow x^2 + (y-3)^2 = 9$

Description of Graph

The graph is the circle with radius 3 and center $(0, 3)$.

□

D. $r^2 = 1/(3 + \cos^2\theta)$

Section 141: Graphs of Polar Equations

Properties

- The graph of the polar equation $r = f(\theta)$ is symmetric with respect to:
  - the line $\theta = 0$ if $f(-\theta) = f(\theta)$ for all $\theta \in \mathbf{R}$;
  - the line $\theta = \pi/2$ if $f(-\theta) = -f(\theta)$ for all $\theta \in \mathbf{R}$.
- The graph of the polar equation $r^2 = f(\theta)$ is symmetric with respect to the pole.

Comments

- The graph of the polar equation $r = f(\theta)$ often will be symmetric with respect to:
  - the line $\theta = 0$ if $f(\theta)$ involves $\cos\theta$ since $\cos(-\theta) = \cos\theta$ for all $\theta \in \mathbf{R}$;
  - the line $\theta = \pi/2$ if $f(\theta)$ involves $\sin\theta$ since $\sin(-\theta) = -\sin\theta$ for all $\theta \in \mathbf{R}$.
- If $f(\theta)$ involves $\cos\theta$ or $\sin\theta$, then we often can draw the complete graph of $r = f(\theta)$ using only points (r, θ) where $\theta \in [0, 2\pi]$ (or any other interval of width 2π); the cosine and sine functions have a period of 2π, so we would retrace the graph with additional points.
- Properties of symmetry allow us to draw graphs accurately with fewer calculations of coordinates of points.

Terminology

Certain shapes of graphs of polar equations have names.

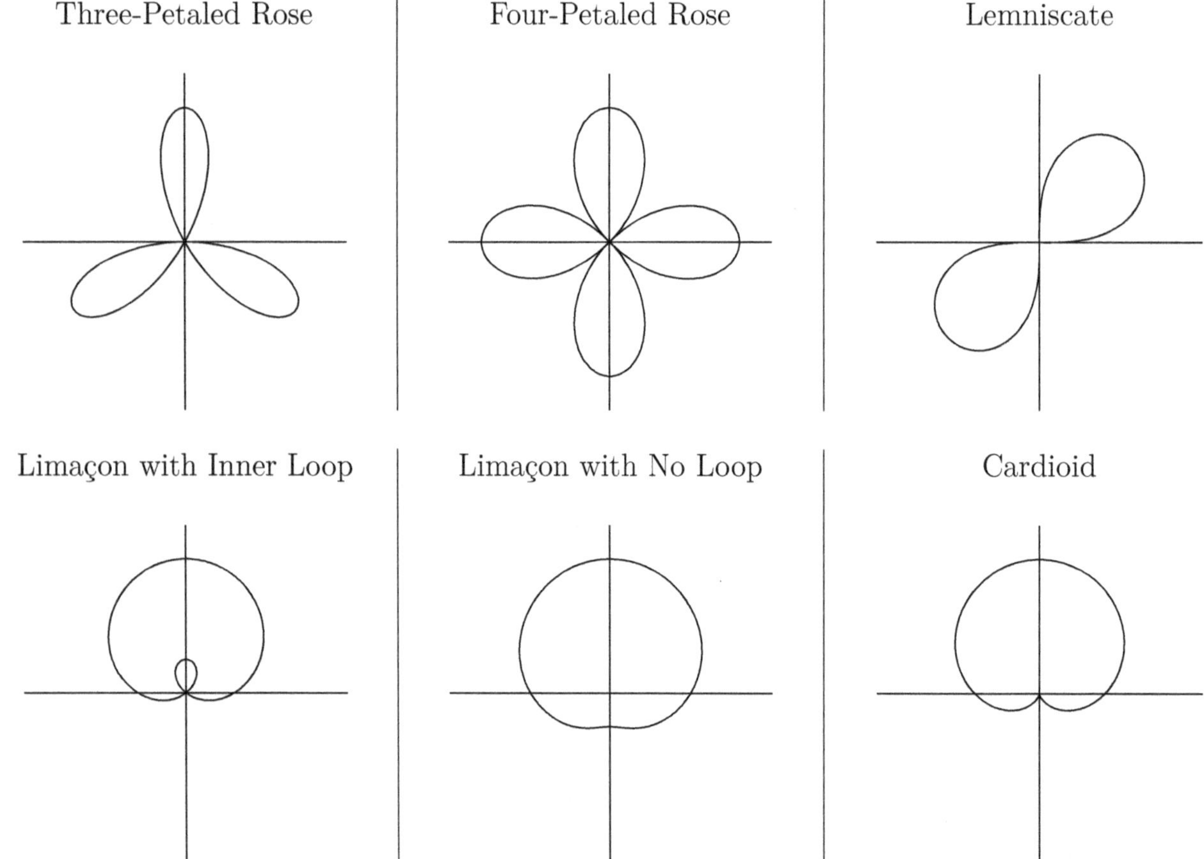

EXAMPLE 1

Draw the graph of each polar equation.

A. $r = 2\cos 3\theta$

SOLUTION

Table of Values

θ	r
0	$2\cos(3\cdot 0) = 2$
$\pi/12$	$2\cos(3\cdot \pi/12) \approx 1.41$
$\pi/6$	$2\cos(3\cdot \pi/6) = 0$
$\pi/4$	$2\cos(3\cdot \pi/4) \approx -1.41$
$\pi/3$	$2\cos(3\cdot \pi/3) = -2$
$5\pi/12$	$2\cos(3\cdot 5\pi/12) \approx -1.41$
$\pi/2$	$2\cos(3\cdot \pi/2) = 0$
$7\pi/12$	$2\cos(3\cdot 7\pi/12) \approx 1.41$
$2\pi/3$	$2\cos(3\cdot 2\pi/3) = 2$
$3\pi/4$	$2\cos(3\cdot 3\pi/4) \approx 1.41$
$5\pi/6$	$2\cos(3\cdot 5\pi/6) = 0$
$11\pi/12$	$2\cos(3\cdot 11\pi/12) \approx -1.41$
π	$2\cos(3\cdot \pi) = -2$
$13\pi/12$	$2\cos(3\cdot 13\pi/12) \approx -1.41$

Graph

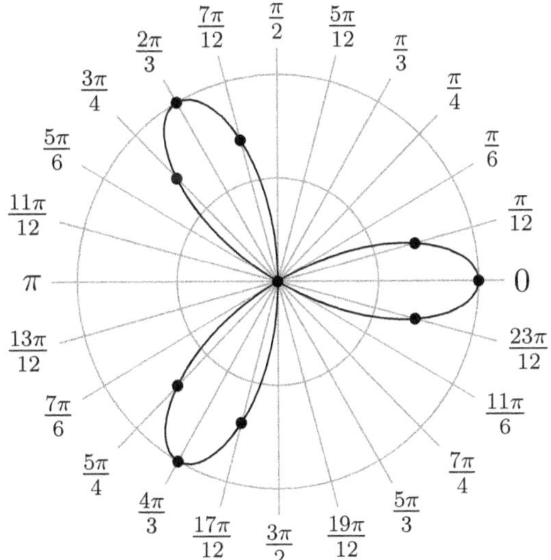

Comment

We terminated our table of values once we realized that we would gain no new points on our graph by evaluating r for further values of θ.

□

B. $r = 4\sin 2\theta$

C. $r = 3$

Solution

Note

We connect points of the form $(3, \theta)$.

Graph

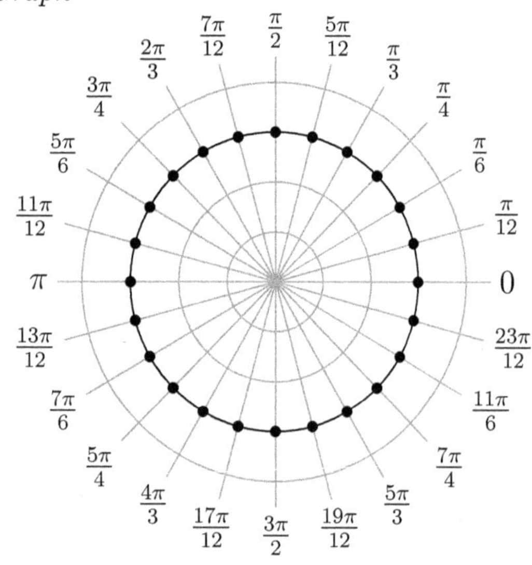

□

D. $r = 4$

E. $\theta = 7\pi/4$

Solution

Note

We connect points of the form $(r, 7\pi/4)$.

Graph

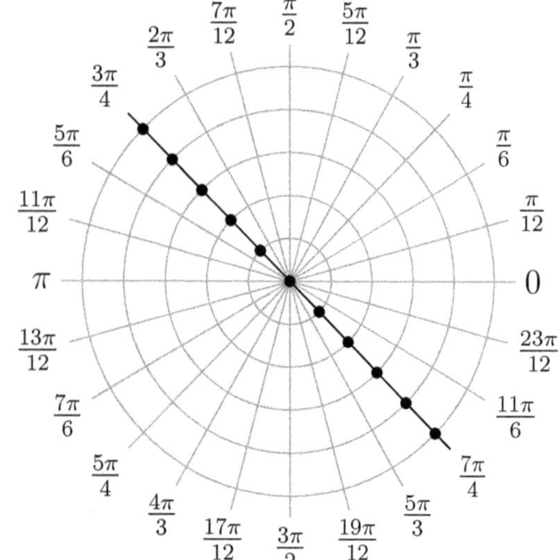

□

F. $\theta = \pi/3$

G. $r = 2 + \sin\theta$

SOLUTION

Notes

- The graph is symmetric about $\theta = \pi/2$.
- We determine points (r, θ) on the graph for which $\theta \in [-\pi/2, \pi/2]$, plot them and their reflections through the line $\theta = \pi/2$, and then connect the dots.
- The reflections will be points (r, θ) for which $\theta \in [\pi/2, 3\pi/2]$.

Table of Values

θ	r
$-\pi/2$	$2 + \sin(-\pi/2) = 1$
$-5\pi/12$	$2 + \sin(-5\pi/12) \approx 1.03$
$-\pi/3$	$2 + \sin(-\pi/3) \approx 1.13$
$-\pi/4$	$2 + \sin(-\pi/4) \approx 1.29$
$-\pi/6$	$2 + \sin(-\pi/6) = 1.5$
$-\pi/12$	$2 + \sin(-\pi/12) \approx 1.74$
0	$2 + \sin 0 = 2$
$\pi/12$	$2 + \sin(\pi/12) \approx 2.26$
$\pi/6$	$2 + \sin(\pi/6) = 2.5$
$\pi/4$	$2 + \sin(\pi/4) \approx 2.71$
$\pi/3$	$2 + \sin(\pi/3) \approx 2.87$
$5\pi/12$	$2 + \sin(5\pi/12) \approx 2.97$
$\pi/2$	$2 + \sin(\pi/2) = 3$

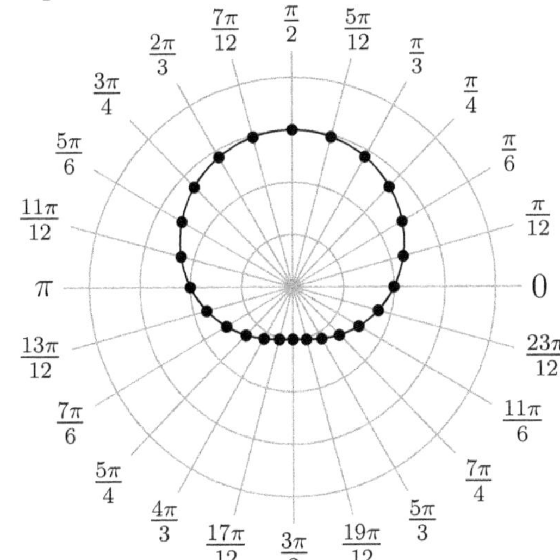

□

H. $r = 1 - 4\cos\theta$

I. $r = -2\sec\theta$

SOLUTION

Notes

- $r = -2\sec\theta \Leftrightarrow r = -2/(\cos\theta)$
- The graph is symmetric about $\theta = 0$.
- We determine points (r, θ) on the graph for which $\theta \in [0, \pi]$, plot them and their reflections through the line $\theta = 0$, and then connect the dots.
- The reflections will be points (r, θ) for which $\theta \in [\pi, 2\pi]$.

Table of Values

θ	r
0	$-2/(\cos 0) = -2$
$\pi/12$	$-2/(\cos(\pi/12)) \approx -2.07$
$\pi/6$	$-2/(\cos(\pi/6)) \approx -2.31$
$\pi/4$	$-2/(\cos(\pi/4)) \approx -2.83$
$\pi/3$	$-2/(\cos(\pi/3)) = -4$
$5\pi/12$	$-2/(\cos(5\pi/12)) \approx -7.73$
$\pi/2$	undefined
$7\pi/12$	$-2/(\cos(7\pi/12)) \approx 7.73$
$2\pi/3$	$-2/(\cos(2\pi/3)) = 4$
$3\pi/4$	$-2/(\cos(3\pi/4)) \approx 2.83$
$5\pi/6$	$-2/(\cos(5\pi/6)) \approx 2.31$
$11\pi/12$	$-2/(\cos(11\pi/12)) \approx 2.07$
π	$-2/(\cos\pi) = 2$

Graph

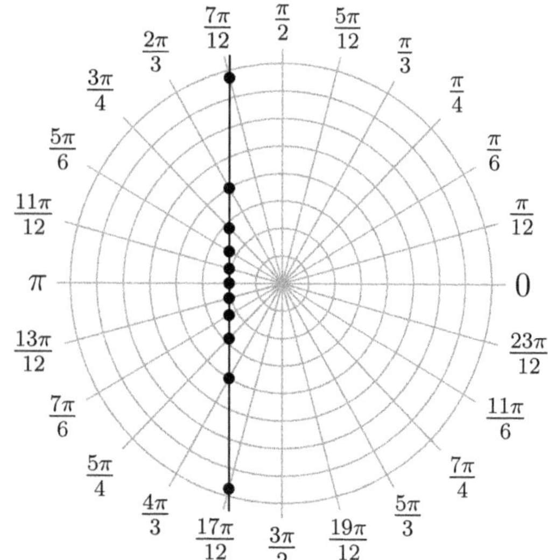

□

J. $r = 3\csc\theta$

K. $r = 5\cos\theta$

Solution

Notes

- The graph is symmetric about $\theta = 0$.
- We determine points (r, θ) on the graph for which $\theta \in [0, \pi]$, plot them and their reflections through the line $\theta = 0$, and then connect the dots.
- The reflections will be points (r, θ) for which $\theta \in [\pi, 2\pi]$.

Table of Values

θ	r
0	$5\cos 0 = 5$
$\pi/12$	$5\cos(\pi/12) \approx 4.83$
$\pi/6$	$5\cos(\pi/6) \approx 4.33$
$\pi/4$	$5\cos(\pi/4) \approx 3.54$
$\pi/3$	$5\cos(\pi/3) = 2.5$
$5\pi/12$	$5\cos(5\pi/12) \approx 1.29$
$\pi/2$	$5\cos(\pi/2) = 0$
$7\pi/12$	$5\cos(7\pi/12) \approx -1.29$
$2\pi/3$	$5\cos(2\pi/3) = -2.5$
$3\pi/4$	$5\cos(3\pi/4) \approx -3.54$
$5\pi/6$	$5\cos(5\pi/6) \approx -4.33$
$11\pi/12$	$5\cos(11\pi/12) \approx -4.83$
π	$5\cos\pi = -5$

Graph

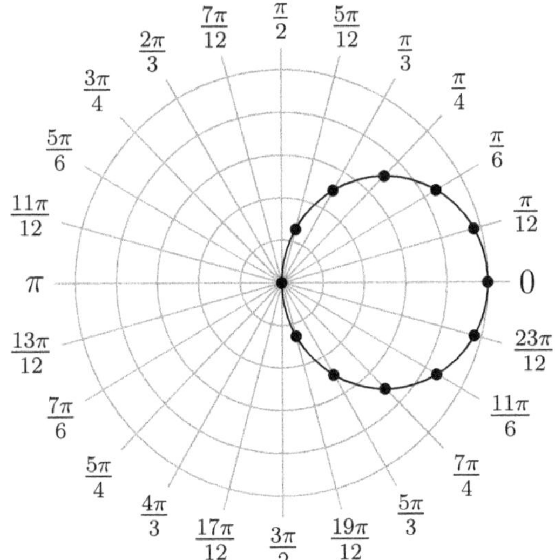

□

L. $r = 4/(2 + \sin\theta)$

M. $r = 3 - 3\sin\theta$

Solution

Notes

- The graph is symmetric about $\theta = \pi/2$.
- We determine points (r, θ) on the graph for which $\theta \in [-\pi/2, \pi/2]$, plot them and their reflections through the line $\theta = \pi/2$, and then connect the dots.
- The reflections will be points (r, θ) for which $\theta \in [\pi/2, 3\pi/2]$.

Table of Values

θ	r
$-\pi/2$	$3 - 3\sin(-\pi/2) = 6$
$-5\pi/12$	$3 - 3\sin(-5\pi/12) \approx 5.90$
$-\pi/3$	$3 - 3\sin(-\pi/3) \approx 5.60$
$-\pi/4$	$3 - 3\sin(-\pi/4) \approx 5.12$
$-\pi/6$	$3 - 3\sin(-\pi/6) = 4.5$
$-\pi/12$	$3 - 3\sin(-\pi/12) \approx 3.78$
0	$3 - 3\sin 0 = 3$
$\pi/12$	$3 - 3\sin(\pi/12) \approx 2.22$
$\pi/6$	$3 - 3\sin(\pi/6) = 1.5$
$\pi/4$	$3 - 3\sin(\pi/4) \approx 0.88$
$\pi/3$	$3 - 3\sin(\pi/3) \approx 0.40$
$5\pi/12$	$3 - 3\sin(5\pi/12) \approx 0.10$
$\pi/2$	$3 - 3\sin(\pi/2) = 0$

Graph

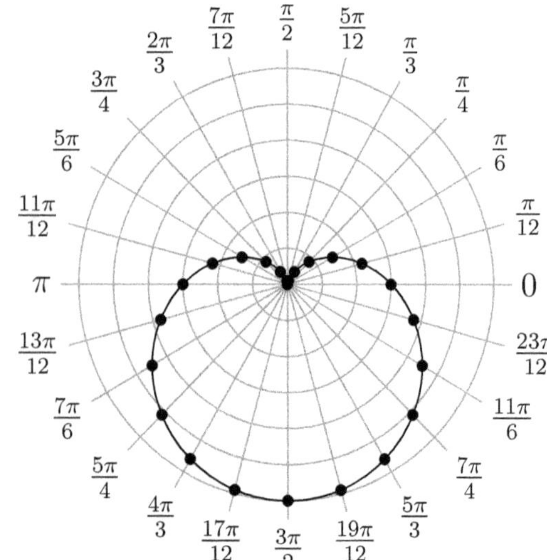

□

N. $r = 2\theta/\pi$

o. $r^2 = 15\sin 4\theta$

Solution

Notes

- $r^2 = 15\sin 4\theta \Rightarrow r = \pm\sqrt{15\sin 4\theta}$
- Values of θ must be such that $\sin 4\theta \geq 0$.
- We will evaluate r for values of θ until we gain no new points on our graph.

Table of Values

θ	r
0	$\pm\sqrt{15\sin(4\cdot 0)} = 0$
$\pi/24$	$\pm\sqrt{15\sin(4\cdot\pi/24)} \approx \pm 2.74$
$\pi/12$	$\pm\sqrt{15\sin(4\cdot\pi/12)} \approx \pm 3.60$
$\pi/8$	$\pm\sqrt{15\sin(4\cdot\pi/8)} \approx \pm 3.87$
$\pi/6$	$\pm\sqrt{15\sin(4\cdot\pi/6)} \approx \pm 3.60$
$5\pi/24$	$\pm\sqrt{15\sin(4\cdot 5\pi/24)} \approx \pm 2.74$
$\pi/4$	$\pm\sqrt{15\sin(4\cdot\pi/4)} = 0$
$\pi/2$	$\pm\sqrt{15\sin(4\cdot\pi/2)} = 0$
$13\pi/24$	$\pm\sqrt{15\sin(4\cdot 13\pi/24)} \approx \pm 2.74$
$7\pi/12$	$\pm\sqrt{15\sin(4\cdot 7\pi/12)} \approx \pm 3.60$
$5\pi/8$	$\pm\sqrt{15\sin(4\cdot 5\pi/8)} \approx \pm 3.87$
$2\pi/3$	$\pm\sqrt{15\sin(4\cdot 2\pi/3)} \approx \pm 3.60$
$17\pi/24$	$\pm\sqrt{15\sin(4\cdot 17\pi/24)} \approx \pm 2.74$
$3\pi/4$	$\pm\sqrt{15\sin(4\cdot 3\pi/4)} = 0$
π	$\pm\sqrt{15\sin(4\cdot\pi)} = 0$
$25\pi/24$	$\pm\sqrt{15\sin(4\cdot 25\pi/24)} \approx \pm 2.74$

Graph

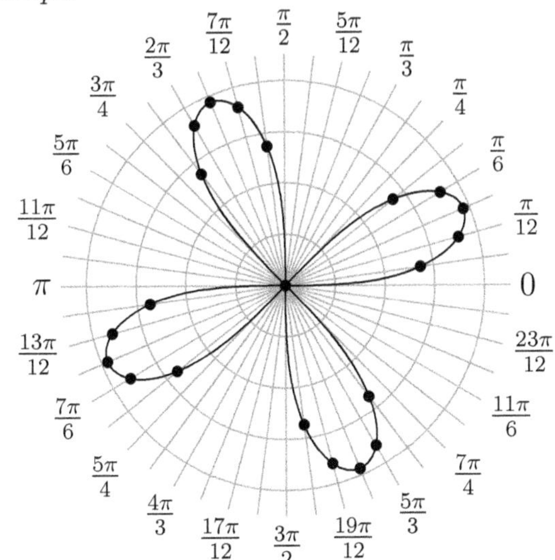

□

p. $r^2 = 9\cos 2\theta$

Section 142: Rotation of Axes

Definition

The *rectangular coordinate system based on axes rotated through the angle of measure θ* is a system in which the point in the plane that corresponds to the ordered pair (A, B) is located $|A|$ units from the line l_x and $|B|$ units from the line l_y, where l_x and l_y represent the lines that the x- and y-axes, respectively, would overlap if rotated about the origin through the angle of measure θ.

Properties

Suppose (x, y) are the rectangular coordinates of a point in the usual system
and (X, Y) are the rectangular coordinates of the point in the system with axes rotated through the angle of measure θ.

Then:

- $X = x\cos\theta + y\sin\theta$
- $Y = -x\sin\theta + y\cos\theta$
- $x = X\cos\theta - Y\sin\theta$
- $y = X\sin\theta + Y\cos\theta$

Proofs (first two properties only)

Let P denote the point with coordinates (x, y) and (X, Y) as described above,
O denote the origin,
r denote the distance from O to P, and
ϕ denote the measure of the angle formed by $\overrightarrow{OP}$ and the x-axis, as shown below.

Note that:

- $x = r\cos\phi$ and $y = r\sin\phi$, since the standard polar coordinates of P are (r, ϕ);
- $X = r\cos(\phi - \theta)$ and $Y = r\sin(\phi - \theta)$, since the polar coordinates of P are $(r, \phi - \theta)$ in the polar coordinate system with polar axis rotated through the angle of measure θ.

- $$\begin{aligned} X &= r\cos(\phi-\theta) \\ &= r[\cos\phi\cos\theta + \sin\phi\sin\theta] \\ &= r\cos\phi\cos\theta + r\sin\phi\sin\theta \\ &= (r\cos\phi)\cos\theta + (r\sin\phi)\sin\theta \\ &= x\cos\theta + y\sin\theta \end{aligned}$$

- $$\begin{aligned} Y &= r\sin(\phi-\theta) \\ &= r[\sin\phi\cos\theta - \cos\phi\sin\theta] \\ &= r\sin\phi\cos\theta - r\cos\phi\sin\theta \\ &= (r\sin\phi)\cos\theta - (r\cos\phi)\sin\theta \\ &= y\cos\theta - x\sin\theta \\ &= -x\sin\theta + y\cos\theta \end{aligned}$$

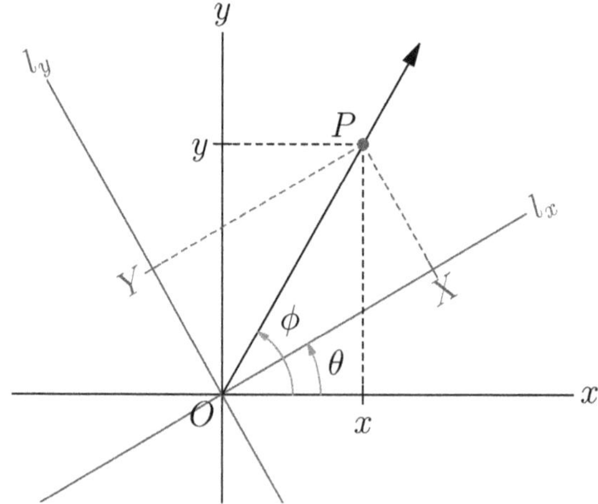

⊡

Property

Suppose (x, y) are the rectangular coordinates of a point in the usual system
and (X, Y) are the rectangular coordinates of the point in the system with axes rotated through the angle of measure θ.

Then the equation $ax^2 + bxy + cy^2 + dx + ey + f = 0$, where a, b, c, d, e, and f are real constants with $b \neq 0$, is equivalent to the equation $AX^2 + CY^2 + DX + EY + F = 0$ for some real constants A, C, D, E, and F if $\cot 2\theta = (a - c)/b$.

Proof

$$
\begin{aligned}
& ax^2 + bxy + cy^2 + dx + ey + f = 0 \\
&\Leftrightarrow a(X\cos\theta - Y\sin\theta)^2 + b(X\cos\theta - Y\sin\theta)(X\sin\theta + Y\cos\theta) \\
&\qquad + c(X\sin\theta + Y\cos\theta)^2 + d(X\cos\theta - Y\sin\theta) + e(X\sin\theta + Y\cos\theta) + f = 0 \\
&\Leftrightarrow a(X^2\cos^2\theta - 2XY\cos\theta\sin\theta + Y^2\sin^2\theta) \\
&\qquad + b(X^2\cos\theta\sin\theta + XY\cos^2\theta - XY\sin^2\theta - Y^2\cos\theta\sin\theta) \\
&\qquad + c(X^2\sin^2\theta + 2XY\cos\theta\sin\theta + Y^2\cos^2\theta) \\
&\qquad + d(X\cos\theta - Y\sin\theta) + e(X\sin\theta + Y\cos\theta) + f = 0 \\
&\Leftrightarrow aX^2\cos^2\theta - 2aXY\cos\theta\sin\theta + aY^2\sin^2\theta \\
&\qquad + bX^2\cos\theta\sin\theta + bXY\cos^2\theta - bXY\sin^2\theta - bY^2\cos\theta\sin\theta \\
&\qquad + cX^2\sin^2\theta + 2cXY\cos\theta\sin\theta + cY^2\cos^2\theta \\
&\qquad + dX\cos\theta - dY\sin\theta + eX\sin\theta + eY\cos\theta + f = 0 \\
&\Leftrightarrow (a\cos^2\theta + b\cos\theta\sin\theta + c\sin^2\theta)(X^2) \\
&\qquad + (-2a\cos\theta\sin\theta + b\cos^2\theta - b\sin^2\theta + 2c\cos\theta\sin\theta)(XY) \\
&\qquad + (a\sin^2\theta - b\cos\theta\sin\theta + c\cos^2\theta)(Y^2) \\
&\qquad + (d\cos\theta + e\sin\theta)(X) + (-d\sin\theta + e\cos\theta)(Y) + f = 0 \\
&\Leftrightarrow AX^2 + BXY + CY^2 + DX + EY + F = 0 \\
&\qquad \text{where } A = a\cos^2\theta + b\cos\theta\sin\theta + c\sin^2\theta, \\
&\qquad\qquad B = -2a\cos\theta\sin\theta + b\cos^2\theta - b\sin^2\theta + 2c\cos\theta\sin\theta, \\
&\qquad\qquad C = a\sin^2\theta - b\cos\theta\sin\theta + c\cos^2\theta, \\
&\qquad\qquad D = d\cos\theta + e\sin\theta, \\
&\qquad\qquad E = -d\sin\theta + e\cos\theta, \\
&\qquad \text{and } F = f
\end{aligned}
$$

But:

$$
\begin{aligned}
B = 0 &\Leftrightarrow -2a\cos\theta\sin\theta + b\cos^2\theta - b\sin^2\theta + 2c\cos\theta\sin\theta = 0 \\
&\Leftrightarrow (-a + c)(2\cos\theta\sin\theta) + b(\cos^2\theta - \sin^2\theta) = 0 \\
&\Leftrightarrow (-a + c)(\sin 2\theta) + b(\cos 2\theta) = 0 \\
&\Leftrightarrow b(\cos 2\theta) = -(-a + c)(\sin 2\theta) \\
&\Leftrightarrow (\cancel{b}\cdot\cos 2\theta)/(\cancel{b}\cdot\sin 2\theta) = (a - c)(\cancel{\sin 2\theta})/(b\cdot\cancel{\sin 2\theta}) \\
&\Leftrightarrow \cot 2\theta = (a - c)/b
\end{aligned}
$$

Therefore:

$\cot 2\theta = (a - c)/b \Rightarrow B = 0$

$\Rightarrow$ the equation $ax^2 + bxy + cy^2 + dx + ey + f = 0$ is equivalent to the equation $AX^2 + CY^2 + DX + EY + F = 0$ for such A, C, D, E, F

⊡

NOTATION

In subsequent graphs we will label the lines denoted by l_x and l_y in the definition of this section as the X-axis and Y-axis, respectively, using the capital letters X and Y to differentiate these axes from the standard x-axis and y-axis.

PROCEDURE

To graph an equation of the form $ax^2 + bxy + cy^2 + dx + ey + f = 0$ (where $b \neq 0$), we:

1. determine the values of $\cos\theta$, $\sin\theta$, and θ where $0° < \theta < 90°$ and $\cot 2\theta = (a - c)/b$;
2. transform the given equation into one of the form $AX^2 + CY^2 + DX + EY + F = 0$ by substituting the expression $X\cos\theta - Y\sin\theta$ for x and $X\sin\theta + Y\cos\theta$ for y;
3. draw the graph of the equation $AX^2 + CY^2 + DX + EY + F = 0$ using the rectangular coordinate system with axes rotated through the angle of measure θ.

COMMENTS

- Since the measure θ in the procedure above will satisfy $0° < \theta < 90°$:
  - the terminal side of the angle of measure 2θ will always lie in the first or second quadrant;
  - the signs of $\cot 2\theta$ and $\cos 2\theta$ will always be the same (both positive or both negative).
- The following identities may prove useful when we apply the procedure above:
  - $\cos^2\theta = \cot^2\theta/(1 + \cot^2\theta)$
  - $\cos\theta = \sqrt{(1 + \cos 2\theta)/2}$ (since $0° < \theta < 90°$)
  - $\sin\theta = \sqrt{(1 - \cos 2\theta)/2}$ (since $0° < \theta < 90°$)
- The graph of an equation of the form $ax^2 + bxy + cy^2 + dx + ey + f = 0$ is a conic section of some kind. If the graph of the equation is not a degenerate conic section, then we can use the values of a, b, and c to identify its shape immediately:
  - if $b^2 - 4ac < 0$, then the graph is an ellipse or a circle;
  - if $b^2 - 4ac = 0$, then the graph is a parabola;
  - if $b^2 - 4ac > 0$, then the graph is a hyperbola.

EXAMPLE 1

Determine the shape of the graph of each equation. Assume that all graphs in this example are nondegenerate conic sections.

A. $5x^2 + 12xy + 10y^2 - 4x + 11y + 7 = 0$

SOLUTION

Notes

- $a = 5$, $b = 12$, and $c = 10$
- $$\begin{aligned} b^2 - 4ac &= (12)^2 - 4(5)(10) \\ &= 144 - 200 \\ &= -56 \\ &< 0 \end{aligned}$$

Conclusion

The graph is an ellipse.

□

B. $9x^2 + 6xy + y^2 - 5x - 13 = 0$

EXAMPLE 2

For each point with the given rectangular coordinates (x, y) in the system with axes unrotated, find the coordinates (X, Y) of the point in the system with axes rotated through the angle of measure θ.

A. $(x, y) = (2, 4)$; $\theta = 60°$

SOLUTION

Values of Coordinates

- $X = (2)(\cos 60°) + (4)(\sin 60°)$
  $= (2)(1/2) + (4)(\sqrt{3}/2)$
  $= 1 + 2\sqrt{3}$
- $Y = (-2)(\sin 60°) + (4)(\cos 60°)$
  $= (-2)(\sqrt{3}/2) + (4)(1/2)$
  $= -\sqrt{3} + 2$
  $= 2 - \sqrt{3}$

Pair of Coordinates

$(X, Y) = (1 + 2\sqrt{3}, 2 - \sqrt{3})$

□

B. $(x, y) = (-7, 3)$; $\theta = 45°$

EXAMPLE 3

For each point with the given rectangular coordinates (X, Y) in the system with axes rotated through the angle of measure θ, find the coordinates (x, y) of the point in the system with axes unrotated.

A. $(X, Y) = (9, -5)$; $\theta = 135°$

SOLUTION

Values of Coordinates

- $x = (9)(\cos 135°) - (-5)(\sin 135°)$
  $= (9)(-\sqrt{2}/2) + (5)(\sqrt{2}/2)$
  $= -9\sqrt{2}/2 + 5\sqrt{2}/2$
  $= -4\sqrt{2}/2$
  $= -2\sqrt{2}$
- $y = (9)(\sin 135°) + (-5)(\cos 135°)$
  $= (9)(\sqrt{2}/2) - (5)(-\sqrt{2}/2)$
  $= 9\sqrt{2}/2 + 5\sqrt{2}/2$
  $= 14\sqrt{2}/2$
  $= 7\sqrt{2}$

Pair of Coordinates

$(x, y) = (-2\sqrt{2}, 7\sqrt{2})$

□

B. $(X, Y) = (-6, -8)$; $\theta = 30°$

EXAMPLE 4

For each equation of the form $ax^2 + bxy + cy^2 + dx + ey + f = 0$, determine the values of $\cos\theta$, $\sin\theta$, and θ if $0° < \theta < 90°$ and $\cot 2\theta = (a - c)/b$.

A. $5x^2 + 12xy + 10y^2 - 4x + 11y + 7 = 0$

SOLUTION

Notes

- $a = 5$, $b = 12$, and $c = 10$
- $\cot 2\theta = (5 - 10)/12$
  $= -5/12$
- $\cos 2\theta < 0$ since $\cot 2\theta < 0$

Value of $\cos 2\theta$

$$\begin{aligned}\cos^2(2\theta) &= \cot^2(2\theta)/(1 + \cot^2(2\theta)) \\ &= (-5/12)^2/(1 + (-5/12)^2) \\ &= (25/144)/(144/144 + 25/144) \\ &= (25/144)/(169/144) \\ &= (25/144) \cdot (144/169) \\ &= 25/169\end{aligned}$$

$$\begin{aligned}\Rightarrow \cos 2\theta &= -\sqrt{25/169} \\ &= -5/13\end{aligned}$$

Value of $\cos\theta$

$$\begin{aligned}\cos\theta &= \sqrt{(1 + \cos 2\theta)/2} \\ &= \sqrt{(1 + (-5/13))/2} \\ &= \sqrt{(13/13 + (-5/13))/2} \\ &= \sqrt{(8/13) \cdot (1/2)} \\ &= \sqrt{4/13} \\ &= 2/\sqrt{13}\end{aligned}$$

Value of $\sin\theta$

$$\begin{aligned}\sin\theta &= \sqrt{(1 - \cos 2\theta)/2} \\ &= \sqrt{(1 - (-5/13))/2} \\ &= \sqrt{(13/13 - (-5/13))/2} \\ &= \sqrt{(18/13) \cdot (1/2)} \\ &= \sqrt{9/13} \\ &= 3/\sqrt{13}\end{aligned}$$

Value of θ

$$\begin{aligned}\theta &= \cos^{-1}(2/\sqrt{13}) \\ &\approx 56.3°\end{aligned}$$

□

B. $9x^2 + 6xy + y^2 - 5x - 13 = 0$

Example 5

Draw the graph of each equation of the form $ax^2 + bxy + cy^2 + dx + ey + f = 0$ with a rotation of axes through the angle of measure θ where $0° < \theta < 90°$ and $\cot 2\theta = (a - c)/b$.

A. $29x^2 - 24xy + 36y^2 = 180$

Solution

Notes

- $\cot 2\theta = (29 - 36)/(-24)$
  $= -7/(-24)$
  $= 7/24$
- $\cos 2\theta > 0$ since $\cot 2\theta > 0$

Value of $\cos 2\theta$

$$\begin{aligned}\cos^2 2\theta &= (7/24)^2/(1 + (7/24)^2)\\ &= (49/576)/(1 + 49/576)\\ &= (49/576)/(625/576)\\ &= (49/576) \cdot (576/625)\\ &= 49/625\\ \Rightarrow \cos 2\theta &= \sqrt{49/625}\\ &= 7/25\end{aligned}$$

Value of $\cos\theta$

$$\begin{aligned}\cos\theta &= \sqrt{(1 + (7/25))/2}\\ &= \sqrt{(25/25 + 7/25)/2}\\ &= \sqrt{(1/2)(32/25)}\\ &= \sqrt{16/25}\\ &= 4/5\end{aligned}$$

Value of $\sin\theta$

$$\begin{aligned}\sin\theta &= \sqrt{(1 - (7/25))/2}\\ &= \sqrt{(25/25 - 7/25)/2}\\ &= \sqrt{(1/2)(18/25)}\\ &= \sqrt{9/25}\\ &= 3/5\end{aligned}$$

Value of θ

$$\begin{aligned}\theta &= \cos^{-1}(4/5)\\ &= \approx 36.9°\end{aligned}$$

Substitutions

- $x = (4/5)X - (3/5)Y$
- $y = (3/5)X + (4/5)Y$

Transformation of Equation

$$\begin{aligned}&29x^2 - 24xy + 36y^2 = 180\\ &\Rightarrow 29((4/5)X - (3/5)Y)^2\\ &\qquad - 24((4/5)X - (3/5)Y)((3/5)X + (4/5)Y)\\ &\qquad + 36((3/5)X + (4/5)Y)^2 = 180\\ &\Rightarrow 29((16/25)X^2 - (24/25)XY + (9/25)Y^2)\\ &\qquad - 24((12/25)X^2 + (7/25)XY - (12/25)Y^2)\\ &\qquad + 36((9/25)X^2 + (24/25)XY + (16/25)Y^2)\\ &\qquad = 180\\ &\Rightarrow (464/25)X^2 - (696/25)XY + (261/25)Y^2\\ &\qquad - (288/25)X^2 - (168/25)XY + (288/25)Y^2\\ &\qquad + (324/25)X^2 + (864/25)XY + (576/25)Y^2\\ &\qquad = 180\\ &\Rightarrow (500/25)X^2 - (0/25)XY + (1125/25)Y^2 = 180\\ &\Rightarrow 20X^2 + 45Y^2 = 180\\ &\Rightarrow (20X^2 + 45Y^2)/180 = (180)/180\\ &\Rightarrow X^2/9 + Y^2/4 = 1\end{aligned}$$

Graph

We draw the graph of $X^2/9 + Y^2/4 = 1$ in reference to axes rotated 36.9°:

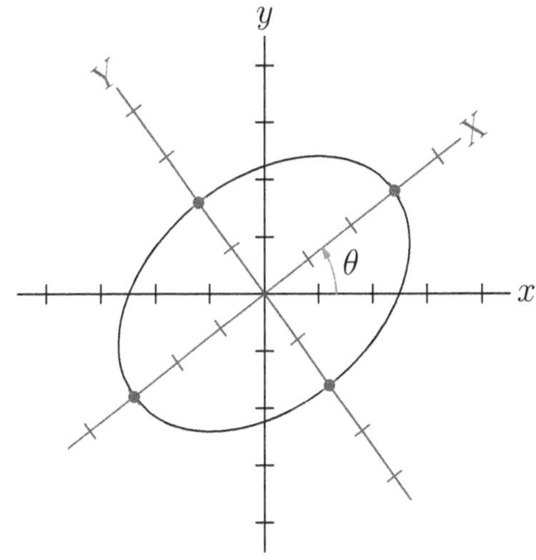

□

B. $2x^2 + 16xy + 32y^2 + (8\sqrt{17})x + (15\sqrt{17})y - 34 = 0$

SOLUTION

Notes

- $\cot 2\theta = (2 - 32)/16$
  $= -15/8$
- $\cos 2\theta < 0$ since $\cot 2\theta < 0$

Value of $\cos 2\theta$

$$\begin{aligned}\cos^2 2\theta &= (-15/8)^2/(1 + (-15/8)^2) \\ &= (225/64)/(1 + 225/64) \\ &= (225/64)/(289/64) \\ &= (225/64) \cdot (64/289) \\ &= 225/289 \\ \Rightarrow \cos 2\theta &= -\sqrt{225/289} \\ &= -15/17\end{aligned}$$

Value of $\cos\theta$

$$\begin{aligned}\cos\theta &= \sqrt{(1 + (-15/17))/2} \\ &= \sqrt{(17/17 - 15/17)/2} \\ &= \sqrt{(1/2)(2/17)} \\ &= \sqrt{1/17} \\ &= 1/\sqrt{17}\end{aligned}$$

Value of $\sin\theta$

$$\begin{aligned}\sin\theta &= \sqrt{(1 - (-15/17))/2} \\ &= \sqrt{(17/17 + 15/17)/2} \\ &= \sqrt{(1/2)(32/17)} \\ &= \sqrt{16/17} \\ &= 4/\sqrt{17}\end{aligned}$$

Value of θ

$$\begin{aligned}\theta &= \cos^{-1}(1/\sqrt{17}) \\ &= \approx 76.0^\circ\end{aligned}$$

Substitutions

- $x = (1/\sqrt{17})X - (4\sqrt{17})Y$
- $y = (4/\sqrt{17})X + (1\sqrt{17})Y$

Transformation of Equation

$$\begin{aligned}&2x^2 + 16xy + 32y^2 + (8\sqrt{17})x \\ &\quad + (15\sqrt{17})y - 34 = 0\end{aligned}$$

$\Rightarrow$ (continued at upper right)

$$\begin{aligned}\Rightarrow\ & 2((1/\sqrt{17})X - (4/\sqrt{17})Y)^2 \\ &+ 16((1/\sqrt{17})X - (4/\sqrt{17})Y) \\ &\qquad \cdot((4/\sqrt{17})X + (1/\sqrt{17})Y) \\ &+ 32((4/\sqrt{17})X + (1/\sqrt{17})Y)^2 \\ &+ (8\sqrt{17})((1/\sqrt{17})X - (4/\sqrt{17})Y) \\ &+ (15\sqrt{17})((4/\sqrt{17})X + (1/\sqrt{17})Y) \\ &- 34 = 0 \\ \Rightarrow\ & 2((1/17)X^2 - (8/17)XY + (16/17)Y^2) \\ &+ 16((4/17)X^2 - (15/17)XY - (4/17)Y^2) \\ &+ 32((16/17)X^2 + (8/17)XY + (1/17)Y^2) \\ &+ 8X - 32Y + 60X + 15Y - 34 = 0 \\ \Rightarrow\ & (2/17)X^2 - (16/17)XY + (32/17)Y^2 \\ &+ (64/17)X^2 - (240/17)XY - (64/17)Y^2 \\ &+ (512/17)X^2 + (256/17)XY + (32/17)Y^2 \\ &+ 8X - 32Y + 60X + 15Y - 34 = 0 \\ \Rightarrow\ & (578/17)X^2 + 68X - 17Y - 34 = 0 \\ \Rightarrow\ & 34X^2 + 68X - 34 = 17Y \\ \Rightarrow\ & (34X^2 + 68X - 34)/17 = (17Y)/17 \\ \Rightarrow\ & Y = 2X^2 + 4X - 2 \\ &\quad = 2(X^2 + 2X) - 2 \\ &\quad = 2(X^2 + 2X + (1)^2 - (1)^2) - 2 \\ &\quad = 2(X^2 + 2X + (1)^2) - 2(1)^2 - 2 \\ &\quad = 2(X + 1)^2 - 4\end{aligned}$$

Graph

We draw the graph of $Y = 2(X + 1)^2 - 4$ in reference to axes rotated 76.0°:

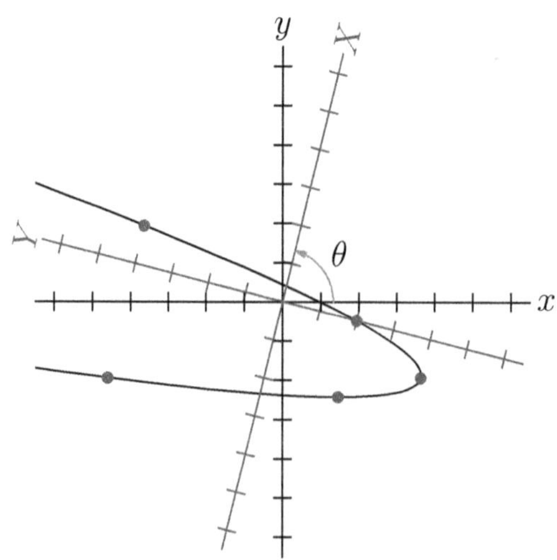

□

EXAMPLE 6

Suppose $X = x\cos\theta + y\sin\theta$ and $Y = -x\sin\theta + y\cos\theta$, where $x, y, X, Y, \theta \in \mathbf{R}$.
Prove each statement.

A. $x = X\cos\theta - Y\sin\theta$

PROOF

Suppose $x, y, X, Y, \theta \in \mathbf{R}$.
Note that:

- $X = x\cos\theta + y\sin\theta$
  $\Rightarrow (X)(\cos\theta) = (x\cos\theta + y\sin\theta)(\cos\theta)$
  $\Rightarrow X\cos\theta = x\cos^2\theta + y\cos\theta\sin\theta$
  $\Rightarrow X\cos\theta - x\cos^2\theta = y\cos\theta\sin\theta$
- $Y = -x\sin\theta + y\cos\theta$
  $\Rightarrow (Y)(\sin\theta) = (-x\sin\theta + y\cos\theta)(\sin\theta)$
  $\Rightarrow Y\sin\theta = -x\sin^2\theta + y\cos\theta\sin\theta$
  $\Rightarrow Y\sin\theta + x\sin^2\theta = y\cos\theta\sin\theta$

Therefore:

$X\cos\theta - x\cos^2\theta = Y\sin\theta + x\sin^2\theta$

$$\begin{aligned}\Rightarrow X\cos\theta - Y\sin\theta &= x\sin^2\theta + x\cos^2\theta \\ &= x(\sin^2\theta + \cos^2\theta) \\ &= x(1) \\ &= x\end{aligned}$$

$\Rightarrow x = X\cos\theta - Y\sin\theta$

$\boxdot$

B. $y = X\sin\theta + Y\cos\theta$

Section 143: Complex Numbers

Definitions

- The *imaginary unit* i is a number whose square is -1.
- A *complex number* is an expression of the form $a + bi$ where a and b are real constants.
- The *real part* of the complex number $a + bi$ is the real number a.
- The *imaginary part* of the complex number $a + bi$ is the real number b.
- A *pure imaginary number* is a complex number $a + bi$ where $a = 0$ and $b \neq 0$.
- The *complex conjugate* of the complex number $a + bi$ is the complex number $a - bi$.
- Two complex numbers $a + bi$ and $c + di$ are *equal* if $a = c$ and $b = d$.
- The *principal square root* of $-b$, denoted $\sqrt{-b}$, is $\sqrt{b}i$ (assuming $b > 0$).

Notation

- We denote the set of complex numbers by $\mathbf{C}$.
- We denote the complex conjugate of a complex number z by $\overline{z}$.
- Complex numbers can take different forms. We can write:
  - $a + 0i$ as a
  - $0 + bi$ as bi
  - $a + bi$ as $a + ib$, $bi + a$, or $ib + a$
  - $a + (-b)i$ as $a - bi$, $a - ib$, $-bi + a$, or $-ib + a$

Operations on Complex Numbers

We define the operations addition $(+)$ and multiplication $(\cdot)$ on $\mathbf{C}$ by:

- $(a + bi) + (c + di) = (a + c) + (b + d)i$
- $(a + bi) \cdot (c + di) = (ac - bd) + (ad + bc)i$

Comments

- A complex number is not a binomial and the imaginary unit i is not a variable; however:
  - the sum of two complex numbers is equivalent to the result yielded by collecting like terms as if i were a variable;
  - the product of two complex numbers is equivalent to the result yielded by multiplying the factors as if they were binomials, substituting -1 for i^2, and collecting like terms as if i were a variable.
- $\sqrt{-1} = i$
- We often write $\sqrt{b}i$ as $i\sqrt{b}$ for clarity because we could easily misread $\sqrt{b}i$ as $\sqrt{bi}$.
- Both $\sqrt{b}i$ and $-\sqrt{b}i$ are square roots of $-b$ since the squares of both numbers equal $-b$, yet only the former is the principal square root of $-b$.
- The property $\sqrt[n]{ab} = \sqrt[n]{a}\sqrt[n]{b}$ is not valid if n is even and a and b are negative real numbers. We should convert all radicals with an even index and a negative radicand to the form bi before we apply arithmetic operations.

Example 1

For each complex number:

- write the number in the form $a + bi$,
- identify the real part of the number, and
- identify the imaginary part of the number.

A. $5i + 2$

Solution

- alternative form: $2 + 5i$
- real part: 2
- imaginary part: 5

□

B. $3i + 8$

C. $8 - (5/2)i$

Solution

- alternative form: $8 + (-5/2)i$
- real part: 8
- imaginary part: $-5/2$

□

D. $1 - 4i$

E. -10

Solution

- alternative form: $-10 + 0i$
- real part: -10
- imaginary part: 0

□

F. i

Example 2

Add or multiply the complex numbers as indicated.

A. $(3 - i) \cdot (7 + 4i)$

Solution

$$
\begin{aligned}
&(3 - i) \cdot (7 + 4i) \\
&\quad = (3)(7) + (3)(4i) + (-i)(7) + (-i)(4i) \\
&\quad = 21 + 12i - 7i - 4i^2 \\
&\quad = 21 + 5i - 4(-1) \\
&\quad = 21 + 5i + 4 \\
&\quad = 25 + 5i
\end{aligned}
$$

□

B. $(10 + i) \cdot (3 + 5i)$

C. $(3-i)+(7+4i)$

Solution

$$\begin{aligned}(3-i)+(7+4i) &= (3+7)+(-i+4i)\\ &= 10+3i\end{aligned}$$

□

D. $(10+i)+(3+5i)$

E. $9\cdot(2-3i)$

Solution

$$\begin{aligned}9\cdot(2-3i) &= (9)(2)-(9)(3i)\\ &= 18-27i\end{aligned}$$

□

F. $6i\cdot(-1-8i)$

G. $(6+5i)\cdot(6-5i)$

Solution

$$\begin{aligned}&(6+5i)\cdot(6-5i)\\ &\quad= (6)(6)-(6)(5i)+(5i)(6)-(5i)(5i)\\ &\quad= 36-\cancel{30i}+\cancel{30i}-25i^2\\ &\quad= 36-25(-1)\\ &\quad= 36+25\\ &\quad= 61\end{aligned}$$

□

H. $(4-2i)\cdot(\frac{4}{5}+\frac{2}{5}i)$

Example 3

Simplify each of the following.

A. i^7

Solution

$$\begin{aligned}i^7 &= (i^6)(i^1)\\ &= (i^2)^3(i)\\ &= (-1)^3(i)\\ &= (-1)(i)\\ &= -i\end{aligned}$$

□

B. i^5

C. i^{102}

Solution

$$\begin{aligned}i^{102} &= (i^2)^{51}\\ &= (-1)^{51}\\ &= -1\end{aligned}$$

□

D. i^{40}

Example 4

Convert each of the following to the form bi.

A. $\sqrt{-9}$

Solution

$$\begin{aligned}\sqrt{-9} &= \sqrt{9}\cdot\sqrt{-1}\\ &= 3i\end{aligned}$$

□

B. $\sqrt{-25}$

C. $\sqrt{-8}$

Solution

$$\begin{aligned}\sqrt{-8} &= \sqrt{8}\cdot\sqrt{-1}\\ &= (\sqrt{4}\sqrt{2})i\\ &= (2\sqrt{2})(i)\\ &= 2i\sqrt{2}\end{aligned}$$

□

D. $\sqrt{-180}$

Example 5

Perform the indicated operations.

A. $\sqrt{-2}\sqrt{-8}$

Solution

$$\begin{aligned}\sqrt{-2}\sqrt{-8} &= (i\sqrt{2})(i\sqrt{8})\\ &= (i^2)(\sqrt{2}\sqrt{8})\\ &= (-1)(\sqrt{16})\\ &= -4\end{aligned}$$

□

B. $\sqrt{-15}\sqrt{-5}$

Example 6

Identify the complex conjugate of each complex number.

A. $2+5i$

B. $4-7i$

C. -10

D. i

E. $-3i-8$

F. $-10+i$

G. 0

H. $5i$

Solutions

A. $2-5i$

B. $4+7i$

C. -10

D. $-i$

□

Section 144: Properties of Operations on Complex Numbers

Properties of Addition $(+)$

- Addition is commutative since $z_1 + z_2 = z_2 + z_1$ for all $z_1, z_2 \in \mathbf{C}$.
- Addition is associative since $(z_1 + z_2) + z_3 = z_1 + (z_2 + z_3)$ for all $z_1, z_2, z_3 \in \mathbf{C}$.
- 0 serves as an identity element for addition since $z + 0 = 0 + z = z$ for all $z \in \mathbf{C}$.
- The inverse of a complex number $a + bi$ under addition is $-a - bi$, since $(a + bi) + (-a - bi) = (-a - bi) + (a + bi) = 0$.

Properties of Multiplication $(\cdot)$

- Multiplication is commutative since $z_1 \cdot z_2 = z_2 \cdot z_1$ for all $z_1, z_2 \in \mathbf{C}$.
- Multiplication is associative since $(z_1 \cdot z_2) \cdot z_3 = z_1 \cdot (z_2 \cdot z_3)$ for all $z_1, z_2, z_3 \in \mathbf{C}$.
- 1 serves as an identity element for multiplication since $z \cdot 1 = 1 \cdot z = z$ for all $z \in \mathbf{C}$.
- The inverse of a complex number $a + bi$ under multiplication is $a/(a^2 + b^2) - (b/(a^2 + b^2))i$, since $(a + bi) \cdot (a/(a^2 + b^2) - (b/(a^2 + b^2))i) = (a/(a^2 + b^2) - (b/(a^2 + b^2))i) \cdot (a + bi) = 1$.

Distributive Property

Suppose $z_1, z_2, z_3 \in \mathbf{C}$.
Then:

- $z_1 \cdot (z_2 + z_3) = z_1 \cdot z_2 + z_1 \cdot z_3$
- $(z_2 + z_3) \cdot z_1 = z_2 \cdot z_1 + z_3 \cdot z_1$

Notation

- We denote the additive inverse of a complex number z by $-z$.
- We denote the multiplicative inverse of a complex number z by z^{-1}.

Comments

- The multiplicative inverse of $a + bi$ does not exist if $a + bi = 0$.
- A number a can represent either the real number a or the complex number $a + 0i$; however, the distinction between these interpretations of a lacks significance since both:
  - a one-to-one correspondence exists between the set $\mathbf{R}$ and the set $\{a + bi \in \mathbf{C} \mid b = 0\}$;
  - properties of operations on real numbers hold for operations on complex numbers.

  Hence for practical purposes, we may say that:
  - $a + 0i \in \mathbf{R}$
  - $\mathbf{R} = \{a + bi \in \mathbf{C} \mid b = 0\}$
  - $\mathbf{R} \subset \mathbf{C}$

Example 1

State the additive inverse of each complex number.

A. $5 - 2i$

Solution

$-5 + 2i$

□

B. $-7 - 9i$

Example 2
State the multipicative inverse of each complex number.

A. $5 - 2i$

Solution
Note
$$\begin{aligned}(5)^2 + (-2)^2 &= 25 + 4 \\ &= 29\end{aligned}$$

Inverse
$$\begin{aligned}(5 - 2i)^{-1} &= (5/29) - (-2/29)i \\ &= 5/29 + (2/29)i\end{aligned}$$

□

B. $-7 - 9i$

Example 3
Prove that $z_1 + z_2 = z_2 + z_1 \ \forall \ z_1, z_2 \in \mathbf{C}$.

Proof
Let $z_1, z_2 \in \mathbf{C}$.
$z_1 = a + bi$ AND $z_2 = c + di$
for some constants $a, b, c, d \in \mathbf{R}$
$$\begin{aligned}\Rightarrow z_1 + z_2 &= (a + bi) + (c + di) \\ &= (a + c) + (b + d)i \\ &= (c + a) + (d + b)i \\ &= (c + di) + (a + bi) \\ &= z_2 + z_1\end{aligned}$$

⊡

Example 4
Prove that $z + 0 = z \ \forall \ z \in \mathbf{C}$.

Example 5
Prove that $z \cdot 1 = z$ for all $z \in \mathbf{C}$.

Proof
Let $z \in \mathbf{C}$.
$z = a + bi$ for some constants $a, b \in \mathbf{R}$
$$\begin{aligned}\Rightarrow z \cdot 1 &= (a + bi) \cdot 1 \\ &= (a + bi) \cdot (1 + 0i) \\ &= (a(1) - b(0)) + (a(0) + b(1))i \\ &= (a - 0) + (0 + b)i \\ &= a + bi \\ &= z\end{aligned}$$

⊡

Example 6
Suppose $a, b \in \mathbf{R}$ with $a \neq 0$ or $b \neq 0$.
Prove that
$$(a + bi) \cdot \left(\frac{a}{a^2 + b^2} - \frac{b}{a^2 + b^2} i\right) = 1.$$

Example 7
Prove that the product of a complex number $a + bi$ and its conjugate is a real number.

Proof

$$\begin{aligned}(a+bi)(a-bi) &= (a)(a) + (a)(-bi) + (bi)(a) + (bi)(-bi)\\ &= a^2 - \cancel{abi} + \cancel{abi} - b^2 i^2\\ &= a^2 - b^2(-1)\\ &= a^2 + b^2\end{aligned}$$

But clearly $a^2 + b^2 \in \mathbf{R}$ since $a \in \mathbf{R}$ and $b \in \mathbf{R}$.
Thus $(a + bi)(a - bi) \in \mathbf{R}$.

$\boxdot$

Example 8
Suppose $z \in \mathbf{C}$.
Prove that $z = \overline{z} \Leftrightarrow z \in \mathbf{R}$.

Proof
Let $z \in \mathbf{C}$.
So $z = a + bi$ for some constants $a, b \in \mathbf{R}$.

$$\begin{aligned}(\Rightarrow)\ z = \overline{z} &\Rightarrow z + (-\overline{z}) = \overline{z} + (-\overline{z})\\ &\Rightarrow a + bi + (-(a - bi)) = 0\\ &\Rightarrow a + bi + (-a + bi) = 0\\ &\Rightarrow (a + (-a)) + (b + b)i = 0\\ &\Rightarrow 0 + \underline{2b}i = 0 + \underline{0}i\\ &\Rightarrow 2b = 0\\ &\Rightarrow b = 0/2\\ &\qquad = 0\\ &\Rightarrow z = a + 0i\\ &\Rightarrow z = a\\ &\Rightarrow z \in \mathbf{R}\end{aligned}$$

$$\begin{aligned}(\Leftarrow)\ z \in \mathbf{R} &\Rightarrow z = a\\ &\Rightarrow z = a + 0i\\ &\Rightarrow \overline{z} = a - 0i\\ &\qquad = a\\ &\qquad = z\end{aligned}$$

$\boxdot$

Example 9
Suppose $z \in \mathbf{C}$.
Show that $\overline{\overline{z}} = z$.

Section 145: Subtraction and Division of Complex Numbers

Operations on Complex Numbers

We define the operations subtraction (−) and division (÷) on **C** by:

- $z_1 - z_2 = z_1 + (-z_2)$
- $z_1 \div z_2 = z_1 \cdot z_2^{-1}$

Procedure A (preferred)

To divide one complex number by another, we:

1. multiply the given quotient by the fraction whose numerator and denominator both equal the complex conjugate of the divisor;
2. simplify the result of step (1).

Procedure B (alternative)

To divide one complex number by another, we:

1. convert the given quotient to the product of its dividend and the multiplicative inverse of its divisor;
2. perform the multiplication indicated in step (1);
3. simplify the result of step (2).

Comments

- The fraction z_1/z_2 of complex numbers is equivalent to the expression $z_1 \div z_2$, just as the fraction a/b of real numbers is equivalent to the expression $a \div b$.
- The rules for addition and multiplication of fractions of real numbers also apply to fractions of complex numbers.
- By following procedure A instead of procedure B, we can:
  - avoid the need to identify the multiplicative inverse of the divisor and
  - exploit the similarity between procedure A and the familiar procedure of rationalizing a denominator that contains a radical.

Example 1

Subtract the complex numbers as indicated.

A. $(-5 + 10i) - (2 - 3i)$

Solution

$$\begin{aligned}&(-5 + 10i) - (2 - 3i)\\ &\quad = -5 + 10i + (-(2 - 3i))\\ &\quad = -5 + 10i + (-2 + 3i)\\ &\quad = (-5 + (-2)) + (10i + 3i)\\ &\quad = -7 + 13i\end{aligned}$$

□

B. $6i - (1 + 8i)$

C. $(7+4i)-(-8)$

Solution

$$\begin{aligned}(7+4i)-(-8) &= 7+4i+(-(-8)) \\ &= 7+4i+8 \\ &= 15+4i\end{aligned}$$

□

D. $6i-(1+8i)$

Example 2

Divide the complex numbers as indicated. Leave the result in the form $a+bi$.

A. $(3-i)\div(7+4i)$

Solution

$$\begin{aligned}&(3-i)\div(7+4i)\\ &\quad=[(3-i)/(7+4i)]\cdot[(7-4i)/(7-4i)]\\ &\quad=[(3-i)(7-4i)]/[(7+4i)(7-4i)]\\ &\quad=(21-12i-7i+4i^2)\\ &\qquad\qquad\div(49-\cancel{28i}+\cancel{28i}-16i^2)\\ &\quad=(21-19i+4(-1))/(49-16(-1))\\ &\quad=(21-19i-4)/(49+16)\\ &\quad=(17-19i)/65\\ &\quad=17/65+(-19/65)i\end{aligned}$$

□

B. $(5-15i)\div(2-i)$

C. $(2+3i)\div(-5i)$

Solution

$$\begin{aligned}&(2+3i)\div(-5i)\\ &\quad=[(2+3i)/(-5i)]\cdot[(5i)/(5i)]\\ &\quad=[(2+3i)(5i)]/[(-5i)(5i)]\\ &\quad=(10i+15i^2)/(-25i^2)\\ &\quad=(10i+15(-1))/(-25(-1))\\ &\quad=(-15+10i)/25\\ &\quad=-15/25+10i/25\\ &\quad=-3/5+(2/5)i\end{aligned}$$

□

D. $(8+2i)\div 4$

Section 146: Complex Solutions of Quadratic Equations

Definition
The *discriminant* of the quadratic equation $ax^2 + bx + c = 0$ is the expression $b^2 - 4ac$.

Theorem (proof omitted)
Suppose $z \in \mathbf{C}$ and $f(x)$ is a polynomial with positive degree and real coefficients.
If $f(z) = 0$, then $f(\overline{z}) = 0$.

Properties
Suppose $ax^2 + bx + c = 0$ where $a, b, c \in \mathbf{R}$ and $a \neq 0$.
Then:

- if $b^2 - 4ac > 0$, then the equation has two distinct real solutions
- if $b^2 - 4ac = 0$, then the equation has one repeated real solution
- if $b^2 - 4ac < 0$, then the equation has two distinct complex (but nonreal) solutions

Comments

- The discriminant of the quadratic equation $ax^2 + bx + c = 0$ is the radicand in the quadratic formula $x = (-b \pm \sqrt{b^2 - 4ac})/(2a)$.
- The theorem above asserts that zeros of polynomials with real coefficients occur in conjugate pairs. In other words, if $a + bi$ is a zero of such a polynomial, then $a - bi$ is also a zero of the polynomial.
- We can solve all quadratic equations over the field of complex numbers.

Example 1
Solve each equation.

A. $x^2 - 2x + 13 = 0$

Solution
$x^2 - 2x + 13 = 0$

$$\begin{aligned}
\Leftrightarrow x &= \frac{-(-2) \pm \sqrt{(-2)^2 - 4(1)(13)}}{2(1)} \\
&= [2 \pm \sqrt{4 - 52}]/2 \\
&= (2 \pm \sqrt{-48})/2 \\
&= (2 \pm i\sqrt{48})/2 \\
&= (2 \pm i\sqrt{16}\sqrt{3})/2 \\
&= (2 \pm 4i\sqrt{3})/2 \\
&= \cancel{2}(1 \pm 2i\sqrt{3})/\cancel{2} \\
&= 1 \pm 2i\sqrt{3}
\end{aligned}$$

Set of solutions: $\{1 - 2i\sqrt{3}, 1 + 2i\sqrt{3}\}$ □

B. $3x^2 + x = -2$

C. $x^2 = -4$

SOLUTION

$$\begin{aligned} x^2 = -4 \Leftrightarrow x &= \pm\sqrt{-4} \\ &= \pm\sqrt{4}i \\ &= \pm 2i \end{aligned}$$

Set of solutions: $\{-2i, 2i\}$

□

D. $3x^2 + 15 = 0$

EXAMPLE 2

Determine the discriminant and the nature of the solutions of each quadratic equation.

A. $9x^2 - 12x + 4 = 0$

SOLUTION

Let $a = 9$, $b = -12$, and $c = 4$.

- discriminant $= b^2 - 4ac$
  $= (-12)^2 - 4(9)(4)$
  $= 144 + 144$
  $= 0$
- Since $b^2 - 4ac = 0$, the equation has one repeated real solution.

□

B. $4x^2 + x + 7 = 0$

C. $x^2 + 3x - 5 = 0$

SOLUTION

Let $a = 1$, $b = 3$, and $c = -5$.

- discriminant $= b^2 - 4ac$
  $= (3)^2 - 4(1)(-5)$
  $= 9 + 20$
  $= 29$
- Since $b^2 - 4ac > 0$, the equation has two distinct real solutions.

□

D. $x^2 = 10x - 25$

Section 147: Complex Zeros of Polynomials

Fundamental Theorem of Algebra (proof omitted)
Every polynomial of positive degree has at least one complex zero.

Complete Factorization Theorem (proof omitted)
Suppose $f(x)$ is a polynomial of positive degree n.
Then $f(x) = a(x - c_1)(x - c_2)\cdots(x - c_n)$ for some constants a, c_1, c_2, …, $c_n \in \mathbf{C}$.

Definition
The *multiplicity* of a zero c of a polynomial $f(x)$ is the frequency of the occurrence of the factor $x - c$ in the factorization of $f(x)$ as $a(x - c_1)(x - c_2)\cdots(x - c_n)$.

Theorem (proof omitted)
A polynomial of positive degree n has exactly n complex zeros, where a zero of multiplicity k is counted k times.

Comments

- The Remainder Theorem and the Factor Theorem (see page 280) remain valid when the constant c in each theorem is a complex number.
- We can use the Factor Theorem and the Rational Zeros Theorem to factorize polynomials over complex numbers as well as over real numbers.
- While every polynomial of positive degree is a product of linear factors over the field of complex numbers, we cannot always discover this factorization easily.

Example 1
Determine the zeros and the multiplicities of the zeros of each polynomial $f(x)$.

A. $f(x) = 2(x - 4)^3(x + 1)^7(x + 5)$

Solution
Factors
$f(x)$ has three unique linear factors:
- the factor $x - 4$ occurs 3 times
- the factor $x + 1$ occurs 7 times
- the factor $x + 5$ occurs 1 time

Zeros
$f(x)$ has three zeros:
- the zero 4 has multiplicity 3
- the zero -1 has multiplicity 7
- the zero -5 has multiplicity 1

□

B. $f(x) = 10x^6(x - 1)(x + 5)^2(x + 3)$

EXAMPLE 2

Determine the polynomial $f(x)$ with the given characteristics. Assume the polynomial has real coefficients and leave it in expanded form.

A. degree: 3
zeros: 2, -3, 7
leading coefficient: 1

SOLUTION

Factors of $f(x)$

$x-2$, $x+3$, $x-7$

Polynomial

$$\begin{aligned} f(x) &= 1(x-2)(x+3)(x-7) \\ &= (x^2+x-6)(x-7) \\ &= x^3+x^2-6x-7x^2-7x+42 \\ &= x^3-6x^2-13x+42 \end{aligned}$$

□

B. degree: 4
zeros: $-3i$, $3i$, 0, 5
leading coefficient: 2

C. degree: 3
zeros: $\frac{1}{3}$, $2-i$
leading coefficient: 6

SOLUTION

Note

$2+i$ is also a zero of $f(x)$ since:

- $2-i$ is a zero of $f(x)$;
- zeros occur in conjugate pairs.

Factors of $f(x)$

$x-\frac{1}{3}$, $x-(2-i)$, $x-(2+i)$

Polynomial

$$\begin{aligned} f(x) &= 6(x-\tfrac{1}{3})(x-(2-i))(x-(2+i)) \\ &= (6x-2)((x-2)+i)((x-2)-i) \\ &= (6x-2)((x-2)^2-\cancel{(x-2)i} \\ &\qquad +\cancel{(x-2)i}-i^2) \\ &= (6x-2)(x^2-4x+4-(-1)) \\ &= (6x-2)(x^2-4x+5) \\ &= 6x^3-24x^2+30x-2x^2+8x-10 \\ &= 6x^3-26x^2+38x-10 \end{aligned}$$

□

D. degree: 3
zeros: 3, -4
leading coefficient: 1
the zero 3 has multiplicity 2

EXAMPLE 3
Factorize each polynomial $f(x)$ completely.

A. $f(x) = 5x^2 + 45$

SOLUTION
Notes
- $f(x)$ has leading coefficient 5
- $f(x) = 0 \Leftrightarrow 5x^2 + 45 = 0$
  $\Leftrightarrow 5(x^2 + 9) = 0$
  $\Leftrightarrow x^2 + 9 = 0$
  $\Leftrightarrow x^2 = -9$
  $\Leftrightarrow x = \pm\sqrt{-9}$
  $= \pm 3i$
- $x - 3i$ and $x + 3i$ are factors of $f(x)$

Factorization
$f(x) = 5(x - 3i)(x + 3i)$

□

B. $f(x) = 2x^2 - 32x + 34$
[Hint: Use the Quadratic Formula.]

C. $f(x) = 3x^3 - 192$

SOLUTION
Notes
- $f(x)$ has leading coefficient 3
- $f(x) = 0$
  $\Leftrightarrow 3x^3 - 192 = 0$
  $\Leftrightarrow 3(x^3 - 64) = 0$
  $\Leftrightarrow 3(x - 4)(x^2 + 4x + 16) = 0$
  $\Leftrightarrow x - 4 = 0$ OR $x^2 + 4x + 16 = 0$
  $\Leftrightarrow x = 4$
  OR $x = (-4 \pm \sqrt{4^2 - 4(1)(16)})/(2(1))$
  $= (-4 \pm \sqrt{16 - 64})/2$
  $= (-4 \pm \sqrt{-48})/2$
  $= (-4 \pm i\sqrt{16 \cdot 3})/2$
  $= (-4 \pm 4i\sqrt{3})/2$
  $= \not{2}(-2 \pm 2i\sqrt{3})/\not{2}$
  $= -2 \pm 2i\sqrt{3}$
- $x - (-2 - 2i\sqrt{3})$, $x - (-2 + 2i\sqrt{3})$, and $x - 4$ are factors of $f(x)$

Factorization
$f(x) = 3 \cdot (x - 4) \cdot (x - (-2 - 2i\sqrt{3}))$
$\cdot (x - (-2 + 2i\sqrt{3}))$

□

D. $f(x) = x^3 - 5x^2 + 7x - 35$

E. $f(x) = x^4 - 5x^3 + 26x^2 + 46x - 68$

SOLUTION

Factors of Coefficients

- constant term (-68): $\pm 1, \pm 2, \pm 4, \pm 17, \pm 34, \pm 68$
- leading coefficient (1): ± 1

Possible Rational Zeros

By the Rational Zeros Theorem, the set A of the possible rational zeros of $f(x)$ is given by:

$$\begin{aligned} A &= \{\pm\tfrac{1}{1}, \pm\tfrac{2}{1}, \pm\tfrac{4}{1}, \pm\tfrac{17}{1}, \pm\tfrac{34}{1}, \pm\tfrac{68}{1}\} \\ &= \{\pm 1, \pm 2, \pm 4, \pm 17, \pm 34, \pm 68\} \end{aligned}$$

Partial Factorization of $f(x)$

We attempt through the process of synthetic division to find an $a_1 \in A$ for which the remainder yielded by dividing $f(x)$ by $x - a_1$ is zero (that is, for which $f(a_1) = 0$):

$$\begin{array}{r|rrrrr} 1 & 1 & -5 & 26 & 46 & -68 \\ & & 1 & -4 & 22 & 68 \\ \hline & 1 & -4 & 22 & 68 & \boxed{0} \end{array}$$

Hence $f(1) = 0$ and $f(x) = (x - 1)(q_1(x))$, where $q_1(x) = x^3 - 4x^2 + 22x + 68$.

Partial Factorization of $q_1(x)$

The set of possible rational zeros of $q_1(x)$ is the same as the set for $f(x)$. We now attempt to find an $a_2 \in A$ for which $q_1(a_2) = 0$:

$$\begin{array}{r|rrrr} -1 & 1 & -4 & 22 & 68 \\ & & -1 & 5 & -27 \\ \hline & 1 & -5 & 27 & \boxed{41} \end{array}$$

$$\begin{array}{r|rrrr} 2 & 1 & -4 & 22 & 68 \\ & & 2 & -4 & 36 \\ \hline & 1 & -2 & 18 & \boxed{104} \end{array}$$

$$\begin{array}{r|rrrr} -2 & 1 & -4 & 22 & 68 \\ & & -2 & 12 & -68 \\ \hline & 1 & -6 & 34 & \boxed{0} \end{array}$$

Hence $q_1(-2) = 0$

$$\begin{aligned} \text{and } q_1(x) &= (x - (-2))(q_2(x)) \\ &= (x + 2)(q_2(x)), \end{aligned}$$

where $q_2(x) = x^2 - 6x + 34$.

Notes for $q_2(x)$

- $q_2(x)$ has leading coefficient 1
- $q_2(x) = 0$

$$\begin{aligned} &\Leftrightarrow x^2 - 6x + 34 = 0 \\ &\Leftrightarrow x = [6 \pm \sqrt{(-6)^2 - 4(1)(34)}\,]/(2(1)) \\ &\quad = (6 \pm \sqrt{36 - 136})/2 \\ &\quad = (6 \pm \sqrt{-100})/2 \\ &\quad = (6 \pm i\sqrt{100})/2 \\ &\quad = (6 \pm i(10))/2 \\ &\quad = \not{2}(3 \pm i(5))/\not{2} \\ &\quad = 3 \pm 5i \end{aligned}$$

- $q_2(x)$ has factors $x - (3 - 5i)$ and $x - (3 + 5i)$

Factorization of $q_2(x)$

$$\begin{aligned} q_2(x) &= (1)(x - (3 - 5i))(x - (3 + 5i)) \\ &= (x - (3 - 5i))(x - (3 + 5i)) \end{aligned}$$

Complete Factorization of $f(x)$

$$\begin{aligned} f(x) &= (x - 1) \cdot (q_1(x)) \\ &= (x - 1) \cdot (x + 2) \cdot (q_2(x)) \\ &= (x - 1) \cdot (x + 2) \cdot (x - (3 - 5i)) \\ &\qquad \cdot (x - (3 + 5i)) \end{aligned}$$

□

Section 148: Graphical Representations of Complex Numbers

DEFINITION
The *complex plane* is a plane in which each point represents a complex number.

PROCEDURE
We construct a portion of the complex plane in the same way that we construct a portion of the Cartesian plane, but:

- we call the horizontal axis the *real axis* and label it "Re";
- we call the vertical axis the *imaginary axis* and label it "Im".

PROCEDURE
We plot in the complex plane the point that corresponds to the complex number $a + bi$ in the same way that we plot in the Cartesian plane the point with coordinates (a, b). Thus the point that corresponds to $a + bi$ should lie:

- $|a|$ units to the right of the imaginary axis if $a \geq 0$ or
- $|a|$ units to the left of the imaginary axis if $a < 0$, and
- $|b|$ units above the real axis if $b \geq 0$ or
- $|b|$ units below the real axis if $b < 0$.

EXAMPLE 1
Construct a portion of the complex plane that will contain the points that correspond to the numbers below and then plot those points in the plane.

A. $z_a = 4 + 2i$
B. $z_b = -2 - 5i$
C. $z_c = -3 + i$
D. $z_d = 3 - 3i$
E. $z_e = -2i$
F. $z_f = 1$

SOLUTION

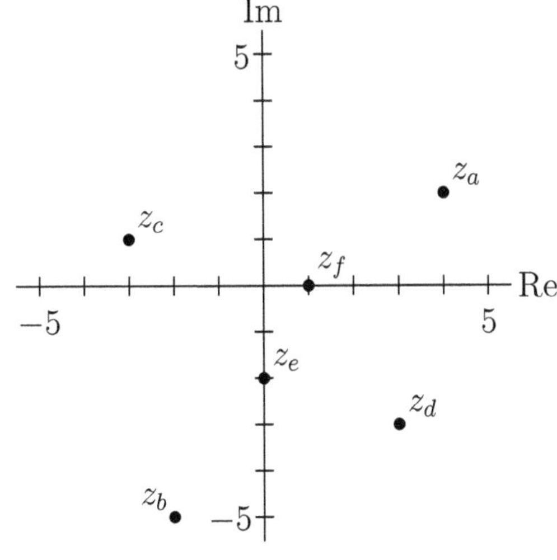

□

EXAMPLE 2
Construct a portion of the complex plane that will contain the points that correspond to the numbers below and then plot those points in the plane.

G. $z_g = -6 - 2i$
H. $z_h = -1$
I. $z_i = 3 - i$
J. $z_j = 5i$
K. $z_k = -4 + 4i$
L. $z_l = 2 + 7i$

Section 149: Trigonometric Form of Complex Numbers

DEFINITIONS

- The *modulus* of a complex number z, denoted by $|z|$, is the distance between 0 and z in the complex plane.
- An *argument* of a complex number z is the measure of an angle formed in the complex plane by the ray with endpoint at 0 that passes through 1 and the ray with endpoint at 0 that passes through z.

PROPERTIES

Suppose a complex number z has modulus r, argument θ, real part a, and imaginary part b. Then:

- $a = r\cos\theta$
- $b = r\sin\theta$
- $\tan\theta = b/a$ (provided $a \neq 0$)
- $r^2 = a^2 + b^2$
- $z = r(\cos\theta + i\sin\theta)$

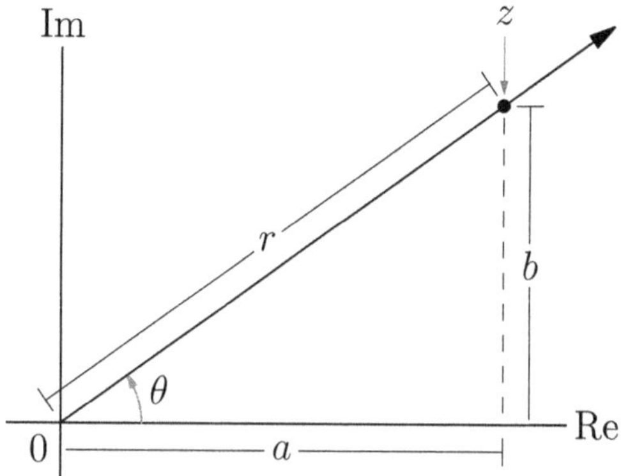

PROOFS (last property only)

$$\begin{aligned} z &= a + bi \\ &= r\cos\theta + (r\sin\theta)i \\ &= r(\cos\theta + i\sin\theta) \end{aligned}$$

(We can prove the first four properties above in the same way that we proved the properties on page 440 that established the relationships between the rectangular coordinates (x, y) and the polar coordinates (r, θ) of a point in the Cartesian plane.)

⊡

TERMINOLOGY

Suppose a complex number z has modulus r, argument θ, real part a, and imaginary part b. Then:

- $a + bi$ is called the *rectangular* form of z;
- $r(\cos\theta + i\sin\theta)$ is called the *trigonometric* or *polar* form of z.

COMMENTS

- As reflected in the notation $|z|$, the definition of the modulus of a complex number resembles closely the definition of the absolute value of a real number.
- The modulus of a complex number (as a distance) is always nonnegative.
- The initial side of the angle described in the definition of argument above is the ray with endpoint at 0 that passes through 1.
- If a ray forms an angle of measure θ with another ray, then it will also form an angle of measure $\theta + 2\pi n$ with that ray for any $n \in \mathbf{Z}$. Therefore, complex numbers do not have unique arguments; however, any two arguments of a complex number will differ by an integral multiple of 2π.
- For any $x \in \mathbf{R}$, $e^{ix} = \cos x + i\sin x$. This equation is known as Euler's formula.

THEOREM

Suppose $z_1, z_2 \in \mathbf{C}$ where $z_1 = r_1(\cos\theta_1 + i\sin\theta_1)$ and $z_2 = r_2(\cos\theta_2 + i\sin\theta_2)$.
Then:

- $z_1 z_2 = r_1 r_2[\cos(\theta_1 + \theta_2) + i\sin(\theta_1 + \theta_2)]$
- $z_1/z_2 = (r_1/r_2)[\cos(\theta_1 - \theta_2) + i\sin(\theta_1 - \theta_2)]$ (provided $z_2 \neq 0$)

PROOFS

Suppose $z_1, z_2 \in \mathbf{C}$ where $z_1 = r_1(\cos\theta_1 + i\sin\theta_1)$ and $z_2 = r_2(\cos\theta_2 + i\sin\theta_2)$.

- $$\begin{aligned} z_1 z_2 &= (r_1(\cos\theta_1 + i\sin\theta_1)) \cdot (r_2(\cos\theta_2 + i\sin\theta_2)) \\ &= r_1 \cdot r_2 \cdot (\cos\theta_1 + i\sin\theta_1)(\cos\theta_2 + i\sin\theta_2) \\ &= r_1 r_2 \cdot (\cos\theta_1\cos\theta_2 + i\cos\theta_1\sin\theta_2 + i\sin\theta_1\cos\theta_2 + i^2\sin\theta_1\sin\theta_2) \\ &= r_1 r_2 \cdot (\cos\theta_1\cos\theta_2 + i\cos\theta_1\sin\theta_2 + i\sin\theta_1\cos\theta_2 + (-1)\sin\theta_1\sin\theta_2) \\ &= r_1 r_2 \cdot (\cos\theta_1\cos\theta_2 - \sin\theta_1\sin\theta_2 + i\sin\theta_1\cos\theta_2 + i\cos\theta_1\sin\theta_2) \\ &= r_1 r_2 \cdot ((\cos\theta_1\cos\theta_2 - \sin\theta_1\sin\theta_2) + i(\sin\theta_1\cos\theta_2 + i\cos\theta_1\sin\theta_2)) \\ &= r_1 r_2[\cos(\theta_1 + \theta_2) + i\sin(\theta_1 + \theta_2)] \end{aligned}$$

- $$\begin{aligned} z_1/z_2 &= [r_1(\cos\theta_1 + i\sin\theta_1)]/[r_2(\cos\theta_2 + i\sin\theta_2)] \\ &= \frac{r_1(\cos\theta_1 + i\sin\theta_1)}{r_2(\cos\theta_2 + i\sin\theta_2)} \cdot \frac{\cos\theta_2 - i\sin\theta_2}{\cos\theta_2 - i\sin\theta_2} \\ &= \frac{r_1}{r_2} \cdot \frac{\cos\theta_1\cos\theta_2 - i\cos\theta_1\sin\theta_2 + i\sin\theta_1\cos\theta_2 - i^2\sin\theta_1\sin\theta_2}{\cos^2\theta_2 - \cancel{i\cos\theta_2\sin\theta_2} + \cancel{i\cos\theta_2\sin\theta_2} - i^2\sin^2\theta_2} \\ &= \frac{r_1}{r_2} \cdot \frac{\cos\theta_1\cos\theta_2 - i^2\sin\theta_1\sin\theta_2 + i\sin\theta_1\cos\theta_2 - i\cos\theta_1\sin\theta_2}{\cos^2\theta_2 - (-1)\sin^2\theta_2} \\ &= \frac{r_1}{r_2} \cdot \frac{\cos\theta_1\cos\theta_2 - (-1)\sin\theta_1\sin\theta_2 + i(\sin\theta_1\cos\theta_2 - \cos\theta_1\sin\theta_2)}{\cos^2\theta_2 + \sin^2\theta_2} \\ &= \frac{r_1}{r_2} \cdot \frac{(\cos\theta_1\cos\theta_2 + \sin\theta_1\sin\theta_2) + i(\sin\theta_1\cos\theta_2 - \cos\theta_1\sin\theta_2)}{1} \\ &= (r_1/r_2)[\cos(\theta_1 - \theta_2) + i\sin(\theta_1 - \theta_2)] \end{aligned}$$

⊡

EXAMPLE 1

Write each complex number in rectangular form.

A. $7(\cos(4\pi/3) + i\sin(4\pi/3))$

SOLUTION

$$\begin{aligned} &7(\cos(4\pi/3) + i\sin(4\pi/3)) \\ &\quad = 7(-1/2 + i(-\sqrt{3}/2)) \\ &\quad = -7/2 - (7\sqrt{3}/2)i \end{aligned}$$

□

B. $3(\cos(3\pi/2) + i\sin(3\pi/2))$

EXAMPLE 2
Write each complex number in trigonometric form with argument between 0 and 2π.

A. $\sqrt{2} - i\sqrt{2}$

SOLUTION
Declarations
Let a denote the real part,
b denote the imaginary part,
θ denote the argument, and
r denote the modulus of $\sqrt{2} - i\sqrt{2}$.

Argument
$a = \sqrt{2}$ AND $b = -\sqrt{2}$ AND $0 \leq \theta < 2\pi$
$\Rightarrow \tan\theta = -\sqrt{2}/\sqrt{2}$ AND $3\pi/2 \leq \theta < 2\pi$
$\Rightarrow \tan\theta = -1$ AND $3\pi/2 \leq \theta < 2\pi$
$\Rightarrow \theta = 7\pi/4$

Modulus
$a = \sqrt{2}$ AND $b = -\sqrt{2}$ AND $r > 0$
$\Rightarrow r^2 = (\sqrt{2})^2 + (-\sqrt{2})^2$ AND $r > 0$
$\Rightarrow r^2 = 2 + 2$ AND $r > 0$
$\Rightarrow r^2 = 4$ AND $r > 0$
$\Rightarrow r = \sqrt{4}$
$= 2$

Trigonometric Form
$2(\cos(7\pi/4) + i\sin(7\pi/4))$

□

B. $\sqrt{3} + i$

EXAMPLE 3
Determine the product z_1z_2 of the complex numbers z_1 and z_2. Leave your answer in trigonometric form.

A. $z_1 = 8(\cos(\pi/6) + i\sin(\pi/6))$;
$z_2 = 2(\cos(5\pi/3) + i\sin(5\pi/3))$

SOLUTION

$$\begin{aligned} z_1z_2 &= (8)(2)\cdot[\cos(\pi/6 + 5\pi/3) \\ &\qquad + i\sin(\pi/6 + 5\pi/3)] \\ &= 16\cdot[\cos(\pi/6 + 10\pi/6) \\ &\qquad + i\sin(\pi/6 + 10\pi/6)] \\ &= 16(\cos(11\pi/6) + i\sin(11\pi/6)) \end{aligned}$$

□

B. $z_1 = 5\sqrt{3}(\cos 245° + i\sin 245°)$;
$z_2 = \sqrt{3}(\cos 100° + i\sin 100°)$

EXAMPLE 4

Determine the quotient z_1/z_2 of the complex numbers z_1 and z_2. Leave your answer in trigonometric form.

A. $z_1 = 8(\cos(\pi/6) + i\sin(\pi/6))$;
$z_2 = 2(\cos(5\pi/3) + i\sin(5\pi/3))$

SOLUTION

$$\begin{aligned} z_1/z_2 &= (8/2) \cdot [\cos(\pi/6 - 5\pi/3) \\ &\qquad + i\sin(\pi/6 - 5\pi/3)] \\ &= 4 \cdot [\cos(\pi/6 - 10\pi/6) \\ &\qquad + i\sin(\pi/6 - 10\pi/6)] \\ &= 4[\cos(-9\pi/6) + i\sin(-9\pi/6)] \\ &= 4[\cos(-3\pi/2) + i\sin(-3\pi/2)] \\ &= 4(\cos(\pi/2) + i\sin(\pi/2)) \end{aligned}$$

□

B. $z_1 = 5\sqrt{3}(\cos 245° + i\sin 245°)$;
$z_2 = \sqrt{3}(\cos 100° + i\sin 100°)$

Section 150: Powers and Roots of Complex Numbers

DeMoivre's Theorem (proof omitted)
Suppose $n \in \mathbf{N}$ and $z \in \mathbf{C}$, where z has modulus r and argument θ.
Then $z^n = r^n(\cos n\theta + i \sin n\theta)$.

Theorem B (proof omitted)
Suppose $n \in \mathbf{N}$ and $z \in \mathbf{C}$, where z has modulus r and argument θ and $n \geq 2$.
Then the set R of the nth roots of z is given by

$R = \{\ \sqrt[n]{r}\ [\cos((\theta + 2\pi k)/n) + i \sin((\theta + 2\pi k)/n)] \mid k \in \{0, 1, 2, 3, \ldots, n-1\}\ \}.$

Theorem C (proof omitted)
Any nonzero complex number has exactly n distinct complex n^{th} roots.

Comments

- When n is even, we can determine the nth roots of the complex number $r(\cos\theta + i \sin\theta)$ by evaluating either:
  - $+\sqrt[n]{r}\ [\cos(\theta + 2\pi k)/n + i \sin(\theta + 2\pi k)/n]$ for $k \in \{0, 1, 2, 3, \ldots, n-1\}$, or
  - $-\sqrt[n]{r}\ [\cos(\theta + 2\pi k)/n + i \sin(\theta + 2\pi k)/n]$ for $k \in \{0, 1, 2, 3, \ldots, n-1\}$.
- The points in the complex plane that correspond to the n distinct n^{th} roots of a complex number with modulus r will be spaced evenly around the circle with radius $\sqrt[n]{r}$ and center 0 (thereby serving as the vertices of a regular polygon with n sides), since:
  - the modulus of each root will be $\sqrt[n]{r}$ and
  - the arguments of the roots will increase in increments of $2\pi/n$.

Example 1
Use DeMoivre's Theorem to evaluate each expression. Leave your answer in the same form as the base of the power.

A. $[\sqrt[3]{4}(\cos 10° + i \sin 10°)]^6$

Solution
$[\sqrt[3]{4}(\cos 10° + i \sin 10°)]^6$
$= (\sqrt[3]{4})^6(\cos(6 \cdot 10°) + i \sin(6 \cdot 10°))$
$= (4)^{6/3}(\cos 60° + i \sin 60°)$
$= 16(\cos 60° + i \sin 60°)$

□

B. $[2(\cos(\pi/3) + i \sin(\pi/3))]^4$

C. $[3(\cos(\pi/6) - i \sin(\pi/6))]^3$

Solution
$[3(\cos(\pi/6) - i \sin(\pi/6))]^3$
$= (3)^3(\cos(3 \cdot \pi/6) - i \sin(3 \cdot \pi/6))$
$= 27(\cos(\pi/2) - i \sin(\pi/2))$

□

D. $[\sqrt{7}(\cos 50° - i \sin 50°)]^4$

E. $(-1 - i\sqrt{3})^5$

SOLUTION

Declarations

Let a denote the real part,
b denote the imaginary part,
θ denote an argument, and
r denote the modulus of $-1 - i\sqrt{3}$.

Argument and Modulus

- $a = -1$ AND $b = -\sqrt{3}$ AND $0 \le \theta < 2\pi$
  $\Rightarrow \tan\theta = -\sqrt{3}/(-1)$ AND $\pi < \theta < 3\pi/2$
  $\Rightarrow \tan\theta = \sqrt{3}$ AND $\pi < \theta < 3\pi/2$
  $\Rightarrow \theta = 4\pi/3$
- $a = -1$ AND $b = -\sqrt{3}$ AND $r > 0$
  $\Rightarrow r^2 = (-1)^2 + (-\sqrt{3})^2$ AND $r > 0$
  $\Rightarrow r^2 = 1 + 3$ AND $r > 0$
  $\Rightarrow r^2 = 4$ AND $r > 0$
  $\Rightarrow r = \sqrt{4}$
  $= 2$

Evaluation

$$
\begin{aligned}
&(-1 - i\sqrt{3})^5 \\
&= [2(\cos(4\pi/3) + i\sin(4\pi/3))]^5 \\
&= (2)^5(\cos(5 \cdot 4\pi/3) + i\sin(5 \cdot 4\pi/3)) \\
&= 32(\cos(20\pi/3) + i\sin(20\pi/3)) \\
&= 32(-1/2 + i(\sqrt{3}/2)) \\
&= -16 + (16\sqrt{3})i
\end{aligned}
$$

□

F. $(-2 + 2i)^3$

EXAMPLE 2

The seventh roots of 128 are:

- 2
- $2(\cos 2\pi/7 + i\sin 2\pi/7)$
- $2(\cos 4\pi/7 + i\sin 4\pi/7)$
- $2(\cos 6\pi/7 + i\sin 6\pi/7)$
- $2(\cos 8\pi/7 + i\sin 8\pi/7)$
- $2(\cos 10\pi/7 + i\sin 10\pi/7)$
- $2(\cos 12\pi/7 + i\sin 12\pi/7)$

We plot the points corresponding to these roots in the complex plane shown at right. Observe that the points would serve as the vertices of a regular heptagon.

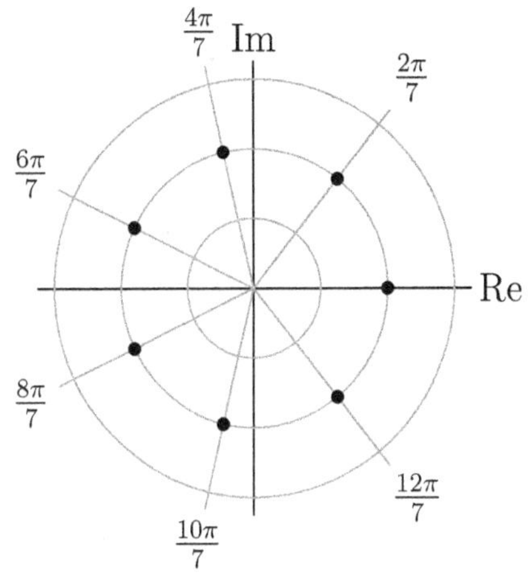

□

Example 3
Find the cube roots of $27i$.

Solution
Declarations
Let w_k denote a cube root of $27i$,
s denote the modulus of any w_k, and
α_k denote an argument of w_k.

Modulus and Argument of $27i$
The point $27i$ in the complex plane lies on the imaginary axis. Clearly:

- the modulus of $27i$ is 27
- an argument of $27i$ is $\pi/2$

Modulus and Arguments of Roots

- $s = \sqrt[3]{27}$
  $= 3$
- $\alpha_k = (\pi/2 + 2\pi k)/3$ for $k \in \{0, 1, 2\}$
  $= \pi/6 + (2\pi/3)k$ for $k \in \{0, 1, 2\}$

Determination of Roots

- $k = 0 \Rightarrow \alpha_0 = \pi/6 + (2\pi/3)(0)$
  $= \pi/6$
  $\Rightarrow w_0 = 3(\cos(\pi/6) + i\sin(\pi/6))$
  $= 3(\sqrt{3}/2 + i(1/2))$
  $= 3\sqrt{3}/2 + (3/2)i$
- $k = 1 \Rightarrow \alpha_1 = \pi/6 + (2\pi/3)(1)$
  $= \pi/6 + 4\pi/6$
  $= 5\pi/6$
  $\Rightarrow w_1 = 3(\cos(5\pi/6) + i\sin(5\pi/6))$
  $= 3(-\sqrt{3}/2 + i(1/2))$
  $= -3\sqrt{3}/2 + (3/2)i$
- $k = 2 \Rightarrow \alpha_2 = \pi/6 + (2\pi/3)(2)$
  $= \pi/6 + 4\pi/3$
  $= 9\pi/6$
  $= 3\pi/2$
  $\Rightarrow w_2 = 3(\cos(3\pi/2) + i\sin(3\pi/2))$
  $= 3(0 + i(-1))$
  $= -3i$

Roots
The three cube roots of $27i$ are $-3i$, $3\sqrt{3}/2 + (3/2)i$, and $-3\sqrt{3}/2 + (3/2)i$.

□

Example 4
Find the fourth roots of -16.

EXAMPLE 5
Solve the equation $z^4 = 1$.

SOLUTION
Declarations
Note that z must be a fourth root of 1.
Let s denote the modulus of z and
α_k denote an argument of z.

Modulus and Argument of 1
The point 1 lies on the real axis. Clearly:
- the modulus of 1 is 1
- an argument of 1 is 0

Modulus and Arguments of Roots
- $s = \sqrt[4]{1}$
  $= 1$
- $\alpha_k = (0 + 2\pi k)/4$ for $k \in \{0, 1, 2, 3\}$
  $= (\pi/2)k$ for $k \in \{0, 1, 2, 3\}$

Determination of Roots
- $k = 0 \Rightarrow \alpha_0 = (\pi/2)(0)$
  $= 0$
  $\Rightarrow z = 1(\cos 0 + i \sin 0)$
  $= 1(1 + i(0))$
  $= 1$
- $k = 1 \Rightarrow \alpha_1 = (\pi/2)(1)$
  $= \pi/2$
  $\Rightarrow z = 1(\cos(\pi/2) + i \sin(\pi/2))$
  $= 1(0 + i(1))$
  $= i$
- $k = 2 \Rightarrow \alpha_2 = (\pi/2)(2)$
  $= \pi$
  $\Rightarrow z = 1(\cos \pi + i \sin \pi)$
  $= 1(-1 + i(0))$
  $= -1$
- $k = 3 \Rightarrow \alpha_3 = (\pi/2)(3)$
  $= 3\pi/2$
  $\Rightarrow z = 1(\cos(3\pi/2) + i \sin(3\pi/2))$
  $= 1(0 + i(-1))$
  $= -i$

Set of Solutions
Set of solutions: $\{1, -1, i, -i\}$

□

EXAMPLE 6
Solve the equation $z^2 = -1/2 - (\sqrt{3}/2)i$.

About the Author

The author was born in North Dakota and resides presently in Oregon. Other books by the author include:

Exercises in Algebra and Trigonometry

Economics

A Study in Romans

Bimetrical Psalter

Index